# Luminescence of Lanthanide Ions in Coordination Compounds and Nanomaterials

# Luminescence of Lanthanide Ions in Coordination Compounds and Nanomaterials

Edited by

ANA DE BETTENCOURT-DIAS
*Department of Chemistry, University of Nevada, Reno, USA*

WILEY

This edition first published 2014


*Registered office*
John Wiley & Sons Ltd, The Atrium, Southern Gate, Chichester, West Sussex, PO19 8SQ, United Kingdom

For details of our global editorial offices, for customer services and for information about how to apply for permission to reuse the copyright material in this book please see our website at www.wiley.com.



***Library of Congress Cataloging-in-Publication Data***

Luminescence of lanthanide ions in coordination compounds and nanomaterials
/ edited by Dr Ana de Bettencourt-Dias.
pages cm
Includes bibliographical references and index.
ISBN 978-1-119-95083-7 (cloth)
1. Nanostructured materials. 2. Luminescence. 3. Rare earth metals–Optical properties. 4. Coordination compounds. I. Bettencourt-Dias, Ana de, editor.
TA418.9.N35L86 2014
546'.41–dc23

2014012258

A catalogue record for this book is available from the British Library.

ISBN: 9781119950837

Set in 10/12 pt TimesLTStd-Roman by Thomson Digital, Noida, India

Printed and bound in Singapore by Markono Print Media Pte Ltd

# Contents


| | |
|---|---|
| ***List of Contributors*** | **xi** |
| ***Preface*** | **xiii** |
| **1 Introduction to Lanthanide Ion Luminescence** | **1** |
| *Ana de Bettencourt-Dias* | |
| 1.1 History of Lanthanide Ion Luminescence | 1 |
| 1.2 Electronic Configuration of the +III Oxidation State | 2 |
| 1.2.1 The 4*f* Orbitals | 2 |
| 1.2.2 Energy Level Term Symbols | 2 |
| 1.3 The Nature of the *f-f* Transitions | 5 |
| 1.3.1 Hamiltonian in Central Field Approximation and Coulomb Interactions | 5 |
| 1.3.2 Spin–Orbit Coupling | 10 |
| 1.3.3 Crystal Field or Stark Effects | 13 |
| 1.3.4 The Crystal Field Parameters $B_q^k$ and Symmetry | 14 |
| 1.3.5 Energies of Crystal Field Split Terms | 18 |
| 1.3.6 Zeeman Effect | 19 |
| 1.3.7 Point Charge Electrostatic Model | 21 |
| 1.3.8 Other Methods to Estimate Crystal Field Parameters | 25 |
| 1.3.9 Allowed and Forbidden *f-f* Transitions | 27 |
| 1.3.10 Induced Electric Dipole Transitions and Their Intensity – Judd–Ofelt Theory | 34 |
| 1.3.11 Transition Probabilities and Branching Ratios | 37 |
| 1.3.12 Hypersensitive Transitions | 38 |
| 1.3.13 Emission Efficiency and Rate Constants | 39 |
| 1.4 Sensitisation Mechanism | 40 |
| 1.4.1 The Antenna Effect | 40 |
| 1.4.2 Non-Radiative Quenching | 44 |
| References | 46 |
| **2 Spectroscopic Techniques and Instrumentation** | **49** |
| *David E. Morris and Ana de Bettencourt-Dias* | |
| 2.1 Introduction | 49 |
| 2.2 Instrumentation in Luminescence Spectroscopy | 52 |
| 2.2.1 Challenges in Design and Interpretation of Lanthanide Luminescence Experiments | 52 |
| 2.2.2 Common Luminescence Experiments | 57 |

2.2.3 Basic Design Elements and Configurations in Luminescence Spectrometers 61
2.2.4 Luminescence Spectrometer Components and Characteristics 63
2.2.5 Recent Advances in Luminescence Instrumentation 67
2.3 Measurement of Quantum Yields of Luminescence in the Solid State and in Solution 69
2.3.1 Measurement Against a Standard in Solution 70
2.3.2 Measurement Against a Standard in the Solid State 71
2.3.3 Absolute Measurement with an Integrating Sphere 72
2.4 Excited State Lifetimes 73
2.4.1 Number of Coordinated Solvent Molecules 73
References 74

**3 Circularly Polarised Luminescence** **77**
*Gilles Muller*

3.1 Introduction 77
3.1.1 General Aspects: Molecular Chirality 77
3.1.2 Chiroptical Tools: from CD to CPL Spectroscopy 78
3.2 Theoretical Principles 79
3.2.1 General Theory 79
3.2.2 CPL Intensity Calculations, Selection Rules, Luminescence Selectivity, and Spectra–Structure Relationship 82
3.3 CPL Measurements 84
3.3.1 Instrumentation 84
3.3.2 Calibration and Standards 88
3.3.3 Artifacts in CPL Measurements 90
3.3.4 Proposed Instrumental Improvements to Record Eu(III)-Based CPL Signals 91
3.4 Survey of CPL Applications 93
3.4.1 Ln(III)-Containing Systems 93
3.4.2 Ln(III) Complexes with Achiral Ligands 94
3.4.3 Ln(III) Complexes with Chiral Ligands 99
3.5 Chiral Ln(III) Complexes to Probe Biologically Relevant Systems 109
3.5.1 Sensing through Coordination to the Metal Centre 109
3.5.2 Sensing through Coordination to the Antenna/Receptor Groups 112
3.6 Concluding Remarks 114
References 115

**4 Luminescence Bioimaging with Lanthanide Complexes** **125**
*Jean-Claude G. Bünzli*

4.1 Introduction 125
4.2 Luminescence Microscopy 127

4.2.1 Classical Optical Microscopy: a Short Survey 127
4.2.2 Principle of Luminescence Microscopy 128
4.2.3 Principle of Time-resolved Luminescence Microscopy 131
4.2.4 Early Instrumental Developments for Time-resolved Microscopy with LLBs 134
4.2.5 Optimisation of Time-resolved Microscopy Instrumentation 140
4.2.6 Commercial Instruments 143
4.3 Bioimaging with Lanthanide Luminescent Probes and Bioprobes 144
4.3.1 β-Diketonate Probes 144
4.3.2 Aliphatic Polyaminocarboxylate and Carboxylate Probes 154
4.3.3 Macrocyclic Probes 163
4.3.4 Self-assembled Triple Helical Bioprobes 171
4.3.5 Other Bioprobes 177
4.4 Conclusions and Perspectives 180
References 184

**5 Two-photon Absorption of Lanthanide Complexes: from Fundamental Aspects to Biphotonic Imaging Applications 197**
*Anthony D'Aléo, Chantal Andraud and Olivier Maury*

5.1 Introduction 197
5.2 Two-photon Absorption, a Third Nonlinear Optical Phenomenon 198
5.2.1 Theoretical and Historical Background 198
5.2.2 Experimental Determination of the 2PA Efficiency of Molecules 199
5.2.3 Two-photon Fluorescence Microscopy for Biological Imaging 200
5.2.4 Molecular Engineering for Multiphonic Imaging 201
5.3 Spectroscopic Evidence for the Two-photon Sensitisation of Lanthanide Luminescence 205
5.3.1 1961: The Breakthrough Experiments 205
5.3.2 Two-photon Excitation of *f-f* Transitions 206
5.3.3 The Two-photon Antenna Effect 207
5.3.4 The Charge Transfer State Mediated Sensitisation Process 209
5.3.5 Optimising Molecular Two-photon Cross Section: the Brightness Trade-off 211
5.3.6 Two-photon Excited Luminescence in Solid Matrix 214
5.3.7 Two-photon Time-gated Spectroscopy 214
5.4 Towards Biphotonic Microscopy Imaging 215
5.4.1 Proof of Concept 215
5.4.2 Towards the Design of an Optimised Bio-probe 217
5.4.3 Design of Lanthanide containing Nano-probes, toward Single-object Imaging 222
5.4.4 Towards NIR-to-NIR Imaging 223
5.5 Conclusions 225
References 226

**6 Lanthanide Ion Complexes as Chemosensors** **231**
*Thorfinnur Gunnlaugsson and Simon J. A. Pope*

6.1 Photophysical Properties of $Ln^{III}$ Based Sensors 231
6.1.1 Emission Based Sensors 231
6.1.2 Luminescence Lifetime 232
6.1.3 Spectral Form, Hypersensitivity and Ratiometric Peaks 233
6.2 Sensor Design Principles 233
6.2.1 The Design of Ln-receptor Sites and Antenna Components 234
6.2.2 Covalent versus Self-assembled Ln-receptor Design 235
6.2.3 Sensors for Cations 237
6.2.4 Sensors for Anions 249
6.3 Interactions with DNA and Biological Systems 260
References 265

**7 Upconversion of $Ln^{3+}$-based Nanoparticles for Optical Bio-imaging** **269**
*Frank C.J.M. van Veggel*

7.1 Introduction 269
7.2 Physical Properties of $Ln^{3+}$ Ions 272
7.3 Basic Principles of Upconversion 272
7.4 Synthesis of Core and Core–Shell Nanoparticles 277
7.4.1 Syntheses in Organic Solvent 277
7.4.2 Syntheses in Aqueous Media 277
7.4.3 Surface Modification 278
7.5 Characterisation 278
7.5.1 Basic Techniques 278
7.5.2 Advanced Techniques 279
7.6 Bio-imaging 283
7.6.1 Basics 283
7.6.2 Cell Studies 283
7.6.3 Animal Studies 287
7.6.4 Discussion 290
7.7 Upconversion and Magnetic Resonance Imaging 293
7.8 Conclusions and Outlook 295
References 295

**8 Direct Excitation Ln(III) Luminescence Spectroscopy to Probe the Coordination Sphere of Ln(III) Catalysts, Optical Sensors and MRI Agents** **303**
*Janet R. Morrow and Sarina J. Dorazio*

8.1 Introduction 303
8.1.1 Luminescence Spectroscopy for Defining the Ln(III) Coordination Sphere 303

8.2 Direct Excitation Lanthanide Luminescence 304
8.2.1 Luminescence Properties of the Lanthanide Ions 304
8.2.2 Ln(III) Excitation Spectroscopy 306
8.2.3 Ln(III) Emission Spectroscopy 307
8.2.4 Time-Resolved Ln(III) Luminescence Spectroscopy 308
8.2.5 Luminescence Resonance Energy Transfer 310
8.3 Defining the Ln(III) Ion Coordination Sphere through Direct Eu(III) Excitation Luminescence Spectroscopy 311
8.3.1 Eu(III) Complex Speciation in Solution: Number of Excitation Peaks 311
8.3.2 Excitation Spectra of Geometric Isomers 311
8.3.3 Innersphere Coordination of Anions 312
8.3.4 Ligand Ionisation 314
8.4 Luminescence Studies of Anion Binding in Catalysis and Sensing 317
8.4.1 Phosphate Ester Binding and Cleavage 317
8.4.2 Sensing Biologically Relevant Anions 318
8.5 Luminescence Studies of Ln(III) MRI Contrast Agents 320
8.5.1 Types of Ln(III) MRI Contrast Agents 320
8.5.2 Luminescence Studies of Ln(III) ParaCEST Agents 322
8.6 Conclusions 326
References 326

**9 Heterometallic Complexes Containing Lanthanides 331**
*Stephen Faulkner and Manuel Tropiano*

9.1 Introduction 331
9.2 Properties of a Heteromultimetallic Complex 332
9.3 Lanthanide Assemblies in the Solid State 335
9.4 Lanthanide Assemblies in Solution 338
9.4.1 Lanthanide Helicates 338
9.4.2 Non-helicate Structures 341
9.5 Heterometallic Complexes Derived from Bridging and Multi-compartmental Ligands 342
9.6 Energy Transfer in Assembled Systems 347
9.7 Responsive Multimetallic Systems 351
9.8 Summary and Prospects 353
References 353

***Index*** **359**

# List of Contributors

**Anthony D'Aléo**, CINaM, UMR 7325 CNRS-Aix Marseille Université, France

**Chantal Andraud**, Laboratoire de chimie, UMR 5281 ENS Lyon-CNRS-Université de Lyon, France

**Ana de Bettencourt-Dias**, Department of Chemistry, University of Nevada, USA

**Jean-Claude G. Bünzli**, Swiss Federal Institute of Technology, Switzerland; and Korea University, Republic of Korea

**Sarina J. Dorazio**, University at Buffalo, State University of New York, USA

**Stephen Faulkner**, Chemistry Research Laboratory, University of Oxford, UK

**Thorfinnur Gunnlaugsson**, Trinity College, University of Dublin, Ireland

**Olivier Maury**, Laboratoire de chimie, UMR 5281 ENS Lyon-CNRS-Université de Lyon, France

**David E. Morris**, Los Alamos National Laboratory, USA

**Janet R. Morrow**, University at Buffalo, State University of New York, USA

**Gilles Muller**, Department of Chemistry, San José State University, USA

**Simon J.A. Pope**, School of Chemistry, Cardiff University, Wales, UK

**Manuel Tropiano**, Chemistry Research Laboratory, University of Oxford, UK

**Frank C.J.M. van Veggel**, Department of Chemistry, University of Victoria, Canada

# Preface

The unique spectroscopic properties of the lanthanide ions prompted Sir William Crookes in his lecture delivered 1887 at the Royal Institution to say: "These elements perplex us in our researches, baffle us in our speculations, and haunt us in our very dreams. They stretch like an unknown sea before us – mocking, mystifying, and murmuring strange revelations and possibilities" (*The Chemical News*, 1887, pp. 83–88). These unique properties, which are line-like absorption and equally narrow emission spectra, played a central role in the separation and identification of the 14 elements. As each lanthanide ion shows a characteristic spectroscopic signature and line-like spectra, they have continued to fascinate researchers through the ages and have led to many applications as well as new fields of research. The interest in spectroscopy and spectroscopic applications of the lanthanide ions has resulted in a growing number of publications. Among these are several books that address one or more areas of lanthanide chemistry and spectroscopy, such as the recent *Rare Earth Coordination Chemistry* edited by Chunhui Huang, Wybourne and Smentek's theoretical treatise on the *Optical Spectroscopy of Lanthanides – Magnetic and Hyperfine Interactions*, or *Lanthanide Luminescence* edited by Hänninen and Härmä. Our new book aims to serve scientists whose primary field of interest is spectroscopy and spectroscopic applications of lanthanide ions, veteran scientists for whom the field is reviewed, as well as new scientists, who can find here information that will help them to get started. Finally, this book is also intended as the basis for an intermediate to advanced course in *f* element spectroscopy.

The first two chapters of this work cover theoretical and practical aspects of the emission process, the spectroscopic techniques and the equipment used to characterize the emission. Chapter 3 introduces and reviews the property of circularly polarized emission, while Chapter 4 reviews the use of lanthanide ion complexes in bioimaging and fluorescence microscopy. Chapter 5 covers the phenomenon of two-photon absorption, its theory as well as applications in imaging, while Chapter 6 reviews the use of lanthanide ions as chemosensors. Chapter 7 introduces the basic principles of nanoparticle upconversion luminescence and its use for bioimaging and Chapter 8 reviews direct excitation of the lanthanide ions and the use of the excitation spectra to probe the metal ion's coordination environment in coordination compounds and biopolymers. Finally, Chapter 9 describes the formation of heterobimetallic complexes, in which the lanthanide ion emission is promoted through the hetero-metal.

I am deeply indebted to all who made this book possible. My thanks to the contributing authors of the nine chapters, without whom this book would not have been possible. They are major driving forces in their respective areas and have contributed chapters that are at once excellent tutorials and thorough reviews of their fields. My heartfelt thanks go also to the publisher and everyone involved with the book at Wiley, who, with their continued patience, encouragement, professionalism and enthusiasm led the project to its successful conclusion.

# 1

# Introduction to Lanthanide Ion Luminescence

*Ana de Bettencourt-Dias*

*Department of Chemistry, University of Nevada, USA*

## 1.1 History of Lanthanide Ion Luminescence

After the isolation of a sample of yttrium oxide from a new mineral by Johan Gadolin in 1794, several of the lanthanides, namely praseodymium and neodymium, as well as cerium, lanthanum, terbium and erbium were isolated in different degrees of purity [1]. It was only after Kirchhoff and Bunsen introduced the spectroscope in 1859 as a means of characterising elements that the remaining lanthanides were discovered and the already known ones could be obtained in pure form [2]. Spark spectroscopy provided the means to finally isolate in pure form the remaining lanthanides [3–5]. As will be discussed below, the $4f$ valence orbitals are buried within the core of the ions, shielded from the coordination environment by the filled $5s$ and $5p$ orbitals, and do not experience significant coupling with the ligands. Therefore, the electronic levels of the ions can be described in an analogous way to the atomic electronic levels with a Hamiltonian in central field approximation with electrostatic Coulomb interactions, spin–orbit coupling and finally crystal field and Zeeman effects added as perturbations. All these perturbations lead to a lifting of the degeneracy of the electronic levels and transitions between these split levels are experimentally observed [6]. These transitions, however, are forbidden by the parity rule, as there is no change in parity between excited and ground state. That the emission was nonetheless seen puzzled scientists for a long time [7]. Only when Judd and Ofelt independently proposed their theory of induced electric dipole

*Luminescence of Lanthanide Ions in Coordination Compounds and Nanomaterials*, First Edition.
Edited by Ana de Bettencourt-Dias.

transitions [8,9] could the appearance of these transitions be satisfactorily explained. As the transitions are forbidden, the direct excitation of the lanthanide ions is also not easily accomplished, and this is why sensitised emission is a more appealing and energy efficient way to promote lanthanide-centred emission. While the ability of the lanthanide salts to emit light was key to their isolation in pure form, sensitised emission was first described by S.I. Weissman only in 1942 [10]. This author realised that when complexes of Eu(III) with salicylaldehyde and benzoylacetonato, as well as other related ligands, were irradiated with light in the wavelength range in which the organic ligands absorb, strong europium-characteristic red emission ensued. Weissman further observed that the emission intensity was temperature and solvent dependent, as opposed to what is seen for europium nitrate solutions [10]. After this seminal work, interest in sensitised luminescence spread through the scientific community, as the potential application of lanthanides for imaging and sensing was quickly recognised [11,12].

## 1.2 Electronic Configuration of the +III Oxidation State

### 1.2.1 The 4*f* Orbitals

The lanthanides' position in the fourth period as the inner transition elements of the periodic table indicates that the filling of the 4*f* valence orbitals commences with them. The electronic configuration of the lanthanides is $[Xe]4f^n6s^2$, with notable exceptions for lanthanum, cerium, gadolinium and lutetium, which have a $[Xe]4f^{n-1}5d^16s^2$ configuration. Upon ionisation to the most common +III oxidation state, the configuration is uniformly $[Xe]4f^{n-1}$. La(III) therefore does not possess any *f* electrons, while Lu(III) has a filled 4*f* orbital. While the 4*f* orbitals are the valence orbitals, they are shielded from the coordination environment by the filled 5*s* and 5*p* orbitals, which are more spatially extended, as shown in Fig. 1.1, which displays the radial charge density distribution for Pr(III) [13]. Therefore, lanthanides bind mostly through ionic interactions and the ligand field perturbation upon the 4*f* orbitals is minimal. Nonetheless, as will be discussed below, symmetry considerations imposed by the ligand field affect the emission spectra of the lanthanide ions.

### 1.2.2 Energy Level Term Symbols

It is usual to describe the configurations of hydrogen-like atoms or ions, that is with only one electron, in terms of the quantum numbers $n$, $l$, $m_l$, $s$ and $m_s$. In polyelectronic atoms and ions, exchange and pairing energies lead to different configurations, or microstates, with different energies, which are described by new quantum numbers, the total orbital angular momentum quantum number $L$ and its projection along the $z$ axis, the total magnetic orbital angular momentum $M_L$, and the total spin angular momentum quantum number $S$, often indicated as the spin multiplicity, $2S+1$, as well as its projection along the $z$ axis, the total magnetic spin quantum number $M_S$. In the case of heavy elements, such as lanthanides, coupling of the spin and angular momenta is seen, and an additional quantum number, $J$, the spin–orbit coupling or Russell–Saunders quantum number, is commonly utilised. As will be mentioned below, intermediate coupling for lanthanides is more correct, but the

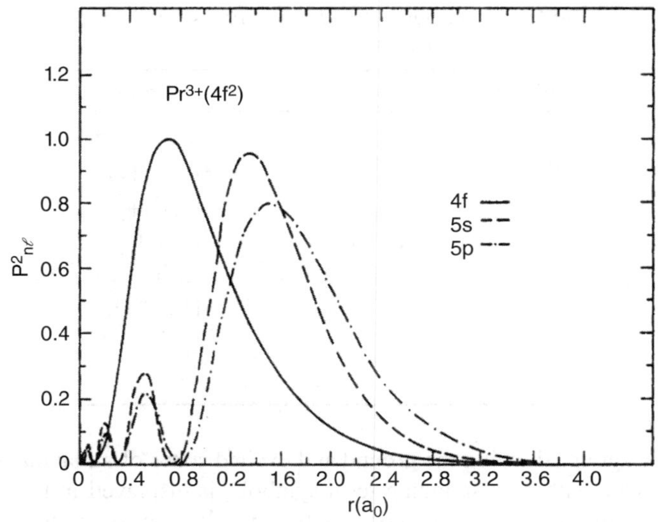


***Figure 1.1*** *Radial charge density distribution of Pr(III). Reproduced from [13] with permission from Elsevier*

Russell–Saunders formalism is simple to use and will be carried through this chapter. Term symbols with the format $^{2S+1}L_J$, which summarise the quantum number information, are assigned to describe the individual microstates. For a polyelectronic atom or ion with $i$ electrons,

$$L = \sum_i l_i, \quad M_L = -L, \ldots, L$$

$$S = \sum_i s_i, \quad s_i = 1/2$$

and

$$J = L + S, L + S - 1, \ldots, |L - S|.$$

Term symbols can be obtained by determining the microstates, or allowed combinations of all electrons described by quantum numbers, of the atom or ion under consideration and methods to do it is can be found in textbooks [14,15]. Since multiple combinations of electrons are allowed, and therefore many microstates are present, Hund's rules are followed for determination of the ground state. The ground state will have the largest spin multiplicity and the largest orbital multiplicity corresponding to the largest value of $L$. Finally, if $S$ and $L$ are equal for two states, the ground state will correspond to the largest value of $J$, if the electron shell is more than half-filled, or an inverted multiplet and the smallest value of $J$, if the orbital is less than half-filled, which is a regular multiplet. The ground state term symbols for $f^n$ ($n =$ number of electrons in the $f$ shell) configurations are shown in Table 1.1.

**Table 1.1** *Ground state term symbols for $f^n$ electronic configurations*

| Configuration | Term |
|---|---|
| $f^0/f^{14}$ | $^1S_0$ |
| $f^1/f^{13}$ | $^2F_{5/2}/^2F_{7/2}$ |
| $f^2/f^{12}$ | $^3H_4/^3H_6$ |
| $f^3/f^{11}$ | $^4I_{9/2}/^4I_{15/2}$ |
| $f^4/f^{10}$ | $^5I_4/^5I_8$ |
| $f^5/f^9$ | $^6H_{5/2}/^6H_{15/2}$ |
| $f^6/f^8$ | $^7F_0/^7F_6$ |
| $f^7$ | $^8S_{7/2}$ |

A complete diagram, showing the ground and excited states of all lanthanide ions in the +III oxidation state with corresponding term symbols, is displayed in Fig. 1.2.

Table 1.2 summarises the most commonly observed emission transitions for the emissive Ln(III) ions.

**Table 1.2** *Most common emissive f-f transitions of $Ln^{3+}$ [16–28]*

| Ln | Transition | λ [nm] |
|---|---|---|
| Pr | $^1D_2 \rightarrow {}^3F_4$ | 1000 |
| | $^1D_2 \rightarrow {}^1G_4$ | 1440 |
| | $^1D_2 \rightarrow {}^3H_J(J = 4, 5)$ | 600, 690 |
| | $^3P_0 \rightarrow {}^3H_J(J = 4-6)$ | 490, 545, 615, 640, |
| | $^3P_0 \rightarrow {}^3F_J(J = 2-4)$ | 700, 725 |
| Nd | $^4F_{3/2} \rightarrow {}^4I_J(J = 9/2-13/2)$ | 900, 1060, 1350 |
| Sm | $^4G_{5/2} \rightarrow {}^6H_J(J = 5/2-13/2)$ | 560, 595, 640, 700, 775 |
| | $^4G_{5/2} \rightarrow {}^6F_J(J = 1/2-9/2)$ | 870, 887, 926, 1010, 1150 |
| Eu | $^5D_0 \rightarrow {}^7F_J(J = 0-6)$ | 580, 590, 615, 650, 720, 750, 820 |
| Gd | $^6P_{7/2} \rightarrow {}^8S_{7/2}$ | 315 |
| Tb | $^5D_4 \rightarrow {}^7F_J(J = 6-0)$ | 490, 540, 580, 620, 650, 660, 675 |
| Dy | $^4F_{9/2} \rightarrow {}^6H_J(J = 15/2-9/2)$ | 475, 570, 660, 750 |
| | $^4I_{15/2} \rightarrow {}^6H_J(J = 15/2-9/2)$ | 455, 540, 615, 695 |
| Ho | $^5S_2 \rightarrow {}^5I_J(J = 8, 7)$ | 545, 750 |
| | $^5F_5 \rightarrow {}^5I_J(J = 8, 7)$ | 650, 965 |
| Er | $^4S_{3/2} \rightarrow {}^4I_J(J = 15/2, 13/2)$ | 545, 850 |
| | $^4F_{9/2} \rightarrow {}^4I_{15/2}$ | 660 |
| | $^4I_J(J = 9/2, 13/2) \rightarrow {}^4I_{15/2}$ | 810, 1540 |
| Tm | $^1D_2 \rightarrow {}^3F_4, {}^3H_4, {}^3F_J(J = 3, 2)$ | 450, 650, 740, 775 |
| | $^1G_4 \rightarrow {}^3H_6, {}^3F_4, {}^3H_5$ | 470, 650, 770 |
| | $^3H_4 \rightarrow {}^3H_6$ | 800 |
| Yb | $^2F_{5/2} \rightarrow {}^2F_{7/2}$ | 980 |

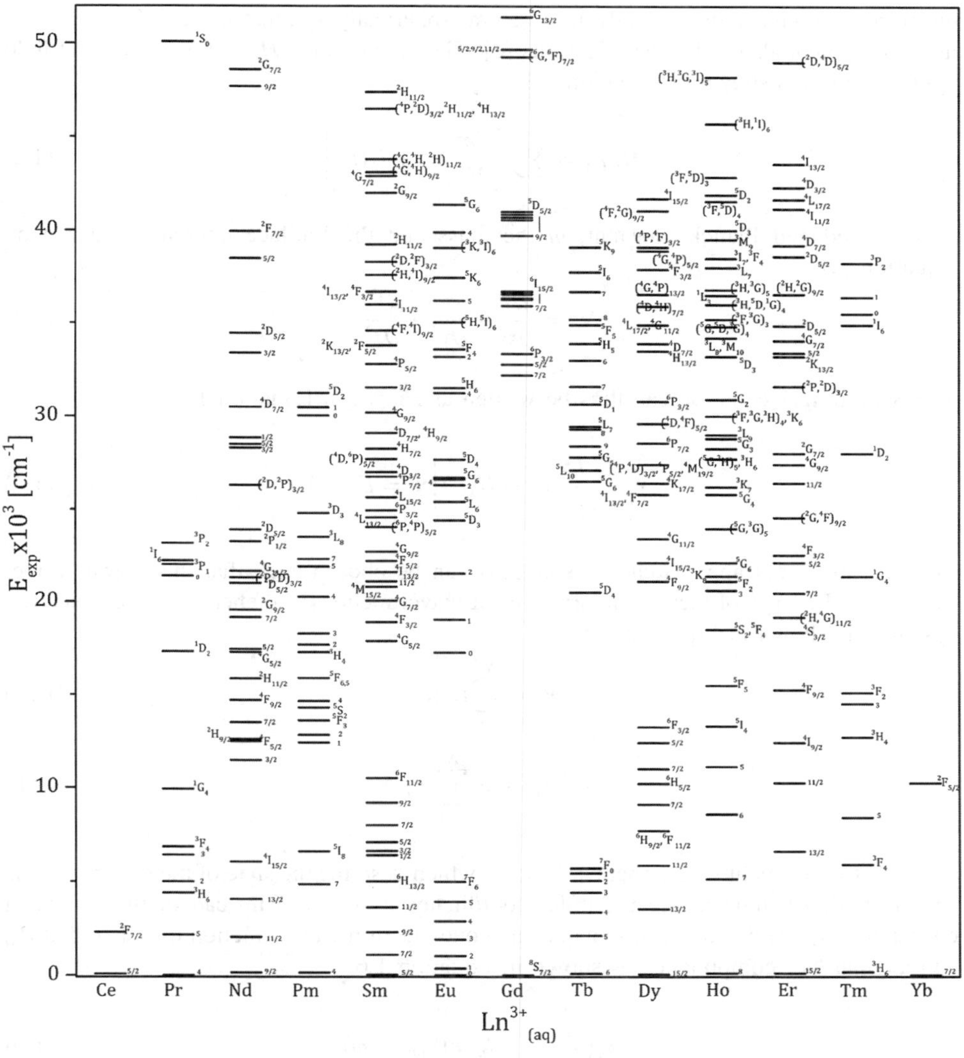


***Figure 1.2*** *Diagram of energy levels with corresponding term symbols for Ln(III) [16]*

## 1.3 The Nature of the *f-f* Transitions

### 1.3.1 Hamiltonian in Central Field Approximation and Coulomb Interactions

The behaviour of an electron is described by the wave function $\psi$, which is a solution of the Schrödinger equation 1.1.

$$H\psi = E\psi \tag{1.1}$$

This equation only has an exact solution for systems with one electron, but for polyelectronic systems with $N$ electrons, the solution can be approximated by considering that each electron is moving independently in a central spherically symmetric field $U(r_i)/e$ of the averaged potentials of all other electrons [6]. The Hamiltonian $H_{CFA}$ for this central field approximation is shown in Equation 1.2.

$$H_{CFA} = \sum_{i=1}^{N} \left[ \frac{-\hbar^2}{2m} \nabla^2 + U(r_i) \right] \tag{1.2}$$

$\hbar$ is the reduced Planck constant, $m$ the mass and the Laplace operator is given by Equation 1.3.

$$\nabla^2 = \frac{\partial^2}{\partial x^2} + \frac{\partial^2}{\partial y^2} + \frac{\partial^2}{\partial z^2} \tag{1.3}$$

The Schrödinger equation can thus be written as shown in Equation 1.4.

$$\sum_{i=1}^{N} \left[ \frac{-\hbar^2}{2m} \nabla^2 + U(r_i) \right] \Psi = E_{CFA} \Psi \tag{1.4}$$

In the central field approximation, solutions can be chosen such that the overall wavefunction and energy of the system are sums of wavefunctions and energies of one-electron systems, as shown in Equation 1.5.

$$\Psi = \sum_{i=1}^{N} \psi_i \left( a^i \right) \tag{1.5a}$$

$$E_{CFA} = \sum_{i=1}^{N} E_i \tag{1.5b}$$

$a^i$ stands for the quantum numbers $n$, $l$ and $m_l$ which describe the state of the electron in the central field. By introducing the polar coordinates $r$, $\theta$ and $\phi$ instead of the Cartesian coordinates $x$, $y$ and $z$, one can separate each one-electron wave function into its radial $R_{nl}$ and angular $Y_{lm_l}$ components, as shown in Equation 1.6.

$$\psi_i \left( a^i \right) = \frac{1}{r} R_{nl}(r) Y_{lm_l}(\theta, \phi) \tag{1.6}$$

Since $R_{nl}$ is a function of $r$ only, it depends on the central field potential $U(r_i)$. A solution to this wave function, shown in Equation 1.7, is approximated and depends on the form of the central field.

$$R_{nl}(r) = -\left[ \left( \frac{2Z}{na_0} \right)^3 \frac{(n-l-1)!}{2n\{n+l\}^3} \right]^{1/2} e^{-\frac{\rho}{2}} \rho^l L_{n+l}^{2l+1}(\rho) \tag{1.7}$$

with $\rho = \frac{2Z}{na_0} r$ and $a_0 = \frac{h^2}{4\pi^2 \mu e^2}$, where $a_0$ is the Bohr radius and $\mu$ the reduced mass. This expression also includes the Laguerre polynomials $L_{n+l}^{2l+1}(\rho)$ shown in Equation 1.8.

$$L_{n+l}^{2l+1}(\rho) = \sum\nolimits_{k=0}^{n-l-1} (-1)^{k+1} \frac{\{(n+l)!\}^2}{(n-l-1-k)!(2l+1+k)!k!}\rho^k \tag{1.8}$$

The angular wave functions, which are Laplacian spherical harmonics, on the other hand, are similar to the one-electron wave function and can thus be solved. Their expression is given in Equation 1.9.

$$Y_{lm_l}(\theta, \phi) = (-1)^m \left[\frac{(2l+1)(l-|m_l|)!}{4\pi(l+|m_l|)!}\right]^{\frac{1}{2}} P_l^{m_l}(\cos\theta)e^{im_l\phi} \tag{1.9}$$

$P_l^{m_l}$ (cos $\theta$) are the Legendre functions shown in Equation 1.10.

$$P_l^{m_l}(\cos\theta) = \frac{(1-\cos^2\theta)^{m_l/2}}{2^l l!} \frac{d^{m_l+l}}{d\cos^{m_l+l}\theta}(\cos^2\theta - 1)^l \tag{1.10}$$

Relativistic corrections to the Schrödinger equation lead to the introduction of a spin function $\delta(m_s, \sigma)$, where $\sigma$ is a spin coordinate and $m_s$ is the magnetic spin quantum number, to the one electron wave function in Equation 1.6, which then takes the shape shown in Equation 1.11.

$$\psi(n, l, m_l, m_s) = \delta(l, n, m_l, m_s)R_{nl}(r)Y_{lm_l}(\theta, \phi) \tag{1.11}$$

Equation 1.5a can now be rewritten as Equation (1.12).

$$\Psi = \sum_{i=1}^{N} \psi_i(\alpha^i) \tag{1.12}$$

While the two equations look similar, in Equation 1.12 $\alpha^i$ stands for the four quantum numbers $n$, $l$, $m_l$ and $m_s$, which describe the state of each $i$ of the $N$ electrons. These permutate to generate equally valid states following Pauli's exclusion principle, to yield anti-symmetric wave functions in the central field, which are solutions to the Schrödinger equation (Equation 1.4).

The lack of perturbations to the Hamiltonian in the central field approximation results in high degeneracy $D$ (Equation 1.13) of the $f$ electron configurations.

$$D = \frac{(4l+2)!}{N!(4l+2-N)!} = \frac{14!}{N!(14-N)!} \quad \text{for } l = 3 \tag{1.13}$$

The Hamiltonian for the perturbation introduced by the potential energy $H_{pot}$ felt by all electrons in the field of the nucleus corrected for the central spherically symmetric field is given by Equation 1.14.

$$H_{pot} = \sum_{i=1}^{N} \left[-\frac{Ze^2}{r_i} - U(r_i)\right] \tag{1.14}$$

$Ze$ is the nuclear charge, $r_i$ the position coordinates of electron $i$ and $U(r_i)$ the spherical repulsive potential of all other electrons experienced by electron $i$ moving independently in the field of the nucleus.

The repulsive Coulomb energy between pairs of electrons is an important perturbation to the central field approximation and its Hamiltonian $H_{Coulomb}$ is given by Equation 1.15.

$$H_{Coulomb} = \sum_{i<j}^{N} \frac{e^2}{r_{ij}} \tag{1.15}$$

$e$ is the charge of the electron and $r_{ij}$ is the distance between electrons $i$ and $j$.

By applying $H_{Coulomb}$ to the wave function of the unperturbed system, it can be shown that the electrostatic repulsion energy $E_{ER}$ of the system is given by Equation 1.16.

$$E_{ER} = \sum_{k=2,4,6} f_k F^k \tag{1.16}$$

Here, $k$ is an integer of values 2, 4 and 6, $f_k$ are the coefficients representing the angular part of the wave function [29] and $F^k$ are the electrostatic Slater two-electron radial integrals given by Equation 1.17.

$$F^k = (4\pi)^2 e^2 \int_0^\infty \int_0^\infty \frac{r_<^k}{r_>^{k+1}} R_{nl}^2(r_i) R_{n'l'}^2(r_j) r_i^2 r_j^2 dr_i dr_j \tag{1.17}$$

$r_<$ is the smaller and $r_>$ the larger of the values of $r_i$ and $r_j$. $F_k$ instead of the Slater integrals are often indicated, for which:

$$F_2 = F^2/225$$
$$F_4 = F^4/1089$$
$$F_6 = F^6/7361.64$$

In the case of hydrogenic wave functions the following relationships are valid [30].

$$F_4 = 0.145\, F_2 \qquad F_6 = 0.0164\, F_2$$

These show that the values of $F_k$ decrease as $k$ increases. Values of $F_2$ for the configurations $f^2$ to $f^{13}$ are tabulated in Table 1.3 and show that they increase with increasing atomic number, as the inter-electronic repulsion is expected to increase.

The $f_k$ angular coefficients are hydrogen-like and can be determined from

$$f_k = \frac{(2l+1)(l-|m_l|)!}{2(l+|m_l|)!} \frac{(2l'+1)(l'-|m_l'|)!}{2(l'+|m_l'|)!} \int_0^\pi \{P_l^{m_l}(\cos\theta_i)\}^2 P_0^k(\cos\theta_i) \sin\theta_i d\theta_i \times \int_0^\pi \{P_{l'}^{m_{l'}}(\cos\theta_i)\}^2 P_0^k(\cos\theta_i) \sin\theta_i d\theta_i \tag{1.18}$$

As above, $P_l^{m_l}$, $P_{l'}^{m_{l'}}$ and $P_0^k$ are Legendre polynomials.

*Table 1.3 Comparison of the average magnitude of perturbations for transition metal and lanthanide ions in cm$^{-1}$ [13]*

| Valence configuration | $H_{Coulomb}$ | $H_{s-o}$ | $H_{cf}$ |
|---|---|---|---|
| $3d^N$ | 70 000 | 500 | 15 000 |
| $4d^N$ | 50 000 | 1000 | 20 000 |
| $5d^N$ | 20 000 | 2000 | 25 000 |
| $4f^N$ | 70 000 | 1500 | 500 |
| $5f^N$ | 50 000 | 2500 | 2000 |

In addition to the Coulomb interactions of electron–electron repulsion and electron–nucleus attraction, further perturbations influence the energy levels of the lanthanide ions, such as the coupling of the spin and angular momenta, commonly designated spin–orbit coupling, the crystal field or Stark effect, and the interaction with a magnetic field or Zeeman effect, which will be described in the following sections.

As illustrated in Fig. 1.3, by comparison to electron–electron repulsion, which leads to energy splits on the order of $10^4\,cm^{-1}$, and spin–orbit coupling, with splits on the order of $10^3\,cm^{-1}$, the crystal field and Zeeman effects are small perturbations, resulting in energy level splitting on the order of $10^2\,cm^{-1}$ at the most [13]. The magnitude of these data compared to the *d* metals is shown comparatively in Table 1.4. In the case of transition metals, the crystal field splitting dominates the spin–orbit coupling. However, for lanthanide ions, the crystal field splitting is almost negligible. The spin–orbit coupling is of increasing

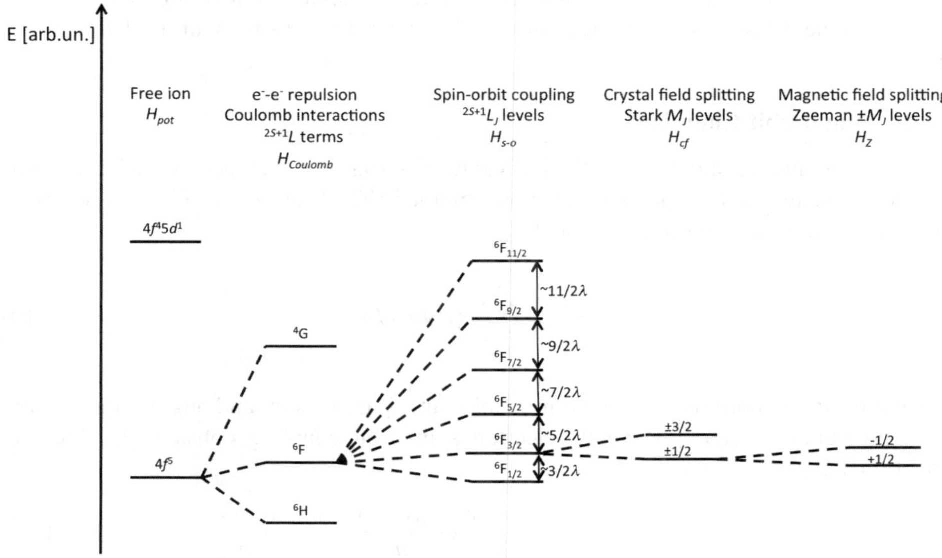


*Figure 1.3 Effect of the perturbations [Coulomb ($H_{Coulomb}$), spin–orbit ($H_{s\text{-}o}$), crystal field ($H_{cf}$), and magnetic field ($H_Z$)] on the electron configuration of an arbitrary Ln(III) Kramers' ion. Energy units are arbitrary and not to scale. λ is described in Section 3.2*

***Table 1.4*** *Spin–orbit radial integral $\zeta_{nl}$, spin-orbit coupling constant $\lambda$ and $F_2$ values for the* $Ln^{3+}_{(aq)}$ *ions [25–28,31]*

| $f^N$ | $\zeta_{nl}$ [$cm^{-1}$] | $\lambda$ [$cm^{-1}$][a] | $F_2$ [$cm^{-1}$][b] |
|---|---|---|---|
| $f^1$ | 625 | 625 | |
| $f^2$ | 740 | 370 | 305 |
| $f^3$ | 884 | 295 | 321 |
| $f^4$ | 1022 | 250 | 338 |
| $f^5$ | 1157 | 231 | 364 |
| $f^6$ | 1326 | 221 | 369 |
| $f^7$ | 1450 | 0 | 384 |
| $f^8$ | 1709 | −285 | 401 |
| $f^9$ | 1932 | −386 | 407 |
| $f^{10}$ | 2141 | −535 | 419 |
| $f^{11}$ | 2380 | −793 | 440 |
| $f^{12}$ | 2628 | −1314 | 461 |
| $f^{13}$ | 2870 | −2880 | 444[c] |

[a] $f^1$ as Ce:$LaCl_3$ [32] and $f^{13}$ as $Yb_3Ga_5O_{12}$ [33].
[b] [16]
[c] [30]

importance for the heavier elements. However, in the case of the lanthanides, it is still approximately an order of magnitude smaller than the Coulomb interactions and one order of magnitude larger than the crystal field splitting; therefore an intermediate coupling scheme, in which *j-j* in addition to Russell–Saunders coupling is also important, is more correct. Nonetheless, as mentioned above, the latter formalism is utilised due to its simplicity.

### 1.3.2 Spin–Orbit Coupling

The spin and angular momenta of the individual electrons couple with each other and this coupling is increasingly important with atomic number. The Hamiltonian $H_{s\text{-}o}$ that describes this perturbation is given in Equation 1.19.

$$H_{s\text{-}o} = \sum_{i=1}^{N} \xi(r_i)(s_i \cdot l_i) \tag{1.19}$$

$r_i$ is the position coordinate of electron $i$, and $s_i$ and $l_i$ are its spin and angular momentum quantum numbers. $\xi(r_i)$, the single electron spin–orbit coupling constant, is given by Equation 1.20.

$$\xi(r_i) = \frac{\hbar^2}{2m^2c^2 r_i}\frac{dU(r_i)}{dr_i} \tag{1.20}$$

In this equation, $c$ is the speed of light in a vacuum and $\hbar$ is the reduced Planck constant. $\xi(r_i)$ is related to the spin–orbit radial integral $\zeta_{nl}$ by equation 1.21.

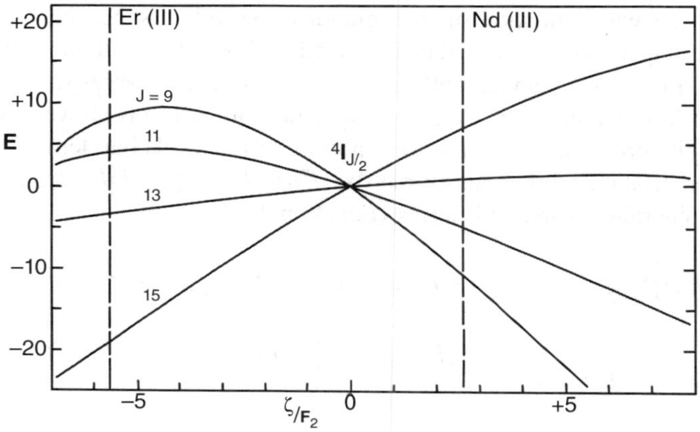


***Figure 1.4*** *The energies and splitting of the $^4I$ level for the $f^3$ and $f^{11}$ configurations as a function of the ratio $\zeta_{nl}/F_2$. The energy levels for the ratios −5.7 for Er(III) and 2.6 for Nd(III) are indicated by the dashed vertical lines. Adapted with permission from [16]. Interscience Publishers: New York, 1968*

$$\zeta_{nl} = \int_0^\infty R_{nl}^2 \xi(r) dr \tag{1.21}$$

and to the many electron spin–orbit coupling constant $\lambda$ by Equation 1.22, for $S \neq 0$.

$$\lambda = \pm \frac{\xi(r)}{2S} \tag{1.22}$$

Values of $\zeta_{nl}$ and $\lambda$ for the hydrated $Ln^{3+}$ ions are summarised in Table 1.4, with $\lambda$ positive for a more than half-filled shell and negative for a less than half-filled shell. It can be seen that $\zeta_{nl}$ increases with increasing number of $f$ electrons, which corresponds to a higher atomic number $Z$ and a stronger spin–orbit interaction, as expected.

$H_{s\text{-}o}$ will permit coupling of $^{2S+1}L$ states for $\Delta S \leq 1$ and $\Delta L \leq 1$. This effect is shown in Fig. 1.4, in which the energy splitting of the $^4I$ level due to spin–orbit coupling is shown as a function of the ratio $\zeta_{nl}/F_2$. The increased curvature of the levels shows the increasing spin–orbit coupling. The energy levels of the reverse multiplet of Er(III) and of the multiplet of Nd(III) are indicated by the vertical dashed lines.

The calculated compositions of the $^4I$ multiplet levels of Nd(III) and of Er(III) are given below.

| Nd(III) | Er(III) |
|---|---|
| $\langle ^4I_{9/2}\| = -0.166[^2H] + 0.984[^4I]$ | $\langle ^4I_{15/2}\| = 0.982[^4I] - 0.186[^2K]$ |
| $\langle ^4I_{11/2}\| = 0.995[^4I]$ | $\langle ^4I_{13/2}\| = 0.995[^4I]$ |
| $\langle ^4I_{13/2}\| = -0.993[^4I]$ | $\langle ^4I_{11/2}\| = 0.133[^4G] - 0.129[^2H] + 0.442[^2H'] + 0.875[^4I]$ |
| $\langle ^4I_{15/2}\| = 0.993[^4I] + 0.118[^2K]$ | $\langle ^4I_{9/2}\| = -0.416[^4F] - 0.342[^2G] + 0.276[^2G'] - 0.219[^2H] + 0.438[^2H'] + 0.627[^4I]$ |

Here, $\langle {}^4I_J|$ is the wave function of the spin–orbit perturbed state and $[{}^4I]$ is the wave function of the unperturbed state; a state indicated by ' is a state with the same $L$ and $S$ but higher energy. Er(III), the heavier lanthanide ion, experiences a larger spin–orbit coupling, as can be seen from the graph as well as composition of the levels above. It can further be inferred that spin–orbit coupling leads to a splitting of the levels into terms with different $J$ values. Diagonalisation of the energy matrix $\langle l^n\alpha LSJ|\sum_i \xi(r_i)s_i l_i|l^n\alpha' L'S'J'\rangle$ allows estimation of the energies of the split terms (Equation 1.23).

$$\langle l^n\alpha LSJ|\textstyle\sum_i \xi(r_i)s_i l_i|l^n\alpha' L'S'J'\rangle = (-1)^{L+S'+j}\zeta_{nl}\sqrt{(2l+1)(l+1)}\delta_{JJ'} \times \begin{Bmatrix} L & S & J \\ S' & L' & 1 \end{Bmatrix} \langle l^n\alpha LS||V^{11}||l^n\alpha' L'S'\rangle \tag{1.23}$$

$\delta_{ij}$ are the Kronecker delta symbols, for which $\delta_{ij}=0$ for $i\neq j$ and $\delta_{ij}=1$ for $i=j$. $\alpha$ stands for all additional quantum numbers which describe the initial and final states of $l^n$. The doubly reduced matrix elements $\langle l^n\alpha LS||V^{11}||l^n\alpha' L'S'\rangle$, containing the spin–orbit operator $V^{11}$, are tabulated [34]. The term between curly brackets is the six-$j$ symbol, which describes the coupling of three momenta, in this case $L$, $S$ and $J$. Online calculators are available to determine these, or they are tabulated [35]. From the 6-$j$ symbol selection rules arise, as it is only non-zero when:

$$\Delta S = 0, \pm 1 \quad \Delta L = 0, \pm 1$$

$$S' + S \geq 1 \quad L' + L \geq 1$$

$$\Delta J = 0$$

The energy of each term with respect to the barycentre of the parent term ${}^{2S+1}L$ can be approximated by Equation 1.24.

$$E_J = {}^1\!/_2\lambda[J(J+1) - L(L+1) - S(S+1)] \tag{1.24}$$

Using this equation, it is possible to estimate that the ${}^3H_5$ energy level of $Pr^{3+}$ ($4f^2$) will be located approximately 370 cm$^{-1}$ or $-1\lambda$ below the barycentre of the ${}^3H$ level, while the ${}^3H_6$ will be $6\lambda$ or 2220 cm$^{-1}$ above and the ${}^3H_4$ level $-5\lambda$ or 1850 cm$^{-1}$ below [16]. From Equation 1.24 it can further be concluded that the energy gap $\Delta E$ between two adjacent levels with $J' = J + 1$ is approximated by Landé's interval rule (see also Fig. 1.3), given in Equation 1.25.

$$\Delta E = \lambda J' \tag{1.25}$$

Landé's interval rule is only strictly obeyed in the case of strong $LS$ coupling and is only approximated in lanthanides, where intermediate coupling, consisting of interaction of levels with the same $J$ but different $L$ and $S$, is more correct. As a consequence, the magnitude of the interval $\Delta E$ determined through Equation 1.25 is usually more accurate for the lower energy levels of the lighter lanthanides. Nonetheless, a good approximation between the experimentally observed gaps and the gaps calculated by Landé's rule is usually seen, especially for ground-state multiplets. In the case of $Pr^{3+}$ the free ion energy levels for $^3H_4$, $^3H_5$ and $^3H_6$ are located at 0, 2152 and 4389 cm$^{-1}$, respectively [16], leading to $\Delta E$ values of 2152 and 2237 cm$^{-1}$ between $J=4$ and 5 and $J=5$ and 6, which reasonably approximate the values of 1850 and 2220 cm$^{-1}$ obtained through Equation 1.25.

### 1.3.3 Crystal Field or Stark Effects

When lanthanide ions are in inorganic lattices or compounds in general, in addition to the Coulomb interactions and the spin–orbit coupling, each electron $i$ also feels the effect of the crystal field generated by the ligands surrounding the metal ion, in analogy to the effect first described by Stark of an electric field on the lines of the hydrogen spectrum [36]. This perturbation lifts the $2J+1$ degeneracy and generates new levels with $M_J$ quantum numbers. Since a potential is generated by the electrons of the $N$ ligands, which is felt by the electrons of the lanthanide ions, the Hamiltonian can be defined by Equation 1.26.

$$H_{cf} = -e\sum_{1}^{N} V(r_i) \tag{1.26}$$

$e$ is the elementary charge, $V(r_i)$ is the potential felt by electron $i$ and $r_i$ its position. Following the same reasoning utilised to derive Equations 1.6 and 1.12 one can express the Hamiltonian as a function of the crystal field parameters $B^k_q$, which are related to the spherical harmonics $Y^k_q$, as shown in Equation 1.27 [37].

$$H_{cf} = \sum_{i,j,k} \left(B^k_q\right)(C^k_q)_i \tag{1.27}$$

The relationships between $B^k_q$ and $Y^k_q$ are shown in Equation 1.28.

$$\begin{aligned} B^k_0 &= \int_0^\infty R^2_{nl}(r) r^k dr \sqrt{\frac{4\pi}{2k+1}} Y^k_0 \sum_L \frac{Z_L e^2}{R_L^{k+1}} \\ B^k_q &= \int_0^\infty R^2_{nl}(r) r^k dr \sqrt{\frac{4\pi}{2k+1}} Re\ Y^k_q \sum_L \frac{Z_L e^2}{R_L^{k+1}} \\ B'^k_q &= \int_0^\infty R^2_{nl}(r) r^k dr \sqrt{\frac{4\pi}{2k+1}} Im\ Y^k_q \sum_L \frac{Z_L e^2}{R_L^{k+1}} \end{aligned} \tag{1.28}$$

$L$ are the ligands responsible for the crystal field at a distance $R_L$, $Z$ their charge and $e$ the elementary charge. Often, instead of $B^k_q$, the equivalent structural parameters $A^q_k$ are utilised as shown below.

$$B^k_q = a \times A^q_k \langle r^k \rangle \tag{1.29}$$

*Table 1.5* Expectation values $\langle r^k \rangle$ in a.u. [38]

| | $\langle r^1 \rangle$ | $\langle r^2 \rangle$ | $\langle r^3 \rangle$ | $\langle r^4 \rangle$ | $\langle r^5 \rangle$ | $\langle r^6 \rangle$ |
|---|---|---|---|---|---|---|
| $Ce^{3+}$ | 0.97 | 1.17 | 1.73 | 3.08 | 6.44 | 15.55 |
| $Pr^{3+}$ | 0.93 | 1.08 | 1.55 | 2.65 | 5.36 | 12.53 |
| $Nd^{3+}$ | 0.90 | 1.01 | 1.39 | 2.31 | 4.53 | 10.31 |
| $Sm^{3+}$ | 0.84 | 0.89 | 1.15 | 1.81 | 3.38 | 7.32 |
| $Eu^{3+}$ | 0.82 | 0.84 | 1.06 | 1.62 | 2.96 | 6.28 |
| $Gd^{3+}$ | 0.79 | 0.79 | 0.98 | 1.46 | 2.61 | 5.45 |
| $Tb^{3+}$ | 0.77 | 0.75 | 0.91 | 1.33 | 2.33 | 4.76 |
| $Dy^{3+}$ | 0.75 | 0.71 | 0.84 | 1.21 | 2.08 | 4.19 |
| $Ho^{3+}$ | 0.74 | 0.68 | 0.79 | 1.11 | 1.87 | 3.71 |
| $Er^{3+}$ | 0.72 | 0.65 | 0.74 | 1.02 | 1.69 | 3.31 |
| $Tm^{3+}$ | 0.70 | 0.62 | 0.69 | 0.94 | 1.54 | 2.97 |
| $Yb^{3+}$ | 0.69 | 0.60 | 0.65 | 0.87 | 1.40 | 2.67 |

$a$ is a constant for each $B^k_q$ and $A^q_k$ pair [29], and $\langle r^k \rangle$ represents the average or expectation value of $r^k$, where $r$ is the nucleus–electron distance of the lanthanide ion, given by

$$\langle r^k \rangle = \int_0^\infty R^2_{nl}(r) r^k dr \tag{1.30}$$

Tabulated values of $\langle r^k \rangle$ for all $Ln^{3+}$ are summarised in Table 1.5.

$(C^k_q)_i$ are the related tensor operators, which transform as the spherical harmonics and are given by

$$(C^k_q)_i = \sqrt{\frac{4\pi}{2k+1}} Y^k_q(i) \tag{1.31}$$

### 1.3.4 The Crystal Field Parameters $B^k_q$ and Symmetry

The integer $k$ runs in the range 0–7 and the parameters containing even values of $k$ are responsible for the crystal field splitting, while those with odd values influence the intensity of the induced electronic dipole transitions (see Section 1.3.10 for more details) [8,9]. $q$ is also an integer and its values depend on the symmetry of the crystal field and the magnitude of $k$, since $|q| \leq k$. The possible combinations of $k$ and $q$ for the crystal field parameters are given in Table 1.6 and the symmetry elements contained in the crystal field parameters are summarised in Table 1.7.

The $B^0_0$ coefficient is notably absent from these tables; since it is spherically symmetric, it acts equally on all $f^N$ configurations. In energy level calculations it can therefore be incorporated into all spherically symmetric interactions and does not need to be considered individually.

**Table 1.6** *Values of q allowed as a function of the symmetry elements of the crystal field [30]*

| Symmetry element present | Values of $q$ allowed |
|---|---|
| $C_n$ (coincident with main axis) | $\lvert q\rvert \leq k$, but is integer of $n$ |
| $\sigma_h$ ($xy$-plane) | Odd $k \Rightarrow q$ odd or $q \neq$ zero |
| | Even $k \Rightarrow q$ even |
| $\sigma_v$ ($xz$-plane) | No imaginary terms |
| $i$ (inversion center) | $k$ even |
| $C_2'$ | Odd $k \Rightarrow q \neq$ zero |
| | Odd $k + \lvert q\rvert \Rightarrow$ no real terms |
| | Even $k + \lvert q\rvert \Rightarrow$ no imaginary terms |
| $S_n$ (coincident with main axis) | Odd $k \Rightarrow q \neq$ zero |
| | Odd $k + \lvert q\rvert \Rightarrow q \neq n$-fold |
| | Even $k + \lvert q\rvert \Rightarrow q \neq (2x+1)n/2$ ($x = 0, 1, 2, \ldots$) |

**Table 1.7** *Symmetry elements of the crystal field parameters $B_q^k$ [30]*

| $q$ | 0 | 1 | 2 | 3 | 4 | 5 | 6 | 7 |
|---|---|---|---|---|---|---|---|---|
| $B_q^1$ | $C_\infty, \sigma_v$ | $\sigma_h, \sigma_v$ | | | | | | |
| $B_q^2$ | $C_\infty, i, C2, S_4, \sigma_h$ | $\sigma_v, i, C2$ | $C_2, i, C_2', \sigma_h, \sigma_v$ | | | | | |
| $B_q^3$ | $C_\infty, \sigma_v$ | $\sigma_h, \sigma_v$ | $C_2, \sigma_v, S_4$ | $C_3, \sigma_h, \sigma_v$ | | | | |
| $B_q^4$ | $C_\infty, i, C_2', S_4, \sigma_h$ | $\sigma_v, i, C_2'$ | $C_2, i, C_2', \sigma_h, \sigma_v$ | $C_3, i, C_2', \sigma_v, S_6$ | $C_4, i, C_2', \sigma_h, \sigma_v, S_4$ | | | |
| $B_q^5$ | $C_\infty, \sigma_v$ | $\sigma_h, \sigma_v$ | $C_2, \sigma_v, S_4$ | $C_3, \sigma_h, \sigma_v$ | $C_4, \sigma_v$ | $C_5, \sigma_h, \sigma_v$ | | |
| $B_q^6$ | $C_\infty, i, C_2', S_4, \sigma_h$ | $\sigma_v, i, C_2'$ | $C_2, i, C_2', \sigma_h, \sigma_v$ | $C_3, i, C_2', \sigma_v, S_6$ | $C_4, i, C_2', \sigma_h, \sigma_v, S_4$ | $C_5, i, C_2', \sigma_v$ | $C_6, i, C_2', \sigma_h, \sigma_v, S_6$ | |
| $B_q^7$ | $C_\infty, \sigma_v$ | $\sigma_h, \sigma_v$ | $C_2, \sigma_v, S_4$ | $C_3, \sigma_h, \sigma_v$ | $C_4, \sigma_v$ | $C_5, \sigma_h, \sigma_v$ | $C_6, \sigma_v$ | $C_7, \sigma_h, \sigma_v$ |
| $B_q'^1$ | | $\sigma_h, \sigma_d, C_2'$ | | | | | | |
| $B_q'^2$ | | $\sigma_d, i$ | $C_2, i, \sigma_h, \sigma_d$ | | | | | |
| $B_q'^3$ | | $\sigma_h, \sigma_d, C_2'$ | $C_2, C_2', \sigma_d, S_4$ | $C_3, C_2', \sigma_h, \sigma_d$ | | | | |
| $B_q'^4$ | | $\sigma_d, i$ | $C_2, i, \sigma_h, \sigma_d$ | $C_3, i, \sigma_d, S_6$ | $C_4, i, \sigma_h, \sigma_d, S_4$ | | | |
| $B_q'^5$ | | $\sigma_h, \sigma_d, C_2'$ | $C_2, C_2', \sigma_d, S_4$ | $C_3, C_2', \sigma_h, \sigma_d$ | $C_4, C_2', \sigma_d$ | $C_5, C_2', \sigma_h, \sigma_d$ | | |
| $B_q'^6$ | | $\sigma_d, i$ | $C_2, i, \sigma_h, \sigma_d$ | $C_3, i, \sigma_d, S_6$ | $C_4, i, \sigma_h, \sigma_d, S_4$ | $C_5, i, \sigma_d$ | $C_6, i, \sigma_h, \sigma_d, S_6$ | |
| $B_q'^7$ | | $\sigma_h, \sigma_d, C_2'$ | $C_2, C_2', \sigma_d, S_4$ | $C_3, C_2', \sigma_h, \sigma_d$ | $C_4, C_2', \sigma_d$ | $C_5, C_2', \sigma_h, \sigma_d$ | $C_6, C_2'$ | $C_7, C_2', \sigma_h, \sigma_d$ |

The expression for $H_{cf}$ varies depending on the symmetry of the crystal field, as shown by the information in Tables 1.3 and 1.4. For example, if the metal ion is in a site of $C_{2v}$ symmetry, which includes two mirror planes and one $C_2$ axis, the expression for $H_{cf}$ becomes:

$$H_{cf} = \sum_{k=2,4,6} B_0^k C_0^k + \sum_{k=2,4,6} B_2^k (C_2^k + C_{-2}^k) + \sum_{k=4,6} B_4^k (C_4^k + C_{-4}^k) + B_6^6 (C_6^6 + C_{-6}^6) \quad (1.32)$$

Complete expressions for the summations for symmetry point groups of interest in coordination chemistry can be found for example in Reference [30].

Further discussion of the crystal field perturbation and crystal field parameters will be continued in Section 1.3.7.

As stated, the crystal field lifts the degeneracy of the $J$ levels. However, in the case of Kramers' ions, which have an odd number of electrons and for which therefore $J$ is half-integer, the degeneracy is not completely removed and each sub-level is two-fold degenerate and therefore a Kramers' doublet [39]. Nonetheless, the lifting of the degeneracy is related to the symmetry around the metal ion, and the number of new $M_J$ sub-levels as a function of symmetry is summarised in Table 1.8.

In the case of the Eu(III) ion, where ground and excited state manifolds are well-separated, this direct dependence of the number of $M_J$ levels on the crystal field symmetry is often utilised to determine the point group symmetry of the metal ion in a complex or solid state material from the emission spectra. This method of descending symmetry is performed with the help of a diagram such as the one shown in Fig. 1.5 [41]. A similar analysis can also be performed on the basis of absorption spectra.

By using the reasoning above, the splitting of the $4f^6$ configuration of $Eu^{3+}$ in $O_h$ symmetry will be as shown in Fig. 1.6.

***Table 1.8*** *Number of new $M_J$ sub-levels for a parent $J$ term split by the crystal field in a given group symmetry [13,40]*

| Symmetry | Integer $J$ | | | | | | |
|---|---|---|---|---|---|---|---|
| | 0 | 1 | 2 | 3 | 4 | 5 | 6 |
| Icosahedral Ih | 1 | 1 | 1 | 2 | 2 | 3 | 4 |
| Cubic $O_h$, $O$, $T_d$, $T_h$, $T$ | 1 | 1 | 2 | 3 | 4 | 4 | 6 |
| Hexagonal $D_{6h}$, $D_6$, $C_{6v}$, $C_{6h}$, $D_{3h}$, $C_{3h}$, $D_{3d}$, $D_3$, $C_{3v}$, $S_6$, $C_3$ | 1 | 2 | 3 | 5 | 6 | 7 | 9 |
| Pentagonal $D_{5h}$, $C_{5h}$, $C_{5v}$, $C_5$, $D_5$ | 1 | 2 | 3 | 4 | 5 | 7 | 8 |
| Tetragonal $D_{4h}$, $D_4$, $C_{4v}$, $C_{4h}$, $C_4$, $D_{2d}$, $S_4$ | 1 | 2 | 4 | 5 | 7 | 8 | 10 |
| Low $D_{2h}$, $D_2$, $C_{2v}$, $C_{2h}$, $C_2$, $C_s$, $S_2$, $C_1$ | 1 | 3 | 5 | 7 | 9 | 11 | 13 |

| | Half-integer $J$ | | | | | | | | |
|---|---|---|---|---|---|---|---|---|---|
| | 1/2 | 3/2 | 5/2 | 7/2 | 9/2 | 11/2 | 13/2 | 15/2 | 17/2 |
| Cubic | 1 | 1 | 2 | 3 | 3 | 4 | 5 | 6 | 6 |
| Other symmetries | 1 | 2 | 3 | 4 | 5 | 6 | 7 | 8 | 9 |

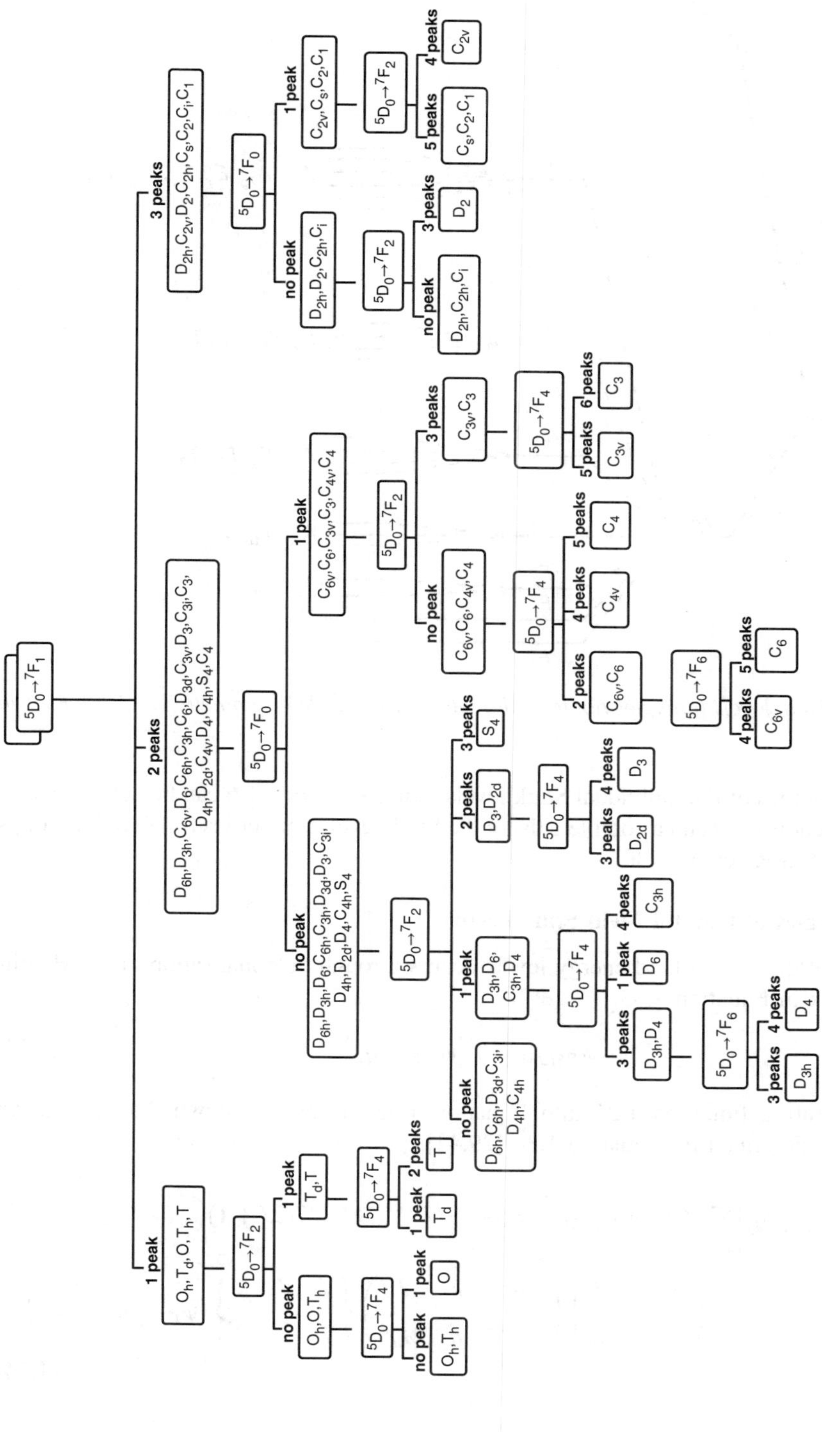


**Figure 1.5** *Method of descending symmetry applied to the J levels of the Eu(III) ion to determine the point group symmetry of the metal ion [41]*

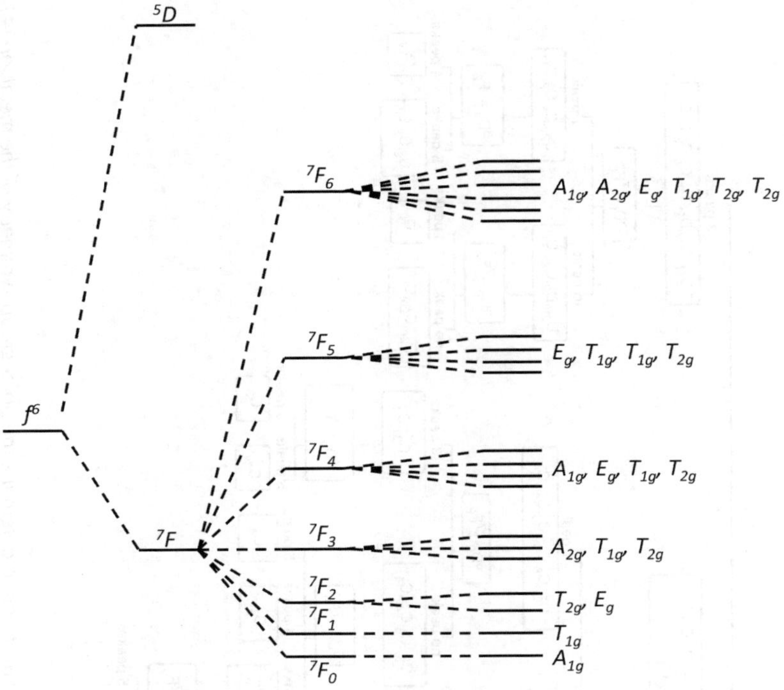


***Figure 1.6*** *Stark levels (energies not to scale) with corresponding symmetry labels for* $Eu^{3+}$ *in* $O_h$ *symmetry*

The symmetries of the individual Stark levels indicated in Fig. 1.6 can be determined as shown in Section 1.3.9 and a complete list of the Stark level symmetries for all point groups can be found in Reference [30].

### 1.3.5 Energies of Crystal Field Split Terms

Estimation of the crystal field energy levels occurs through diagonalisation of the Hamiltonian matrix in Equation 1.33.

$$\langle \psi_{l^n SLJM_J} | H_{cf} | \psi_{l^n S'L'J'M_J'} \rangle \tag{1.33}$$

After substituting Equation 1.26 into Equation 1.33, it can be shown that the matrix elements are described by Equation 1.34 [29,42].

$$\sum\nolimits_{k,q} B_q^k \left\langle \psi_{l^n SLJM_J} \middle| \sum\nolimits_i (C_q^k)_i \middle| \psi_{l^n SLJM_J'} \right\rangle = (-1)^{2J-M_J+S+L+k+3} 7(2J+1)$$
$$\times \begin{pmatrix} 3 & k & 3 \\ 0 & 0 & 0 \end{pmatrix} \begin{pmatrix} J & k & J \\ -M_J & q & M_J' \end{pmatrix} \begin{Bmatrix} J & J & k \\ L & L & S \end{Bmatrix} \langle \psi_{l^n SL} \| U^k \| \psi_{l^n SL} \rangle \tag{1.34}$$

The terms in parentheses are 3-*j* symbols and the term in braces is a 6-*j* symbol. The first 3-*j* symbol indicates the coupling of the angular momenta $l=3$ between final and initial states, the second 3-*j* and the 6-*j* symbol denote the coupling of two, $J$ and $M_J$, and three, $J$, $L$ and $S$, angular momenta, respectively. These symbols are tabulated in [35] or can be calculated in *Mathematica* [43] or by using several calculators available online. While the general formulas for the symbols are complicated, it can be shown that for the first 3-*j* symbol to be non-zero the following relationship must hold:

$$|3-3| \leq k \leq 3+3 \text{ or } k = 0-6.$$

The second 3-*j* symbol will be non-zero for: $-M_J + q + M_J' = 0 \Leftrightarrow M_J - M_J' = q$ and $|J-J| \leq k \leq J+J$ or $k \leq 2J$ and $|q| \leq k$.

These conditions for the 3-*j* symbols dictate the selection of $k=0, 2, 4, 6$ for the crystal field splitting.

The 6-*j* symbol will be non-zero for $|J-k| \leq J' \leq J+k$ and $|L'-S| \leq L \leq L'+S$.

The doubly reduced matrix elements $\langle \psi_{l^n SL} \| U^k \| \psi_{l^n S'L'} \rangle$ are specific to a given lanthanide ion. It was shown that the crystal field does not affect them substantially. Therefore, instead of calculating them for each system, the values tabulated by Nielson and Koster can be used [34].

With all of these tabulated values, only the $B_q^k$ needs to be determined to evaluate the energy level splitting based on the crystal field and its symmetry.

For $Pr^{3+}$, which has the valence configuration $f^2$, in $D_{3h}$ symmetry the reduced matrix elements $\langle U^4 \rangle$ and $\langle U^6 \rangle$ vanish (Table 1.9) and only $\langle U^2 \rangle$ needs to be evaluated. Therefore, for the ${}^3P_2$ term, with $L=S=0$, $J=2$, $q=0$, $k=2$ (see above for the discussion of the values of $q$ and $k$ depending on symmetry), and $M_J=0, \pm 1, \pm 2$, it can be shown with Equation 1.34 above that

$$\langle f^2\,{}^3P_2\,20 | H_{cf} | f^2\,{}^3P_2\,20 \rangle = -\frac{1}{5} B_0^2$$

$$\langle f^2\,{}^3P_2\,21 | H_{cf} | f^2\,{}^3P_2\,21 \rangle = -\frac{1}{10} B_0^2$$

$$\langle f^2\,{}^3P_2\,22 | H_{cf} | f^2\,{}^3P_2\,22 \rangle = \frac{1}{5} B_0^2$$

$$\langle f^2\,{}^3P_2\,20 | H_{cf} | f^2\,{}^3P_2\,00 \rangle = \frac{\sqrt{2}}{5} B_0^2$$

$$\langle f^2\,{}^3P_2\,21 | H_{cf} | f^2\,{}^3P_2\,11 \rangle = \frac{3}{10} B_0^2$$

etc.

Similar reasoning for the ${}^3P_0$ and ${}^3P_1$ terms allows the drawing of the diagram shown in Fig. 1.7, which represents the relative energies of the crystal field split terms.

### 1.3.6 Zeeman Effect

As mentioned above, in the case of Kramers' ions with non-integer value of $J$, the crystal field does not completely lift the degeneracy of the $J$ levels. This degeneracy can however

**Table 1.9** *Selected doubly reduced matrix elements* $\langle \psi_{f^n SL} \| U^k \| \psi_{f^n S'L'} \rangle$ *for the* $f^2$ $Pr^{3+}$ *ion [34]*

| $Pr^{3+}$, $f^2$ | $\langle \psi_{f^n SL} \| U^2 \| \psi_{f^n S'L'} \rangle$ | $\langle \psi_{f^n SL} \| U^4 \| \psi_{f^n S'L'} \rangle$ | $\langle \psi_{f^n SL} \| U^6 \| \psi_{f^n S'L'} \rangle$ |
|---|---|---|---|
| $^3P/^3P$ | $-3/\sqrt{14}$ | 0 | 0 |
| $^3P/^3F$ | $\sqrt{6/7}$ | $-\sqrt{11/21}$ | 0 |
| $^3F/^3F$ | $-1/3$ | $-1/3$ | $-1/3$ |
| $^3F/^3H$ | $2/3\sqrt{11/7}$ | $\sqrt{65/63}$ | $-1/3\sqrt{14}$ |

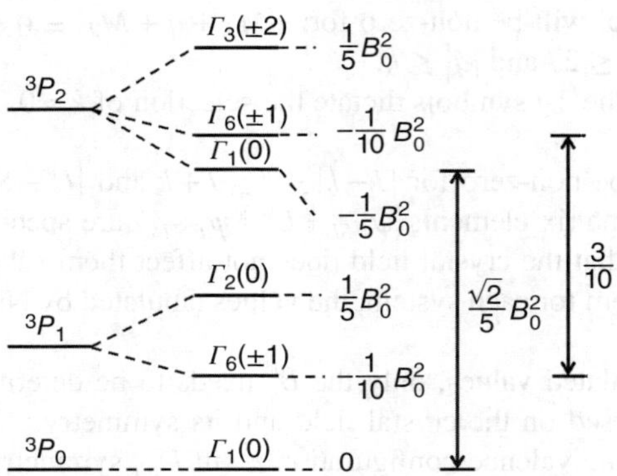


***Figure 1.7*** *Relative energies (not to scale) of the crystal field split* $^3P_J$ *(J = 0, 1, 2) terms of* $Pr^{3+}$ *in* $D_{3h}$ *symmetry*

be lifted in the presence of an external magnetic field and is often referred to as the Zeeman effect [44]. The magnitude of the splitting is proportional to the strength of the applied magnetic field. The expression for the Hamiltonian $H_z$ is given in Equation 1.35 [42].

$$H_z = g\,\mu_B B \cdot J \tag{1.35}$$

$\mu_B$ is the Bohr Magneton; $B$ is the magnitude of the external magnetic field and $g$ is Landé's factor (Equation 1.36) in the $LS$ coupling scheme.

$$g = 1 + \frac{J(J+1) - L(L+1) + S(S+1)}{2J(J+1)} \tag{1.36}$$

The effective magnetic moment $\mu$ is given by Equation 1.37.

$$\mu = -\mu_B g J \tag{1.37}$$

If the applied magnetic field is parallel to the $z$-axis of a crystal, the splitting energy $E_Z$ is given by

$$E_Z = \mu_B g B \langle JM_z | J_z | JM_z \rangle \tag{1.38}$$

$M_Z$ is the quantum number associated with this perturbation. If, on the other hand, the Zeeman interaction is anisotropic, the Hamiltonian in Equation 1.35 should be re-written as

$$H_Z = \mu_B (g_x B_x J_x + g_y B_y J_y + g_z B_z J_z) \tag{1.39}$$

with the components along the $x$, $y$ and $z$ axes designated by the appropriate subscripts.

### 1.3.7 Point Charge Electrostatic Model

Qualitative estimation of the $B_q^k$ parameters and therefore of the symmetry around the metal ion and the positions of the energy levels can be done utilising the point charge electrostatic model (PCEM), which assumes that the electric field acting on the central metal ion is generated by the ligands as negative point charges, and all ligands have the same charge. These point charges are arranged according to the correct symmetry of the metal ion site. Taking into account Equations 1.27 and 1.30, and considering that $N$ ligands will be at a distance $R_A$ and $M$ ligands at a distance $R_B$ ($R_A < R_B$), Equation 1.28 can be re-written as Equation 1.40.

$$\begin{aligned}
B_0^k &= \sum_{A=1}^{N} Z_L e^2 \frac{\langle r^k \rangle}{R_A^{k+1}} \sqrt{\frac{4\pi}{2k+1}} Y_0^k(\theta_A, \phi_A) + \sum_{B=1}^{M} Z_L e^2 \frac{\langle r^k \rangle}{R_B^{k+1}} \sqrt{\frac{4\pi}{2k+1}} Y_0^k(\theta_B, \phi_B) \\
B_q^k &= \sum_{A=1}^{N} Z_L e^2 \frac{\langle r^k \rangle}{R_A^{k+1}} \sqrt{\frac{4\pi}{2k+1}} (-1)^q \mathrm{Re}\, Y_q^k(\theta_A, \phi_A) \\
&\quad + \sum_{B=1}^{M} Z_L e^2 \frac{\langle r^k \rangle}{R_B^{k+1}} \sqrt{\frac{4\pi}{2k+1}} (-1)^q \mathrm{Re}\, Y_q^k(\theta_B, \phi_B) \\
B_q'^k &= \sum_{A=1}^{N} Z_L e^2 \frac{\langle r^k \rangle}{R_A^{k+1}} \sqrt{\frac{4\pi}{2k+1}} (-1)^q \mathrm{Im} Y_q^k(\theta_A, \phi_A) \\
&\quad + \sum_{B=1}^{M} Z_L e^2 \frac{\langle r^k \rangle}{R_B^{k+1}} \sqrt{\frac{4\pi}{2k+1}} (-1)^q \mathrm{Im} Y_q^k(\theta_B, \phi_B)
\end{aligned} \tag{1.40}$$

$\theta_L$ and $\phi_L$ are coordinates of the ligand point charges. Since the positions of the ligands are important to indicate the angles, and therefore the sign of the crystal field parameters, the choice of molecular axes should follow the usual convention, in which the axis of highest rotational symmetry coincides with the $z$-axis. Also, with judicial choice of the molecular axes, it is possible in some cases to set $B_q'^k = 0$ [30]. Finally, since the radial parts of the parameters $B_0^k$ and $B_q^k$ are the same, the ratio $B_q^k/B_0^k$ does not have a radial component and will be a constant for a given symmetry. Therefore, only the $B_0^k$ with $k = 2$, 4 and 6 needs to be fit.

For example, if the metal is situated in a hypothetical $D_{3h}$ symmetry environment with three equidistant ligands, the parameters of interest are given by the expression for $H_{cf}$

below, as demonstrated in Section 1.3.4:

$$H_{cf} = B_0^2 C_0^2 + B_0^4 C_0^4 + B_0^6 C_0^6 + B_6^6 (C_{-6}^6 + C_6^6)$$

Since the ligands span a triangle and their position angles $\theta/\phi$ are equal to 90°/90°, 90°/120° and 90°/330°, with the help of the tabulated spherical harmonics, and substituting in Equation 1.40, it follows that:

$$B_0^2 = -\frac{3}{2} Ze^2 \frac{\langle r^2 \rangle}{R^3}$$

$$B_0^4 = \frac{9}{8} Ze^2 \frac{\langle r^4 \rangle}{R^5}$$

$$B_0^6 = -\frac{15}{16} Ze^2 \frac{\langle r^6 \rangle}{R^7}$$

$$B_6^6 = -\frac{3\sqrt{231}}{32} Ze^2 \frac{\langle r^6 \rangle}{R^7}$$

By utilising a similar procedure, it can be shown that for a general prismatic polyhedron with $p$ axial ligands, $n$ equatorial ligands and $m$ ligands in the base of the prism [30]

$$\begin{aligned} B_0^2 &= Ze^2 \frac{\langle r^2 \rangle}{R^3} \left[ p - \frac{n}{2} + m(3\cos^2\theta - 1) \right] \\ B_0^4 &= Ze^2 \frac{\langle r^4 \rangle}{R^5} \left[ p + \frac{3n}{8} + \frac{m}{4}(35\cos^4\theta - 30\cos^2\theta + 3) \right] \\ B_0^6 &= Ze^2 \frac{\langle r^2 \rangle}{R^7} \left[ p - \frac{5n}{16} + \frac{m}{8}(231\cos^6\theta - 315\cos^4\theta + 105\cos^2\theta - 5) \right] \end{aligned} \tag{1.41}$$

Some polyhedra are relatively common for lanthanide ion complexes, and they will be discussed here in more detail. They are the square antiprism, the tricapped trigonal prism and the monocapped square antiprism, shown below in Fig. 1.8. The metal ion is situated in the centre of the polyhedron and the ligands, as point charges, are located at the vertices. The proper axis of highest symmetry is chosen to coincide with the $z$-axis, as mentioned before.

The previous section described the expressions for the crystal field parameters for regular polyhedra. However, most frequently the geometry around the lanthanide ion departs more or less drastically from the regular geometry. Of the crystal field parameters, $B_0^k$ depends on $\theta$, while $B_q^k$ depends both on $\theta$ and $\phi$. Therefore, distortions of both angles will affect both sets of parameters, while changes in $\phi$ will affect only the latter. When fitting the parameters, it is common to do an initial fit at a higher symmetry and then refine at the lower, coordination compound-appropriate symmetry.

The locations of the point charges are given in Table 1.10 and different coordination polyhedra and their parameters are described below.

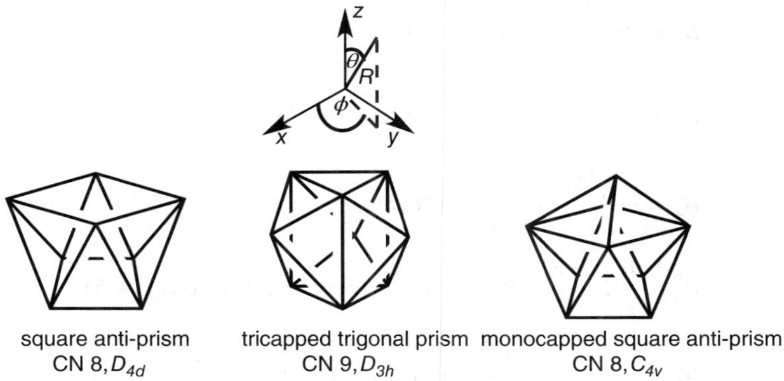


***Figure 1.8*** *Common coordinat ion polyhedra for lanthanide ions, shown with the Cartesian and polar coordinate system*

### *1.3.7.1 Square Antiprism*

This coordination geometry is often encountered when the metal ion has the coordination number (CN) 8, for which $p=n=0$ and $m=4$. The point group symmetry is $D_{4d}$, with four ligands spanning a square above and four ligands the other square below the central ion. An $S_8$ improper axis of rotation coincides with the $z$-axis and passes through the metal ion and in between all eight ligand point charges. The square antiprism is obtained by 22.5° clockwise and counter-clockwise of the top and bottom faces with respect to the metal ion from a cube. The square antiprism will be elongated when $\theta > 54.74°$ and compressed when $\theta < 54.74°$. The angle $\varphi$ can distort from $2\times 22.5°$ by $\varphi$, upon which the symmetry will decrease to $D_4$. Substituting into Equation 1.40 yields the following $B_q^k$ parameters for a distorted square antiprism.

***Table 1.10*** *Angular coordinates of the ligands as point charges for the three coordination polyhedra and distorted coordination polyhedra*[a]

| $\theta/\varphi$ [°] | $D_{4d}\rightarrow D_4$ | $D_{3h}\rightarrow D_3$ | $C_{4v}\rightarrow C_4$ |
|---|---|---|---|
| Atoms spanning top face | $\theta/45+\varphi$ | $\theta/90+\varphi$ | $125.7/45+\varphi$ |
| | $\theta/135+\varphi$ | $\theta/210+\varphi$ | $125.7/135+\varphi$ |
| | $\theta/225+\varphi$ | $\theta/330+\varphi$ | $125.7/225+\varphi$ |
| | $\theta/315+\varphi$ | | $125.7/315+\varphi$ |
| Atoms spanning bottom face | $180-\theta/45-\varphi$ | $180-\theta/90-\varphi$ | $70.1/45-\varphi$ |
| | $180-\theta/135-\varphi$ | $180-\theta/210-\varphi$ | $70.1/135-\varphi$ |
| | $180-\theta/225-\varphi$ | $180-\theta/330-\varphi$ | $70.1/225-\varphi$ |
| | $180-\theta/315-\varphi$ | | $70.1/315-\varphi$ |
| Capping atoms | — | 90/30 | 0/0 |
| | | 90/150 | |
| | | 90/270 | |

[a] $\varphi$ is the angle by which the bottom and top faces distort from the regular polyhedron, past the 45° angle in the case of the square antiprism and monocapped square antiprism and 0° angle in the case of the tricapped trigonal prism.

$$B_0^2 = Ze^2 \frac{\langle r^2 \rangle}{R^3} 4(3\cos^2\theta - 1)$$

$$B_0^4 = Ze^2 \frac{\langle r^4 \rangle}{R^5} (35\cos^4\theta - 30\cos^2\theta + 3)$$

$$B_4^4 = -\frac{\sqrt{70}}{2} Ze^2 \frac{\langle r^4 \rangle}{R^5} \sin^4\theta\cos 4\varphi$$

$$B_0^6 = \frac{1}{2} Ze^2 \frac{\langle r^6 \rangle}{R^7} (231\cos^6\theta - 315\cos^4\theta + 1 - 5\cos^2\theta - 5)$$

$$B_4^6 = \frac{21}{2\sqrt{14}} Ze^2 \frac{\langle r^6 \rangle}{R^7} \sin^4\theta(11\cos^2\theta - 1)\cos 4\varphi$$

As can be seen from the equations above, parameters $B_4^4$ and $B_4^6$ have the distortion angle $\varphi$ in their expression and are therefore indicators of the magnitude of the distortion of the square antiprism. Along similar lines, $(3\cos^2\theta - 1)$ vanishes for 54.74°, the cubic angle, and therefore the presence of the parameter $B_0^2$ is an indication of the distortion of the square antiprism from the higher symmetry cube.

#### *1.3.7.2 Tricapped Trigonal Prism*

The tricapped trigonal prism is a commonly encountered coordination polyhedron for CN 9, although often distorted. For this polyhedron $p = 0$ and $n = m = 3$. The symmetry is $D_{3h}$ and the polyhedron has three atoms at the top spanning a triangular face and three atoms at the bottom spanning another triangular face, eclipsed with the top face. A $C_3$ axis contains the metal ion and is collinear with the *z*-axis. The point charges spanning the top and bottom faces will be at a distance $R_A$, while the capping charges will be at a distance $R_B$. When the top and base faces twist by $2 \times \varphi$, the symmetry decreases to $D_3$.

$$B_0^2 = Ze^2 \frac{\langle r^2 \rangle}{R^3} \frac{9}{2} (2\cos^2\theta - 1)$$

$$B_0^4 = Ze^2 \frac{\langle r^4 \rangle}{R^5} \frac{3}{8} (70\cos^4\theta - 60\cos^2\theta + 9)$$

$$B_3^4 = Ze^2 \frac{\langle r^4 \rangle}{R^5} \frac{3\sqrt{35}}{2} (\sin^3\theta \cos\theta \sin 3\varphi)$$

$$B_0^6 = Ze^2 \frac{\langle r^6 \rangle}{R^7} \frac{3}{16} (462\cos^6\theta - 630\cos^4\theta + 210\cos^2\theta - 15)$$

$$B_3^6 = Ze^2 \frac{\langle r^6 \rangle}{R^7} \frac{3\sqrt{105}}{8} \sin^3\theta(11\cos^3\theta - 3\cos\theta)\sin 3\varphi$$

$$B_3^6 = Ze^2 \frac{\langle r^6 \rangle}{R^7} \frac{3\sqrt{231}}{32} (2\sin^6\theta \cos 6\varphi + 1)$$

### *1.3.7.3 Monocapped Square Antiprism*

This coordination polyhedron is a special case of the square antiprism, with $p = 1$, $n = 0$ and $m = 4$, as a capping atom is added on top of the top face of the antiprism, which results in a top face with a larger area. The symmetry around the metal ion is $C_{4v}$ with CN 9, with a $C_4$ axis passing through the metal ion and the capping point charge.

$$B_0^2 = Ze^2 \frac{\langle r^2 \rangle}{R^3} \left[12\cos^2\theta - 11\right]$$

$$B_0^4 = Ze^2 \frac{\langle r^4 \rangle}{R^5} \left[35\cos^4\theta - 30\cos^2\theta + 4\right]$$

$$B_0^6 = Ze^2 \frac{\langle r^6 \rangle}{R^7} \left[-\frac{1}{2} + \frac{1}{2}(231\cos^6\theta - 315\cos^4\theta + 105\cos^2\theta)\right]$$

The crystal field parameters $B_4^4$ and $B_4^6$, which account for the distortions from $C_{4v}$ symmetry, are analogous to the square antiprismatic case.

## 1.3.8 Other Methods to Estimate Crystal Field Parameters

The point-charge electrostatic model is useful in illustrating how symmetry influences the signs of the crystal field parameters $B_q^k$. However, it does not usually result in accurate determinations of their magnitude and therefore other methods have been developed that lead to a better estimation. One such approach is based on the angular overlap model AOM developed and expanded to the *f* elements by Jørgensen [45]. Another approach is the simple overlap model SOM, proposed by Malta [46].

### *1.3.8.1 Angular Overlap Model*

The angular overlap model (AOM) considers the existence of weak covalent interactions between the Ln(III) orbitals and the ligand orbitals which perturb the metal-based orbitals. The perturbation is proportional to the overlap of the metal and ligand orbitals and is evaluated in terms of the anti-bonding energy $E^*$ of the *f* orbital considered, as the anti-bonding orbital is mostly metal in character. For a complex with $N$ ligands L, $ML_N$, this energy is described by Equation 1.42 [47].

$$E^* = e_\lambda \sum_{j=1}^{N} (F_\lambda^l)^2 \tag{1.42}$$

$e_\lambda$ are quantities that depend on the radial functions of the metal M and the ligands and on the bond distances between the two and the nature of the interaction [$\lambda = 0$ ($\sigma$), 1 ($\pi$)], $l$ is the angular momentum quantum number, and $F_\lambda^l$ are the angular overlap integrals. For an $ML_9$ system with $D_{3h}$ symmetry, for which was shown above that $H_{cf} = B_0^2C_0^2 + B_0^4C_0^4 + B_0^6C_0^6 + B_6^6(C_{-6}^6 + C_6^6)$, the energies of the $f$ orbitals in units of $e_\lambda$ are shown in Table 1.11. The energies for other symmetries can be found in Reference [47].

**Table 1.11** *Energies of the f orbitals in an $ML_9$ complex in units of $e_\lambda$ for $D_{3h}$ symmetry [47]*

| f Orbital | Symmetry | $e_\sigma$ | $e_\pi$ |
|---|---|---|---|
| $z^3$ | $a_2''$ | 3/16 | 117/32 |
| $xz^2$ | e′ | 117/64 | 213/128 |
| $yz^2$ | | | |
| $z(x^2-y^2)$ | e″ | 45/32 | 195/64 |
| $xyz$ | | | |
| $x(x^2-3y^2)$ | $a_1'$ | 75/32 | 45/64 |
| $y(3x^2-y^2)$ | $a_2'$ | 0 | 135/32 |

Based on these energies, the splitting pattern shown in Fig. 1.9 for the $f$ orbitals of complex $ML_9$ in $D_{3h}$ symmetry can be deduced.

The $B_q^k C_q^k$ parameters can be expressed in terms of the AOM $e_\lambda$ [48], and for the $B_0^2 C_0^2$ in $D_{3h}$ symmetry [49].

$$B_0^2 C_0^2 = \frac{3}{7}\sqrt{5\pi}\left[\frac{(2f_1+3)}{2f_1}\right] \times (-1+4f_2-6f_2\sin^2\theta)e_{\sigma_1}$$

with $f_1 = \frac{e_{\sigma 1}}{e_{\pi 1}} = \frac{e_{\sigma 2}}{e_{\pi 2}} = \frac{e_\sigma}{e_\pi}$ and $f_2 = \frac{e_{\sigma 2}}{e_{\sigma 1}} = \frac{e_{\pi 2}}{e_{\pi 1}}$, for the ligands at distances $R_1$ and $R_2$.

Using these considerations, for Pr(III) in $LaCl_3$, Urland calculated $B_0^2 C_0^2 = 164\ \mathrm{cm}^{-1}$, which compares well with the experimental value of $150\ \mathrm{cm}^{-1}$ [49].

### 1.3.8.2 Simple Overlap Model

In the simple overlap model (SOM), the crystal field experienced by the 4$f$ electrons is generated by point charges located in small regions around the middle distance between the lanthanide ion and its ligands. These regions of charge are not necessarily positioned exactly in between the lanthanide ion and the ligands; therefore, a correction factor, $\beta_L$, which is given by Equation 1.43, is introduced to account for this asymmetry.

$$\beta_L = \frac{1}{1 \pm \rho_L} \tag{1.43}$$

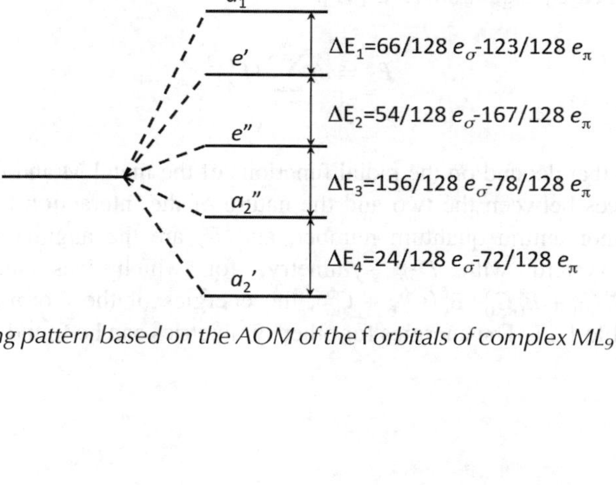


**Figure 1.9** *Splitting pattern based on the AOM of the f orbitals of complex $ML_9$ in $D_{3h}$ symmetry*

In this equation, $\rho_L$ is an additional correction factor to the position of the region of charge. A positive sign indicates that this region is shifted towards the ligand, which is usually observed for small electronegative species such as O and F. The negative sign will be used when the ligand is bulkier and contains atoms such as N and Cl and the region of charge is displaced towards the lanthanide ion. In initial publications on $LaCl_3:Nd^{3+}$ $\rho_L$ was set at 0.05 and therefore $\beta_L = 1.1$ with the region of charge towards the lanthanide [46], and on $YOCl:Eu^{3+}$ $\beta_O = 0.95$ and $\beta_{Cl} = 1.05$, showing the centroid shifted towards O for the Ln–O interaction and towards Ln for the Ln–Cl interaction, respectively [50]. It was later suggested that $\rho_L$ can be estimated using Equation 1.44 [51].

$$\rho_L = 0.05\left(\frac{R_0}{R_L}\right)^{3.5} \tag{1.44}$$

$R_0$ corresponds the shortest M–L bond and $R_L$ is the M–L bond distance for the ligand L of interest.

The charge of the region between the ligand and the lanthanide is equal to $-g_L e\rho$, where $\rho$ is proportional to the magnitude of orbital overlap between lanthanide and ligand, $e$ is the electron charge and $g_L$ is the ligand charge factor. Instead of the ligand charge $Z_L$, which is used in the PCEM, in the SOM $g_L$ is used in the estimation of $B_q^k$ (Equation 1.40). Both crystal field parameters are related by Equation 1.45.

$$B_q^k(\mathrm{SOM}) = \rho(2\beta)^{k+1} B_q^k(\mathrm{PCEM}) \tag{1.45}$$

The success of the SOM for estimation of the crystal field parameters can be seen from the data summarised in Table 1.12, which shows the phenomenological $B_q^k$ compared to the numbers obtained through the PCEM, SOM, and through AOM for an acetylacetonato complex of Nd(III).

### 1.3.9 Allowed and Forbidden *f-f* Transitions

Optical transitions within the 4*f* shell arise due to interaction of the electric and magnetic components of the electromagnetic radiation with the *f* electrons and are therefore either electric or magnetic dipole transitions (ED or MD), respectively. The electric dipole transition can be considered a linear displacement of charge and its direction is reversed through inversion, meaning it has odd parity. The magnetic dipole transition, on the other

***Table 1.12*** *Comparison of experimental $B_q^k$ [cm$^{-1}$] with values calculated with PCEM, AOM and SOM for Nd(btmsa)$_3$ [52]*

| $B_q^k$ | Phenomenological | PCEM | AOM | SOM |
|---|---|---|---|---|
| $B_0^2$ | −2912 | −4220 | −1834 | −2964 |
| $B_0^4$ | +920 | +370 | +1391 | +821 |
| $B_0^6$ | −516 | −61 | −304 | −427 |
| $B_6^6$ | ±331 | ±154 | ±512 | ±1083 |

hand, corresponds to a rotational displacement of charge, which, upon inversion, does not change its sense of rotation and therefore has even parity. Which transitions are allowed and which are forbidden is determined by selection rules.

### *1.3.9.1 Empirical Parameters Describing Optical Transitions*

In absorption spectra the peak maxima is reported either as absorbance $A$ or as molar absorptivity $\varepsilon$, which is independent of sample concentration, and they are related as shown in Equation 1.46.

$$A = \varepsilon c l \tag{1.46}$$

$c$ is the sample concentration and $l$ the path length covered by the light through the sample.

In the case of narrow absorption bands, due to signal-to-noise considerations, it is better to report the area of the peak $\int \varepsilon(\overline{\nu})d\overline{\nu}$ ($\overline{\nu}$ are wavenumbers in $cm^{-1}$), which is proportional to the peak intensity. Two other quantities, which can be reported instead of $\varepsilon$, are the transition dipole strength $D$ and the oscillator strength $P$, to which they are related by the following two equations [53].

$$D = 9 \times 10^{-39} \int \frac{\varepsilon(\overline{\nu})}{\overline{\nu}} d\overline{\nu} \tag{1.47}$$

$$P = 4.32 \times 10^{-9} \int \varepsilon(\overline{\nu})d\overline{\nu} = 4702 \times 10^{29} \times \overline{\nu}_0 \times D \tag{1.48}$$

$\overline{\nu}_0$ is the wavenumber at absorption maximum.

For transitions forbidden by selection rules, $P \ll 1$ and, in the case of the Ln(III), $P \sim 10^{-6}\ esu^2\ cm^2$ for MD allowed and ED forbidden transitions, with $\varepsilon_{max} < 10\,cm^{-1}\ M^{-1}$, which is habitually on the order of $1\ cm^{-1}\ M^{-1}$.

Correction factors for the dielectric medium, represented by its refractive index $n$, and for the degeneracy $2J+1$ of the terms are included, since both influence the transition dipole and oscillator strengths. The corrected expressions are given in Equations 1.49 and 1.50.

$$D_{corr.} = \frac{\chi}{2J+1} D \tag{1.49}$$

$$P_{corr.} = \frac{\chi}{2J+1} P \tag{1.50}$$

$\chi$ are the dielectric medium correction factors for ED and MD transitions. For absorption, these are given by Equation 1.51.

$$\chi_{ED} = \frac{(n^2+2)^2}{9n} \qquad \chi_{MD} = n \tag{1.51}$$

For emission, the equations are:

$$\chi_{ED} = \frac{n^2(n^2+2)^2}{9}$$
$$\chi_{MD} = n^3 \tag{1.52}$$

Since many *f-f* transitions of the Ln(III) have both ED and MD components, the overall experimental dipole strength is given by Equation 1.53.

$$D_{\exp} = \frac{1}{2J+1}(\chi_{MD}D_{MD} + \chi_{ED}D_{ED}) \tag{1.53}$$

### *1.3.9.2 The Spin Rule*

For a transition to be spin allowed, the spins of excited and ground state need to be the same. In the case of several *f-f* transitions of the lanthanide ions, these are accompanied by a change in spin and are therefore strictly forbidden. However, as mentioned previously, due to the spin–orbit coupling the total spin quantum number *S* is no longer completely valid, and therefore this rule is relaxed [53].

### *1.3.9.3 The Parity Rule*

The parity or Laporte rule states that for an optical transition to be allowed, the parity between final and initial states needs to change [53]. Utilising group theory arguments, it can be concluded that, for the transition to be allowed, the totally symmetric irreducible representation $\Gamma_1$ (Bethe's symbol, equivalent to Mulliken's symbols $A_1$) needs to be contained in the direct product shown in Equation 1.54.

$$\Gamma_i \otimes \Gamma \otimes \Gamma_f \tag{1.54}$$

$\Gamma_i$ and $\Gamma_f$ are the symmetry labels of the initial and final states and $\Gamma$ is the symmetry of the operator of the transition being considered. In the case of an electric dipole transition, in which interaction of the electric part of the electromagnetic radiation induces a linear displacement of electric charge in the ion or atom, the operator $O_{ED}$ transforms as *x*, *y* and *z*. It is given by:

$$O_{ED} = -e\sum_i \vec{r_i} \tag{1.55}$$

Since both the operator and the *f* orbitals have *u* (*ungerade*, odd) symmetry, electric dipole transitions are forbidden by the parity rule (see section below on Judd–Ofelt theory and induced electric dipole transitions). The selection rules for these transitions are summarised in Table 1.13.

The magnetic dipole operator $O_{MD}$ induces a rotational displacement of electric charge in the ion or atom due to interaction with the magnetic component of the electromagnetic

***Table 1.13*** *Selection rules for f-f transitions*

| Transition | $\Delta S$ | $\Delta L$ | $\Delta J$ |
|---|---|---|---|
| ED | 0 | ≤6 | ≤6 (2,4 or 6 if $J$ or $J'$ = 0) |
| MD | 0 | 0 | 0, ±1 |

radiation. It transforms as $R_x$, $R_y$ and $R_z$. Its expression is:

$$O_{MD} = -\frac{e\hbar}{2mc}\sum_i \vec{l}_i + 2\vec{s}_i \tag{1.56}$$

Since $O_{MD}$ has $g$ (*gerade*, even) symmetry and the $f$ orbitals $u$, magnetic dipole transitions are allowed in centrosymmetric and noncentrosymmetric point groups. However, the selection rules $\Delta J = 0$, 1 (but not $0 \leftrightarrow 0$) are followed (Table 1.13), and so few magnetic dipole transitions, such as the $Eu^{3+}$ $^5D_0 \rightarrow {}^7F_1$ transition, are known.

Electric quadrupole transitions are also possible; despite the even parity of the electric quadrupole operator, the intensity of these transitions is low, and they are less relevant for the lanthanide ions. They will therefore not be addressed here [29].

### *1.3.9.4 Symmetries of the Terms*

To help decide which transitions between energy levels of lanthanide ions are electric or magnetic dipole allowed, depending on the symmetry in which the lanthanide ion is located, it is useful to determine the symmetries of the terms split by the crystal field. This can be done with the rotation formula (Equation 1.57).

$$\chi(C_n) = \frac{\sin\left(J + \frac{1}{2}\right)\alpha}{\sin\frac{\alpha}{2}} \tag{1.57}$$

In this equation, $\chi$ is the character of the symmetry operation $C_n$, and $\alpha$ the rotation angle. For the identity operation $\chi(E) = 2J + 1$, which corresponds to the degeneracy of the term. For the operations $\sigma$ and $i$ the characters are $\chi(\sigma) = -\chi(C_2)$ and $\chi(i) = -\chi(E)$, respectively. Using these formulas it is possible to determine the reducible representation $\Gamma$ associated with each term and, upon reduction into its irreducible components, utilising the reduction formula (Equation 1.58), the symmetries of the individual crystal field split terms.

$$a_i = \frac{1}{h}\sum_R g_R \cdot \chi_r^{(R)} \cdot \chi_i^{(R)} \tag{1.58}$$

$a_i$ is the number of times the reducible representation $\Gamma_r$ is contained in the irreducible representation $\Gamma_i$, $h$ is the order of the group, $g_R$ is the number of operations in each class $R$ of symmetry operations and $\chi$ is the character of the irreducible $\Gamma_i$ or reducible representation $\Gamma_r$ for each class $R$.

**Table 1.14** *Multiplication table showing the selection rules for the* $O_h$ *point group [30]*

| $O_h$ | $\Gamma_1^+$ | $\Gamma_2^+$ | $\Gamma_3^+$ | $\Gamma_4^+$ | $\Gamma_5^+$ |
|---|---|---|---|---|---|
| $\Gamma_1^+$ | — | — | — | $R_x, R_y, R_z$ | — |
| $\Gamma_2^+$ | — | — | — | — | $R_x, R_y, R_z$ |
| $\Gamma_3^+$ | — | — | — | $R_x, R_y, R_z$ | $R_x, R_y, R_z$ |
| $\Gamma_4^+$ | $R_x, R_y, R_z$ | — | $R_x, R_y, R_z$ | $R_x, R_y, R_z$ | $R_x, R_y, R_z$ |
| $\Gamma_5^+$ | — | $R_x, R_y, R_z$ | $R_x, R_y, R_z$ | $R_x, R_y, R_z$ | $R_x, R_y, R_z$ |

If $J$ is a half-integer, character tables for double groups are utilised for the reduction, which contain the additional symmetry element $R$, defined as rotation by $2\pi$ and $\chi(\alpha + 2\pi) = -\chi(\alpha)$ [54]. In addition to the symmetry elements of the common character tables, the additional symmetry elements $C_2R$ (same character as $C_2$), $C_n^{n-1}$ (same character as $C_n$) and $C_n^{n-m}$ (same character as $C_n^m$) are present.

For example, Eu(III) has the electronic configuration $4f^6$ with the ground state multiplet $^7F_0$, $^7F_1$, $^7F_2$, $^7F_3$, $^7F_4$, $^7F_5$ and $^7F_6$, in order of increasing energy. Under octahedral symmetry $O$ (order of the group $h = 24$) and utilising the rotation formula (Equation 1.57) it can be shown that $^7F_3$ transforms as the reducible representation $\Gamma_r$ shown below.

| $O$ | $E$ | $6C_4$ | $3C_2$ | $8C_3$ | $6C_2$ |
|---|---|---|---|---|---|
| $\Gamma_r$ | 7 | −1 | −1 | 1 | −1 |

Using the reduction formula (Equation 1.58), it can further be shown that $\Gamma_r = A_2 + T_1 + T_2$ or $\Gamma_r = \Gamma_2 + \Gamma_4 + \Gamma_5$ using Bethe's notation. If the $O_h$ group symmetry is used instead, $\Gamma_r = A_{2g} + T_{1g} + T_{2g}$, or $\Gamma_r = \Gamma_2^+ + \Gamma_4^+ + \Gamma_5^+$ using Bethe's notation. The term splits into seven levels, one of symmetry $A_2$ (or $A_{2g}$), three degenerate ones of symmetry $T_1$ (or $T_{1g}$) and three of symmetry $T_2$ (or $T_{2g}$), consistent with the $2J+1 = 7$ degeneracy of the parent term. A similar exercise for the $^5D_0$ term shows that its $\Gamma_r = \Gamma_1^+$. Inspection of the multiplication tables indicates that the only transitions allowed are the magnetic dipole transitions, which occur between terms with irreducible representation, which contain the $O_{MD}$, as required by Equation 1.54 and shown in Table 1.14. Therefore the $^5D_0 - {}^7F_1$ transition has a magnetic dipole allowed component. Other transitions are observed for Eu(III) nonetheless, and those include induced electric dipole transitions, which will be discussed in the following paragraph.

Complete multiplication tables can be found in the literature [30].

Similarly, it can be shown that the $^2F_{5/2}$ term of Yb(III) transforms in $D_3$ as the reducible representation $\Gamma_r$ displayed in Table 1.15.

**Table 1.15** *Reducible representation of the* $^2F_{5/2}$ *term of Yb(III) in* $D_3$ *symmetry*

| $D_3'$ [a] | $E$ | $R$ | $C_3$ | $C_3^2$ | $3C_2$ | $3C_2R$ |
|---|---|---|---|---|---|---|
| | | | $C_3^2R$ | $C_3R$ | | |
| $\Gamma_r$ | 6 | −6 | 0 | 0 | 0 | 0 |

[a] The double group $D_3'$ is used instead of $D_3$, as Yb(III) is a Kramer's ion.

**Table 1.16** *Multiplication table showing the selection rules for the $D_3'$ point group [30]*

| $D_3'$ | ED | | MD | |
|---|---|---|---|---|
| | $\Gamma_4$ | $\Gamma_{5,6}$ | $\Gamma_4$ | $\Gamma_{5,6}$ |
| $\Gamma_4$ | $\alpha, \pi, \sigma$ | $\alpha, \sigma$ | $\alpha, \sigma, \pi$ | $\alpha, \pi$ |
| $\Gamma_{5,6}$ | $\alpha, \sigma$ | $\pi$ | $\alpha, \pi$ | $\sigma$ |

After reduction $\Gamma_r = 2\Gamma_4 + \Gamma_{5,6}$, which corresponds to two doubly degenerate levels with symmetry $\Gamma_4$ and two degenerate levels with symmetries $\Gamma_5$ and $\Gamma_6$, consistent with an overall sixfold degeneracy which is not completely lifted by the crystal field, as Yb(III) is a Kramers' ion. Analogously, the $^2F_{7/2}$ term transforms as the reducible representation $\Gamma_r = 3\Gamma_4 + \Gamma_{5,6}$.

Inspection of Table 1.16 above shows that transitions between these two terms are both electric and magnetic dipole allowed as well as polarised, since they are allowed only in certain directions. The transition between levels with $\Gamma_{5,6}$ symmetry is allowed only in $\pi$ polarised spectra. A $\pi$-polarised spectrum is measured with the magnetic field perpendicular and the electric field parallel to the crystallographic $c$ axis. An $\alpha$-spectrum, on the other hand, is measured with both the magnetic and electric field vectors perpendicular to the crystallographic $c$ axis. Measurement of the $\sigma$-spectrum occurs with the magnetic field parallel and the electric field perpendicular to the $c$ axis. Polarised spectra yield useable results only in uniaxial crystals, which have trigonal, tetragonal and hexagonal unit cells.

### 1.3.9.5 *Intensity of the MD Transitions*

The magnetic dipole strength of a transition between the initial and final states $\psi_i$ and $\psi_f$ can be theoretically evaluated by determining the matrix elements of the dipole moment operator $O_{MD}$ coupling the two states [40]. By re-writing Equation 1.56 without the summation over all $i$ electrons, $O_{MD}$ is given by

$$O_{MD} = -\frac{e\hbar}{2m_e c}(\hat{L} + 2\hat{S}) \tag{1.59}$$

Therefore, the strength of the magnetic dipole transition $D_{MD}$ is given by Equation 1.60.

$$D_{MD} = \langle \psi_i | O_{MD} | \psi_f \rangle^2 = \frac{e^2\hbar^2}{4m_e^2c^2} \left| \langle l^N SLJ | O_{MD} | l^N S'L'J' \rangle^2 \right. \tag{1.60}$$

After separating the quadratic term into expressions containing individually the total orbital angular momentum operator $\hat{L}$ and the total spin angular momentum operator $\hat{S}$, it can be shown that [55]

$$\begin{aligned} \langle l^N SLJ | \hat{L} | l^N S'L'J' \rangle &= \delta_{SS'} \delta_{LL'} (-1)^{S+L+J+1} \begin{Bmatrix} L & J & S \\ J & L & 1 \end{Bmatrix} [(2L+1)(2J+1)^2 L(L+1)]^{1/2} \\ \langle l^N SLJ | \hat{S} | l^N S'L'J' \rangle &= \delta_{SS'} \delta_{LL'} (-1)^{S+L+J+1} \begin{Bmatrix} S & J & L \\ J & S & 1 \end{Bmatrix} [(2S+1)(2J+1)^2 S(S+1)]^{1/2} \end{aligned} \tag{1.61}$$

Therefore, for MD transitions the following selection rules arise.

$$\Delta S = 0$$

$$\Delta L = 0$$

$$\Delta J = 0, \pm 1, \text{ but } 0 \leftrightarrow 0 \text{ forbidden}$$

As there are three possible values for $\Delta J$, it can be shown that for:

1. $\Delta J = 0$, $J = J'$ and

$$\langle l^n SLJ | \hat{L} + 2\hat{S} | l^n S'L'J' \rangle = g[J(J+1)(2J+1)]^{1/2}$$

$$g = 1 + \frac{J(J+1) - L(L+1) + S(S+1)}{2J(J+1)}$$

   $g$, the Landé factor, describes the magnetic moment of the ion.
2. $J' = J - 1$

$$\langle l^n SLJ | \hat{L} + 2\hat{S} | l^n S'L'(J-1) \rangle = \left[ \frac{1}{4J} (S+L+J+1)(S+L+J-1)(J+S-L)(J+L-S) \right]^{1/2}$$

3. $J' = J + 1$

$$\langle l^n SLJ | \hat{L} + 2\hat{S} | l^n S'L'(J+1) \rangle = \left[ \frac{1}{4J+1} (S+L+J+2)(S+J+1-L)(L+J+1-S)(S+L-J) \right]^{1/2}$$

Representative values for dipole and oscillator strengths for selected MD allowed Ln(III) transitions are summarised in Table 1.17.

Since MD allowed transitions are relatively independent of the geometry surrounding the lanthanide ions, both ligand identity as well as coordination polyhedra do not influence transition intensity appreciably.

***Table 1.17*** *Dipole D and oscillator P strengths for MD allowed transitions for select Ln(III)*

| Ln(III) | MD allowed transition | $E$ [cm$^{-1}$] | $P_{MD}$ [$10^{-8}$ Debye$^2$][a] | $D_{MD}$ [$10^{-6}$ Debye$^2$] |
|---|---|---|---|---|
| Pr(III) | $^3H_5 \leftarrow {}^3H_4$ | 2300 | 9.76 | 90 |
| Nd(III) | $^4I_{11/2} \leftarrow {}^4I_{9/2}$ | 2000 | 14.11 | 15 |
| Eu(III) | $^5D_0 \leftarrow {}^7F_1$ | 16 900 | 7.47 | 9.4 |

[a] 1 Debye = $1 \times 10^{-18}$ esu cm

### 1.3.10 Induced Electric Dipole Transitions and Their Intensity – Judd–Ofelt Theory

Since the electric dipole operator has odd parity, electric dipole transitions are allowed only between states of different parity, as mentioned before. Therefore, the electric dipole operator must be combined with other odd-parity operators to enable transitions within the *f* shell [56]. If the metal is present in a symmetry site without inversion symmetry, the parity rule is no longer applicable and therefore electric dipole transitions can be observed. However, even in centrosymmetric point groups, electric dipole transitions can be observed, as vibronic coupling, which is coupling of the electronic and vibrational wave functions, lifts the site symmetry. Alternatively, admixture of electronic states of opposite parity, such as *d* orbitals, mediated by the crystal field, which was proposed concurrently and independently by Brian Judd [8] and George Ofelt [9], can be invoked to explain the intensity of these transitions. While arbitrary, a small participation of the *d* orbitals can be seen as a small perturbation to the system, which results in new perturbed wave functions $\varphi_a$ and $\varphi_b$ for the initial and final states, where the wave functions $\psi$ designated with $\alpha$ and $\beta$ correspond to the orbitals of higher energy with opposite parity.

$$\langle\varphi_a| = \langle\psi_a| + \sum_{\beta} \frac{\langle\psi_a|H_{cf}|\psi_\beta\rangle}{E_a - E_\beta} \langle\psi_\beta|$$

$$|\varphi_b\rangle = |\psi_b\rangle + \sum_{\beta} \frac{\langle\psi_\beta|H_{cf}|\psi_b\rangle}{E_b - E_\beta} |\psi_\beta\rangle$$

By utilising these new wave functions, the dipole strength $D_{ED}$ of the line corresponding to the transition between the states $a$ and $b$ is given by Equation 1.62.

$$\begin{aligned} D_{ED} &= \langle\varphi_a|O_{ED}|\varphi_b\rangle^2 \\ &= \left[\sum_{\beta}\left\{\frac{\langle\psi_a|H_{cf}|\psi_\beta\rangle\langle\psi_\beta|O_{ED}|\psi_b\rangle}{E_a - E_\beta} + \frac{\langle\psi_a|O_{ED}|\psi_\beta\rangle\langle\psi_\beta|H_{cf}|\psi_b\rangle}{E_b - E_\beta}\right\}\right]^2 \end{aligned} \tag{1.62}$$

In this expression, since $O_{ED}$ is odd-parity and the states $a$ and $b$ have opposite parity to $\alpha$ and $\beta$, those terms will not vanish. To prevent the terms involving the crystal field Hamiltonian from vanishing, the $B^k_q$ parameters have to be odd, with $k = 1, 3, 5$ and $q \leq k$. The crystal field operator $H_{cf}$ and electric dipole operator $O_{ED}$ can be re-written as shown below.

$$H_{cf} = \sum_i r_i C^1_q(\theta_i, \phi_i) \tag{1.63}$$

$$O_{ED} = \sum_{t,p} A^t_p \sum_i r^t_i (C^t_p)_i \tag{1.64}$$

$C^k_q$ are the tensor operators introduced in Equation 1.31, which transform like the spherical harmonics.

Substitution of these expressions into Equation 1.62 leads to Equation 1.65.

$$D_{ED} = \left[ \begin{array}{l} \sum\limits_{p,t,\text{ even }\lambda} (-1)^{J-M_J+p+q}(2\lambda+1)A_p^t \begin{pmatrix} 1 & \lambda & t \\ q & -p-q & p \end{pmatrix} \begin{pmatrix} J & \lambda & J' \\ -M_J & -p-q & M_J' \end{pmatrix} \begin{Bmatrix} 1 & \lambda & t \\ l & l' & l \end{Bmatrix} \\ \times \langle \psi_a \parallel U^\lambda \parallel \psi_b \rangle \times \Xi(t,\lambda) \end{array} \right]^2$$

with

$$\Xi(t,\lambda) = 2\sum_{nl}(2l+1)(2l'+1)(-1)^{l+l'} \times \begin{Bmatrix} 1 & \lambda & t \\ l & l' & l \end{Bmatrix} \begin{pmatrix} l & 1 & l' \\ 0 & 0 & 0 \end{pmatrix} \begin{pmatrix} l & t & l' \\ 0 & 0 & 0 \end{pmatrix} \times \frac{(nl|r|n'l')(nl|r^t n'l')}{\Delta E_{nl}} \tag{1.65}$$

In this expression $\lambda = 1 + t$, $t$ is odd, $A_p^t = \frac{B_p^t}{\langle r \rangle^t}$ (Equation 1.29) are the static crystal field parameters, with $t \rightarrow k$ and $p \rightarrow q$ and, for the $3j$ and $6j$ symbols to be different from zero and therefore the dipole moment does not vanish, the following parameter values and selection rules arise for induced electric dipole transitions:

$\lambda = 2, 4, 6$

$t = 1, 3, 5$

$\Delta J \leq 6$ (no $0 \leftrightarrow 0$)

$\Delta J = 2, 4, 6$ when $J$ or $J' = 0$, otherwise $\Delta J = 0, \pm 1$

$\Delta L \leq 6$

$\Delta S = 0$

Following some simplifications, such as the assumption that all Stark levels of the ground state manifold are equally populated, and that the emission is isotropic, the dipole strength can be simplified to Equation 1.66.

$$D_{ED} = e^2 \sum_{\lambda=2,4,6} \Omega_\lambda |\langle \psi_a || U^\lambda || \psi_b \rangle|^2 \tag{1.66}$$

$U^\lambda$ are the tensors of the electric dipole operator of rank $\lambda = 2, 4, 6$, and the terms in brackets are the doubly reduced matrix elements for intermediate coupling, which have been determined by Carnall for all lanthanide aqua ions [25–28]. Some are summarised in Table 1.18.

$\Omega_\lambda$ are the Judd–Ofelt parameters in $cm^{-1}$ [8,9,55].

$$\Omega_\lambda = (2\lambda+1)\sum_p \sum_{t=1,3,5} \frac{|A_p^t|^2}{(2t+1)} \Xi^2(t,\lambda) \tag{1.67}$$

***Table 1.18*** *Selected squares of doubly reduced matrix elements* $|\langle\psi_a||U^\lambda||\psi_b\rangle|^2$ *for Eu(III) [26]*

| | $\lambda=2$ | $\lambda=4$ | $\lambda=6$ |
|---|---|---|---|
| $^7F_0 \rightarrow$ | | | |
| $^5D_0$ | 0 | 0 | 0 |
| $^5D_1$ | 0 | 0 | 0 |
| $^5D_2$ | 0.0008 | 0 | 0 |
| $^5L_6$ | 0 | 0 | 0.0155 |
| $^7F_1 \rightarrow$ | | | |
| $^5D_0$ | 0 | 0 | 0 |
| $^5D_1$ | 0.0026 | 0 | 0 |
| $^5D_2$ | 0.0001 | 0 | 0 |
| $^5L_6$ | 0 | 0 | 0.0090 |

The expression above shows that odd-order components of the crystal field and radial integrals of $4f^n$ wavefunctions and of perturbing wavefunctions of opposite parity comprise these parameters. These parameters appear in the definition of the oscillator strength $P_{ED}$ for a particular induced electric dipole transition between and $a$ and $b$, as shown in the equation below [57].

$$P_{ED}(a \leftrightarrow b) = \frac{8\pi^2 m_e \nu (n^2+2)^2}{27hn(2J+1)} \sum_{\lambda=2,4,6} \Omega_\lambda |\langle\psi_a||U^\lambda||\psi_b\rangle|^2 \tag{1.68}$$

The Judd–Ofelt parameters can, in principle, be calculated; however, it is more common to treat them as phenomenological parameters, which can be fitted from absorption or emission spectra, through measurement of the experimental dipole strength as defined in Equation 1.48 and the use of the tabulated doubly reduced matrix elements with Equation 1.68. The Judd–Ofelt parameters, as determined experimentally, are often given in units of $10^{-20}$ cm$^2$. $\Omega_2$ is strongly affected by dynamic coupling between the ligands and the lanthanide ion, which is related to changes in the ligand-generated field due to the incident light. This changing field in turn induces *f-f* transitions. $\Omega_6$ is strongly influenced by the rigidity of the host the lanthanide is embedded in. In this static coupling model, the ligands generate a crystal field potential of odd parity, which in turn induce 4*f* states of mixed parity. The electric dipole component of the incident photons then induces transitions between these states, but the ligands themselves are not affected by the incident radiation. The $\Omega_4$ parameters do not show specific trends and involve both dynamic and static coupling mechanisms [57,58]. In general, if the host matrix is the same and the lanthanide ions are in isostructural positions, a decrease of the $\Omega_\lambda$ parameters along the series is expected, and is indeed observed for $\Omega_6$, due to static coupling prevalence (Tables 1.19 and 1.20). In the case of $\Omega_4$, the trend is less consistent, and, for $\Omega_2$ no trend is observed, as dynamic coupling is prevalent.

The Judd–Ofelt theory does not estimate well transition intensities for Pr(III), as it is likely that the $4f^1 5d^1$ configuration contributes significantly to the perturbation described in $\varphi_a$ and $\varphi_b$.

***Table 1.19*** *Judd–Ofelt parameters for aqueous Ln(IIII) ions in acidic solutions [31]*

| Ln | $\Omega_2$ [$10^{-20}$ cm$^2$] | $\Omega_4$ [$10^{-20}$ cm$^2$] | $\Omega_6$ [$10^{-20}$ cm$^2$] |
|---|---|---|---|
| Pr | 32.6 | 5.7 | 32.0 |
| Nd | 0.93 | 5.00 | 7.91 |
| Sm | 0.91 | 4.13 | 2.70 |
| Eu | 1.46 | 6.66 | 5.40 |
| Gd | 2.56 | 4.70 | 4.73 |
| Tb | 0.004 | 7.19 | 3.45 |
| Dy | 1.50 | 3.44 | 3.46 |
| Ho | 0.36 | 3.14 | 3.07 |
| Er | 1.59 | 1.95 | 1.90 |
| Tm | 0.80 | 2.08 | 1.86 |

***Table 1.20*** *Judd–Ofelt parameters for Nd(III) in selected environments [57]*

| Nd | $\Omega_2$ [$10^{-20}$ cm$^2$] | $\Omega_4$ [$10^{-20}$ cm$^2$] | $\Omega_6$ [$10^{-20}$ cm$^2$] |
|---|---|---|---|
| Aqua | $0.93 \pm 0.3$ | $5.0 \pm 0.3$ | $7.9 \pm 0.4$ |
| Nitrate | $9.2 \pm 0.4$ | $5.4 \pm 0.3$ | $7.7 \pm 0.45$ |
| Acetylacetonate in DMF | 24.5 | 0.71 | 9.1 |
| Acetylacetonate in MeOH/EtOH | 15.7 | 0.73 | 7.4 |
| Dibenzoylacetonate in MeOH/EtOH | 34.1 | 2.5 | 9.1 |
| Bromide | 180 | 9 | 9 |
| Iodide | 275 | 9 | 9 |
| YAG | 0.2 | 2.7 | 5.0 |

### 1.3.11 Transition Probabilities and Branching Ratios

The probability of an individual transition $J \leftrightarrow J'$ to occur is given by Equation 1.69.

$$A(J,J') = \frac{64\pi^4 e^2 \bar{\upsilon}^3}{2h(2J+1)}\left[\chi_{ED} D_{ED} + \chi_{MD} D_{MD}\right] \tag{1.69}$$

The correction factors for dielectric medium, $\chi$, used in this equation will depend on the transitions being absorption or emission. Further, since individual transitions will have different probabilities, it is possible to define a radiative branching ratio $\beta_R(J,J')$, given by Equation 1.70.

$$\beta_R(J,J') = \frac{A(J,J')}{\sum_J A(J,J')} = \frac{E(J,J')}{\sum_J E(J,J')} \tag{1.70}$$

The radiative branching ratio can be calculated through the probabilities of the transitions, or, in the case of emission, it can be determined experimentally from the emission spectra, where $E(J,J')$ is the integrated emission spectrum of transition $J \leftrightarrow J'$ and $\sum_J E(J,J')$ is the integrated emission spectrum over all transitions.

In the case of emission, $A(J,J')$ is also known as Einstein's coefficient of spontaneous emission, and the sum of all probabilities for all radiative transitions is equal to the inverse of the radiative rate constant, $k_R$, in turn the reciprocal of the emissive state lifetime, $\tau_R$.

$$\beta_R(J,J') = \frac{A(J,J')}{\sum_J A(J,J')} = \frac{A(J,J')}{k_R} = \tau_R A(J,J') \tag{1.71}$$

Werts and co-workers [59] demonstrated that for the purely magnetic dipole transition of $Eu^{3+}$, $^5D_0 \rightarrow {}^7F_1$, the equation above can be rearranged to

$$\frac{1}{\tau_R} = A(^5D_0 \rightarrow {}^7F_1)n^3 \left(\frac{E(^5D_0 \rightarrow {}^7F_J)}{E(^5D_0 \rightarrow {}^7F_1)}\right) \tag{1.72}$$

and $A(^5D_0 \rightarrow {}^7F_1) = 14.65\ \mathrm{s}^{-1}$ is the spontaneous emission probability of the purely magnetic dipole transition *in vacuo*, which was calculated through Equation 1.69 above.

For other lanthanide ions, Equation 1.72 can be re-written as Equation 1.73.

$$\frac{1}{\tau_R} = 2303 \frac{8\pi c n^2 \overline{\nu}^2}{N_A} \frac{(2J+1)}{(2J'+1)} \int \varepsilon(\overline{\nu}) d\overline{\nu} \tag{1.73}$$

$\int \varepsilon(\overline{\nu}) d\overline{\nu}$ is the integrated absorption spectrum of the transition in molar absorptivity as a function of wavenumber.

Werts and co-workers tested the validity of these calculations by comparing experimental and calculated parameters for well-studied systems, the europium salts of dipicolinic acid [59]. They found an error of about 15% between experimentally determined and calculated branching ratios and radiative lifetimes.

### 1.3.12 Hypersensitive Transitions

Hypersensitive transitions are electric dipole transitions whose shape and intensity display large dependence on the point group symmetry of the metal ion, as well as on the pH, temperature, and ligand type. These transitions obey the following selection rules.

$$|\Delta S| = 0 \qquad |\Delta L| \leq 2 \qquad |\Delta J| \leq 2$$

Table 1.21 below summarises hypersensitive transitions observed for the different lanthanides. These transitions are usually observed for systems that display large values of $U^2$ and therefore $\Omega_2$ and comparatively small and symmetry-independent values of $\Omega_4$ and $\Omega_6$ [13].

While over the years several explanations have been proposed for hypersensitivity, including symmetry arguments, vibronic, charge-transfer and electric-quadrupole transition contributions [57,58], the most successful to date has been the dynamic coupling mechanism proposed by Mason et al. [60]. These authors suggest that the intensity of the hypersensitive transitions results from a non-zero electric dipole transition, which arises from an electric dipole in the ligand induced by the *f* orbitals of the metal ion. By analogy

***Table 1.21*** *Hypersensitive transitions of the Ln(III) ions observed in absorption or emission spectra [13,57]*

| Ln(III) | Transition | $\lambda$ [nm] |
|---|---|---|
| Pr | $^3F_2 - ^3H_4$ | 1920 |
| Nd | $^4G_{5/2} - ^4I_{9/2}$ | 578 |
| | $^2H_{9/2}, ^4F_{5/2} - ^4I_{9/2}$ | 806 |
| | $^4G_{7/2}, ^3K_{13/2} - ^4I_{9/2}$ | 521 |
| Sm | $^4F_{1/2}, ^4F_{3/2} - ^6H_{5/2}$ | 1560 |
| Eu | $^5D_2 - ^7F_0$ | 465 |
| | $^5D_1 - ^7F_1$ | 535 |
| | $^5D_0 - ^7F_2$ | 613 |
| Gd | $^6P_{5/2}, ^6P_{7/2} - ^8S_{7/2}$ | 308 |
| Dy | $^6F_{11/2} - ^6H_{15/2}$ | 1300 |
| | $^4G_{11/2}, ^4I_{15/2} - ^6H_{15/2}$ | 427 |
| Ho | $^3H_6 - ^5I_8$ | 361 |
| | $^5G_6 - ^5I_8$ | 452 |
| Er | $^4G_{11/2} - ^4I_{15/2}$ | 379 |
| | $^2H_{11/2} - ^4I_{15/2}$ | 521 |
| Tm | $^1G_4 - ^3H_6$ | 469 |
| | $^3H_4 - ^3H_6$ | 787 |
| | $^3F_4 - ^3H_6$ | 1695 |

with Equation 1.62 it is possible to write Equation 1.74 for the electric dipole moment $\mu$.

$$\langle A_0 B_0 | \mu | A_a B_0 \rangle = \sum_b \frac{2E_b}{(E_b^2 - E_a^2)} \langle A_0 A_a | V | B_0 B_b \rangle \mu_{0b} \tag{1.74}$$

$\langle A_0|$ and $|Aa\rangle$ are the ground and excited states of the metal ion, respectively, which perturb the $\langle B_0|$ ground and $|B_b\rangle$ excited states of the ligand. $\mu_{0b}$ is the electric dipole moment of the ligand transition influenced by the Coulombic interaction $V$ between the $f$ orbitals and the ligand $B$. The dynamic coupling contributes then to the $\Omega_2$ Judd–Ofelt parameter, if the crystal field potential has crystal field terms with $k$ odd ($k = 3$).

### 1.3.13 Emission Efficiency and Rate Constants

The emission efficiency is given by the quantum yield of luminescence $\phi$, which is simply the ratio of emitted photons, $p_{em}$, to photons absorbed by the sample, $p_{abs}$, as shown in Equation 1.75.

$$\phi = \frac{p_{em}}{p_{abs}} \tag{1.75}$$

In the case of lanthanides, following direct excitation of the metal ion, the efficiency of emission is called the intrinsic emission efficiency $\phi_{Ln}^{Ln}$, which is directly related to the overall rate at which the emissive state is depopulated through radiative $R$ and non-radiative $NR$ pathways, $k_{exp} = k_R + k_{NR}$, and the radiative rate constant, $k_R$, or their corresponding lifetimes, $\tau_{exp}$ and $\tau_R$, as shown in Equation 1.76.

$$\phi_{Ln}^{Ln} = \frac{k_R}{k_R + k_{NR}} = \frac{k_R}{k_{exp}} = \frac{\tau_{exp}}{\tau_R} \tag{1.76}$$

The experimentally observed excited state lifetime reflects the contribution of all processes, which lead to the deactivation of the excited state, both radiative and non-radiative.

Equation 1.72 above relates the radiative lifetime to the magnetic dipole-allowed transition of Eu(III). It is therefore straightforward to determine experimentally, for this ion, $\tau_R$ and therefore the intrinsic emission efficiency.

## 1.4 Sensitisation Mechanism

### 1.4.1 The Antenna Effect

Due to the forbidden nature of the *f-f* transitions, for many applications the direct excitation of the lanthanide ion is not desirable, as it requires the use of high intensity sources. It is therefore more efficient to promote the emission through an appended sensitiser or antenna. The antenna effect, illustrated in Fig. 1.10, relies on a sensitiser to harvest energy, for example through photon absorption, which leads to population of an excited single state $^1S$. In the presence of a heavy atom, intersystem crossing ISC is facilitated, which leads to population of a triplet excited state $^3T$. This state is usually long-lived and is capable therefore of energy transfer ET to the lanthanide ion's emissive $f^*$ excited state. If the emissive state is too close in energy to the triplet state, back energy transfer BT occurs.

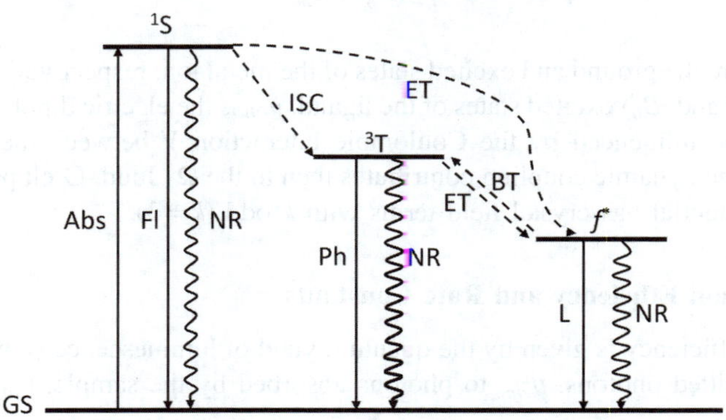


*Figure 1.10* Modified Jablonski diagram illustrating the antenna effect. Abs – absorption, Fl – fluorescence, Ph – phosphorescence, L – luminescence, ISC – intersystem crossing, ET – energy transfer, BT – back energy transfer, NR – non-radiative deactivation, $^1S$ – first excited singlet state, $^3T$ – lowest excited triplet state, GS – ground state, $f^*$ – emissive f excited state

Fluorescence Fl and phosphorescence Ph compete for deactivation of the singlet and triplet states, respectively, and non-radiative processes NR can also lead to deactivation of all excited states. Although it is usually assumed that the energy transfer occurs from the triplet excited state due to the favourable intersystem crossing in the presence of the heavy atom [61], several cases have been reported in which the energy is transferred from the singlet excited state directly to the emissive *f* level [62–66]. It has also been shown that sensitisation through metal to ligand charge-transfer states of coordinated transition metal complexes or intra-ligand charge-transfer states are viable pathways to promote lanthanide-centred emission [67].

The overall efficiency of sensitised emission $\phi_{Ln}^{L}$ is given by Equation 1.77.

$$\phi_{Ln}^{L} = \phi_{ISC} \times \phi_{ET} \times \phi_{Ln}^{Ln} = \phi_{sens} \times \phi_{Ln}^{Ln} \tag{1.77}$$

The efficiency of intersystem crossing $\phi_{ISC}$ and efficiency of energy transfer $\phi_{ET}$ combine to give the efficiency of sensitisation $\phi_{sens}$. In the case of europium, where $\tau_r$ (Equation 1.72) can be obtained from the integrated emission spectrum and $\phi_{Ln}^{Ln}$ can be estimated with Equation 1.76, it is therefore possible to experimentally determine the efficiency of sensitisation $\phi_{sens}$.

### *1.4.1.1 Singlet and Triplet States and Intersystem Crossing*

The relative positions of the singlet and triplet states are important, as they can help influence the efficiency of the intersystem crossing as well as the energy transfer and magnitude of back energy transfer. Work done by Latva et al. [68] indicates that the position of the lowest triplet state relative to the emissive state of the Ln(III) ion is important for the efficiency of the energy transfer. From their survey of 41 different ligands, these authors conclude that for Eu(III) good energy transfer happens if the triplet to $f^*$ gap is in the range 2500–4000 $cm^{-1}$, with emission efficiencies in the range 15–38%. Nonetheless, an emission efficiency of 12% was observed for a complex with $\Delta E \approx 9000\,cm^{-1}$, and an emission efficiency of 11% was observed for another complex with $\Delta E \approx 1000\,cm^{-1}$. In the case of Tb(III), emission efficiencies in the range 21–58% were observed for ΔE in the range 2000–4300 $cm^{-1}$. For this metal ion, with its emissive $^5D_4$ state at 20 450 $cm^{-1}$, substantial energy back-transfer was observed if $\Delta E \leq 1850\,cm^{-1}$.

In the course of the pioneering work on sensitised luminescence, Yuster and Weissman studied the promotion of intersystem crossing due to spin–orbit coupling with heavy atoms [69]. They found that dibenzoylmethanide coordinated to the non-emissive La, Lu and Gd displayed different efficiencies of intersystem crossing, as seen by different ratios of fluorescence versus phosphorescence intensity and changes in phosphorescence lifetimes, directly related to spin–orbit coupling and to the magnetic moment of the lanthanide ion.

The lifetime $\tau(^1S)$ of the singlet excited state is given by Equation 1.78.

$$\tau(^1S) = \frac{1}{k_{NR} + k_{Fl} + k_{ISC}} \tag{1.78}$$

$k_{NR}$ is the non-radiative decay rate, while $k_{Fl}$ is the radiative decay rate and $k_{ISC}$ the rate of intersystem crossing (see Fig. 1.10). The lifetime $\tau(^3T)$ of the triplet excited state is given by Equation 1.79.

$$\tau(^3T) = \frac{1}{k'_{NR} + k_{Ph}} \tag{1.79}$$

$k'_{NR}$ is the non-radiative decay from the triplet state back to the singlet ground state and $k_{Ph}$ is the radiative phosphorescence decay rate.

The emission efficiencies for the fluorescence $\phi_{Fl}$ and phosphorescence $\phi_{Ph}$ can then be defined by Equations 1.80 and 1.81, respectively.

$$\phi_{Fl} = \frac{k_{Fl}}{k_{NR} + k_{ISC} + k_{Fl}} \tag{1.80}$$

$$\phi_{Ph} = \frac{k_{Ph} + k_{ISC}}{(k_{NR} + k_{Ph})_{Ph}(k_{ISC} + k_{Fl} + k_{NR})_{Fl}} \tag{1.81}$$

#### *1.4.1.2 Ligand to Metal Ion Energy Transfer Mechanisms*

As mentioned above, sensitised luminescence is the process by which lanthanide ions emit following absorption of light by a coordinated ligand. During this process the absorbed energy has to be transferred to the lanthanide ion. In general, energy transfer between a donor *D* and an acceptor *A* will follow one of two mechanisms, one involving electron transfer, also called Dexter energy transfer [70], and the other involving dipole–dipole exchange, also known as Förster energy transfer [71], represented in Fig. 1.11 [72].

Due to the lack of overlap of the 4*f* orbitals with the ligand orbitals, it is generally accepted that the energy transfer occurs through the latter for lanthanide ions and so only the Förster mechanism will be briefly discussed in the following. However, in the case of lanthanide ions such as europium and ytterbium, which are easily reduced, electron transfer and phonon-assisted mechanisms have been invoked [73,74].

***Förster or Dipole–Dipole Energy Transfer*** For systems in which molecules are present that are capable of absorbing and emitting energy and energy transfer is observed between donors and acceptors, Förster postulated that the decrease in excitation of the donor occurs in parallel with the increase in emission of the acceptor, which is consistent with a dipole–dipole exchange mechanism. This is different from a sequential process in which the donor emits and its emission wavelengths are reabsorbed by the acceptor [71].

When donors and acceptors are present in the system and the former are de-excited while the latter absorb energy, the rate constant for energy transfer $k_{D\text{-}A}$ between donor *D* and acceptor *A* is given by Equation 1.82.

$$k_{D\text{-}A} = \frac{161.9\,\kappa^2}{\pi^6 n^4 N \tau_0 R^6} \int_0^\infty f_D(\nu)\varepsilon_A(\nu)\frac{d\nu}{\nu^4} = \frac{161.9\,\kappa^2}{\pi^6 n^4 N \tau_0 R^6} J(\nu) \tag{1.82}$$

*R* is the distance between donor *D* and acceptor *A*, *n* is the refractive index of the solution, $\tau_0$

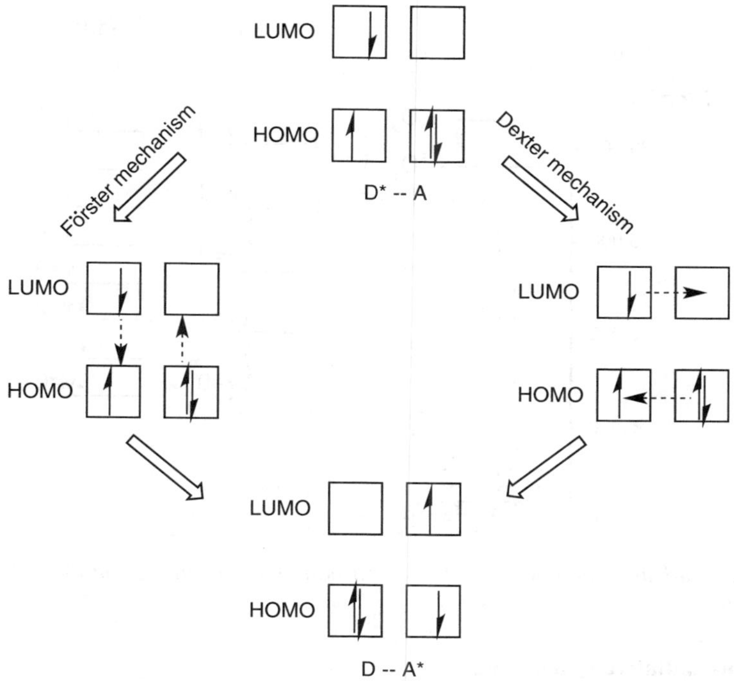


***Figure 1.11*** *Pictorial representation of energy transfer between donor* D *and acceptor* A *through the Förster and Dexter mechanisms*

is the natural radiative lifetime of the donor in absence of the acceptor and $N$ is Avogadro's number. The overlap integral $J(\nu)$ is determined through integration of the absorption spectrum of the acceptor $\varepsilon_A(\nu)$, in units of $cm^{-1}$ $M^{-1}$, and the emission spectrum of the donor $f_D(\nu)$, in units of $cm^{-1}$, normalised to an area of 1.0. Finally, since dipole–dipole interactions depend on molecular orientation, the orientation factor $\kappa$ is present in this equation [71]. For solutions in which molecular positions are averaged due to random motion, $\kappa^2$ is usually equal to two-thirds [75]. From this equation it follows that the energy transfer depends on $R^{-6}$, and therefore Förster energy transfer is a long-range interaction which requires overlap of emission spectrum of the donor and absorption spectrum of the acceptor, but does not require physical contact between the donor and acceptor. The Förster distance $R_0$, which is the maximum donor–acceptor distance for which energy transfer is possible, is given by Equation 1.83 [71].

$$R_0 = \sqrt[6]{\frac{3000\tau_0 c^2 J(\nu)}{8\pi^4 n^2 N^2 \nu_0^2}} \tag{1.83}$$

Here, $c$ is the speed of light in a vacuum and $\nu_0$ is the frequency at which absorption and emission spectra intersect.

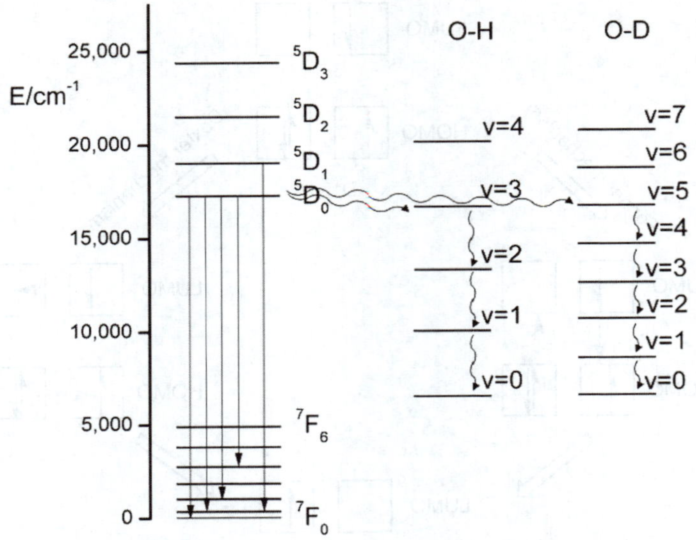


*Figure 1.12 Radiative transitions of Eu(III) and non-radiative quenching through O–H and O–D bonds*

### 1.4.2 Non-Radiative Quenching

As shown in Fig. 1.11, non-radiative quenching pathways are present at several places of the sensitisation process and lead to overall low emission efficiency of sensitised emission. In the case of intrinsic Ln(III) ion emission, the smaller the gap between the emissive state and the highest sublevel of the ground-state manifold, the easier it is to non-radiatively quench the emission. This is shown in Fig. 1.12 for Eu(III), where the gap between the $^5D_0$ and the $^7F_6$ states is approximately 12 000 cm$^{-1}$. It follows that the quenching of the emissive state is easily accomplished through three vibrational quanta of the O–H bond, with a vibrational energy of 3600 cm$^{-1}$. If instead the bond is O–D, with a vibrational energy of 2200 cm$^{-1}$, five vibrational quanta are needed, and the process is less efficient.

Table 1.22 summarises the number of phonons necessary to bridge the gap $\Delta E$ between an excited Ln(III) state and the highest energy sub-level of the ground-state manifold for $H_2O$ and $D_2O$.

*Table 1.22 Vibrational bridging of the $\Delta E$ gap for different Ln(III) ions by O–H and O–D bonds and typical radiative lifetimes $\tau$*

| Ln | $\Delta E$ [cm$^{-1}$] | Number of phonons | | $\tau$ [ms] | |
|---|---|---|---|---|---|
| | | OH | OD | $H_2O$ | $D_2O$ |
| Gd | 32 100 | 9 | 15 | 2.3 | n.a. |
| Tb | 14 800 | 4 | 7 | 0.47 | 3.8 |
| Eu | 12 300 | 3–4 | 5–6 | 0.11 | 4.1 |
| Dy | 8850 | 2–3 | 3–4 | 0.002 | 0.06 |
| Sm | 7400 | 2 | 3 | 0.002 | 0.08 |

Horrocks and co-workers took advantage of this and derived an empirical equation, which allows the determination of the number of water molecules coordinated to Eu(III) and Tb(III) by comparing emission lifetimes in water and deuterated water. This will be further discussed in Chapter 2 [76,77].

In addition to O–H other bonds, such as N–H ($3300\,cm^{-1}$) and C–H ($2900\,cm^{-1}$), also contribute to vibrational quenching of Ln(III)-centred emission.

Since the non-radiative quenching processes are vibrationally mediated, they are temperature dependent and the temperature dependence of the rate constants is described by an Arrhenius-like Equation 1.84.

$$\ln\left(k_{exp} - k_0\right) = \ln A - \frac{E_A}{RT} \tag{1.84}$$

$k_0$ is the rate constant at 0 K, which can be approximated by the rate constant measured experimentally at 4 K or 77 K and $E_A$ is the activation energy for the quenching process. This can be used, for example, to determine the activation energy for energy back-transfer from the emissive Ln(III) state to the triple state. Charbonnière et al. determined an activation energy of $180\,cm^{-1}$ for the back-transfer from the $^5D_4$ level of Tb(III) to the triplet state of a calixarene located at $2400\,cm^{-1}$ [78]. The authors hypothesise that the back-transfer is mediated by Ln–O vibrations in the complex, which occur at around $220\,cm^{-1}$.

## Acknowledgement

Support through the National Science Foundation grant CHE-1058805 is gratefully acknowledged.

## Abbreviations

| | |
|---|---|
| $^1S$ | first excited singlet state |
| $^3T$ | lowest excited triplet state |
| A | acceptor |
| AOM | angular overlap model |
| BT | back energy transfer |
| CN | coordination number |
| D | donor |
| ED | electric dipole |
| ET | energy transfer |
| Fl | fluorescence |
| GS | ground state |
| ISC | intersystem crossing |
| L | luminescence |
| Ln | lanthanide |
| MD | magnetic dipole |
| NR | non-radiative |

PCEM point charge electrostatic model
Ph phosphorescence
R radiative
SOM simple overlap model

## References


1. Evans, C. H. (ed.) *Episodes from the History of the Rare Earth Elements;* Cambridge Unversity Press: Cambridge, **1996**.
2. Kirchhoff, G.; Bunsen, R. *Quart. J., Chem. Soc., London* **1861**, *13*, 270–289.
3. von Welsbach, C. A. *Justus Liebigs Ann., Chem.* **1907**, *351*, 458–466.
4. von Welsbach, C. A. *Monatsh. Chem.* **1883**, *4*, 630–642.
5. von Welsbach, C. A. *Monatsh. Chem.* **1884**, *5*, 508–522.
6. Condon, E. U.; Shortley, G. H. *The Theory of Atomic Spectra;* Cambridge Unversity Press: Cambridge, **1996**.
7. Van Vleck, J. H. *J. Phys. Chem.* **1937**, *41*, 67–80.
8. Judd, B. R. *Phys. Rev.* **1962**, *127*, 750–761.
9. Ofelt, G. S. *J. Chem. Phys.* **1962**, *37*, 511–520.
10. Weissman, S. I. *J. Chem. Phys.* **1942**, *10*, 214–217.
11. Bünzli, J.-C. G.; Choppin, G. R. (eds) *Lanthanide Probes in Life, Chemical and Earth Sciences – Theory and Practice;* Elsevier: Amsterdam, **1989**.
12. Bünzli, J. C. G. In *Spectroscopic Properties of Rare Earths in Optical Materials;* Liu, G., Jacquier, B., (eds); Springer: Berlin, 2005; Vol. 83, pp. 462–499.
13. Sastri, V. S.; Bünzli, J. C.; Perumareddi, J. R.; Rao, V. R.; Rayudu, G. V. S. (eds) *Modern Aspects of Rare Earths and Their Complexes* Elsevier: Amsterdam, **2003**.
14. Huheey, J. E.; Keiter, E. A.; Keiter, R. L. *Inorganic Chemistry: Principles of Structure and Reactivity;* 4th edn; Harper Collins: New York, **1993**.
15. Reddy, J. V. *Symmetry and Spectroscopy of Molecules;* New Age Science: Maidstone, **2010**.
16. Dieke, G. H. *Spectra and Energy Levels of Rare Earth Ions in Crystals;* Interscience: New York, **1968**.
17. Dieke, G. H.; Sarup, R. *J. Chem. Phys.* **1958**, *29*, 741–745.
18. Varsanyi, F.; Dieke, G. H. *J. Chem. Phys.* **1962**, *36*, 2951–2961.
19. Thomas, K. S.; Singh, S.; Dieke, G. H. *J. Chem. Phys.* **1963**, *38*, 2180–2190.
20. DeShazer, L. G.; Dieke, G. H. *J. Chem. Phys.* **1963**, *38*, 2190–2199.
21. Dieke, G. H.; Pandey, B. *J. Chem. Phys.* **1964**, *41*, 1952–1969.
22. Crosswhite, H. M.; Dieke, G. H.; Carter, W. J. *J. Chem. Phys.* **1965**, *43*, 2047–2054.
23. Dieke, G. H.; Leopold, L. *J. Opt. Soc. Am.* **1957**, *47*, 944–954.
24. Dieke, G. H.; Singh, S. *J. Opt. Soc. Am.* **1956**, *46*, 495–499.
25. Carnall, W. T.; Fields, P. R.; Rajnak, K. *J. Chem. Phys.* **1968**, *49*, 4424–4442.
26. Carnall, W. T.; Fields, P. R.; Rajnak, K. *J. Chem. Phys.* **1968**, *49*, 4450–4455.
27. Carnall, W. T.; Fields, P. R.; Rajnak, K. *J. Chem. Phys.* **1968**, *49*, 4447–4449.
28. Carnall, W. T.; Fields, P. R.; Rajnak, K. *J. Chem. Phys.* **1968**, *49*, 4443–4446.
29. Wybourne, B. G. *Spectroscopic Properties of Rare Earths;* Wiley Interscience: New York, 1965.
30. Görller-Walrand C.; Binnemans K. In: *Handbook on the Physics and Chemistry of Rare Earths;* Karl A. Gschneidner, J. L. E. G. H. L., Choppin, G. R. (eds), **1996**; Vol. 23, p 121–283
31. Carnall W. T. In: *Handbook on the Physics and Chemistry of Rare Earths* **1979**; Vol. 3, pp. 171–208
32. Hellwege K. H.; Orlich E.; Schaack G. *Phys. Kondens. Mater.* **1965**, *4*, 196–206.

33. Pearson J. J.; Herrmann G. F.; Wickersheim K. A.; Buchanan R. A. *Phys. Rev.* **1967**, *159*, 251–261
34. NIelson C. W.; Koster G. F. *Spectroscopic Coefficients for the* $p^n$, $d^n$ *and* $f^n$ *Configurations;* MIT Press: Cambridge, **1963**.
35. Rotenberg M. *3-J and 6-J Symbols;* MIT Press: Cambridge, **1959**.
36. Stark J. *Ann. Phys.* **1914**, *43*, 965–982.
37. Rajnak K.; Wybourne B. G. *J. Chem. Phys.* **1964**, *41*, 565–569
38. Wybourne B. G.; Smentek L. *Optical Spectroscopy of Lanthanides: Magnetic and Hyperfine Interactions;* CRC Press: New York, **2007**.
39. Kramers H. A. *Proc. K. Ned. Akad. Wet.* **1930**, *33*, 959–972.
40. Bünzli J.-C. G.; Eliseeva S. V. *Springer Ser. Fluoresc.* **2011**, *7*, 1–46.
41. Tanner P. A. *Springer Ser. Fluoresc.* **2011**, *7*, 183–234.
42. Liu G.; Jacquier B. (eds) *Spectroscopic Properties of Rare Earths in Optical Materials;* Springer: Heidelberg, **2005**.
43. Sanz J.; Cases R.; Alcala R. *J. Non-Cryst. Solids* **1987**, *93*, 377–386
44. Zeeman P. *Nature* **1897**, *55*, 347.
45. Jørgensen C. K.; Pappalardo R.; Schmidtke H. H. *J. Chem. Phys.* **1963**, *39*, 1422–1430
46. Malta O. L. *Chem. Phys. Lett.* **1982**, *87*, 27–29.
47. Warren K. D. *Inorg. Chem.* **1977**, *16*, 2008–2011.
48. Urland W. *Chem. Phys.* **1976**, *14*, 393–401.
49. Urland W. *Chem. Phys. Lett.* **1978**, *53*, 296–299.
50. Malta O. L. *Chem. Phys. Lett.* **1982**, *88*, 353–356.
51. Albuquerque R. Q.; Rocha G. B.; Malta O. L.; Porcher P. *Chem. Phys. Lett.* **2000**, *331*, 519–525.
52. Jank S.; Reddmann H.; Amberger H. D. *Spectrochim. Acta, Part A* **1998**, *54A*, 1651–1657.
53. Turro N. J.; Ramamurthy V.; Scaiano J. C. *Modern Molecular Photochemistry of Organic Molecules;* University Science Books: Sausalito, **2010**.
54. Cotton F.A. *Chemical Applications of Group Theory;* 3rd edn.; John Wiley & Sons Inc.: New York, **1990**.
55. Walsh B. M. In: *Advances in Spectroscopy for Lasers and Sensing;* Di Bartolo, B., Forte O. (eds); Springer: Dordrecht, **2006**, pp. 403–433.
56. Wybourne B. G. *Mol. Phys.* **2003**, *101*, 899–901
57. Peacock R. D. In: *Structure and Bonding;* Dunitz, J. D., Hemmerich, P., Holm, R. H., Ibers, J. A., Joergensen, K., Neilands, J. B., Reinen, D., Williams, R. J. P. (eds); Springer: Heidelberg, **1975**; Vol. 22.
58. Görller-Walrand C.; Binnemans K. In: *Handbook on the Physics and Chemistry of Rare Earths;* Karl A. Gschneidner, J., Eyring, L. (eds), **1998**; Vol. 25, pp. 101–264.
59. Werts M. H. V.; Jukes R. T. F.; Verhoeven J. W. *Phys. Chem. Chem. Phys.* **2002**, *4*, 1542–1548.
60. Mason S. F.; Peacock R. D.; Stewart B. *Chem. Phys. Lett.* **1974**, *29*, 149–153.
61. Rudzinski C. M.; Engebretson D. S.; Hartmann W. K.; Nocera D. G. *J. Phys. Chem. A* **1998**, *102*, 7442–7446.
62. Hebbink G. A.; Klink S. I.; Grave L.; Oude Alink P. G. B.; Van Veggel F. C. J. M. *ChemPhysChem* **2002**, *3*, 1014–1018.
63. Lamture J. B.; Zhou Z. H.; Kumar A. S.; Wensel T. G. *Inorg. Chem.* **1995**, *34*, 864–869.
64. Nah M.-K.; Cho H.-G.; Kwon H.-J.; Kim Y.-J.; Park C.; Kim H. K.; Kang J.-G. *J. Phys. Chem. A* **2006**, *110*, 10371–10374.
65. Thorne J. R. G.; Rey J. M.; Denning R. G.; Watkins S. E.; Etchells M.; Green M.; Christou V. *J. Phys. Chem. A* **2002**, *106*, 4014–4021.
66. Ha-Thi, M.-H.; Delaire J. A.; Michelet V.; Leray I. *J. Phys. Chem. A* **2010**, *114*, 3264–3269.
67. D'Aleo A.; Pointillart F.; Ouahab L.; Andraud C.; Maury O. *Coord. Chem. Rev.* **2012**, *256*, 1604–1620.

68. Latva M.; Takalo H.; Mukkala V.-M.; Matachescu C.; Rodriguez-Ubis J. C.; Kankare J. *J. Luminescence* **1997**, *75*, 149–169.
69. Yuster P.; Weissman S. I.; *J. Chem. Phys.* **1949**, *17*, 1182–1188.
70. Dexter D. L. *J. Chem. Phys.* **1953**, *21*, 836–850.
71. Förster T. *Ann. Physik* **1948**, *2*, 55–75.
72. Balzani, V.; Campagna, S. (eds) *Photochemistry and Photophysics of Coordination Compounds I*, *[In: Top Curr. Chem., 2007; 280]*, **2007**.
73. Reinhard, C.; Güdel, H. U., *Inorg. Chem.* **2002**, *41*, 1048–1055.
74. Pope, S. J. A.; Burton-Pye, B. P.; Berridge, R.; Khan, T.; Skabara, P. J.; Faulkner, S. *Dalton Trans.* **2006**, 2907–2912.
75. Lakowicz, J. R. *Principles of Fluorescence Spectroscopy;* Plenum: New York, 1999.
76. Horrocks, W. D. Jr.; Sudnick, D. R. *Acc. Chem. Res.* **1981**, *14*, 384–392.
77. Horrocks, W. D. Jr.; Sudnick, D. R. *J. Am. Chem. Soc.* **1979**, *101*, 334–340.
78. Charbonnière, L. J.; Balsiger, C.; Schenk, K. J.; Bünzli, J.-C. G. *J. Chem. Soc., Dalton Trans.* **1998**, 505–510.

# 2

# Spectroscopic Techniques and Instrumentation

*David E. Morris*[1] *and Ana de Bettencourt-Dias*[2]

[1]*Los Alamos National Laboratory, USA*
[2]*Department of Chemistry, University of Nevada, USA*

## 2.1 Introduction

Of the many scientifically challenging and technologically relevant physical properties of the lanthanide elements and their compounds, their luminescence properties are certainly among the most interesting and most exploited for applications. The electronic structure and photophysical properties of the lanthanide ions that give rise to luminescence are well summarised in Chapter 1. These represent an incredibly fertile landscape around which there continues to be enormous activity and innovation in chemistry, biology, materials science and at the interfaces between these disciplines (e.g., nanoscience) to design, control, and optimise the energies and intensities of luminescence from these ions. Of particular consequence to this chapter, studies of luminescent lanthanide systems also present many substantial experimental challenges. The intrinsic luminescence from the *f-f* state(s) in $Ln^{3+}$ ions is generally quite weak (e.g., Fig. 2.1) and can be easily masked by radiative decay (emission) from other constituents in the molecule, solid, or matrix under investigation.

The energy range [ultraviolet (UV) to near infrared (NIR)] over which lanthanide luminescence is observed as one traverses the lanthanide series is substantial and cannot be covered readily by a single experimental combination of excitation source and detector.

*Luminescence of Lanthanide Ions in Coordination Compounds and Nanomaterials*, First Edition.
Edited by Ana de Bettencourt-Dias.

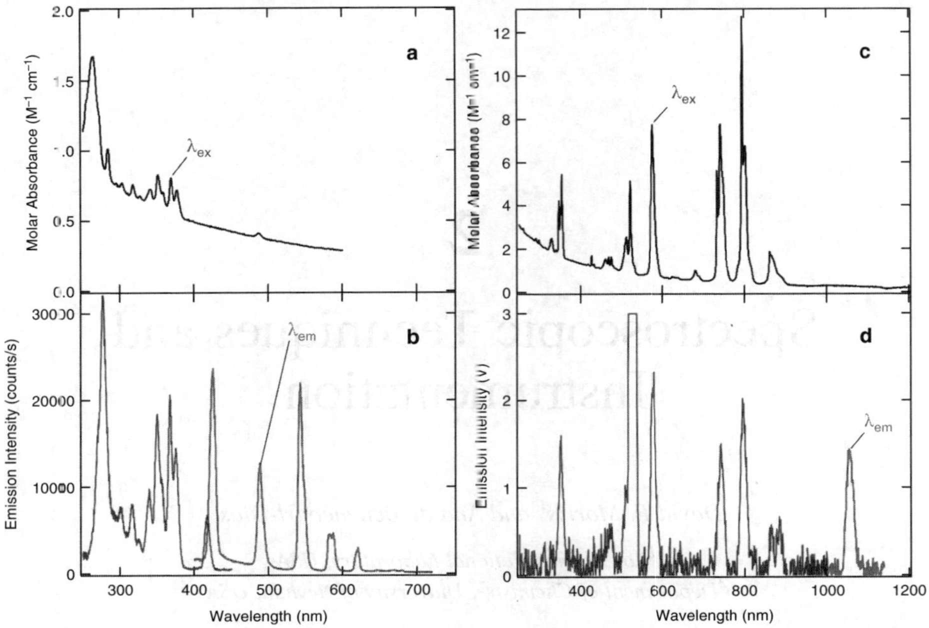


***Figure 2.1*** *Typical absorption and emission spectra for $Ln^{3+}$ ions in dilute aqueous solution: (black lines) absorption data; (red lines) emission data; (blue lines) emission excitation data. (a, b) $Tb^{3+}$. (c, d) $Nd^{3+}$. $[Tb^{3+}]$ = 1.3 mM for (b); $[Nd^{3+}]$ = 27.6 mM for (d). Emission and emission excitation data were collected using continuous wave excitation from a Xe arc lamp. Wavelengths for exciting the emission ($\lambda_{ex}$) and monitoring the emission excitation ($\lambda_{em}$) spectra are indicated. The detector in B was a photomultiplier tube. The detector in d was a single-element InGaAs device. © David E. Morris, Los Alamos National Laboratory (See colour plate section)*

This point is illustrated by example in Fig. 2.2 showing typical luminescence spectral data from several $Ln^{3+}$ ions.

The physical properties and characteristics of samples of interest (aqueous and non-aqueous solutions, powders, single crystals, glasses, and most recently a range of nanoscale and nanostructured materials) dictate a range of experimental considerations to optimally couple light into and out of the sample under interrogation and to exercise control of external variables including temperature and pressure. Finally, a critical comparative assessment of luminescence properties for a given lanthanide ion in differing environments, among lanthanide ions in similar environments, or between lanthanides and other light emitting chromophores, also requires very careful attention to experimental detail in measuring luminescent intensities and lifetimes to extract absolute luminescence quantum yields, as these are the best basis for performance comparison in light-emission applications.

In this chapter we consider the most general elements of luminescence spectroscopy practice and instrumentation with a focus, where appropriate, on experimental aspects of luminescence spectroscopy that are unique to or specific to lanthanide-ion systems. In particular, we emphasise recent advances in spectroscopic hardware and data-processing

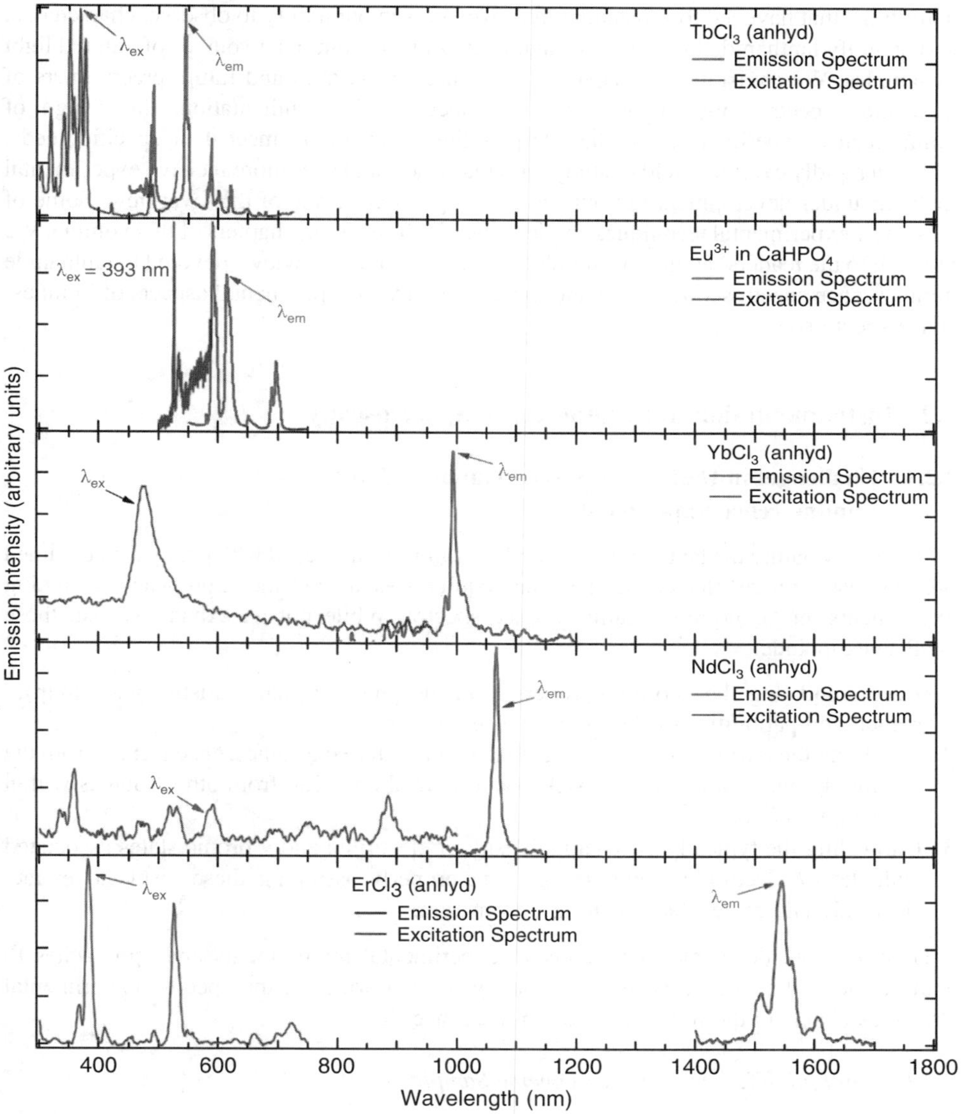


***Figure 2.2*** *Emission (red) and emission excitation (blue) spectral data for a series of $Ln^{3+}$ ion in solid-state matrices illustrating the range of excitation and emission wavelengths over which data need to be collected to characterise these materials. All data were collected using continuous wave excitation from a Xe arc lamp. Wavelengths for exciting the emission ($\lambda_{ex}$) and monitoring the emission excitation ($\lambda_{em}$) spectra are indicated. The detector for the $TbCl_3$ and $Eu^{3+}$ data was a photomultiplier tube. The detector for the $YbCl_3$, $NdCl_3$, and $ErCl_3$ data was a single-element InGaAs device. © David E. Morris, Los Alamos National Laboratory (See colour plate section)*

algorithms that have led to substantial improvements in our ability to observe, characterise, and quantify lanthanide luminescence and distinguish it from other sources of emitted light in samples. We envision this chapter as a resource for current and future practitioners of lanthanide spectroscopy to assist in the selection of instrumentation, the design of luminescence experiments, and the interpretation of results to meet their specific needs. This is a rapidly evolving field of study with many new and ever more incisive experimental methods under development but outside the scope of coverage of this overview. Some of these new experimental techniques are described in the ensuing chapters of this volume. We also call to the readers' attention a number of excellent recent reviews devoted to lanthanide luminescence spectroscopy [1–6] and more generally to experimental aspects of luminescence spectroscopy [7,8].

## 2.2 Instrumentation in Luminescence Spectroscopy

### 2.2.1 Challenges in Design and Interpretation of Lanthanide Luminescence Experiments

By their very nature the lanthanide ions and the sample matrices in which they are contained present many interesting challenges that require special consideration when designing experiments and interpreting luminescence spectra. While not an exhaustive list, these challenges include:

1. Accommodating the enormous diversity in the physical characteristics and environmental compatibilities of the samples to be interrogated.
2. Resolving the intrinsically low intensity lanthanide-based luminescence signals from the commonly much more intense background optical emission from other sources within the sample matrix.
3. Unraveling the typically very large density of electronic and vibronic states associated with the *f-f* electronic configurations and properly assigning these levels to extract electronic and molecular structural information.

Here we consider some of the special experimental requirements and approaches to address these challenges, as these ultimately have bearing on the specific experimental strategies to obtain the most incisive luminescence data.

#### *2.2.1.1 Adapting Experiments for Diverse Samples*

Samples bearing luminescent lanthanides are among the most diverse in all of science. Inorganic, organometallic, and materials chemistries of the lanthanide ions give rise to samples that range from discrete molecules that can be isolated as single crystals and dissolved in a range of solvents, to complex coordination polymers that are only isolable for characterisation as powders, to solid state materials including glasses and ceramics. This broad array of sample states has important implications for experimental luminescence spectroscopy for at least two reasons. First, the physical state of the sample and its environmental sensitivity (e.g., the need for an inert atmosphere) impact flexibility in the choice of sample containment. While most commercial instruments are designed to accommodate standard fluorescence cuvettes, many samples are better suited to containment in capillary tubes or thin-film cells, or as self-supporting films or blocks. For samples other than optically transparent solutions in cuvettes it is common to have to adapt a commercial sample compartment to accommodate the sample container while ensuring that optical characteristics of the instrument system are retained. Second, there are many important refinements to standard luminescence experiments to extract more detailed information on the sample (see below) that can only be carried out with specific instrumental configurations. Examples here include quantum yield and emission polarisation measurements that typically require optically transparent samples such as solutions or glasses, and quantum yield determinations for opaque materials like powders that typically require an integrating sphere detector. Most instrument manufacturers now offer a range of sample holders and instrument configurations to facilitate experiments with a variety of different materials and sample containers to enable the greatest variety of emission experiments.

#### 2.2.1.2 *Resolving Intrinsically Weak Lanthanide Emission from Intense Background Emission*

Optical emission spectroscopy is one of the most sensitive and incisive tools in analytical science. It is even capable of single-molecule detection under optimal conditions [9,10]. However, it suffers from a prevalent shortcoming, namely molecular excited-state emission is a ubiquitous phenomenon particularly in matrices comprised of organic molecules or containing organic impurities, and this typically very broad background emission signal can readily mask the much weaker luminescence signal intrinsic to lanthanide ions under typical conditions. This short-coming is less pronounced for the $Ln^{3+}$ ions that have their dominant emission signature in the NIR (e.g., $Nd^{3+}$, $Yb^{3+}$, $Er^{3+}$) because the lanthanide luminescence is spectrally well resolved from the background emission signal (typically in the UV and visible region). However, for several of the more important luminescent lanthanide ions ($Tb^{3+}$ and $Eu^{3+}$ in particular), the principal luminescent band(s) are in the visible region of the spectrum and very susceptible to being masked by background emission.

This problem can be most readily overcome through experimental means by taking advantage of the typically much longer-lived luminescent signal of the lanthanide ions (measured emission lifetimes in the millisecond regime are common) relative to the nanosecond and shorter lifetimes common to most (organic) fluorescence and phosphorescence signals. Thus, by using a sufficiently short duration pulsed light source and time-resolved detection scheme (see below for details), it is possible to temporally resolve the lanthanide luminescence signal from the background signal. An example of this time-resolved approach to isolating the luminescence from the lanthanide ion is presented in Fig. 2.3 for $Eu^{3+}$ absorbed on a cellulosic substrate.

In some cases the photophysical characteristics of the substrate can even be used to advantage to engender luminescence from the lanthanide ions. As discussed in Chapter 1, under optimised conditions the electronically excited organic substrate can transfer its energy to the lanthanide ion to induce electronic excitation at the $Ln^{3+}$ centre from which luminescence can occur. This mechanism is most prevalent for systems in which the organic chromophore is itself in intimate contact with the lanthanide ion via coordination (e.g., the so-called antenna effect) [2,3,5,11], although longer-range energy transfer processes can also induce lanthanide emission, even between differing lanthanide ions in the same matrix. These processes lead to a more readily discernible lanthanide luminescence signature

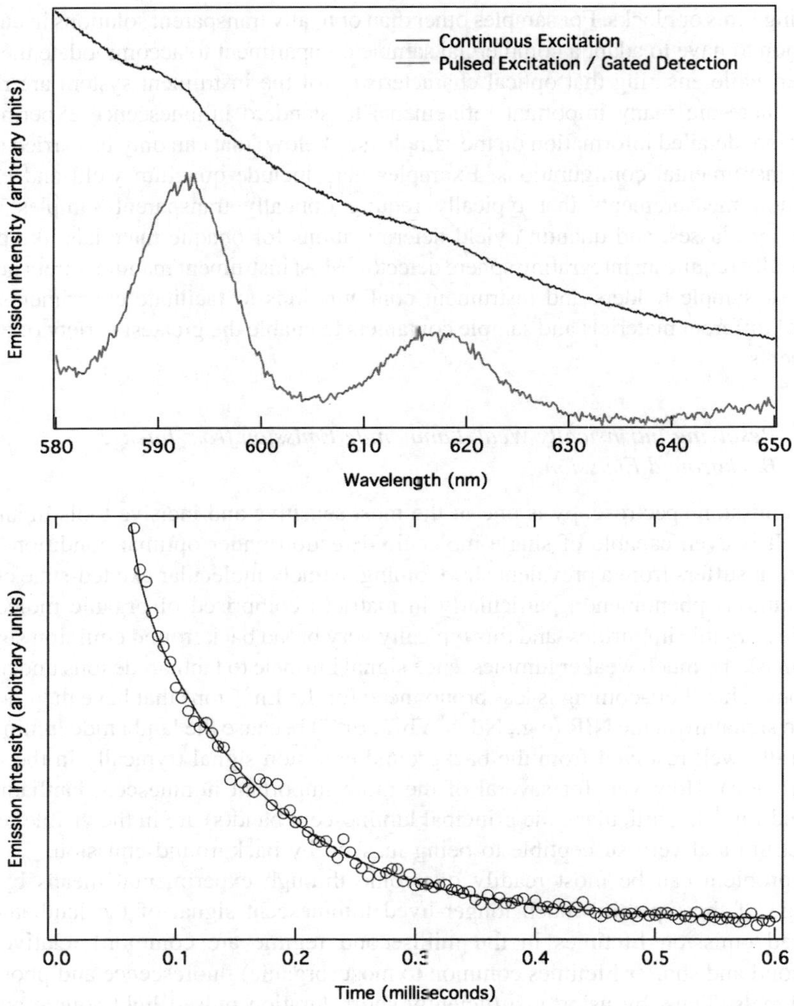


***Figure 2.3*** *Top: Comparison of continuous-wave (cw) excitation versus pulsed excitation/time-delayed gated detection in the emission spectrum of $Eu^{3+}$ adsorbed on Whatman CF-1 cellulose. The excitation source was a Xe arc lamp for the cw spectrum and a Xe flash lamp for the pulsed excitation spectrum. The excitation wavelength for both spectra was 393 nm. The detection in the pulsed spectrum entailed a 50 μs delay and a 10 ms gate width. The strong background emission tail seen in the cw spectrum originates from the cellulose matrix. This signal has a nanosecond lifetime and as illustrated the contribution from this background can be discriminated from the signal for the $Eu^{3+}$ species using gated detection with a delay time longer than the background emission lifetime. Bottom: Emission decay curve for this $Eu^{3+}$/cellulose sample using 393 nm pulsed excitation. Open circles are the experimental data. The green line is an unconstrained double-exponential fit to the data with lifetime values of 7 μs and 110 μs. The short lifetime component required for this fit is an experimental artifact from the residual intensity in the flashlamp pulse. © David E. Morris, Los Alamos National Laboratory (See colour plate section)*

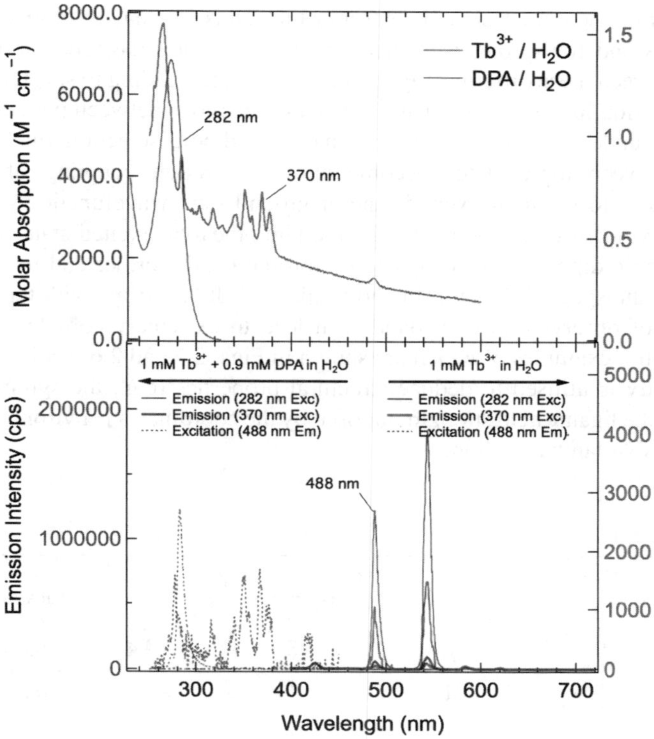


***Figure 2.4*** *Example of the "antenna" (sensitiser) effect on the emission properties of* $Tb^{3+}$. Top: *UV-visible absorption spectra of* $Tb^{3+}$ *(red) and pyridine-2,6-dicarboxylic acid (dipicolinic acid, DPA, green) in* $H_2O$ *illustrating the much greater absorption cross-section for the DPA sensitiser relative to that of the fully aquated* $Tb^{3+}$ *ion.* Bottom: *Continuous-wave emission [solid lines] and emission excitation [dashed lines] spectra for the aquated* $Tb^{3+}$ *ion (red; right hand intensity scale) and the* $Tb^{3+}$*/DPA complex (green; left hand intensity scale). Experimental parameters (slit widths and integration times) are identical for both samples. Excitation in resonance with the strong DPA absorption band (282 nm) leads to an enormous increase (~×1250) in the* $Tb^{3+}$ *emission, and the correspondence between the excitation spectrum for the* $Tb^{3+}$*/DPA sample and the DPA absorption spectrum clearly reflects that the DPA excited state is strongly coupled to emissive* $Tb^{3+}$ *states. Tuning off this ligand absorption band (370 nm) leads to a dramatic decrease in emission intensity. © David E. Morris, Los Alamos National Laboratory (See colour plate section)*

through enhancement of the lanthanide emission intensity rather than through suppression of the background emission signal. Typical examples of this effect are shown in Fig. 2.4.

#### *2.2.1.3 Resolving and Assigning Bands in Systems with High Densities of Spectral Bands*

As described in Chapter 1, the electronic structure derived from the occupied *f*-orbitals of all open-shell lanthanide ions is quite complex with a host of excited states of differing spin and orbital angular momentum designations from which emission may be observed. Further, the selection rules and transition mechanisms (electric dipole, magnetic dipole, etc.) differ within and between these differing angular momenta states leading to substantially differing observed transition intensities. In many instances transitions between pure electronic states are forbidden under both electronic and magnetic dipole selection rules, but radiative transitions between these states become allowed through coupling of one or more vibrational states to the total excited- and/or ground-state wavefunction (e.g., mixing of a non-totally symmetric metal–ligand vibration into the total excited-state wavefunction to enable an electric dipole transition). These vibronic bands further add to the density and complexity in the spectral domain. In short, this overall landscape with the potential for a high number of observed spectral bands can lead to extremely complex experimentally observed spectra, examples of which are shown in Figs. 2.5 and 2.6, yet in many cases this very complexity is utilised to deduce structural properties (e.g., the symmetry about the metal ion and the ligand-field strength imposed by the coordination environment) associated with the emissive lanthanide ion.

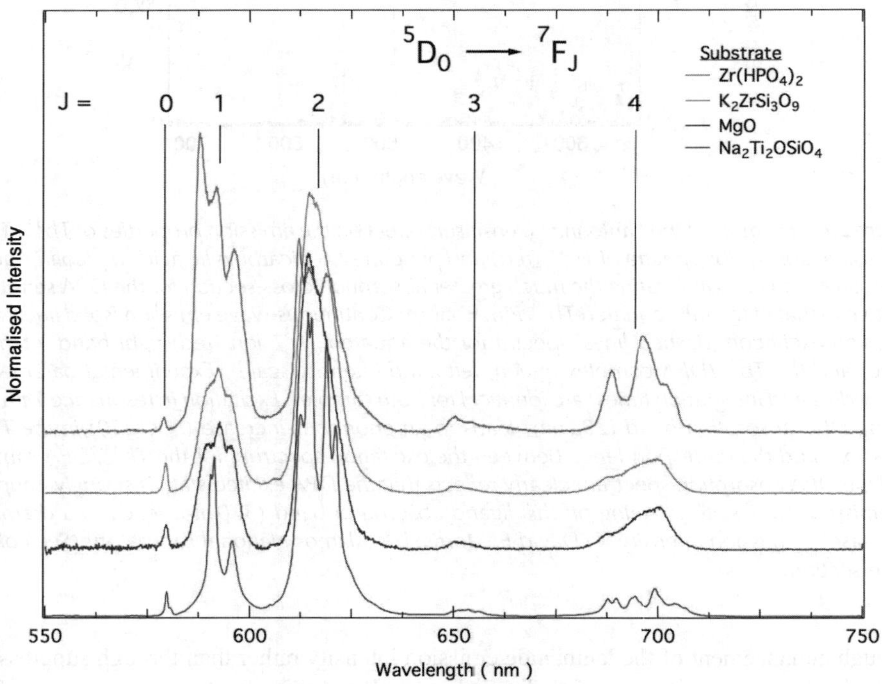


***Figure 2.5*** *Emission spectra for $Eu^{3+}$ ions in a series of solid-state inorganic ion exchange matrices illustrating the often high density of observable spectral transitions from the $^5D_0$ excited state into the manifold of lower-lying $^7F_J$ spin–orbit states and the variation in the relative intensities and degree of resolution in these bands. All spectra were acquired using 393 nm continuous-wave excitation. © David E. Morris, Los Alamos National Laboratory (See colour plate section)*

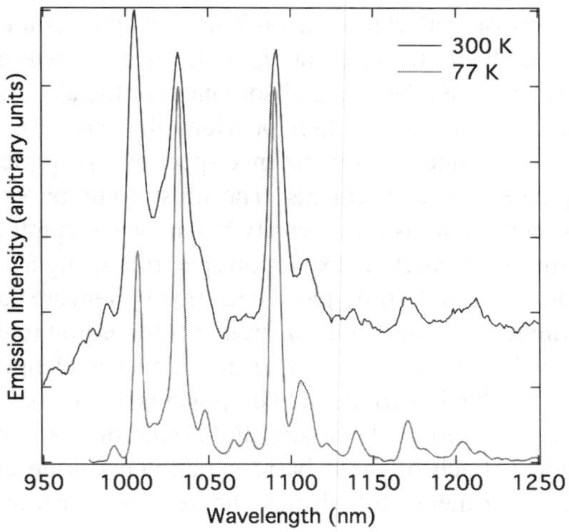


***Figure 2.6*** *Comparison of room-temperature versus liquid-nitrogen temperature emission spectra for* $[(C_5Me_5)_2Yb(2,2'\text{-bipyridine})]^+$ *in* $CH_2Cl_2$ *solution illustrating the substantial improvement in band resolution at cryogenic temperature. These vibronically resolved emission bands for the* $4f^{13}$ $Yb^{3+}$ *ion arise from transitions between the* $^2F_{7/2}$ *excited state and the* $^2F_{5/2}$ *ground state manifolds. The excitation in both spectra occurred at 440 nm from a continuous wave Xe arc lamp. © Ryan E. Da Re and David E. Morris, Los Alamos National Laboratory (See colour plate section)*

The experimental ramifications of this propensity for very complex spectral domain data include: (i) ensuring adequate instrumental spectral resolution to distinguish often closely spaced, narrow spectral features and (ii) manipulating experimental conditions, most notably temperature, to achieve enhanced spectral resolution. As described in greater detail in the ensuing discussion, the instrumental resolution is determined by a number of hardware considerations including the quantity and characteristics of the light dispersion element(s) (most commonly gratings) employed and the spectral bandwidth admitted into the instrument via adjustable slits. In addition, in some newer systems containing multi-element array detectors [e.g., charge-coupled device (CCD) detectors] the density of detector elements can influence the instrumental resolution. The most commonly employed experimental perturbation to influence spectral resolution is to subject the sample to a low temperature environment during data collection. This is most commonly done using cryogens such as liquid nitrogen or liquid helium by placing the sample in a Dewar or cryostat vessel. The principal influence of the low temperature is to diminish contributions to spectral line broadening from environmental interactions and to eliminate thermal populations of low-lying vibrational levels in the ground and/or thermally equilibrated excited electronic-state manifolds. An example of this thermal effect on spectral resolution is provided in Fig. 2.6.

### 2.2.2 Common Luminescence Experiments

The most common starting point in the characterisation of luminescent lanthanide systems is UV-visible-NIR electronic absorption spectroscopy. The UV-visible-NIR spectrum, in which absorption (or diffuse reflectance in optically opaque samples) is measured as a function of the wavelength of the incident light, is used to determine the wavelength (energy) and optical density (i.e., the molar absorbance) of the electronic absorption bands in the sample of interest. This information provides the basis on which luminescence spectral data acquisition is planned. Luminescence spectroscopy typically entails a number of different yet complementary experiments. The most common experiments employed to characterise luminescent systems are intensity versus wavelength (or energy) scans of: (i) the emission spectrum and/or (ii) the excitation spectrum using a continuous wave light source. If a pulsed light source and a time-gated detection system are part of the instrumental arrangement, then luminescence lifetime measurements (emission intensity versus time for a single excitation wavelength and at a single emission wavelength) are also possible. Similarly, if these time-resolved instrumental components are available, it is also common to measure and compare time-resolved emission and excitation scans to those collected with a continuous light source. Each of these basic experiments is described with examples below. This suite of experiments establishes the required instrumental hardware and configurations to perform these investigations.

#### *2.2.2.1 Luminescence Emission Spectra*

In the emission spectral mode the wavelength of the exciting light ($\lambda_{ex}$) is selected to coincide with an electronic absorption band in the sample and is held constant at this wavelength. The emission spectrum is generated by measuring the intensity of the emitted light as a function of wavelength (see Fig. 2.1). Most commonly the spectral range for an emission spectrum is chosen to begin at a wavelength longer than the excitation wavelength and to extend to even longer wavelengths to cover the region in which emission from the excited state(s) of the luminescent lanthanide are anticipated. It is frequently of value to obtain these emission spectral data at multiple excitation wavelengths corresponding to differing optical absorption bands in the sample to evaluate the degree of electronic coupling between the differing electronic excited states (see Chapter 1; see Fig. 2.4). In particular, the excitation wavelength could coincide with: (i) direct population of the emissive *f-f* manifold, (ii) higher lying states derived from the *f*-electronic configuration, (iii) charge-transfer states, and/or (iv) ligand localised states. Commonly the measured intensity in an emission spectral band as a function of the excitation wavelength should correlate with the magnitude of the absorbance in the optical absorption spectrum at the excitation wavelengths indicating that there is good electronic communication among the different states in the excited state manifold(s). Conversely, if a relatively more intense absorption band gives rise to an emission spectrum that is weaker than one obtained by exciting into a less strongly absorbing band, this suggests that the more strongly absorbing excited state is not strongly coupled to the emissive excited state. This can happen, for example, when systems of lanthanides contain multiple discrete chromophores that are either spatially or electronically isolated from each other.

#### *2.2.2.2 Luminescence Excitation Spectra*

In an excitation spectrum a single emission detection wavelength ($\lambda_{em}$) is chosen that coincides with an observed band in the emission spectrum. The excitation source is then scanned through the wavelength region in which the sample absorbs light, and the intensity of emission at the single selected wavelength is recorded versus the excitation wavelength. The wavelength range of the excitation spectrum is typically chosen to begin at the short wavelength end of the instrument's capability and extend to a wavelength just short of that of the emission band being monitored (see Fig. 2.1). The excitation spectrum is analogous to a UV-visible-NIR absorption spectrum, but instead of measuring the diminution in transmitted light as the samples absorbs energy, the excitation spectrum measures the light given off from an emissive excited state. Thus, the excitation spectrum not only maps out the electronic absorption bands in the sample, but through relative intensity variations as noted above between the absorption spectrum and the excitation spectrum, it also provides information about the extent to which the higher lying, absorbing excited states are coupled to the emissive excited state(s). The acquisition of an excitation spectrum requires a tunable broadband light source such as an arc lamp or flash lamp (see below). Further, because excitation sources typically have a variable intensity output versus wavelength, it is essential that the excitation spectrum be corrected for this intensity variation so that any observed peak intensity variations in the excitation spectrum reflect effects intrinsic to the electronic excited state processes in the sample and not simple instrumental factors attributable to source intensity variations. This correction is most easily and commonly done by diverting a very small fraction of the excitation source intensity (using a simple beam splitter such as a glass slide) onto an ancillary detector. The output from this ancillary detector, which corresponds to the intensity spectrum of the source, is then used to normalise the emission intensity signal from the sample.

#### *2.2.2.3 Luminescence Lifetime Measurements*

Luminescence lifetime measurements are an essential ingredient in characterising the luminescence properties of lanthanide ions. They form the basis for a number of analytically important determinations regarding electronic structure and coordination environment in lanthanide systems including quantum yields for radiative and non-radiative transitions (see below), estimation of the number of coordinated water molecules and other ligands [12–14], and enumeration of multiple structurally distinct emissive species in complex samples [15][1]. Lifetime measurements require a pulsed excitation source such as a flash lamp or pulsed laser with a pulse duration that is sufficiently short to enable discrimination of the light emitted from the sample from that due to the excitation source itself. In practice, with commercially available sources, the pulse duration ranges from the microsecond regime for flash lamps to the picosecond regime for pulsed laser sources, while measured lanthanide luminescence lifetimes in the range from milliseconds to nanoseconds are common. The actual experiment typically entails the use of an excitation wavelength ($\lambda_{ex}$) that corresponds to an electronic absorption band in the sample of interest. The decay in the intensity of the emitted light is measured as a function of time from arrival of the light pulse on the sample to

[1] The most common elaboration of this technique for lanthanide luminescence studies is carried out in the time domain (i.e., a pulsed light source is employed, and emission intensity is recorded vs. time), although instrumentation is available to make these measurements in the frequency domain as well using a modulated light source. In the ensuing discussion the focus is on time domain experiments.

generate the emission decay curve. In most cases the emitted light is dispersed and the wavelength selected to coincide with a specific band in the emission spectrum ($\lambda_{em}$). For a pure single emissive species the decay in luminescence intensity $I(t)$ should follow simple first-order exponential decay kinetics, described by Equation 2.1.

$$I(t) = I_0 \exp\left(-^{t}/_{\tau}\right) \tag{2.1}$$

$\tau$ is the measured luminescence lifetime and $I_0$ is the intensity at $t=0$. In general the emission decay curves are analysed according to this model to extract the measured lifetime values. Deviations from this simple, single-exponential decay behaviour then reflect the presence of multiple emissive species in the sample, sample inhomogeneity, and/or more complex excited state decay processes. An example of a luminescence lifetime determination is provided in Fig. 2.3.

#### 2.2.2.4 *Time-resolved Luminescence Emission and Excitation Spectra*

Time-resolved spectral data acquisition can be regarded as an elaboration on the luminescence lifetime determination described above, and it is usually the case that hardware for measuring lifetimes (pulsed excitation sources and gated detection electronics) can be adapted to the measurement of time-resolved spectral-domain data. However, the utility of time-resolved spectral domain data acquisition transcends that of lifetime measurements alone. In particular, time-resolved spectral acquisition (both emission and excitation scans) is the principal means to resolve spectral domain band structure associated specifically with the luminescent lanthanide ion of interest (i.e., the *f-f* bands) from background optical emission from other components in the sample matrix. One example of this effect and the benefit derived is illustrated in Fig. 2.3. In practice, time-resolved spectral data are normally acquired in a manner analogous to emission and excitation spectral data described above under steady-state illumination conditions, but with a pulsed light source (wavelength tunable for excitation spectra) and a gated detection system. A number of commercially available instruments have gated detection electronics incorporated into the systems, but this can also be accomplished with common detectors such as photomultiplier tubes and a boxcar averager to provide the detection gate. Unlike the case in luminescent lifetime measurements for which the emitted light is measured from time of arrival of the excitation pulse until the emission intensity decays to zero, for time-resolved spectra one typically choses a set delay time from the excitation pulse and a set gate duration. These parameters are then held constant as one scans the wavelength range of interest. These additional adjustable parameters, gate delay and duration, when chosen properly, enable one to discriminate between multiple emission sources in a sample. This is especially important in discriminating background or impurity emission that is usually fluorescent in nature with nanosecond or shorter lifetimes from lanthanide luminescence that is commonly microsecond-lived or longer. In these cases an essentially pure representation of the lanthanide ion spectrum can be obtained (e.g., Fig. 2.3). This approach can also be used to bias the relative spectral contributions in a very complex measured spectrum arising from multiple species with comparable lifetimes (i.e., differing by factors of two or three instead of orders of magnitude). In this way it may be possible to utilise chemometric analyses of multiple data sets with differing gate delay/duration settings to reconstruct the spectra from the individual pure components [16,17].

### 2.2.3 Basic Design Elements and Configurations in Luminescence Spectrometers

The demands of fully characterising the luminescence properties of most lanthanide ions contained in frequently complex sample matrices dictate a need for both flexibility and modularity in the design of the spectrometer system. In this section the basic elements in the spectrometer system are considered, and the most common system configuration is defined. This common system configuration and its hardware elements are available currently from a number of commercial sources, or can be assembled from individual components (see component details below) to address specific experimental needs. Whether one selects a complete commercial system or assembles a system from components, attention to the need for and importance of design flexibility and compatibility in incorporating or exchanging system components is strongly recommended.

There are four distinct modules to consider in most optical emission spectrometer systems: the excitation module, the sample module, the emission module, and the instrument control/data acquisition module. The elements comprising each module (and variations therein) are captured in Fig. 2.7 and described in greater detail in the ensuing section.

Note that some instrument configurations do not require each element in the module. For example, if a laser excitation source with an intrinsically narrow-bandwidth monochromatic output is employed, then an excitation monochromator is not required and the exciting light is typically coupled directly into the sample compartment via an optical fibre or light pipe. Some of the more elaborate commercial instrument systems also explicitly incorporate

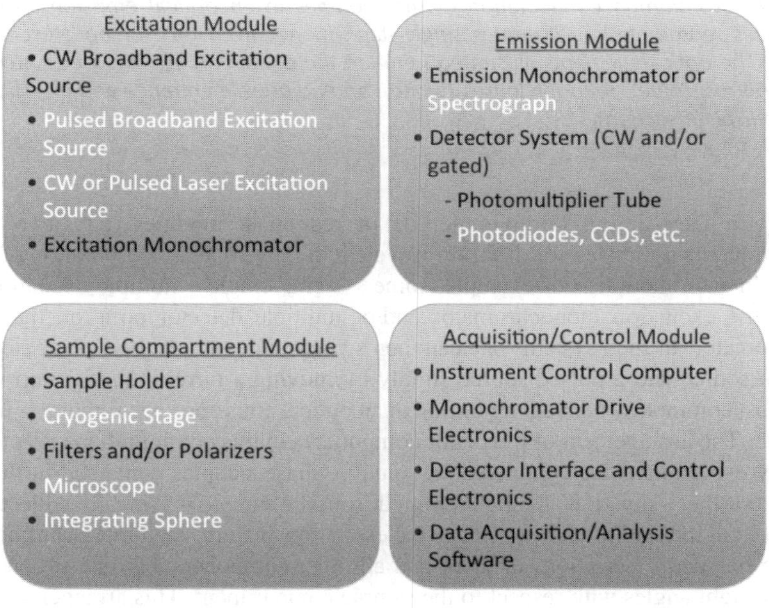


*Figure 2.7* *The four instrument modules common to most optical emission spectrometer systems. Components in black are the standard elements in a UV-visible emission system. Components in white are optional accessories that enhance the versatility of the system and extend the wavelength range for excitation and/or emission. (See colour plate section)*

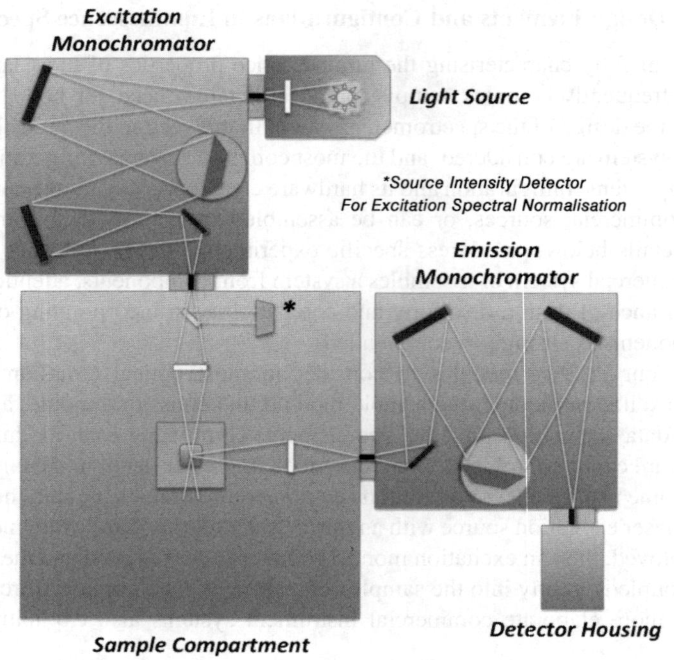


*Figure 2.8 A common configuration of the modules in an optical emission spectrometer system. This representation illustrates single-stage monochromators for both excitation and emission. The optical elements illustrated here are the diffraction gratings (red), steering and focusing mirrors (blue), focusing lenses (white), and adjustable entrance and exit slits (black). (See colour plate section)*

flexibility in their design through the use of redundant modules (e.g., two emission monochromators optimised for differing wavelength ranges) and incorporation of multiple elements in each module. For example, some systems employ multiple excitation source ports on the excitation monochromator and/or multiple detector ports on the emission monochromator. In such systems one can then switch, for example, between a steady-state excitation source and a pulsed source simply by moving a turret-mounted mirror.

The most common system configuration in an optical emission spectrometer is illustrated in Fig. 2.8. The arrangement of the major components in this configuration reflects the fact that many experiments are conducted on solution-phase samples contained in fluorescent cuvettes. For these situations the light emitted from the sample is typically collected at 90° with respect to the excitation beam. Thus the excitation and emission monochromators (see below for detailed descriptions of these and all other components) are most conveniently situated at right angles with respect to the sample compartment. This arrangement reduces the amount of scattered light directed into the emission monochromator (thereby reducing background signal from the excitation source) and facilitates interpretation of, for example, polarisation data when polarised light is used to excite the sample and polarisation analysers are used to filter the emitted light [18].

There are a number of common variants on this basic instrument configuration. For example, optical microscopes are frequently interfaced to optical emission instruments to perform emission-based imaging of samples or to couple light into and out of samples of very small dimensions. Integrating spheres are another common component that are externally interfaced. These are specially designed optical cavities with a reflective interior coating that are used to collect nearly a full $4\pi$ steradians of light from an emissive sample. These are used for very precise quantum yield determinations (see below) on all types of samples, and provide the only practical route to quantum yield determinations for opaque samples such as powders. For both microscope and integrating sphere adaptations to the standard instrument configuration shown above the light is usually coupled into (and sometimes out of) these external components via fibre optics, so in many respects the microscope or the integrating sphere represents a variation on the sample compartment module as defined above. Frequently, however, these components will have detection elements integrated directly into their design.

### 2.2.4 Luminescence Spectrometer Components and Characteristics

This section describes and discusses the principal optical and electronic components that comprise a luminescence spectrometer system. In some cases (e.g., monochromators) a single design dominates all commercial systems, and this design provides the basis for the ensuing discussion. In many other cases there is a variety of possible components in the category. In these cases an attempt has been made to be comprehensive with regard to discussion of current commercially available hardware. However, in the case of both excitation sources and detectors, development and deployment of new technologies and new products are proceeding at a remarkable pace. Thus, one can anticipate an ever-expanding range of capabilities covering (or improving) broader wavelength ranges, enhanced sensitivity, and better spatial and/or temporal resolution.

#### *2.2.4.1 Monochromators*

Monochromators are the essential light dispersion components in a spectrometer system. Monochromators are comprised of: (i) adjustable dimension entrance and exit slits, (ii) the wavelength dispersive element (which is in nearly all cases a diffraction grating, see below), and (iii) a series of mirrors to capture, focus, and steer the light from the entrance slit to the grating to the exit slit. There are two high-level variants in this category, but the functions are quite similar. The first is a *scanning monochromator* in which only a relatively narrow band of light (referred to as the bandpass, typically a fraction of 1 nm in this case) is transmitted through the exit slit, and the dispersive element is scanned through a wavelength range of interest to produce the spectral domain data. The second is a *spectrograph* in which a much broader band of light (typically tens to hundreds of nanometres depending on design and wavelength range) is present at the plane of the exit slit. Spectrographs have the advantage of capturing an entire spectral window in a single shot without scanning of the wavelength dispersive element. These were originally designed for use with photographic films as the detection/recording element, and are now used extensively with multichannel or array detectors such as CCD detectors. For luminescence applications the scanning monochromator is the dominant choice because of enhanced spectral resolution and because luminescence spectra typically cover a broader spectral window than can be conveniently captured in a single shot with a spectrograph.

Nearly all commercial monochromator systems employ diffraction gratings for light dispersion, and the Czerny–Turner optical design has become the standard. This design consists of two concave mirrors and a planar grating. It is the design illustrated in the excitation and emission monochromators shown in Fig. 2.8. While single-stage monochromators are used in most luminescence spectrometer applications, some systems are designed with double monochromators on either or both emission and excitation modules. A double monochromator is merely two tandem monochromators operated in series and in wavelength synchronisation. The advantage of a double monochromator lies in significant improvement in stray-light rejection that concomitantly enhances the signal-to-noise ratio in a spectrum. One of the more important figures of merit for monochromators is their f/# (or its inverse, the numerical aperture). As with simple lenses, the smaller the f/# (e.g., f/4 relative to f/10) the greater the light throughput (also frequently referred to as "optical speed") through the instrument.

As noted previously, diffraction gratings are the primary light dispersion elements in modern monochromators. Gratings are characterised by the groove density (in number of grooves per millimetre) that determines the resolving power of the monochromator; a higher groove density produces greater resolving power[2]. Diffraction gratings are of two types; mechanically ruled gratings or holographic gratings. The former, as the name suggests, are prepared by burnishing grooves using a diamond stylus into a coated layer (typically aluminum) deposited on a substrate (glass or metal). This mechanical approach limits the groove density to ~1800 grooves/mm. Holographic gratings are produced by exposing a photoresist deposited on a substrate to a laser interference pattern. The photoresist is then developed to etch out the exposed areas leaving a periodically grooved surface. Holographically produced gratings can have much higher groove densities and they provide a much simpler, more robust route to larger surface area gratings. Mechanically ruled gratings have faceted grooves with facet angles (relative to the surface normal) determined by the burnishing process. Holographic gratings have sinusoidal grooves. Both mechanically ruled and holographic gratings can be prepared with specific blaze wavelengths. The blaze wavelength refers to the wavelength of optimal grating efficiency; it is an adjustable parameter in the fabrication process and is a function of the detailed geometry of the grooves.

Finally, the adjustable-dimension slits at the entrance to and exit from the monochromator are integral elements of this system component. The slit height is typically not a continuously adjustable parameter, but can frequently by modified over a coarse range of settings using an adjustable mask. In general, the slit height will impact only the total amount of light admitted into/out of the monochromator without impacting resolution. The slit width in contrast is a continuously adjustable parameter that is normally set to achieve the desired spectral bandpass or resolution. Note however that the slit width setting will also affect the light throughput through the monochromator which ultimately determines signal strength, so there will always be tradeoffs to consider in adjusting slit width settings. Narrower slit width settings result in smaller bandpass and better resolution, but also

[2] Resolving power is related to the theoretical ability to separate two spectral features, and should not be confused with resolution or bandpass, which are practical measures of the performance of the entire monochromator system.

decreased signal strength. In general, one adjusts the slit width to achieve optimal signal consistent with required spectral resolution. As a practical matter, since a typical luminescence spectrometer has both excitation and emission monochromators, the slit settings need to be independently considered and optimised for the experiment. If one is interested in the highest resolution emission spectrum, then the emission monochromator slit widths need to be adjusted to provided suitable resolution whereas the excitation monochromator slit widths can be increased to provide greater excitation light throughput to increase the emission intensity from the sample. Conversely, a high resolution excitation spectrum requires that the excitation monochromator slit widths be adjusted to achieve this high resolution, but the emission monochromator slits widths can be increased to allow greater throughput of the emitted light.

#### *2.2.4.2 Excitation Sources*

*Continuous Sources:* The most common continuous wave light sources for optical emission spectrometer applications are high-pressure arc lamps (xenon and mercury) and tungsten halogen lamps. The arc lamp sources have very high radiance in the UV and visible regions, with the mercury lamp in particular having excellent UV output. Tungsten halogen lamps have less radiance in the UV region, but excellent radiance into the NIR region. These lamps are available in a variety of powers (with concomitantly higher radiance) from ~50 W to >1 kW. The higher power arc lamps do represent a source of ozone generation that may require special ventilation requirements. Continuous wave lasers (e.g., argon or krypton ion lasers) and cw-pumped dye lasers can also been used as excitation sources in emission spectrometer applications, but such applications are relatively rare and are usually associated with, for example, the use of a Raman spectrometer system to collect emission data.

*Pulsed Sources:* The number of different pulsed light sources has proliferated dramatically in the last several decades. The three most general categories of pulsed sources are flashlamps, semiconductor-based light emitting and laser diodes, and pulsed lasers. Within each category there is a broad range of performance characteristics, but the figures of merit of greatest importance include output wavelength(s), output power, pulse duration, and repetition rate. In principle any pulsed laser system represents a viable pulsed excitation source for an optical emission spectrometer system, but here the discussion will focus on laser systems designed specifically for use as an excitation source for emission spectrometers.

Xenon flashlamps are the simplest and most readily wavelength tunable pulsed light sources. These lamps typically have sufficient output radiance from ~250 to ~800 nm for excitation applications. The pulse width depends on a number of factors including output power, but pulse widths of 1–2 ms are common. Pulse repetition rates of ~100 Hz are also fairly common for these sources.

Semiconductor-based devices, including light-emitting diodes (LED) and laser diodes, are the newest and most versatile excitation sources for emission spectroscopy applications. These devices are relatively inexpensive and compact. Their biggest drawback is that they are intrinsically single wavelength sources (although some devices can be tuned over narrow ranges using temperature control). Single wavelength pulsed LED sources are now available with centre wavelengths ranging from ~250 nm to ~600 nm, bandwidths of ~25 nm, and with pulse powers of ~1 mW. The pulse duration for these sources is nominally ~1 ns, and they can be operated at repetition rates from single-shot to in excess of 1 MHz. Pulsed laser diodes are available in an even broader range of wavelengths from ~370 nm into the NIR, and with greater power (tens of mW is possible) than pulsed LEDs. These, too, can be operated over a range of repetition rates from single-shot to >10 MHz. A significant advantage of pulsed laser diodes relative to LEDs is the reduced pulse durations. Laser diodes typically have pulse widths of 50–100 ps enabling investigations of very short-lived processes.

A variety of other pulsed laser light sources have been incorporated into optical emission spectrometer systems. The two most common ones are based on either a pulsed nitrogen laser or a frequency-upconverted pulsed Nd:YAG laser as either a single line source (337 nm for the nitrogen laser, 266, 355, or 532 nm for the Nd:YAG laser) or used as a pump laser for a tunable secondary laser device such as a dye laser or an optical parametric oscillator (OPO). The ultimate wavelength tunability in such systems depends on the characteristics of the secondary laser and can be substantial in the case of the OPO, but at a substantial price in both cost and complexity of operation and maintenance. Nonetheless, these sources can have excellent temporal characteristics (sub-nanosecond pulse durations) and high per pulse energies.

### *2.2.4.3 Detectors*

The simplest means to delineate the options for detectors for optical emission spectroscopy applications is one based on the wavelength range over which the detector has optimal sensitivity. By this standard there are two spectral regimes to consider; the UV-visible region (which extends into the NIR in most cases for detectors) and the NIR region. Within these two broad categories, there are two additional means to distinguish available detector technologies; single element vs. array detectors and those capable of time-resolved detection (gateable devices) versus steady-state detection.

In the UV-visible spectral region photomultiplier tubes (PMTs) are the most versatile single-element detection option. PMTs are available with a range of optimal wavelength sensitivities (e.g., those that are only sensitive in the UV-visible versus those that have some sensitivity into the near infrared) that span the range from the deep UV to ~900 nm. PMTs are also very sensitive yet have a large dynamic range (accommodating very low to moderately high light levels), and they have generally very fast response times making them ideal for time-resolved detection applications. Multielement or array detector options in the UV-visible region include diode arrays and CCD detectors. Both of these detector options can be obtained with time-resolved performance at the nanosecond scale and longer. CCD chips, the functional element in CCD detectors, now dominate this market. These are manufactured with a variety of specialty modifications (back-thinned, deep depletion, etc.) that enhance sensitivity in select wavelength ranges including "near infrared" chips with significant sensitivity in the 800–1000 nm range.

Recent advances in semiconductor manufacturing have led to a variety of new options in NIR detectors. Most NIR detectors are based on InGaAs semiconductor architectures. Single element detectors include diodes and, more recently, NIR photomultiplier tubes. These typically have good response from the visible region to ~1700 nm although their use is typically focused in the longer wavelength (~1000–1700 nm) range since traditional PMTs have superior performance in the visible range. The NIR photomultiplier tubes are also now available with excellent response times for use in time-resolved spectral applications. Tubes with rise times on the order of 2–3 ns are currently available opening up exciting new opportunities for systems containing NIR luminescent lanthanides. Multi-element detectors are also now in production for the NIR spectral region based on a linear array of InGaAs diodes. These have performance characteristics similar to the single element diode detectors.

#### 2.2.4.4 *Ancillary Components*

There are a host of additional components and accessories to enhance the performance and range of applications of both commercial and custom optical emission spectrometer systems. A number of optical components are essential adjuncts to any optical emission spectrometer system. The most important of these are optical filters including long pass, band pass, and neutral density filters. The band pass filters are essential because the diffraction gratings diffract higher orders of light than just the first order that is most commonly used. While the intensity in these higher orders is diminished relative to that in first order, it nonetheless can be substantial. Thus, for example, if one selects 300 nm light in the excitation monochromator, there will also be a strong component of 600 nm light (second order from the grating) that passes directly through the exit slit of the excitation monochromator and illuminates the sample. This 600 nm light (and all higher orders) can be filtered out using a 300 nm band pass filter at the entrance to the sample compartment. Similarly, long pass filters can be placed at the entrance to the emission monochromator that will only admit light of wavelengths equal to or longer than the short-wavelength side of the emission band(s) of interest (i.e., reject stray light from the exciting wavelength). Neutral density filters can be used to diminish, for example, excitation light intensity impinging on the sample without altering a desired spectral bandpass established by slit settings. Linear polarisation optics are also frequently of value when detailed information about the polarisation characteristics of the emissive system are sought to more thoroughly map out the symmetries and selection rules of the transition matrix elements. This will most commonly be the case for fluid solution samples in which molecular orientations are totally random or for optically transparent oriented single crystals for which molecular axes remain fixed. Since many excitation sources are not *a priori* polarised, the polarisation experiments require sets of polarising films to first select the linear polarisation of the excitation light, and then to analyse the polarisation of the emitted light.

Among the other more common accessories are sample temperature regulators/controllers. These range from simple jacketed cuvette holders supplied by closed-loop recirculating baths, to thermoelectric coolers integrated directly into the cuvette holder, to liquid nitrogen Dewar assemblies fabricated with optical windows, to more complex nitrogen and helium cryostats. Finally, for studies of reaction dynamics involving luminescent reactants and/or products there are also stop–flow cells and titrator accessories available for most commercial instruments.

### 2.2.5 Recent Advances in Luminescence Instrumentation

The biggest advances in luminescence spectroscopy instrumentation are clearly being driven in large part by the exciting new applications areas for lanthanides in the biosciences and materials sciences. In the biosciences the versatile chelation chemistry of the lanthanides enables specific and very site-directed incorporation of lanthanide ions into functional sites within biomolecules. This versatile chemistry coupled with the intrinsically useful optical properties of the lanthanides is leading to increasing use of lanthanides as luminescent labels for identifying specific features and functions in cellular environments as well as for real-time imaging of intracellular trafficking. This translates into many exciting developments in microscopy and imaging hardware and software that take full advantage of both the breadth of wavelengths and temporal characteristics of lanthanide luminescence [19–23]. The field of bioimaging with lanthanides is considered in greater detail in several subsequent chapters.

Another area of explosive growth for lanthanide applications is in the realm of nanoscience (which has some significant overlap with the biological imaging applications noted above). Lanthanide-based nanoparticles and nanostructured materials as well as other nanoscale materials that incorporate lanthanides as dopants for emergent functionality or merely as luminescent or magnetic tags have proliferated over the past decade as evidenced by the huge increase in citations for publications relating to these applications. While many of the luminescence techniques described above are generally applicable to ensemble samples of nanoparticles or two- and three-dimensional arrays formed by these particles, all of these methods suffer from the physical limits imposed by light diffraction when using conventional optical elements. Spatial resolution using optical probes is limited to approximately one-half the wavelength of the light used for interrogation. Thus, for visible to near infrared probes the intrinsic resolution (from ~300 to 500 nm) falls short of that needed to spatially resolve a single nanoparticle by as much as an order of magnitude. However, there are many unique structural and functional features of these nanomaterials that can only be captured and understood at the single particle spatial scale.

A number of new approaches to overcoming this fundamental diffraction-limit obstacle in optical emission spectroscopy are emerging. These generally fall into two categories; near-field (also called "tip enhanced") optical microscopies that are hardware focused [24,25], and super-resolution techniques that can be either hardware and/or algorithmic approaches to collecting and/or treating confocal microscope data [26–29]. None of these approaches have gained a strong foothold in the lanthanide luminescence research community to date, but one can anticipate that as the need for greater spatial resolution from optical probes increases these approaches will become more prevalent. In the near-field microscopy methods the light is coupled into and out of the system under interrogation by a very small aperture probe (or metal tip in "apertureless" mode) placed at a distance from the materials significantly less than the wavelength of the light being used. Spatial resolution is determined principally by the dimension of the aperture or tip. The probes are typically incorporated into atomic force microscopy stages to achieve suitable manipulation of the tip with respect to the sample surface. This approach to characterisation of lanthanide nanomaterials has been described in at least one recent report of emissive lanthanide doped $Lu_2O_3$ nanorods [30]. Super-resolution approaches include the hardwared-based methods [26,29] known as structured illumination microscopy (SIM) and stimulated emission depletion (STED) and software-based methods [27–29] known as photoactivation localisation microscopy (PALM) and stochastic optical reconstruction microscopy (STORM). The reader is referred to the above references for details. These approaches are all ripe for exploitation in the field of lanthanide luminescence in nanomaterials.

A final relatively recent development in luminescence instrumentation to note here is the time-resolved detection scheme known as time-correlated single photon counting (TCSPC) [31,32]. This is a variation on a theme described above for the measurement of luminescent lifetimes, but requires an additional array of detection electronics for signal processing. In this approach the decay period is discretised into a number of equal time channels, and emitted photons are detected and subsequently binned into these channels. This creates a histogram that defines the conventional exponential (or more complex) decay curve. It has not been extensively utilised in lanthanide luminescence spectroscopy [32], but has as a commending asset its applicability to weakly emitting samples because of its true single-photon counting basis. The method does typically utilise a high repetition rate pulsed light source, since not every excitation pulse results in the binning of an emitted photon, and this high repetition rate could lead to saturation in pumping the electronic excited states thereby making it inconsistent with the long measured lifetimes in some lanthanide systems.

## 2.3 Measurement of Quantum Yields of Luminescence in the Solid State and in Solution

From Fig. 1.10 in Chapter 1 it can be seen that the overall efficiency of the sensitised luminescence process $\Phi_L^{Ln}$ is given by

$$\Phi_L^{Ln} = \varepsilon_L(\lambda_{exc}) \times \phi_L^{Ln} = \varepsilon_L(\lambda_{exc}) \times {}^{I_E}\!/_{I_{abs}} = \varepsilon_L(\lambda_{exc}) \times \eta_{ISC} \times \eta_{ET} \times \phi_{Ln}^{Ln} \tag{2.2}$$

$I_E$ is the integrated intensity of emission, $I_{abs}$ is the total number of photons absorbed, $\varepsilon_L(\lambda_{exc})$ is the molar absorption coefficient at the excitation wavelength $\lambda_{exc}$, $\phi_L^{Ln}$ is the quantum yield of the metal-centred luminescence upon ligand excitation, in turn a product of the efficiency of intersystem crossing $\eta_{ISC}$ from the singlet to the triplet state of the ligand, the efficiency $\eta_{ET}$ of the energy transfer from the triplet state to the lanthanide ion excited state and the intrinsic quantum yield $\phi_{Ln}^{Ln}$ of the lanthanide ion emission upon direct excitation.

The intensity of emitted light $I_E$ is given by

$$I_E = I_{abs} \times \phi_L^{Ln} = I_0 \times \left[1 - 10^{-A(\lambda_{exc})}\right] \times \phi_L^{Ln} \tag{2.3}$$

with the intensity of absorbed light $I_{abs} = I_0 - I = I_0\left[1 - 10^{-A(\lambda_{exc})}\right]$. $I_0$ is the intensity of excitation light and $I$ the intensity of the transmitted light.

Most fluorimeter components have a wavelength-dependent response, which can be accounted for with $k(\lambda)$, the instrumental function; therefore, when doing emission spectroscopy $I_{obs}$ is measured.

$$I_{obs} = k(\lambda) \times I_E \tag{2.4}$$

It is difficult to determine the instrumental function, and therefore absolute values of quantum yield are not routinely measured. Instead, one measures the intensity of observed light $I_{obs}$ for a standard of known quantum yield and compares one's sample to the standard, as $k(\lambda)$ will be the same for both samples, as is illustrated in the following.

### 2.3.1 Measurement Against a Standard in Solution

The most commonly utilised way to measure the unknown emission efficiency of a lanthanide ion complex $x$ is by comparison to a standard $ST$ of known emission efficiency.

$$\frac{\phi_x}{\phi_{ST}} = \frac{I^x_{obs}}{I^{ST}_{obs}} \times \frac{I^{ST}_0\left[1 - 10^{-A_{ST}(\lambda_{exc})}\right]}{I^x_0\left[1 - 10^{-A_x(\lambda_{exc})}\right]} \tag{2.5}$$

and

$$\phi_x = \phi_{ST} \times \frac{I^x_{obs}}{I^{ST}_{obs}} \times \frac{I^{ST}_0\left[1 - 10^{-A_{ST}(\lambda_{exc})}\right]}{I^x_0\left[1 - 10^{-A_x(\lambda_{exc})}\right]} \tag{2.6}$$

$\left[1 - 10^{-A(\lambda_{exc})}\right]$ can be expanded into a Taylor series as follows

$$2.3A(\lambda_{exc}) - \frac{[2.3A(\lambda_{exc})]^2}{2!} + \frac{[2.3A(\lambda_{exc})]^3}{3!} + \ldots \tag{2.7}$$

For $A < 0.05$ the terms involving the factorials are below 5%, so that for dilute solutions with very low absorbance they can be safely ignored and Equation 2.6 simplifies to

$$\phi_x = \phi_{ST} \times \frac{I^x_{obs}}{I^{ST}_{obs}} \times \frac{I^{ST}_0}{I^x_0} \times \frac{A_{ST}}{A_x} \tag{2.8}$$

If the emission quantum yields are measured with sample and standard in different solvents, the differences in refractive index should be corrected by using Equation 2.9.

$$\phi_x = \phi_{ST} \times \frac{I^x_{obs}}{I^{ST}_{obs}} \times \frac{I^{ST}_0}{I^x_0} \times \frac{A_{ST}}{A_x} \times \frac{n_x^2}{n_{ST}^2} \tag{2.9}$$

Common standards used for these measurements are summarised in Table 2.1.

If measuring at $A \leq 0.05$ is impractical, either because the metal ion decomplexes at the concentration needed, or sensitisation is not observed at $\lambda_{exc}$ for this low absorbance, measurements of a series of solutions of different concentrations can be done instead. Samples with different concentrations are prepared and the integrated emission spectra for each are plotted as a function of sample absorbance. Similarly, integrated emission spectra for a series of solutions of standard at different concentrations are plotted as a function of their absorbance. The slopes $m$ of the obtained lines are then used to calculate the emission efficiency of the sample, following Equation 2.10.

$$\phi_x = \phi_{ST} \times \frac{m_x}{m_{ST}} \times \frac{n_x^2}{n_{ST}^2} \tag{2.10}$$

*Table 2.1* Common standards used for emission efficiency measurements

| Standard | Φ (%) | References |
|---|---|---|
| $Cs_3[Eu(dpa)_3]$ – air saturated water | 24 | [33,34] |
| $Cs_3[Tb(dpa)_3]$ – air saturated water | 22 | [33,34] |
| Sodium salicylate – solid state in KBr | 53 | [35] |
| Pyrene – solid state in PMMA | 61 | [36] |
| Anthracene – solid state in PMMA | 22 | |
| $[Ru(bpy)_3]Cl_2$ | 2.8 (air saturated water, room temperature)<br>37.6 (methanol at 77 K) | [37,38] |
| $5 \times 10^{-6}$ M quinine sulfate in 0.1 N $H_2SO_4$ – air saturated | 54.6 | [39] |
| 1 μM $Sm^{3+}$ in Wallac DELFIA enhancement solution | 1.5 | [40] |
| Rhodamine 6G – air saturated water | 81 | [41] |
| Eu(III) chelate of N,N,N′,N′-[4′-phenyl-2, 2′:6′,2″-terpyridine-6,6″-diyl]bis-(methylenenitrilo)tetrakis(acetate) – deoxygenated – air saturated water | 16 | [42] |
| $Yb(tta)_3(H_2O)_2$ | 0.35 (deoxygenated toluene)<br>0.55 (solid state in PMMA) | [43] |

### 2.3.2 Measurement Against a Standard in the Solid State

Measuring emission efficiencies in the solid state against standards is less commonly seen for lanthanide ion-containing materials, but is nonetheless possible. Potential solid-state standards are summarised in Table 2.1. To reduce concentration quenching and assume Lambertian behaviour for the emitting materials, both samples and standards should be dissolved in a non-emissive solid, a matrix, such as PMMA or KBr [35,44]. Emission spectra should be done with the solid state sample holder in the front surface position; in this position, the sample is oriented towards excitation beam and emission detector at an angle of 30° or 60°, to prevent reflected and scattered excitation photons from reaching the emission detector. The powders should be deposited with enough thickness onto the solid-state holder to avoid holder signal.

Bril [45] showed that the diffuse reflectance $R$ and the integrated emission spectrum $I$ for both sample and standard can be used to calculate the emission efficiency of the solid sample with Equation 2.11.

$$\phi_x = \frac{1 - R_{ST}}{1 - R_x} \times \frac{I_x}{I_{ST}} \times \phi_{ST} \tag{2.11}$$

The diffuse reflectance can be determined with the fluorimeter by scanning the emission monochromator through the excitation range and calibrating with the diffuse reflectance of the solid-state matrix. To that effect, the emission spectrum of PMMA or KBr in the excitation range is measured; the integrated area of that spectrum has a corresponding reflectance value of the matrix $R_{ST}$ at that wavelength, which can be measured in an independent step with a diffuse reflectance accessory in an absorption spectrometer. The integrated area of the spectrum of the sample in the same excitation range is then compared with the area of the matrix to obtain $R_x$.

As can be seen from Table 2.1, the reported solid-state standards in the visible region of the spectrum do not contain lanthanide ions. Their emission wavelengths do not necessarily overlap with the lanthanide wavelengths, and similarly the excitation range might be different. Since the instrumental function depends on the wavelengths, this technique gives only approximate results. Therefore, the ability to do absolute measurements, as described in the following, is important.

### 2.3.3 Absolute Measurement with an Integrating Sphere

The most accurate method for determining emission efficiencies is the absolute measurement with an integrating sphere. This method was described by de Mello et al. [46] in 1997 and more recently revisited by Porrès et al. [47] and Bünzli et al. [48]. The emission efficiency is the ratio of absorbed and emitted photons. With a regular cuvette, the detector only counts the photons that are emitted in its direction. The integrating sphere, on the other hand, is coated with a material that shows close to 100% reflectance, regardless of the photon wavelength; the light that is emitted by the sample, either isotropically or anisotropically, will be reflected by the surface of the sphere and the photons that reach the detector will be proportional to the total number of emitted photons [49]. A combination of measurements of the light intensity passing through the integrating sphere empty and with sample enables then the determination of the amount of absorbed and emitted photons, and calculation of the emission efficiency. Equation 2.12 is used for this calculation.

$$\phi_x = \frac{I_{ss} - (1-\alpha)I_{es}}{A_e\alpha} \tag{2.12}$$

In this equation, $I_{ss}$ is the integrated emission spectrum of the sphere with the excitation bean hitting the sample, $I_{es}$ is the integrated emission spectrum of the sphere with the sample with the excitation beam hitting the sphere wall but not the sample, and $A_e$ is the integrated excitation spectrum of the integrating sphere with the sample holder. The absorption coefficient $\alpha$ of the sample is defined as

$$\alpha = \left(1 - \frac{A_x}{A_w}\right) \tag{2.13}$$

Here, $A_w$ is the integrated excitation spectrum of the sphere with the sample with the excitation beam hitting the sphere wall but not the sample and $A_x$ is the integrated excitation spectrum of the sphere loaded with the sample and the excitation beam hitting the sample directly.

## 2.4 Excited State Lifetimes

As mentioned in Section 2.2 of this chapter, the determination of excited state lifetimes relays important information regarding the coordination environment of the lanthanide ion. Since the lifetime of the excited state is strongly influenced by non-radiative quenching pathways, the number of oscillators leading to non-radiative quenching in the coordination sphere of the lanthanide ion can be determined with lifetime measurements, as detailed below.

### 2.4.1 Number of Coordinated Solvent Molecules

Horrocks and co-workers measured emission lifetimes of aqueous solutions of complexes of Eu(III) and Tb(III) with known numbers of water molecules in their first coordination sphere and compared them to the lifetimes measured in $D_2O$ [12,50]. They were able to derive the empirical Equation 2.14 from these measurements, which is now regularly used to determine the number of water molecules $q$ coordinated to Eu(III) and Tb(III) in other compounds.

$$q = A(\tau_{\text{solvent}}^{-1} - \tau_{\text{deutero-solvent}}^{-1} - B) \tag{2.14}$$

The coefficients $A$ and $B$ were empirically determined and are summarised in Table 2.2. Since methanol contains one O–H oscillator, the equation can be extended for this solvent. This equation has been expanded by several authors, to include quenching effects by second sphere solvent molecules, as well as other oscillators, N–H in amines and amides and C–H [51–53].

***Table 2.2*** *Coefficients* A *and* B *for the determination of the number of coordinated solvent molecules*

| Ln(III) | Solvent | *A* | *B* | References |
|---|---|---|---|---|
| Eu | Water | 1.05 | 0.0 | [50] |
| | Methanol | 2.1 | 0.0 | [54] |
| Tb | Water | 4.2 | 0.0 | [50] |
| | Methanol | 8.4 | 0.0 | [54] |
| Sm | Water | 0.0225 | 31.5 | [40] |
| Yb(III) | Methanol | 2.0 | 0.05 | [52] |
| | | | 0.1 | [55] |
| | Water | 1.0 | 0.2 | [52] |
| Nd(III) | Water | 130 | 0.4 | [55,56] |

## Acknowledgements

This effort was supported in part by the United States Department of Energy, Office of Basic Energy Sciences through the Center for Integrated Nanotechnologies and the Heavy Element Chemistry program of the Division of Chemical Sciences, Geosciences, and Biosciences (DEM) and the National Science Foundation grant NSF-CHE 1058805 (AdBD).

## References

1. Bünzli, J.-C. G.; Choppin, G. R. (eds) *Lanthanide Probes in Life, Chemical and Earth Sciences – Theory and Practice;* Elsevier: Amsterdam, **1989**.
2. Choppin, G. R.; Peterman, D. R. *Coord. Chem. Rev.* **1998**, *174*, 283–299.
3. Bünzli, J.-C. G.; Piguet, C. *Chem. Soc. Rev.* **2005**, *34*, 1048–1077.
4. Hemmilae, I.; Laitala, V. *J. Fluoresc.* **2005**, *15*, 529–542.
5. Armelao, L.; Quici, S.; Barigelletti, F.; Accorsi, G.; Bottaro, G.; Cavazzini, M.; Tondello, E. *Coord. Chem. Rev.* **2010**, *254*, 487–505.
6. Bünzli, J.-C. G.; Comby, S.; Chauvin, A.-S.; Vandevyver, C. D. B. *Journal of Rare Earths* **2007**, *25*, 257–274.
7. Winefordner, J. D.; Schulman, S. G.; O'Haver, T. C. *Luminescence Spectrometry in Analytical Chemistry*, **1972**; Vol. *38*.
8. Lancas, F. M.; Carrilho, E. In *Ewing's Analytical Instrumentation Handbook;* 3rd ed.; Cazes, J., Ed.; Marcel Dekker: New York, 2005, pp. 141–162.
9. Weiss, S., *Science* **1999**, *283*, 1676–1683.
10. Kumbhakar, M.; Nath, S.; Mukherjee, T.; Mittal, J. P.; Pal, H., *J. Photochem. Photobiol., C* **2004**, *5*, 113–137.
11. Barela, T. D.; Sherry, A. D., *Anal. Biochem.* **1976**, *71*, 351–357.
12. Horrocks, W. D. Jr.; Sudnick, D. R. *J. Am. Chem. Soc.*, **1979**, *101*, 334–340.
13. Bünzli, J.-C. G.; Mabillard, C.; Yersin, J. R., *Inorg. Chem.* **1982**, *21*, 4214–4218.
14. Bünzli, J.-C. G.; Yersin, J. R., *Inorg. Chim. Acta* **1984**, *94*, 301–308.
15. Bender, J. L.; Corbin, P. S.; Fraser, C. L.; Metcalf, D. H.; Richardson, F. S.; Thomas, E. L.; Urbas, A. M. *J. Am. Chem. Soc.* **2002**, *124*, 8526–8527.
16. van Stokkum, I. H. M.; Larsen, D. S.; van Grondelle, R. *Biochim. Biophys. Acta, Bioenerg.* **2004**, *1657*, 82–104.
17. Divya, O.; Mishra, A. K. *Appl. Spectrosc.* **2008**, *62*, 753–758.
18. Reifenberger, J. G.; Snyder, G. E.; Baym, G.; Selvin, P. R., *J. Phys. Chem. B* **2003**, *107*, 12862–12873.
19. Vereb, G.; Jares-Erijman, E.; Selvin, P. R.; Jovin, T. M., *Biophys. J.* **1998**, *74*, 2210–2222.
20. Faulkner, S.; Pope, S. J. A.; Burton-Pye, B. P. *Appl. Spectrosc. Rev.* **2005**, *40*, 1–31.
21. Bünzli, J.-C. G. *Chem. Rev.* **2010**, *110*, 2729–2755.
22. Tanke, H. J. In *Lanthanide Luminescence: Photophysical, Analytical and Biological Aspects;* Hänninen, P., Härmä, H., Eds.; Springer: Heidelberg, 2010; Vol. 7.
23. Nam, S. H.; Bae, Y. M.; Park, Y. I.; Kim, J. H.; Kim, H. M.; Choi, J. S.; Lee, K. T.; Hyeon, T.; Suh, Y. D. *Angew. Chem., Int. Ed.* **2011**, *50*, 6093–6097, S6093/6091–S6093/6015.
24. Novotny, L.; Stranick, S. J. *Annu. Rev. Phys. Chem.* **2006**, *57*, 303–331.
25. Kim, J.; Song, K.-B. *Micron* **2007**, *38*, 409–426.
26. Patterson George, H. *Semin. Cell Dev. Biol.* **2009**, *20*, 886–893.
27. Patterson, G.; Davidson, M.; Manley, S.; Lippincott-Schwartz, J. *Annu. Rev. Phys. Chem.* **2010**, *61*, 345–368.

28. Wolter, S.; Schuettpelz, M.; Tscherepanow, M.; van de Linde, S.; Heilemann, M.; Sauer, M. *J. Microsc.* **2010**, *237*, 12–22.
29. Galbraith, J. A.; Galbraith, C. G. *Wiley Interdiscip. Rev. Nanomed. Nanobiotechnol.* **2011**, *3*, 247–255.
30. Barrera, E. W.; Pujol, M. C.; Diaz, F.; Choi, S. B.; Rotermund, F.; Park, K. H.; Jeong, M. S.; Cascales, C. *Nanotechnology* **2011**, *22*, 075205/075201–075205/075215.
31. Felekyan, S.; Kuehnemuth, R.; Kudryavtsev, V.; Sandhagen, C.; Becker, W.; Seidel, C. A. M. *Rev. Sci. Instrum.* **2005**, *76*, 083104/083101–083104/083114.
32. Sergent, N.; Levitt, J. A.; Green, M. A.; Suhling, K. *Proc. SPIE–Int. Soc. Opt. Eng.* **2007**, *6771*, 67710X/67711–67710X/67718.
33. Chauvin, A.-S.; Gumy, F.; Imbert, D.; Bünzli, J.-C. G. *Spectroscopy Lett.* **2007**, *40*, 193.
34. Chauvin, A.-S.; Gumy, F.; Imbert, D.; Bünzli, J.-C. G. *Spectroscopy Lett.* **2004**, *37*, 517–532.
35. Wrighton, M. S.; Ginley, D. S.; Morse, D. L. *J. Phys. Chem.* **1974**, *78*, 2229–2233.
36. Melhuish, W. H. *J. Opt. Soc. Am.* **1964**, *54*, 183–186.
37. Nakamaru, K. *Bull. Chem. Soc. Jpn.* **1982**, *55*, 2697–2705.
38. Crosby, G. A.; Demas, J. N. *J. Am. Chem. Soc.* **1971**, *93*, 2841–2847.
39. Zucchi, G.; Scopelliti, R.; Bünzli, J.-C. G. *Dalton Trans.* **2001**, 1975–1985.
40. Hakala, H.; Liitti, P.; Peuralahti, J.; Karvinen, J.; Mukkala, V.-M.; Hovinen, J. *J. Lumin.* **2005**, *113*, 17–26.
41. Olmsted, J. III *J. Phys. Chem.* **1979**, *83*, 2581–2584.
42. Latva, M.; Takalo, H.; Mukkala, V.-M.; Matachescu, C.; Rodriguez-Ubis, J. C.; Kankare, J. *J. Luminescence* **1997**, *75*, 146–169.
43. Meshkova, S. B.; Topilova, Z. M.; Bolshoy, D. V.; Beltyukova, S. V.; Tsvirko, M. P.; Venchikov, V. Y. *Acta Phys. Pol. A* **1999**, *95*, 983–990.
44. Allison, R.; Burns, J.; Tuzzolino, A. J. *J. Opt. Soc. Am.* **1964**, *54*, 747–751.
45. Bril, A.; De Jager-Veenis, A. W. *J. Res. Nat. Bureau of Standards A* **1976**, *80A*, 401–407.
46. De Mello, J. C.; Wittmann, F. H.; Friend, R. H. *Advanced Materials* **1997**, *9*, 230–232.
47. Porrès, L.; Holland, A.; Pålsson, L.-O.; Monkman, A. P.; Kemp, C.; Beeby, A. *J. Fluoresc.* **2006**, *16*, 267–273.
48. Bünzli, J.-C. G.; Eliseeva, S. V. *Springer Ser. Fluoresc.* **2011**, *7*, 1–46.
49. Gaigalas, A. K.; Wang, L. *J. Res. Natl. Inst. Stand. Technol.* **2008**, *113*, 17–28.
50. Horrocks, W. D. Jr.; Sudnick, D. R., *Acc. Chem. Res.*, **1981**, *14*, 384–392.
51. Supkowski, R. M.; Horrocks, W. D. Jr. *Inorg. Chim. Acta* **2002**, *340*, 44–48.
52. Beeby, A.; Clarkson, I. M.; Dickins, R. S.; Faulkner, S.; Parker, D.; Royle, L.; de Sousa, A. S.; Williams, J. A. G.; Woods, M. *J. Chem. Soc., Perkin Trans. 2* **1999**, 493–504.
53. Hemmila, I.; Mukkala, V. M.; Takalo, H. *J. Fluoresc.* **1995**, *5*, 159–163.
54. Holz, R. C.; Chang, C. A.; Horrocks, W. D. Jr. *Inorg. Chem.* **1991**, *30*, 3270–3275.
55. Beeby, A.; Burton-Pye, B. P.; Faulkner, S.; Motson, G. R.; Jeffery, J. C.; McCleverty, J. A.; Ward, M. D. *Dalton Trans.* **2002**, 1923–1928.
56. Faulkner, S.; Beeby, A.; Carrie, M. C.; Dadabhoy, A.; Kenwright, A. M.; Sammes, P. G. *Inorg. Chem. Comm.* **2001**, *4*, 187–190.

# 3

# Circularly Polarised Luminescence

*Gilles Muller*

*Department of Chemistry, San José State University, USA*

## 3.1 Introduction

### 3.1.1 General Aspects: Molecular Chirality

Molecular chirality is the property whereby two mirror images of a molecule cannot be superimposed on each other and is crucial to modern drug research. While the difference between chiral structures may *seem* trivially small [1], the slight change in a compound's three-dimensional structure profoundly alters the given compound's interaction with its surroundings. For example, in the 1960s, racemic thalidomide was widely used to treat morning sickness. One of the enantiomers was effective at reducing morning sickness, but unfortunately the drug's other enantiomer caused birth defects [2]. For this reason, it is easy to understand why single-enantiomer drugs are attractive, and researchers are looking at them as possible treatments for cancer, cardiovascular disease, and central nervous system (CNS) defects [2]. To fully understand how chiral molecules produce their biological effects, science needs to develop reliable tools and/or effective probes.

During the last decade, a growing interest in the development of molecular recognition of guests by artificial hosts has emerged [3–15]. This can be explained by the increasing importance of the host–guest chemistry, since it represents a fundamental process for a variety of chemical and biological phenomena and for the regulation of various functions in living systems. For these reasons, synthetic receptors have become an attractive target for chiral recognition studies. In particular, these studies have been focused on the development of receptors having an increased selectivity and/or binding affinities, since such selectivity and affinities can provide valuable information for a better understanding of the interactions

*Luminescence of Lanthanide Ions in Coordination Compounds and Nanomaterials*, First Edition.
Edited by Ana de Bettencourt-Dias.

between molecules. Moreover, the continued efforts to improve the design of targeted artificial hosts may also lead to the development of useful practical applications including catalysis, separation processes, sensing, or transport through membranes. Among the large variety of synthetic hosts designed, considerable attention has been devoted to the chiral recognition of amino acids, since they are the basic building blocks of many biologically important molecules. It is often the chiral nature of amino acid R-groups that governs structure–function relationships of peptides and proteins in aqueous media (i.e. chiral discrimination in active sites or dictate the folding/unfolding of proteins). Since there are a large variety of biomolecules, this research field has generated considerable interest and is still growing with the continuous discovery of new processes and functions involving proteins. Thus, new tools in chiral separation and resolution must be found to meet this growing demand [16].

### 3.1.2 Chiroptical Tools: from CD to CPL Spectroscopy

Currently, the most commonly used tools in drug discovery, structural resolution, and elucidation include circular dichroism (CD), NMR, X-ray diffraction, mass spectrometry, HPLC, and molecular imprinting methods [16–21]. X-ray diffraction is the most commonly used technique in resolving absolute structures; however the caveat for this approach is the difficulty of obtaining suitable crystals from small amounts of material. NMR methods for determining stereochemical activity based on modified Mosher's methods [22,23] or the Trost-type approach [24,25] are commonly used in the resolution of primary amines, secondary alcohols, and carboxylic acids. A study by Marathias et al. [26], where Ibuprofen's enantiomers were identified using NMR and residual dipolar couplings, serves as an example of the shift to *ab initio* methods. This is due to the complexity in resolving ambiguous proton environments through routine NMR methods. More recently, CD has been used extensively when the structure and specific optical rotation values are known [27]. This technique serves as a staple in current DNA structure analysis, where the absorption bands and intensity can yield substantial information on the structural activity. The use of CD has given rise to newly modified techniques including the spectroscopic approach of comparing observed and calculated vibrational circular dichroism (VCD) spectra, exciton chirality, electronic circular dichroism (ECD), or optical rotation values, which have gained popularity due to commercially manufactured software and supercomputers, powerful PCs and Linux clusters [17,20,28,29]. Although these conventional tools including NMR, chiral HPLC, capillary electrophoresis, optical rotation, or CD are widely used, each may have its own limitations (i.e. time consuming, large sample preparation, high cost, not being sensitive enough to distinguish one enantiomer over another, etc.) or be more appropriate for specific chiral biomolecules. For instance, CD and molecular mechanics calculations rarely provide unambiguous data [30] as corroborated by the revised assignment of the helical handedness initially proposed of peptide nucleic acid double helices [31–33]. It is also interesting to note that Polavarapu recently summarised the importance to use several chiroptical spectroscopic tools for obtaining chiral molecular structural information (i.e. independent verification, single methods may only provide partial information or give ambiguous conclusions, etc.) [34,35].

Consequently, an attractive complementary tool is the use of Ln(III) luminescence spectroscopy, and especially circularly polarised luminescence (CPL) spectroscopy, the emission analog to CD. CD allows one to probe the chiral structure of the ground state, whereas CPL is a probe of the chirality of the excited state. It is worth noting that the measurement of net circular polarisation in luminescence has been referred to as circularly polarised luminescence (CPL – its most common acronym) circularly polarised emission (CPE), or other names in the literature over the years.

## 3.2 Theoretical Principles

### 3.2.1 General Theory

Since the complete details of the CPL theory have been the subject of numerous reviews [36–42], only a brief overview of the theoretical principles is highlighted here. CPL spectroscopy allows one to measure the difference in the luminescence intensity ($\Delta I$) of left circularly polarised light ($I_L$) versus right circularly polarised light ($I_R$). By convention this difference is defined as

$$\Delta I \equiv I_L - I_R \tag{3.1}$$

Because of the difficulty in measuring absolute emission intensities, it is common to report the degree of CPL in terms of the luminescence dissymmetry ratio (or factor)

$$g_{\text{lum}} = \frac{\Delta I}{\frac{1}{2}I} = \frac{I_L - I_R}{\frac{1}{2}(I_L + I_R)} \tag{3.2}$$

which represents the ratio of the difference in intensity divided by the average total luminescence intensity. The extra factor of one-half in this equation is included to make the definition of $g_{\text{lum}}$ consistent with the definition of the related quantity in CD, namely,

$$g_{\text{abs}} = \frac{\Delta\varepsilon}{\varepsilon} = \frac{\varepsilon_L - \varepsilon_R}{\frac{1}{2}(\varepsilon_L + \varepsilon_R)} \tag{3.3}$$

where $\varepsilon$ has always been explicitly defined as an average quantity, and $\varepsilon_L$ and $\varepsilon_R$ indicate, respectively, the molar absorption coefficients for left and right circularly polarised light.

Since CPL spectroscopy probes molecular properties, one must consider that the connection between molecular properties and the relationship of $I$ and $\Delta I$ is through the appropriate expressions for polarised luminescence transition probabilities (Fermi's golden rule). That is, one must express the probability of emitting a left or right circularly polarised photon at a time $t$ following the excitation of the luminescing species to the excited or emitting state [41]. Whereas organic fluorescence chromophores usually possess short-lived excited state lifetimes, the emission lifetimes of organic phosphorescences or, more particularly, Ln(III)-containing systems are normally long (μs to ms), one can assume that in solution the orientational distribution of emitting species is completely random. This relies on the fact that any photo-induced internal molecular structural changes are complete and the orientational distribution is completely isotropic by the time of emission.

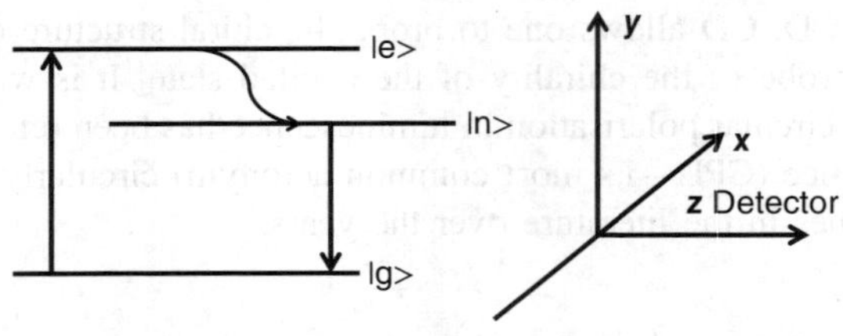


*Figure 3.1* Schematic energy level diagram for absorption and emission transitions (left) and laboratory coordinate system (right)

Since this chapter is solely devoted to the CPL of luminescent Ln(III)-containing systems, we will only consider two limiting cases for the orientational distribution; the so-called "frozen" limit and the isotropic state. The former is often considered for crystals, glasses, rigid sol-gels, or solutions of very high viscosities, whereas the latter is appropriate for solutions where one can assume a random orientational distribution of the emitting species. It is worth noting that no intermediate cases have been reported for such Ln(III)-based systems [38]. As mentioned above, the long radiative lifetimes of the luminescent Ln(III) ions usually ensure that, in solution, the emitting species have had sufficient time to scramble any initial photoselected orientational distribution. Taking into account these considerations, one can express the steady-state differential intensity of left ($L$) minus right ($R$) circularly polarised light for a transition from an initial emitting state $n$ to a final state $g$ by

$$\Delta I(\lambda) = (\hbar c/\lambda) N_n \Delta W_{gn} f_\sigma(\lambda) \tag{3.4}$$

where the differential transition probability

$$\Delta W_{gn} \equiv W_{gn}^L - W_{gn}^R \tag{3.5}$$

represents a lineshape function, $f_\sigma(\lambda)$, and $N_n$ describes the population of the emitting state (see Fig. 3.1). In Equation 3.4 we ignored the time and orientation dependence of $N_n$ and $\Delta W_{gn}$ for the reasons given above and for simplicity.

The probability of emitting a right or left circularly polarised photon may be related in the usual way to molecular transition matrix elements through Fermi's golden rule. Under the assumption that the emitted light is being detected in the laboratory $z$ direction (see Fig. 3.1) and allowing for only electric dipoles and magnetic dipoles in the expansion of the molecule–radiation Hamiltonian interaction, we obtain the following expressions

$$W_{gn}^L = K(\lambda^3)[|\mu_1^{gn}|^2 + |\mu_2^{gn}|^2 + |m_1^{gn}|^2 + |m_2^{gn}|^2 - 2i(\mu_1^{gn} m_1^{gn} + \mu_2^{gn} m_2^{gn})] \tag{3.6}$$

$$W_{gn}^R = K(\lambda^3)[|\mu_1^{gn}|^2 + |\mu_2^{gn}|^2 + |m_1^{gn}|^2 + |m_2^{gn}|^2 + 2i(\mu_1^{gn} m_1^{gn} + \mu_2^{gn} m_2^{gn})] \tag{3.7}$$

where $\boldsymbol{x}$ and $\boldsymbol{y}$ refer to laboratory axes and the electric dipole transition moment $\mu^{gn}$ and the imaginary magnetic dipole transition moment $m^{gn}$ are defined as follows

$$\begin{aligned} \mu_1^{gn} &\equiv < g|\mu_1|n > \\ m_1^{gn} &\equiv < g|m_1|n > \end{aligned} \tag{3.8}$$

Therefore the differential transition rate is

$$\Delta W_{gn} = -K(\lambda^3)[4i(\mu_1^{gn} m_1^{gn} + \mu_2^{gn} m_2^{gn})] \tag{3.9}$$

Note that the $i$ in Equation 3.9 results in the differential transition probability being a real number. The condition for this quantity to be non-zero is that the chromophore of interest must have a non-zero magnetic and electric transition dipole moment along the same molecular direction. In the absence of perturbing external fields, this is only true for molecules that are chiral.

Finally, the relationship between the experimental parameters and the molecular properties needs to account for the contribution of the direction and polarisation of the excitation beam and the direction of detection on the orientational distribution of the emitting molecules. As already indicated above, we only consider the two limiting cases in which the sample of interest is in a "frozen" or isotropic state. Thus, the orientation distribution of emitting molecules in the "frozen" state arises from the distribution generated by the excitation beam whereas, for the latter, it results from a random orientational distribution of the emitting molecules that have had enough time between absorption and emission to completely scramble any orientational distribution created by the excitation beam. It is important to point out that it is always possible to select an experimental geometry that ensures that the molecules that have been excited and are subsequently emitting are isotropically distributed in the plane ($\boldsymbol{x}$, $\boldsymbol{y}$) perpendicular to the direction of emission detection ($z$-axis). This is a key point since it is necessary to ensure that there is no linearly polarised component in the emission. This will be discussed in some detail in Section 3.3.3.

Expressed in the laboratory ($\boldsymbol{x}$ $\boldsymbol{y}$ $\boldsymbol{z}$) coordinate system, the total luminescence transition rate may be obtained by adding Equations 3.6 and 3.7

$$W_{gn} = 2K(\lambda^3)[|\mu_1^{gn}|^2 + |\mu_2^{gn}|^2 + |m_1^{gn}|^2 + |m_2^{gn}|^2] \tag{3.10}$$

Under the assumptions that the lineshapes for CPL and total luminescence are identical (this is appropriate for the usually sharp isolated pure electronic transitions that are often target of CPL measurements), and that the number of molecules in the emitting state is independent of their orientation, the luminescence dissymmetry ratio can then be related to the molecular transition matrix elements as follows

$$g_{\text{lum}} = \frac{\Delta I}{\frac{1}{2}I} = \frac{I_L - I_R}{\frac{1}{2}(I_L + I_R)} = -4i\frac{\mu_1^{gn} m_1^{gn} + \mu_2^{gn} m_2^{gn}}{|\mu_1^{gn}|^2 + |\mu_2^{gn}|^2 + |m_1^{gn}|^2 + |m_2^{gn}|^2} \tag{3.11}$$

The direct connection to molecular structure relies on relating the transition matrix elements from laboratory to molecular coordinate systems (see Scheme 3.1). For the case of a randomly oriented emitting distribution the orientational averaging yields the following general result

$$g_{\text{lum}}(\lambda) = 4\,\text{Re}\left[\frac{\vec{\mu}^{\,gn} \cdot \vec{m}^{\,gn}}{|\vec{\mu}^{\,gn}|^2 + |\vec{m}^{\,gn}|^2}\right] \tag{3.12}$$

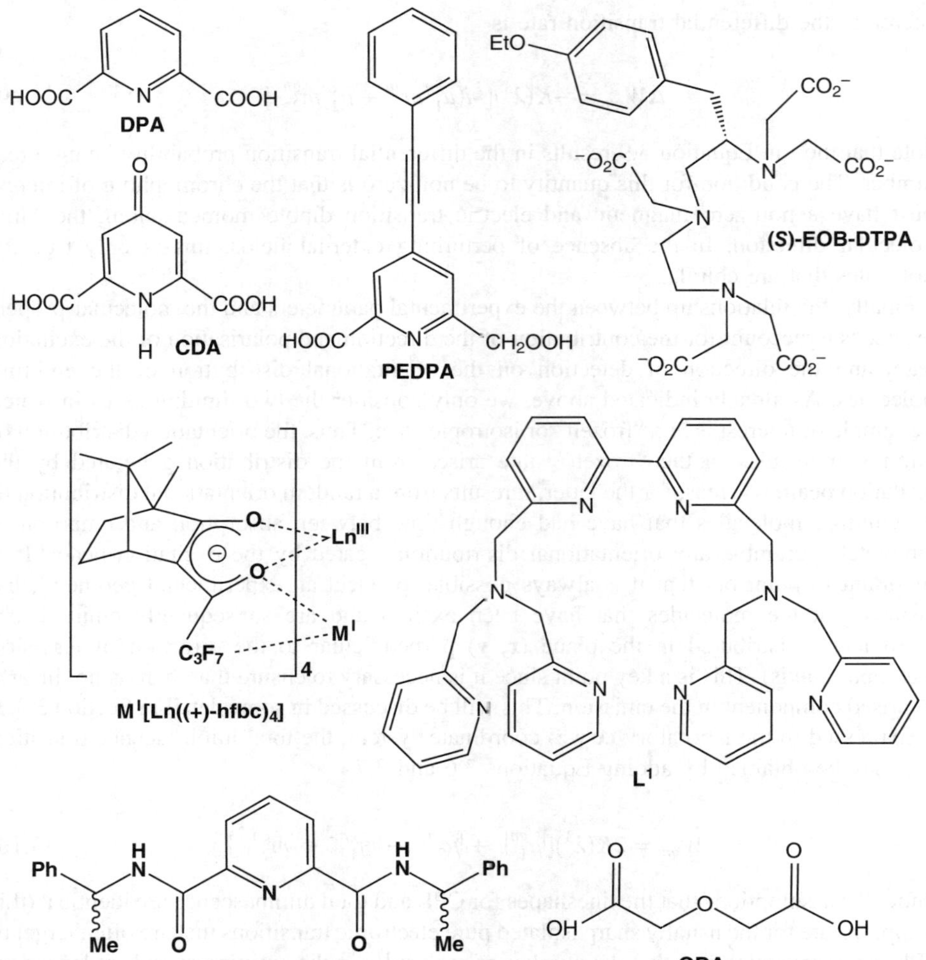


*Scheme 3.1* Chemical structures of ligands and complexes

### 3.2.2 CPL Intensity Calculations, Selection Rules, Luminescence Selectivity, and Spectra–Structure Relationship

The emphasis in much of the CPL-based work developed over the past decade is a result of the unusual spectral characteristics of the luminescent chiral Ln(III) complexes (see Chapter 1) and the technical advantages resulting from the use of CPL. One can expect the measurement of larger $g_{\mathrm{lum}}$ values when the transitions involved are inherently weak. This can be seen from the form of Equation 3.12 where one may expect a large $g_{\mathrm{lum}}$ value when the transition considered is electric dipole forbidden, but magnetic dipole allowed. Since the magnetic dipole transition moments are typically much smaller than the electric dipole terms, the denominator in Equation 3.12 will be dominated by the first term, $|\mu|^2$. Transitions

***Table 3.1*** *Selected magnetic dipole allowed emissive transitions of Ln(III) ions*

| Ln(III) Ion | Emission transition | Approximate wavelength (nm) |
|---|---|---|
| Sm(III) | $^4G_{5/2} \rightarrow {}^6H_{7/2}/{}^4G_{5/2} \rightarrow {}^6H_{5/2}$ | 580/560 |
| Eu(III) | $^5D_0 \rightarrow {}^7F_1$ | 590 |
| Gd(III) | $^8S_{7/2} \rightarrow {}^6P_{7/2}$ | 312 |
| Tb(III) | $^5D_4 \rightarrow {}^7F_3/{}^5D_4 \rightarrow {}^7F_4/{}^5D_4 \rightarrow {}^7F_5$ | 625/580/540 |
| Dy(III) | $^4F_{9/2} \rightarrow {}^6H_{11/2}$ | 650 |
| Yb(III) | $^2F_{5/2} \rightarrow {}^2F_{7/2}$ | 980 |

that satisfy the formal selection rules of $\Delta J = 0, \pm 1$ (except $0 \leftrightarrow 0$) are magnetic dipole allowed, where $J$ stands for the total angular momentum quantum number found in the definition of the term symbol, $^{2S+1}L_J$, describing the electronic states of lanthanides [43]. As a result, the considerable increase in the use of CPL is most probably originated from the discovery that luminescence from intraconfigurational *f-f* transitions of the Ln(III) ions obeying magnetic dipole selection rules often showed large circular polarisation for the reasons outlined earlier [44]. Unfortunately, the paradox of this requirement is that $g_{lum}$ may be quite large (up to 1.38) for such selected transitions [45,46], but these transitions have usually very low intensity, and therefore are more difficult to measure. A value of 0 for $g_{lum}$ corresponds to no circular polarisation, while the absolute maximum value is 2. A list of selected magnetic-allowed emissive transitions of the luminescent Ln(III) ions is given in Table 3.1. It is worth noting that the same arguments apply to CD spectroscopy, where it is often easier to study the $n \rightarrow \pi^*$ transitions of chiral ketones than $\pi \rightarrow \pi^*$ transitions.

Unlike CD spectroscopy, where all the absorbing chromophores contribute to the observed differential absorption measurement, CPL spectroscopy is limited to luminescent species. The former is mainly used for determining macromolecular information about protein secondary and tertiary structures, whereas CPL rather reflects local chiral structure changes associated with the local environment surrounding the emitting chromophore. Lunkley et al. [45,46] recently showed the importance of using CPL for selectively studying only luminescent chromophores present in the systems of interest, in contrast to CD, which is affected by most chromophores and/or equilibrium mixtures in an additive manner. Of special importance, a concentration-dependent exciton CD due to the dissociation of the tetrakis(3-heptafluoro-butylryl-(+)-camphorato = (+)-hfbc) Eu(III) compound into the tris (+)-hfbc Eu(III) complex was shown but, more importantly, the former species exhibited the highest CPL activity ever observed for any CPL studies of Ln(III)-containing systems in solution ($g_{lum}$ values amounted to +1.38 or +1.32 at 595 nm in $CHCl_3$ or EtOH, respectively). The CPL activity of the latter compound, however, was negligible.

This example shows that one needs to be careful on how to interpret CPL results with respect to the distribution of complex species present in solution. It is essential to know what the sample solution contains to avoid misinterpretation, since the CPL and total luminescence intensity from solutions of Ln(III) complexes with a mixture of ligands (i.e. some of which may be chiral or not, or in situations in which one chiral ligand may form multiple solution species) may come from different species [38]. For instance, such a solution of Tb(III) with dipicolinic acid (DPA) and L-malic acid showed that the total luminescence came almost exclusively from tris $Tb(DPA)_3^{3-}$, whereas the CPL originated from some unidentified mixture of Tb(III), L-malic acid, DPA, and water [47]. As a result, the measurement of CPL from these so-called "mixed" solutions may only suggest that the average solution structure is chiral on the luminescence time scale. However, it is not possible to interpret the results in terms of a single solution species.

Finally, since the availability of commercial CD instrumentation in the 1960s, a considerable effort was devoted to the development of reliable rules relating CD sign patterns to absolute configurations or conformations of coordination compounds [48–52]. Unlike CD spectroscopy, there were only a few attempts to develop reliable empirical relationships between chiral structures and CPL sign patterns [53–56]. It was shown that the sign and magnitude of CPL are mainly affected by the degree of helical twist of the complex, the nature of the ligand field, and by modulating the polarisability [57,58] of the axial donor group solvation [36,38]. In other words, the CPL signal reflected the time-averaged local helicity around the Ln(III) ion (i.e. the magnitude of $g_{lum}$ increased with an increase in the degree of conformational rigidity of the complex). However, such a development may be limited to chiral metal complexes containing Ln(III) ions since the difference in absorption or emission of circularly polarised light may approach 50%, that is, if one studies the intraconfigurational *f-f* transitions that obey the magnetic dipole transition selection rules [43,44]. However, it is very unlikely that chiral organic systems will show a difference in absorption or emission of circularly polarised light greater than 0.1% [36,38,49]. In addition to the fact that the nature of the 4*f* electronic states makes the calculation of the observed CPL (or CD) very difficult, one must envisage the use of chiral high symmetric Ln(III)-based systems which would lead to strong CPL activities. This would be a necessary step to make reliable predictions. The recent findings of such studies aimed at the development of an empirical spectra–structure correlation based upon chiroptical spectral information will be discussed in some detail in Section 3.4.

## 3.3 CPL Measurements

### 3.3.1 Instrumentation

Unlike CD, the measurement of CPL is still mainly dependent on the use of custom-made instruments that have been designed, developed and improved by a limited number of research groups around the world over the last three decades [36,38,41,42,45,52,59–63]. However, the growing interest in developing chiral luminescent probes, and in particular Ln(III)-based systems, has resulted in the advertising and some availability of commercial instrumentation. The first commercial CPL spectrometer, which essentially consists of two CD spectrometers with the second one used as the emission spectrometer, was manufactured by JASCO Inc., the JASCO CPL-200. More recently, the company OLIS Inc. developed its Polarisation Toolbox to support fluorescence, polarisation of fluorescence, anisotropy, CPL, CD, and FDCD (fluorescence detected circular dichroism) measurements for its CD instrumentation. As of February 2012, only CPL-related studies using the JASCO CPL-200 instrument have appeared in the literature [64–67].

Although there is still a limited use of commercial CPL instruments, the technique has continued to be developed to a point where the detection of CPL can be performed with a

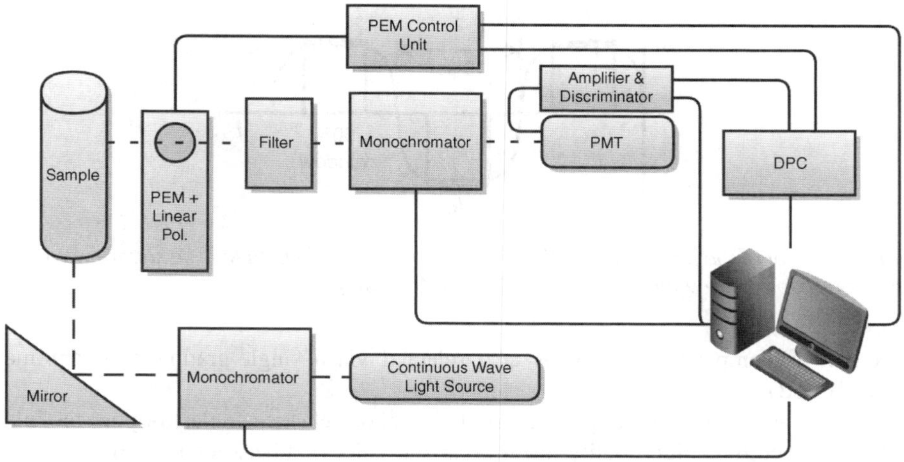


***Figure 3.2*** *Schematic diagram for instrumentation used to perform CPL measurements*

high degree of sensitivity (~1 part in $10^4$–$10^5$) and reliability. It is worth noting that the basic design of custom-made CPL spectrometers is most likely based on the technological approach used for the instrument in operation in our laboratory at San José State University. That is, the instrument is utilised in photon-counting detection with various technological approaches of gated-photon counting [36,38,41,68]. The CPL instrumentation as used in our laboratory follows the schematic diagram shown in Fig. 3.2. It is designed to measure "steady-state" CPL.

The light source for indirect excitation is a continuous wave 1000 W xenon arc lamp (OBB KiloArc). The Spex Fluorolog-2 spectrofluorometer is equipped with excitation and emission monochromators with dispersions of 4 nm/mm (SPEX 1681B). Selective excitation (or direct excitation) of Ln(III) is accomplished with either a tunable dye-laser (Coherent-599) pumped by an argon-ion laser, a Coherent Sabre TSM 15, or the argon-ion laser without the tunable dye-laser. The choice of the excitation source is driven by the type of information one is interested in getting from a CPL measurement. In general, one uses a laser excitation when either high-intensity, or wavelength or polarisation selectivity is required, but also if one is concerned with potential photochemical degradation upon long UV exposure when an indirect sensitisation through the ligand's absorption bands is employed. This indirect sensitisation phenomenon (or the so-called indirect excitation) is what happens when one uses a xenon arc lamp as the source of excitation. Examples of different types of applications of CPL spectroscopy are described in Section 3.4.

Aside from the excitation monochromator, the rest of the instrumentation setup is similar for the laser configuration. In the case of the tunable laser configuration, the excitation wavelength selection is accomplished via a computer-controlled stepping motor, whereas the one for the argon-ion laser is a property of the specific laser. However, the emission wavelength selection for laser and xenon arc-lamp sources is performed via a double monochromator (SPEX 1680) and a single-grating monochromator (SPEX 1681B), respectively. Both monochromators are controlled by the computer. The excitation wavelength in

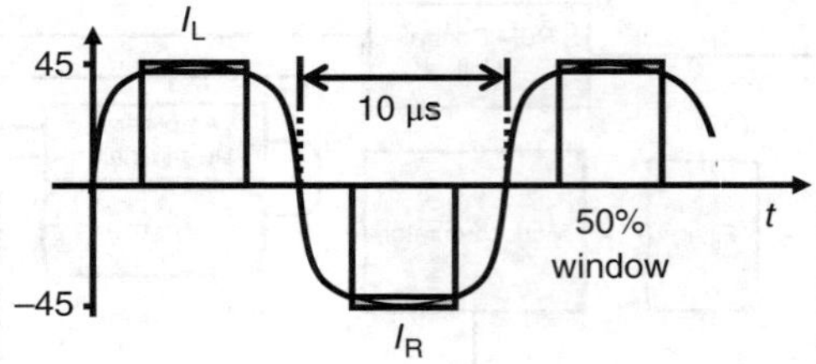


***Figure 3.3*** *Plot of the phase modulation versus time for a 50 kHz PEM. The square wave signal corresponds to a time window equal to 50% of the half-period*

the xenon arc-lamp configuration is also selected via a single-grating monochromator (SPEX 1681B).

A circular analyser, consisting of an oscillating photo-elastic (or elasto-optic) modulator (PEM) followed by a high-quality linear polariser, is located between the emitting sample and the emission monochromator. This circular analyser allows the detection of the net circular polarisation in the luminescence. The PEM control unit provides a reference signal to the differential photon counter (DPC) that is used to define a fixed time window for gated counting, one that alternates and corresponds to the passage of left and right circularly polarised light (Fig. 3.3). It should be noted that the PEM acts as a dynamic quarter-wave device, converting the circularly polarised light emitted from the sample to linearly polarised light sent to the emission monochromator. For the PEM to function in this manner, it must be configured for the emission wavelength. This is done via the computer, which sends a command to the PEM control unit to apply an appropriate voltage to the optical head unit. This is realised using a digital unit (Hinds PEM-100) operating at 50 kHz. In other words, by individually converting either the left or right polarised light to linearly polarised light, it is possible to measure one polarity at time. The PEM performs this in time with a high frequency clock, which is also provided as a reference signal to the DPC to ascertain which polarity is being detected and, thus, should be counted at a given moment. Each clock cycle is represented by a periodic square wave having a high and low state (Fig. 3.3). Each state in turn represents the passage of left or right circularly polarised light. Being periodic, the time window for each polarity is fixed and equal, thus ensuring equal sampling of both polarities. The time window is usually set to 50% of the modulation cycle and all the photons detected during this time window are counted as either left or right circularly polarised. These two sets of counted photons correspond to $I_L$ and $I_R$, respectively. Since the PEM is not a perfect optical element, the phase shift is not exactly a quarter-wave. This is the reason why the time window is set to 50% to allow counting for the slight error (i.e. a theoretical error of less than 5% is calculated for a 50% window). Although a correction factor could be applied to the experimental results, it is standard practice to simply report the experimental results as measured, since the various elements including linear polarisers, samples, containers, filters, mirrors, PEM, and statistical errors associated with the photon counting may lead to systematic sources of error.

Detection of the alternating emitted left and right polarised light is accomplished by a thermoelectrically cooled photomultiplier tube (PMT), operating in a differential photon-counting mode. The photon pulses corresponding to the intensity of left and right circularly polarised emitted light are collected by two separate digital counters. In other words, the light passing through the emission monochromator is detected by the PMT and converted to TTL-level pulses by the amplifier-discriminator for counting by the DPC. The DPC maintains a count of the differential between left and right photon pulses, as well as the total number of photon pulses, to determine a $g_{lum}$ value. In our laboratory, the custom-built DPC is based on a Field Programmable Gate Array (FPGA) design (being a software design, it is easy to modify and replicate) [69]. For our implementation, a SPARTAN-3AN FPGA board was used. This board features a serial port, enabling communication with most computers without the use of an expensive I/O board. The widespread availability of serial port programming interfaces and terminal applications simplifies development and testing of the DPC as well. Commands supported by the FPGA-based DPC include Count, Reset, and Toggle Mode. The latter command enables switching between the standard photon counting mode ("steady-state" mode) and a pulse mode. Although the sample size of the FPGA-based DPC-in its current form is fixed, in the sense that it is a constant value that is set in the FPGA code, one can easily change it by reloading the FPGA board with modified code. Moreover, several FPGA codes with different fixed values can be loaded into the auxiliary on-board, non-volatile memory and then loaded onto the FPGA at any time for greater flexibility. This fixed value can be any value from 1 to 4 294 967 296.

When a count has been issued, the FPGA-based design does not return results in the form of a final $g_{lum}$ value, but rather returns the individual left and right photon counters ($I_R$ and $I_L$, respectively). The benefits in terms of knowledge gained from the left/right counter values is unclear at this time, but may be present itself in the future. What is clear is that we are given more data with which to work or analyse, and flexibility in when or where the final $g_{lum}$ value is calculated. This latter point proves an added benefit by enabling the calculation of $g_{lum}$ values across increments of the sample size (i.e. for a fixed sample of size of a thousand, it would be easy to make counts for any positive integer multiple of a thousand without the need to modify code and reprogram the FPGA). Finally, the inclusion of a pulse mode in the FPGA-based DPC makes it possible to perform photon count based on the duration of a pulse, rather than for a fixed quantity of photons as it is done in the "steady-state" mode. This can be realised with the use of a chopper that would break the continuous wave excitation beam of our laser light sources into pulses.

Although the steady-state CPL instrumentation uses a continuous excitation light source, it is possible to switch to a pulsed excitation light source with the main difference that the collected $g_{lum}$ values would be determined at a series of times after the excitation pulse. This type of measurement can be referred to as time-resolved CPL spectroscopy, as this configuration measures the time-dependence of $g_{lum}$. It is worth noting that the rest of the system including the optical components is similar to the one described above. However, there is a requirement that the time measurement window needs to be a multiple of the phase modulation. This will ensure an equal sampling of the two half cycles of modulation, which correspond to the emission of left and right circular polarisation. This is the reason why the excitation pulses need to be coupled to the polarisation modulation cycle or be set to occur randomly throughout the modulation cycle [70,71].

To conclude, it must be emphasised that it is important to minimise the sources of depolarisation. As a result, no optical elements should be placed between the sample compartment and the PEM. Doing so ensures that no linearly polarised luminescence would be detected, since this latter is 10 to 100 times more intense than circularly polarised emitted light. Additional information is provided in Section 3.3.3. In addition to the use of a right-angle detection (90° excitation mode), it is recommended that the emitted light, once it has passed through the PEM and the linear polariser, travels through an appropriate filter. This ensures that no scattered excitation or other stray light is detected by the PMT. In addition, the polarisation of the laser beam, when a laser excitation source is used, needs to be aligned along the direction of emission (laboratory $z$ direction) in order to minimise any polarisation in the $xy$ plane for the reasons discussed in Section 3.3.3.

Since the CPL technique is based on a photon-counting method, it is possible to calculate the standard deviation $\sigma_d$ in the measurement of the $g_{\text{lum}}$ directly from the total number of photon counts, $N$.

$$\sigma_d = \sqrt{2/N} \tag{3.13}$$

This is possible since the function $\sigma_d$ follows a Poisson distribution [60]. One can see that the determination of accurate $g_{\text{lum}}$ for transitions associated with large $g_{\text{lum}}$ values of highly luminescent systems can be done in a short time. However, a longer time of collection is necessary for transitions associated with small $g_{\text{lum}}$ values of weakly luminescent species to achieve the same percentage error. As the time required for measuring a CPL spectrum is dependent on the intensity of the luminescence of the system of interest and the "chirality" of the transition analysed, it is standard practice that the photon pulses are collected for the same amount of time at each wavelength. Thus, the relative error (or signal-to-noise ratio) at each of these wavelengths is the same in the CPL spectrum measured.

### 3.3.2 Calibration and Standards

Since there is a considerable increase in the use of CPL spectroscopy that mainly relies on homemade instruments, it is important to make sure that these apparatuses are regularly tested and calibrated to ensure the determination of accurate $g_{\text{lum}}$ values. This is a key step even for spectrometers that are based on an analog detection of CPL. The $\Delta I$ and $I$ data, which are proportional to the output of the lock-in amplifier and a DC output voltage, respectively, are typically acquired from two independent measurements. As a result, the determination of a $g_{\text{lum}}$ necessitates the use of a calibration standard. However, the CPL instrumentation based on the photon counting method does not require the use of an independent calibration since, in this case, the $g_{\text{lum}}$ is determined directly. However, it is always important to make sure that the magnitude and sign of the CPL signal are measured accurately. This is even more important for the accurate detection of small degrees of circular polarisation in the total emitted light intensity.

Although various standardisation and calibration methods have been utilised (i.e. use of variable quarter-wave plates or passing unpolarised light through solutions of known CD) [61,72], the most popular approach consists in the use of a calibrant. The idea is that one measures the CPL activity of a solution containing a chiral species of known CPL [36,38]. To date, the commercially available (Aldrich) NMR chiral shift reagent tris[3-trifluoro-acetyl-(+)-camphorato]europium(III), Eu[(+)-facam]$_3$, in DMSO is still the most used compound as a CPL standard [73]. Although this complex is available in high purity at a modest cost and can be excited either by a UV or argon-ion laser source, its use as a CPL calibration standard is still an area of argument and discussion [60,74]. The $g_{\text{lum}}$ values of a DMSO solution of Eu[(+)-facam]$_3$ at, for example, 588.2, 595.2, and 613.5 nm are −0.25,

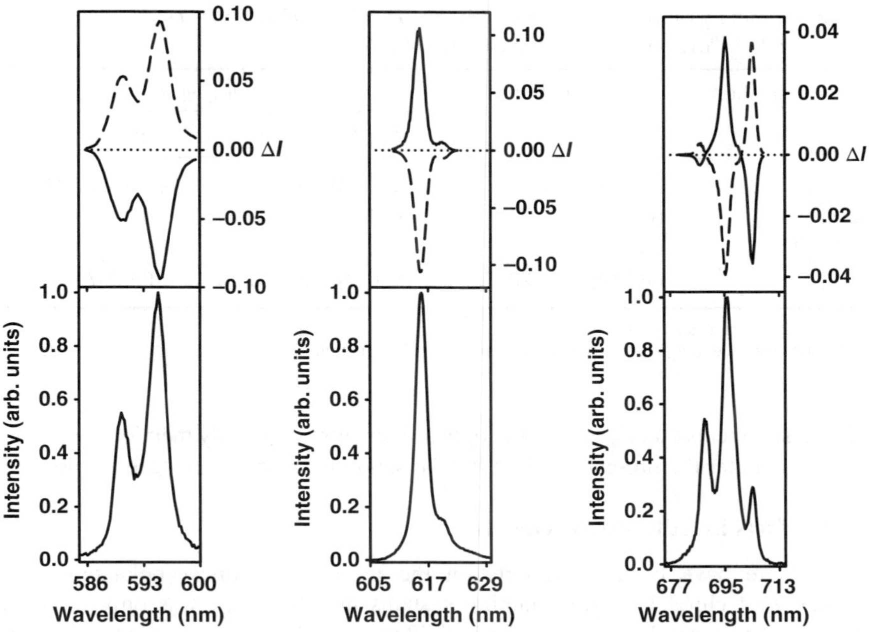


***Figure 3.4*** *CPL (upper curves) and total luminescence (lower curves) spectra for the* $^5D_0 \rightarrow {}^7F_1$ *(left),* $^5D_0 \rightarrow {}^7F_2$ *(middle), and* $^5D_0 \rightarrow {}^7F_3$ *(right) transitions of* $[Eu\{(R,R)\ -\mathbf{1}\}_3]^{3+}$ *(solid lines) and* $[Eu\{(S,S)\ -\mathbf{1}\}_3]^{3+}$ *(dashed lines) in* $6.67 \times 10^{-3}$ *M MeCN at 295 K, following excitation at 308 nm. Reproduced with permission from [75]. Royal Society of Chemistry*

−0.78, and +0.072 in the spectral range of the $^5D_0 \rightarrow {}^7F_1$ and $^5D_0 \rightarrow {}^7F_2$ transitions, respectively. In addition to its water sensitivity, the high cost of Eu[(−)-facam]$_3$ with the other enantiomeric form of facam limits its use as an effective and reliable CPL standard for routine tests.

Working along these lines, Bonsall et al. [75] reported the use of a more suitable CPL calibrating agent based on optical isomers of *N,N′*-bis(1-phenylethyl)-2,6-pyridinecarboxamide (**1**) coordinated to Eu(III) ions in a Eu:**1** ratio of 1:3 (Fig. 3.4 and Table 3.2). In particular, the advantages of these systems are: (i) the ease of the ligand synthesis, (ii) their complex solution stability (i.e. several months), and (iii) the lack of a noticeable photochemical degradation under continuous UV excitation (i.e. 70 hours, $\lambda_{exc} = 308$ nm). A $g_{lum}$ value measured at 595.3 nm amounted to −0.18 for a $[Eu\{(R,R)\ -\mathbf{1}\}_3]^{3+}$ complex solution in MeCN left on the shelf and measured seven months apart.

Finally, it should be mentioned that the use of a calibrant to conduct instrumental routine tests also requires that one performs the necessary modifications to the apparatus in order to correct the source of the error and until the expected $g_{lum}$ value is obtained. However, it is not appropriate to apply a factor of correction [76,77] to the experimental values of $\Delta I$, $I$, or $g_{lum}$ without some justification [36]. This is not a suitable approach knowing that the accuracy and precision of CPL measurements are sample-specific, but also wavelength-dependent. In addition, each sample may respond differently to the various sources of error

***Table 3.2*** *Summary of CPL results for [Eu{(R,R) −**1**}$_3$]$^{3+}$ and [Eu{(S,S) −**1**}$_3$]$^{3+}$ in 6.67 × 10$^{-3}$ M MeCN at 295 K, following an excitation at 308 nm [75]*

| Electronic transition | Wavelength (nm) | $g_{lum} \pm \sigma_d$ [a] | |
|---|---|---|---|
| | | [Eu{(*R*,*R*) −**1**}$_3$]$^{3+}$ | [Eu{(*S*,*S*) −**1**}$_3$]$^{3+}$ |
| $^5D_0 \rightarrow {}^7F_1$ | 590.5/595.3 | −0.19/−0.18 | +0.19/+0.18 |
| $^5D_0 \rightarrow {}^7F_2$ | 615.6 | +0.21 | −0.21 |
| $^5D_0 \rightarrow {}^7F_3$ [b] | 649.6 | −0.22 | +0.23 |
| $^5D_0 \rightarrow {}^7F_4$ | 688.8/696.0/704.1 | +0.001/+0.07/−0.24 | −0.001/−0.07/+0.25 |

[a] Values from Fig. 3.4 and $\sigma_d$ = 0.01.
[b] Very low luminescence that prevents recording a complete CPL spectrum.

in CPL measurements (i.e. electronic and optical elements), as already mentioned earlier. One may also encounter the presence of linear polarisation in the emission (see Section 3.3.3).

### 3.3.3 Artifacts in CPL Measurements

Although there have been no reports on the presence of linear polarisation in the luminescence of chiral Ln(III)-containing systems in solution (it was only reported the observation of linear polarisation in the individual crystal field components of an Eu(III) complex prepared in a very viscous solution) [78], one may not exclude such a source of artifacts. Indeed, it has been recognised for a long time that the principal source of artifacts in the measurement of CPL is the linear polarisation in the luminescent beam. This phenomenon originates from the passing of linearly polarised light through the very slightly birefringent PEM [41,79]. Since the intensity of a linear polarisation signal is much larger than that of a true CPL signal, it may lead to signals of comparable magnitudes, even if the birefringence is usually small (<5%). As an example, luminescence that is 10% linearly polarised may be converted by this effect to circular polarisation of about 0.5%, which is often larger than the true CPL signal.

Although the presence of linear polarisation in the luminescence beam may lead to difficulties in terms of measuring true CPL signals, it is possible to considerably limit and/or eliminate its effect, if any may be present in the luminescence beam. As mentioned earlier, the advantage of studying long-lived luminescent chiral Ln(III)-containing systems over short-lived fluorescent chiral organic and/or metal-based compounds ensures that, in ordinary liquid solution, the emitted species have had sufficient time between excitation and emission to scramble any initial photoselected orientational distribution. This is what is described as the isotropic state. But, the study of short-lived fluorescent chiral organic and/or metal-based compounds, crystals, glasses, rigid sol-gels, or solutions of very high viscosities of chiral molecules such as long-lived Ln(III)-based systems may require the use of specific experimental geometries to ensure that no linear polarisation in the luminescence beam is detected. These types of systems are often defined by the so-called "frozen" case. Examples of such experimental approaches include the rotation of the PEM and the linear polariser in a way that the birefringent axis of the PEM is parallel to the plane of polarisation of the emitted light, or the use of combinations of excitation/emission geometries and linear polarisers that are positioned at the so-called "magic-angles" [80]. The key point is that the sample is isotropic in the plane perpendicular to the direction of emission detection. This ensures that no linear polarisation is detected in the (***xy***) plane. For example, this is the experimental approach used in the measurement of CPL from chiral triarylamine helicenes [81]. The excitation beam was linearly polarised along the laboratory $z$ axis and, thus, resulting in the fact that no linear polarisation was detected in the (***xy***) plane. Another example from the CPL study of chiral derivatised perylene aggregates showed that these latter samples were excited at 0° in an epiluminescence measurement using a depolarised argon-ion laser excitation beam [82]. This, again, ensured that no linear polarisation in the luminescence was detected. It should be emphasised that the most likely best way to ensure that no linear polarisation is observed in the luminescence beam, or at least in the plane perpendicular to the emission detection, is by experiment. As it is the case for the conduct of routine standardisation and calibration tests of CPL instrumentation, this is another important technical aspect of the measurements if one wants to report accurate CPL results.

To conclude this section, it may be of interest to the readers to know that the measurement of CPL from samples classified in the "frozen" category needs to fulfill specific requirements to ensure that no sources of artifacts are present. In addition to the fact that the emitting species must be chiral, a crystal system of, for example, a luminescent Ln(III)-containing compound must be at least uniaxial with the optic axis oriented along the direction of emission detection. This ensures that the CPL is propagated through the crystal without scattering which typically leads to depolarisation. In other words, the index of refraction of the crystal needs to be uniform in directions perpendicular to the direction of emission detection. The same argument applies to randomly oriented "frozen" solutions of chiral molecules. This implies that the orientational distribution of the molecules is isotropic and, also, independent of the direction of emission detection.

To date, CPL measured from crystals containing Ln(III) complexes is still limited. The first one for which it was possible to detect a CPL signal was for the $Na_3$[Eu(ODA = oxydiacetate)$_3$]·$2NaClO_4$·$6H_2O$ complex that crystallises in the cubic space group R32 [53,55,56]. It should be mentioned that no CPL or CD was observed from crystals containing $[Eu(DPA)_3]^{3-}$ and differing amounts of included water molecules, since these crystals are monoclinic or triclinic [83–85]. More recently, Tsumatori et al. [59] reported CPL spectra from single crystals of a series of Ln(III) complexes. However, it is unclear why the authors observed significant variations in the $g_{lum}$ values between the crystalline and solution complex samples. One hypothesis suggested was that there may have been a contribution of the linear polarisation component in the luminescence detected. This is in line with the observation that the CPL signal was dependent on the lattice planes of the crystals.

### 3.3.4 Proposed Instrumental Improvements to Record Eu(III)-Based CPL Signals

Since most of the strategies to improve the measurement of weak CPL signals focus on the design of Ln(III)-based systems with improved stability, efficiency and, in particular, selectivity by choosing selected sensitising and discriminating moieties into the ligand molecules, the limitations for such approaches result from the excessive synthetic efforts required for the preparation of the organic ligands and, also, from the various energy conversion processes occurring in the Ln(III)-containing systems that may not be optimum. Since CPL measurements have limited sensitivity due to the low absorption of the

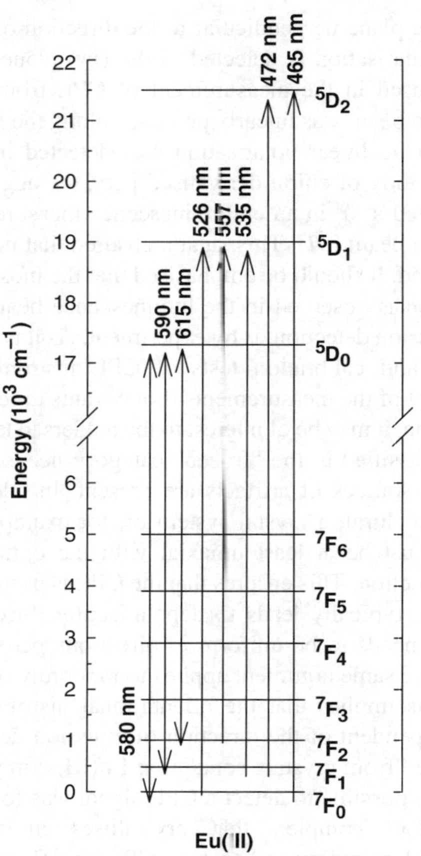


***Figure 3.5*** *Approximate partial energy level diagram for Eu(III)*

luminescent Ln(III) complexes because $f \leftrightarrow f$ transitions are forbidden by the Laporte Rule, one should consider absorption and emission transitions that will exhibit large intrinsic $g_{\text{lum}}$ values. It has been shown that such transitions are often those that obey the magnetic dipole selection rules ($\Delta J = 0, \pm 1$, except $0 \leftrightarrow 0$). Working along these lines, Do et al. [86] reported the use the magnetic dipole allowed $^5D_1 \leftarrow {}^7F_0$ transition (525–529 nm) in absorption and the $^5D_0 \rightarrow {}^7F_1$ transition in emission. It should be noted that this work was limited to the study of Eu(III) systems, due to the fact that: (i) the (Eu) spectroscopy is probably the most useful tool for probing the speciation, solvation numbers, excited-state energetic, or solution dynamics of luminescent Ln(III) edifices, (ii) Eu(III) possesses several absorption transitions and also one emission transition that meet the appropriate selection rules, and (iii) the limited choice of magnetic dipole allowed absorptions of the luminescent Ln(III) ions for which CPL has been reported (Fig. 3.5 and Tables 3.1 and 3.3). The selected absorption $^5D_1 \leftarrow {}^7F_0$ transition will facilitate the detection of CPL and also be more selective and precise in selecting a circular polarised excitation wavelength. This is possible since this latter transition has a maximum of three individual crystal field levels versus 15 for the

*Table 3.3* Selected magnetic dipole allowed absorptions of the luminescent Ln(III) ions

| Ln(III) ion | Absorption transition | Approximate wavelength (nm) |
|---|---|---|
| Nd(III) | $^4G_{7/2} \leftarrow {}^4I_{9/2}/{}^4G_{9/2} \leftarrow {}^4I_{9/2}/{}^4G_{11/2} \leftarrow {}^4I_{9/2}$ | 526/514/465 |
| Sm(III) | $^4G_{5/2} \leftarrow {}^6H_{5/2}$ | 560 |
| Eu(III) | $^5D_1 \leftarrow {}^7F_0/{}^5D_1 \leftarrow {}^7F_1/{}^5D_1 \leftarrow {}^7F_2$ | 526/535/557 |
| | $^5D_2 \leftarrow {}^7F_2/{}^5D_2 \leftarrow {}^7F_1$ | 488/472 |
| Dy(III) | $^4I_{15/2} \leftarrow {}^6H_{15/2}$ | 457 |
| Yb(III) | $^2F_{5/2} \leftarrow {}^2F_{7/2}$ | 980 |

usually employed $^5D_1 \leftarrow {}^7F_2$ absorption transition (~555 nm). Although the $^5D_1 \leftarrow {}^7F_2$ absorption transition satisfies the selection rules, its weak population of the initial state (1%) generates an important problem in maximising emission intensity, which cannot be easily overcome.

It is worth mentioning that the proposed experimental advantage that may be obtained in chiral Eu(III) systems by being able to scan in the spectral region corresponding to the $^5D_1 \leftarrow {}^7F_0$ absorption was illustrated by measuring the emission detected circular dichroism (EDCD) and CPL spectra of a perturbed chiral equilibrium involving $[Eu(DPA)_3]^-$ and (+)-dimethyl-L-tartrate, and also the chiral $[Eu\{(R,R)\ \mathbf{-1}\}_3]^{3+}$ complex [86,87].

## 3.4 Survey of CPL Applications

This section illustrates the utility of CPL to provide specific and unique information concerning its use as a probe of luminescent Ln(III) complexes as sensory systems for anion binding in aqueous media and as a test for the existence of chiral Ln(III) structures (i.e. predominant isomer in solution or if the solution of a complex containing an achiral ligand is indeed a racemix mixture). CPL is also an indicator of changes in chiral structure (i.e. importance of the helical wrapping of the ligand strand contribution, and therefore its influence on the diastereomeric induction). In addition, information concerning metal-ion environments and the associated chiral structures of metal-containing biological systems can be obtained through the measurement of CPL. Because the focus of this chapter is to highlight the use of CPL spectroscopy to study Ln(III)-containing systems, the reader is referred to the numerous and extensive reviews that are devoted to CPL for a complete discussion of the importance of this chiroptical technique [15,36–42,49,62,88]. However, we will attempt to be as complete as possible concerning Ln(III)-based CPL work appearing in the literature since the last full review.

### 3.4.1 Ln(III)-Containing Systems

In order to detect a CPL signal, the emitting species must be chiral. The key point of introducing chirality into luminescent Ln(III) complexes is to develop strongly luminescent probes that also possess chiral properties with the goal of improving the selectivity toward biological systems. One of the enantiomers/diastereoisomers of the chiral Ln(III) complex will preferentially interact with an optically active molecule and quantitatively reflect the result of the interaction with the chiral environment of the molecule of interest. This new property will add considerable power (i.e. flexibility, applicability, sensitivity, specificity, selectivity, etc.) to chiroptical luminescent probes, since it can provide valuable information concerning molecular recognition mechanisms in biological materials. The judicious selection of chiral and/or achiral ligands in the preparation of chiroptically active functional Ln(III)-based materials may be used to mimic complex biological systems and provide useful information on their mechanisms, or enhance their chiral recognition and sensing properties. Since the development of applications for the chiral luminescent Ln(III) compounds with chiral properties depends on how the chirality feature was introduced in these systems (i.e., chiral structure resulting from a chiral arrangement of the ligand molecules about the Ln(III) ion or presence of asymmetric carbon atoms in the ligand molecules), the selected examples have been grouped by type of ligands present in chiral systems containing Ln(III) ions, achiral (non-chiral) and chiral (non-racemic) ligands.

### 3.4.2 Ln(III) Complexes with Achiral Ligands

#### *3.4.2.1 CPL from Racemic Mixtures following Circularly Polarised Excitation*

The use of achiral ligands usually leads to the formation of Ln(III) complexes existing as racemic mixtures of complexes with Δ and Λ helicity in solution (Fig. 3.6). It is worth noting that there is only one example for which CPL has been conducted on a chiral luminescent Ln(III) complex containing achiral ligands that has been resolved and studied in solution [ΛΛ-(−)EuCr($\mathbf{L^8}$)$_3^{6+}$] [89].

One unique aspect of the CPL spectroscopy is that it can identify whether a Ln(III) complex solution contains a mixture of diastereomeric/enantiomeric species or whether this complex solution is a racemic mixture. If only one emitting species contributes to the luminescence observed in the sample, then the CPL results should be independent of excitation polarisation. Variation of the excitation polarisation from left to right circular polarisation will produce an oppositely signed CPL result of equal magnitude for a racemic

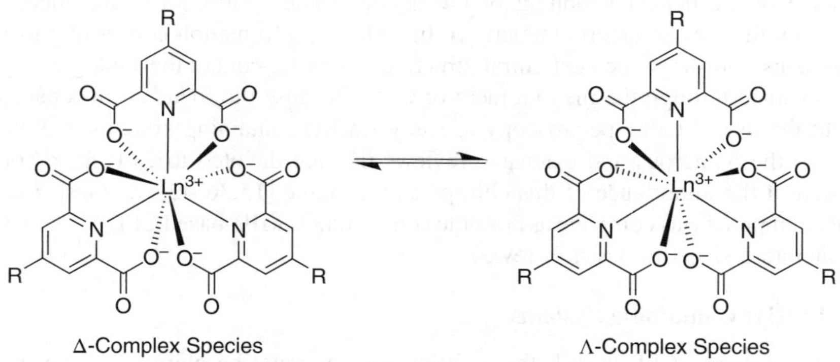


*Figure 3.6* 9-Coordinate tris-terdentate compounds of Ln(III) with achiral, for example, DPA derivative ligands (e.g., R = H, O$^-$) existing as racemic mixtures of complexes with Δ and Λ helicity in solution

mixture. However, no CPL will be observed if such a complex containing an achiral ligand does not exist as a racemic mixture in solution. It is worth noting that no signal will be detected in either case by CD, which shows that this technique cannot be used in such experiments. However, one cannot exclude the lack of CPL detection for some Ln(III) systems on the luminescence time scale. As an example, this experimental approach was used to investigate the chiral nature of the Ln(III) complex solutions with the polypyridine ligand 6,6′-bis[bis(2-pyridylmethyl)aminomethyl]-2,2′-bipyridine (**L$^1$**) [90]. Although no CPL was observed for the Eu(III) and Tb(III) compounds following left and right circularly polarised excitation, an oppositely signed CPL spectrum was observed for the corresponding Dy(III) complex, confirming the chiral nature of this complex and, though, that it exists as a racemic mixture in solution. The lack of CPL detection for the Eu(III) and Tb(III) compounds was explained by an efficient racemisation during the longer lifetime of these latter species compared to the related Dy(III) compound (~ms range vs. 0.1 ms range). Taking into account this result, one may anticipate that a control of the experimental conditions (i.e. temperature, liquid vs. sol-gel systems) may play an important role in the detection of CPL signals for systems that may racemise in solution and, thus, gives one the ability to probe individual enantiomers by circularly polarised excitation followed by CPL. For instance, it was shown that the tris Eu(III) complex with 4-phenylethynyl-2,6-pyridinedicarboxylic acid (PEDPA), a derivative of DPA, that racemises quickly in aqueous solutions (a temperature dependence of its CPL signal was observed due to racemisation occurring during the excited state lifetime) [91], may be imbedded in sol-gel to minimise and/or inhibit racemisation without significantly altering the overall complex structure [92]. A completely opposite CPL spectrum of approximately equal magnitude was measured for its sol-gel sample, confirming the racemic nature of this complex species. Similarly, it can be envisaged that CPL measurements at lower temperatures than, for example, RT may minimise the risk of racemisation or any isomer interconversion on the experimental time scale. This was the case for two diastereoisomers of (*S*)-$[Eu\text{-}EOB\text{-}DTPA(H_2O)]^{2-}$ which were in slow exchange at RT on the NMR and emission time scales [93]. The CPL data obtained at 282 K confirmed that each 9-coordinate isomer, which only differed in the absolute configuration of the central N atom, had a Λ helicity (similar spectra). Although this latter example is from a Eu(III) complex with a chiral ligand, similar findings would be expected for racemic Ln(III) complexes with the assumption that the overall complex structure and the excited state lifetime are independent of temperature, and that any decrease in $g_{lum}$ is due to the increased racemisation at elevated temperatures. As indicated earlier, this was the case for the aqueous solution of the racemic Eu(III) complex containing the terdentate ligand PEDPA [91]. As temperature increased, the magnitude of the $g_{lum}$ value decreased due to a faster racemisation process.

#### *3.4.2.2 CPL from Racemic Mixtures: Perturbation of the Equilibrium*

In addition to verify if a Ln(III) complex with achiral ligands exists as a racemic mixture in solution, the CPL spectroscopy is also often used as a probe of specific molecular chirality. It was shown that the luminescent *tris*-terdentate Ln(III) complexes of ligands derived from DPA are good candidates for such studies. Of special importance is that the modular nature of the DPA backbone allows for easy synthesis of derivatives with tunable photophysical and/or chiroptical properties [94]. This is the reason why they are often used for chirality sensing [15,36]. Since these 9-coordinate Ln(III) compounds exist as racemic mixtures of complexes with Δ and Λ helicity in solution, one can use them toward the molecular recognition of chiral biological substrates. This is possible due to the observation of induced chirality in the latter upon outer sphere coordination with a variety of optically active organic molecules such as tartrate substrates, amino acid or sugar derivatives (*AA**) [15,36,38,87,95–99], a phenomenon referred to as the "Pfeiffer effect" [100]. The three relevant equilibrium expressions are defined as

$$\Delta\text{-Ln(DPA)}_3^{3-} \rightleftarrows \Lambda\text{-Ln(DPA)}_3^{3-} \quad (K_{\text{rac}} = 1) \tag{3.14}$$

$$\Delta\text{-Ln(DPA)}_3^{3-} + AA^* \rightleftarrows \Delta\text{-Ln(DPA)}_3^{3-} : AA^* \quad (K_1) \tag{3.15}$$

$$\Lambda\text{-Ln(DPA)}_3^{3-} + AA^* \rightleftarrows \Lambda\text{-Ln(DPA)}_3^{3-} : AA^* \quad (K_2) \tag{3.16}$$

where the outer sphere association complex is denoted by a colon (:).

Of special importance, it was shown that the chiral recognition of such biomolecular substrates can be modulated by the nature of the ligand interface of the racemic 9-coordinate Ln(III) complexes. Moussa et al. showed that the H-bond character of the negatively charged hydroxyl group in CDA (chelidamic acid), the hydroxylated derivative of DPA, led to a larger "Pfeiffer effect" with L-amino acids (e.g. L-proline or L-arginine) susceptible to form hydrogen bonds with negatively charged groups, while these H-bonding effects were less important with DPA [101]. More recently, Kosareff et al. [15,102] reported that the chiroptical CPL technique has potential for the chiral recognition of optical isomers of a given amino acid (i.e. L- and D-serine), even if one of the two enantiomers of the latter was in a high excess in solution (Fig. 3.7) [15,102]. This was possible because CPL is responsive to the nature and amount of the added chiral probes present in solution. It is worth noting that these findings are in agreement with earlier studies where it was pointed out that the sign and magnitude of the CPL signal were dependent upon several factors and not only from the chirality of the enantiomerically pure amino acid [36,38,103]. In such "Pfeiffer effect" studies, it is imperative to take into account the effect of various non-covalent chiral discriminatory interactions such as hydrogen bonding, Coulombic forces, $\pi$-stacking, hydrophobic effects, experimental conditions (i.e. pH, temperature, ratio of the system of interest to amino acid) and steric effects on the CPL sign and magnitude.

Unlike the Pfeiffer-induced systems described above, where the non-racemic emitting state is produced by the disturbance of the ground state racemic equilibrium by an added chiral molecule, one may envisage a situation in which the non-racemic emitting state is the result of a time-dependent optically enriched excited state. This is generated through enantioselective quenching by an added chiral quencher molecule (Scheme 3.2). Although this is commonly referred to as an enantioselective quenching process, a more appropriate description of it would be a diastereomeric interaction. The discriminatory interactions between the excited enantiomers of a racemic Ln(III) complex and a chiral quencher molecule result in a difference in excited state populations due to the quenching of one enantiomer over the other by the added chiral quencher. In terms of the experimental procedure, an unpolarised beam excites the racemic mixture to an emitted state leading to the presence of equal concentrations of the two excited enantiomers. Then, the chiral

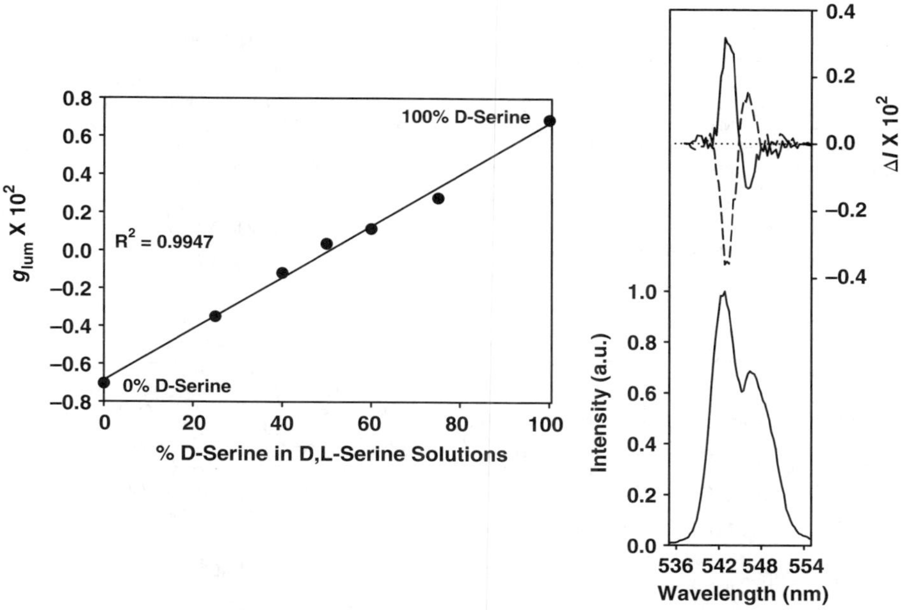


***Figure 3.7*** *Left: Plot of the $g_{lum}$ values for the 543 nm component of the $^5D_4 \rightarrow {}^7F_5$ transition of 0.01 M $[Tb(DPA)_3]^{3-}$ after addition of 0.40 M of L- and D-serine as a function of the percentage of D-serine in the D,L-serine solution. Right: CPL (upper curves) and total luminescence (lower curves) spectra for the $^5D_4 \rightarrow {}^7F_5$ transition of 0.01 M $[Tb(DPA)_3]^{3-}$ after addition of 0.40 M of L- and D-serine (dashed and solid lines), respectively ($\lambda_{exc}$ = 293 nm, pH 7, 295 K). Reproduced with permission from [15]. Royal Society of Chemistry*

quencher, also present in the solution, interacts with the excited enantiomers in such a way that one of the enantiomers is quenched at a faster rate than the other. The result of this differential excited-state quenching process is that the excited state becomes chiral (or a non-racemic emitting state) over time, which can be analysed using the time-resolved feature of the CPL instrumentation. Since the complete mechanistic details of the quenching processes were described numerous times in the literature and that only one review has appeared since the last full review [37], only a brief overview of the findings is highlighted here. Of special importance is that the studies showed that the degree and sense of enantiomeric preference in these quenching processes are governed by the electronic and stereochemical properties of the optically active quenchers. Small structural changes in the chiral quencher systems can result in large changes in the chiral recognition/discrimination properties of the systems of interest. For instance, replacing the amide groups on the side chains of the corrin ring in vitamin $B_{12}$ derivatives by ester moieties resulted in a considerable decrease in the magnitude of the enantioselectivity in the quenching parameter ($E_q$ values went from −0.20 to almost zero). This result suggested that the presence of the amide groups in $B_{12}$ vitamins is critical for effective hydrogen bonding-type intermolecular interactions between its amide protons and the carboxylate groups of the DPA ligands of $[Tb(DPA)_3]^{3-}$. In fact, these hydrogen bonding interactions

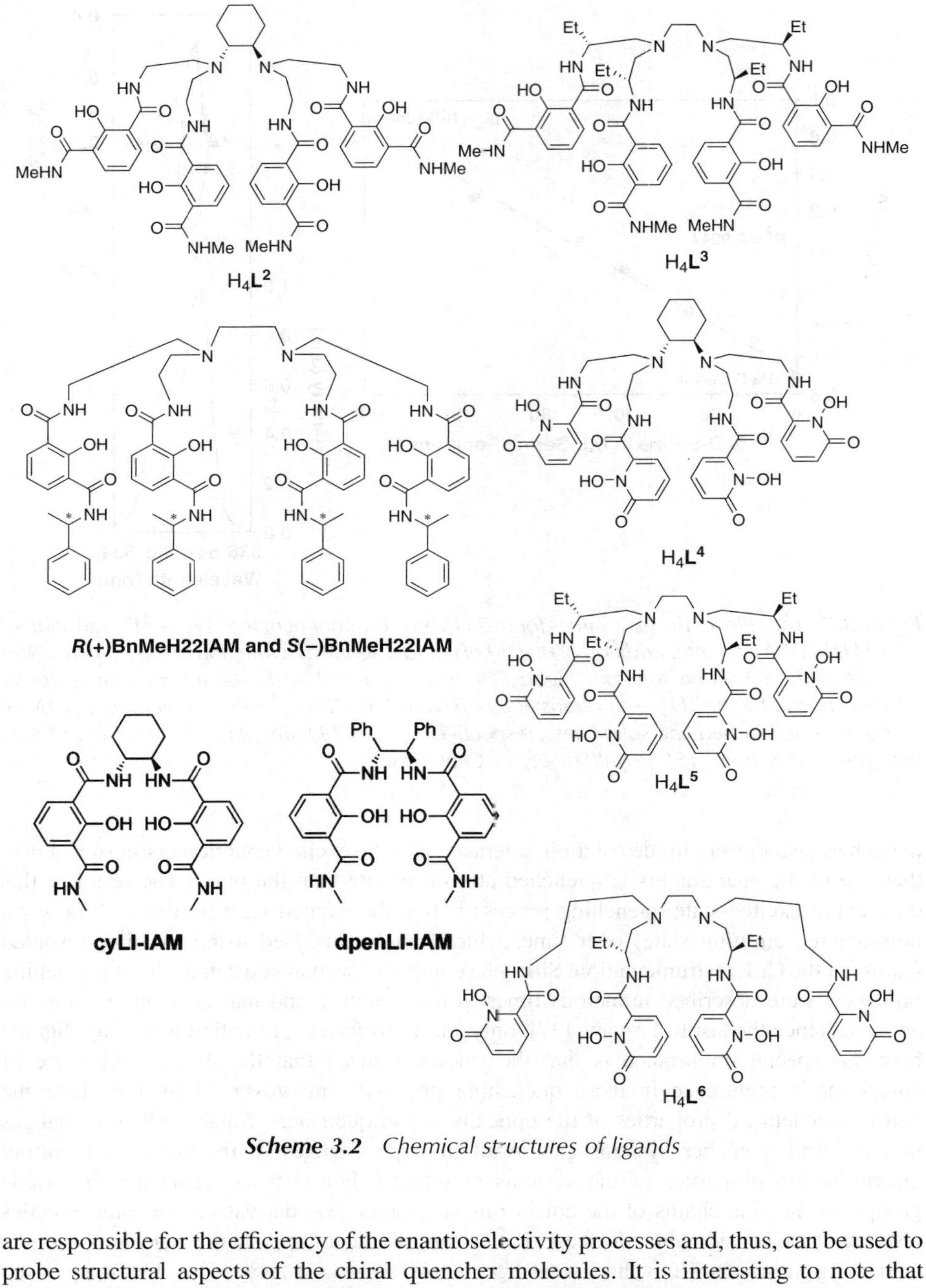


***Scheme 3.2*** *Chemical structures of ligands*

are responsible for the efficiency of the enantioselectivity processes and, thus, can be used to probe structural aspects of the chiral quencher molecules. It is interesting to note that Meskers and Dekkers also envisioned that such quenching reaction studies could be used to investigate the enantioselectivity of metal-containing membrane proteins in the near future [104].

### 3.4.3 Ln(III) Complexes with Chiral Ligands

Luminescent Ln(III) complex-based molecular probes are commonly employed for analytical and biomedical applications, where there are used either as diagnostic or as therapeutic tools. Although it is now commonplace to use them in fields like biochemistry, biology, medicine, and related biomedical disciplines, there is still a considerable surge of interest to develop luminescent Ln(III) compounds possessing chiral properties in addition to their interesting spectral features. These unusual spectral characteristics often consist of, for example, long excited-state lifetimes, line-like emission bands that are easily recognisable and well separated from the broad fluorescence bands of the organic fluorophores, a large Stokes shift, sensitivity to the local environment, or time-resolved separation between Ln(III) luminescence and short-lived background fluorescence. The combination of these Ln(III)-based luminescence properties with CPL activity would result in the development of bioanalytical applications where the detection would be based on the CPL signal.

#### *3.4.3.1 Chiral Ligands with the 1,2-HOPO and IAM Units*

Since most of the Ln(III) complexes are optimised with respect to either luminescent properties or CPL activities, the ideal approach would be to design a system that combines a large luminescence quantum yield with a strong CPL activity. Another requirement is that such a system should be soluble and, more importantly, stable in an aqueous medium if one wants to use it for bioanalytical application purposes.

Working along these lines, the research group of Raymond investigated the possibility of developing such compounds. Petoud et al. [105] reported a new family of Ln(III) complexes formed with octadentate ligands incorporating 2-hydroxyisophthalamide (IAM) as the chelating unit. An interesting and potentially useful result on these chiral IAM ligand-based systems, *R*(+) and *S*(−)BnMeH22IAM, was that the same ligand architecture was used to sensitise four different Ln(III) ions, namely Sm, Eu, Tb, and Dy. Since each of these cations has characteristic sharp luminescence bands in the visible region which in turn yields distinctive CPL activities, these Ln(III) complexes may be suitable for chiral multiplex detection and/or for assays based on CPL measurements. This would be possible since each of these Ln(III) systems can be discriminated from each other.

Although the four Ln(III) compounds displayed strong CPL activity in MeOH solution ($|g_{lum}|$ values of up to 0.3), the insolubility of these systems in aqueous media considerably limits their use for analytical applications. As a result, a new family of water-soluble Ln(III) complexes with chiral octadentate 1-hydroxy-2-pyridinone (1,2-HOPO) and IAM-based ligands was developed for the chelation of Eu(III) ($H_4\mathbf{L}^4$, $H_4\mathbf{L}^5$, and $H_4\mathbf{L}^6$) and Tb(III) ($H_4\mathbf{L}^2$ and $H_4\mathbf{L}^3$), respectively [106,107]. The interesting feature of these compounds is that the chiral stereocentres in the ligand were incorporated in two different positions of the ligand hexamine backbone (central *trans*-1,2-diaminocyclohexane and ethyl substituents in the pheripheral ethylene arms), whereas the former MeOH-soluble ligands had the chiral information introduced in the sensitiser units. The separation of the chiral information from the chromophore presents a significant advantage with respect to the versatility and flexibility of ligand design. In addition to the good luminescent properties (i.e. high quantum yields, long lifetimes) and the solubility in physiologically relevant media, the significant CPL values [$|g_{lum}|$ values of up to 0.2 and 0.4 for Eu(III) and Tb(III) compounds] of these water-soluble systems open new perspectives in terms of the development of practical CPL probes. It is interesting to note that the CPL-active Tb(III) complex with the chiral IAM ligand $H_4\mathbf{L}^2$ showed one of the highest luminescence quantum yields of Ln(III) edifices in aqueous solution ($\phi$ ~60%), but its CPL was relatively modest ($|g_{lum}|$ < 0.08) [106,107]. Since it is well established that the CPL activity may increase with the conformational rigidity of the complex [36,38], Samuel et al. [108] studied tetradentate chiral ligands derived from the chiral IAM ligands, but with an increased conformational rigidity and, thus, would yield a larger CPL activity. As expected, these Tb(III)-containing compounds with chiral tetrapodal octadentate ligands containing IAM antenna chromophores and utilising diaminocyclohexane (cYLI) and diphenylethylenediamine (dpenLI) backbones resulted in a substantial increase in the CPL activity compared to the parent compounds with the chiral IAM ligand $H_4\mathbf{1}$ ($|g_{lum}|$ values up to 0.2 vs. < 0.08), but also maintained their brightness ($\phi$ ~32–60% vs. ~60%). In addition to these findings, the diastereopurity of these compounds was assessed by CPL.

As already discussed in Section 3.4.1, one can use CPL spectroscopy to investigate solutions containing luminescent complexes to determine whether or not the source of the luminescence is one particular complex or a mixture of species. If only one emitting species contributes to the luminescence observed in the sample, then the CPL results should be independent of excitation polarisation (i.e. left-, right-, or plane-polarised light). But, had the complex solution contained a mixture of species, the CPL signal would be dependent on the polarisation of the excitation beam. Thus, the CPL measurements confirmed the presence of a single and multiple emissive species in solution for the 2:1 cyLI-Tb and dpenLI-Tb compounds, respectively [108]. It was shown that the CPL signal of the former complex solution was similar whether a direct or indirect excitation was used (consistent with one single emitting species in solution), whereas the $g_{lum}$ values of the latter complex solution were dependent on the polarisation of the excitation beam, and on whether a direct or indirect excitation was used (Scheme 3.3). These findings were confirmed by the lifetime data, which indicated the presence of two emitting species for dpenLI-Tb that differed in their hydration number, and, consequently, in their coordination environment. It should be noted that NMR is often used for assessing the diastereopurity of chiral systems. In addition to the sample requirement for NMR measurements, it is often the case that Ln(III) complexes, namely Eu(III) and Tb(III), cannot be investigated in this way due to their strong paramagnetic nature. This is the reason why the CPL technique is suitable for a detailed examination of the diastereopurity of chiral Ln(III) systems, as demonstrated in this study.

#### *3.4.3.2 Chiral Ligands Derived from Pybox Moieties*

Chiroptical structures of luminescent tris[$\beta$-diketonate (HFA, 1,1,1,5,5,5-hexafluoropentane-2,4-dione)] Eu(III) and Tb(III) complexes containing a chiral (*R*)-bis(oxazolinyl) pyridine derivative ligand (Ph-, *i*-Pr-, and Me-Ph-pybox) have been characterised by CPL [52]. In particular, CPL spectra of [{$Eu^{III}$(*R*)-Ph-pybox}$(HFA)_3$], [{$Eu^{III}$(*R*)-Ph-pybox}$(HFA)_3$], and [{$Eu^{III}$(*R*)-Ph-pybox}$(HFA)_3$] in $CD_3CN$ solution have revealed large CPL activities in the spectral range of the $^5D_0 \rightarrow {}^7F_1$ transition ($|g_{lum}|$ = 0.15–0.46) but, more importantly, that the induced chirality at the Ln(III) centre in these systems is governed by non-covalent ligand–ligand interactions between the achiral HFA and chiral pybox derivative ligands. The X-ray structural analyses of these various complexes revealed that

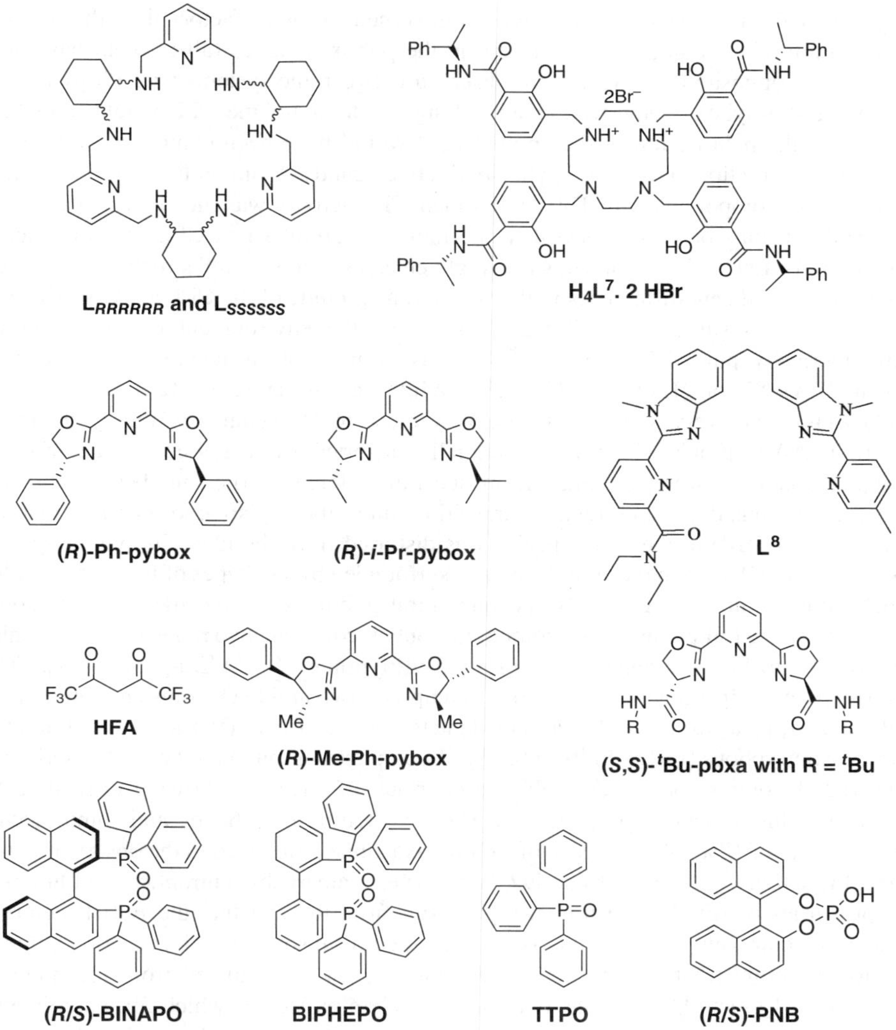


***Scheme 3.3*** *Chemical structures of ligands*

these ligand–ligand interactions resulted in an asymmetric arrangement of the three HFA ligands around the Ln(III) ion. Additionally, it was also shown that each (*R*)-pybox ligand (Ph-, *i*-Pr-, and Me-Ph-pybox) was involved in different non-covalent interactions ($\pi$-$\pi$ stacking, CH/F, and CH/$\pi$ interactions) with the adjacent HFA ligand, respectively. Although the local environment of the metal centre in the three complexes reveals some differences in solution as pointed out by the spectral variations observed (i.e. crystal-field splitting of the $^7F_J$, magnitudes of the $g_{lum}$ values, and/or relative transition intensities), and may be different than what was observed from the X-ray crystal structures, the CPL results suggest that the polarisation of the light is governed by other factors than the absolute configuration of the chiral centres present in the pybox moieties (all of them have an *R* chirality, but opposite CPL signs were observed between the systems with Ph-pybox and *i*-Pr- or Me-Ph-pybox). Otherwise, identical sign patterns for the CPL would have been observed if the polarisation of the emitted light would have been controlled only by the absolute configuration of the chiral pybox derivative ligands. A similar finding was obtained for the parent compounds with Tb(III), but their CPL activity was much weaker ($|g_{lum}| =$ 0.04–0.08). It must be emphasised that one must be careful when CPL spectra:structure correlation for Ln(III) is attempted, since slight changes in the overall structure of a system may lead to significant differences in the sign and magnitude of the CPL activity. This was the finding of a study where it was shown that the environment of Eu(III) in the enantiomerically pure ΛΛ-(−)EuCr($\mathbf{L^8}$)$_3^{6+}$ was clearly not sufficiently similar enough to $[Eu(DPA)_3]^{3-}$ or $[Eu(ODA)_3]^{3-}$ (53). While the structure of $[Eu(DPA)_3]^{3-}$ and $[Eu(ODA)_3]^{3-}$ are very similar, and are very close to $D_3$ symmetry, the symmetry of Eu(III) in ΛΛ-(−)EuCr($\mathbf{L^8}$)$_3^{6+}$ is substantially distorted from a 9-coordinate trigonal-tricapped prism $D_3$ structure. Significant spectral differences (i.e. number of observed crystal field transitions and relative transition intensities) also corroborated that the symmetry of Eu(III) in the latter species was distorted from the ideal $D_3$ symmetry.

While the Ln(III) ion in the complexes was surrounded by two types of ligand molecules, namely three achiral HFA ligands and one chiral pybox derivative ligand in the above example [52], the following Ln(III) compounds only incorporate two molecules of the chiral 2,6-bis[(4*S*)-*tert*-butyl-carbamoyl-2-oxazolin-2-yl]pyridine, [(*S*,*S*)-$^t$Bu-pbxa] [77]. This pybox derivative ligand with amide groups comprises two amide O atoms and three pybox N atoms and, as a result, acts as a pentadentate chelator for Ln(III) ions. In addition to forming isomorphous [Ln{(*S*,*S*)-$^t$Bu-pbxa}$_2$]$^{3+}$ complexes through all the Ln(III) series (all 14 crystals belong to the orthorhombic space group $P2_12_12$) and showing interesting Eu (III)-centred luminescence properties in MeOH (lifetime of 1.66 ms and luminescence quantum yield of 17%), the presence of a chiral oxazoline moiety led to the observation of a large CPL activity ($g_{lum} = -0.18$ at 592.2 nm). These remarkable chiroptical and luminescent properties of [Eu{(*S*,*S*)-$^t$Bu-pbxa}$_2$]$^{3+}$ result from the fact that the Eu(III) centre is situated in a chiral environment that is well shielded from solvent molecules (a $q$ value of 0.03 for the number of coordinated water molecules was calculated from the lifetimes measured in $H_2O$ and $D_2O$) by the two (*S*,*S*)-$^t$Bu-pbxa molecules which form a cryptand-like dimer. This structural feature was confirmed by X-ray crystallography. The X-ray structures of [Ln{(*S*,*S*)-$^t$Bu-pbxa}$_2$]$^{3+}$ revealed that all complexes possess crystallographic $C_2$ axes passing through the Ln(III) ion and, thus, form left-handed double-helical structures. This solid state finding was also supported by the NMR study since the $^1$H-NMR signals pointed to a high symmetry of the diamagnetic La(III) and Lu(III) complex species in MeOD, but also that no fast diastereoisomeric equilibrium occurred on the NMR time scale at room temperature. Additionally, this NMR study confirmed that the CH-$\pi$ interactions observed in the solid state were maintained in solution.

### *3.4.3.3 Chiral Ligands with Phosphine Oxide or Phosphate Units*

Section 3.4.1 described a series of experiments concerning the measurement of CPL from luminescent tris[$\beta$-diketonate (HFA)] Eu(III) and Tb(III) complexes containing a chiral

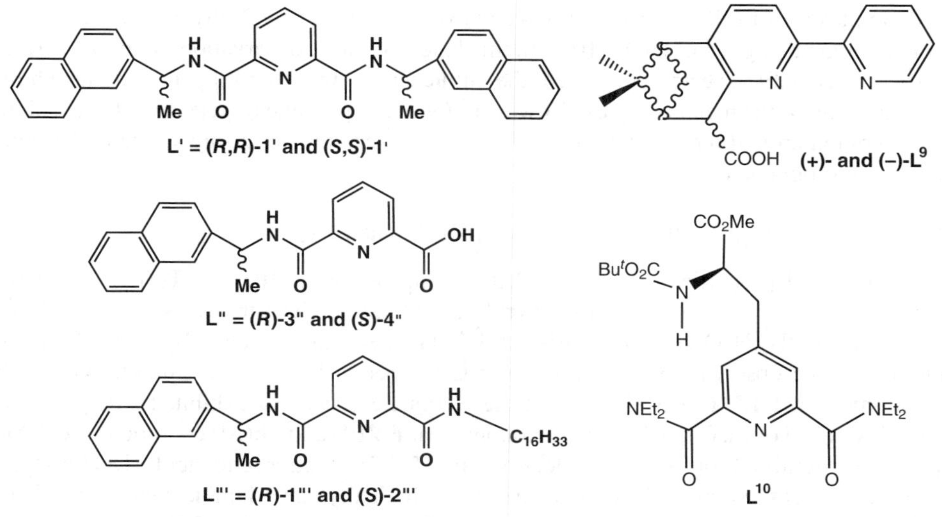


***Scheme 3.4*** *Chemical structures of ligands*

(*R*)-pybox derivative ligand. Here the luminescent Eu(III) compounds combine achiral or chiral *β*-diketonate (HFA or *D*-facam) and point- or axis-chiral or achiral phosphineoxide ligands [(*R* or *S*)-BINAPO, BIPHEPO, or TPPO] (Scheme 3.3) [64,65]. The general formulas of these systems are [Eu{(*R*)-BINAPO}$(HFA)_3$], [Eu{(*R* or *S*)-BINAPO}(*D*-facam)$_3$], [Eu(BIPHEPO)(*D*-facam)$_3$], and [Eu$(TPPO)_2$(*D*-facam)$_3$] where *D*-facam, (*R*/*S*)-BINAPO, BIPHEPO, and TPPO stand for 3-trifluoroacetyl-*d*-camphor, (*R*/*S*)-2,2′-bis(diphenylphosphoryl)-1,1′-binaptthyl, 2,2′-bis(diphenylphosphoryl)-1,1′-biphenyl, and triphenylphosphine oxide, respectively. CPL investigations on these Eu(III)-containing compounds in acetone-$d_6$ have shown that the strong measured CPL ($|g_{lum}| = 0.44$, 0.34, 0.24, and 0.47 in the spectral range of the $^5D_0 \rightarrow {}^7F_1$ transition of [Eu{(*R* or *S*)-BINAPO}(*D*-facam)$_3$], [Eu(BIPHEPO)(*D*-facam)$_3$], and [Eu$(TPPO)_2$(*D*-facam)$_3$]) were mainly dependent on the chiroptical properties induced by the presence of the chiral *D*-facam ligands in these systems. Similar CPL spectral shapes were observed between [Eu{(*S*)-BINAPO}(*D*-facam)$_3$] or [Eu(BIPHEPO)(*D*-facam)$_3$] and [Eu{(*R*)-BINAPO}(*D*-facam)$_3$]. It was suggested that the difference in the $|g_{lum}|$ values between [Eu{(*R*)-BINAPO}(*D*-facam)$_3$] and [Eu{(*S*)-BINAPO}(*D*-facam)$_3$] (0.44 vs. 0.34) was due to slight changes in the local structural environment of Eu(III) in these two. However, one must be cautious with such an interpretation, since it is well established that the measured $g_{lum}$ value represents the sum of all the CPL components present in the studied systems. One can assume that the sign of the $g_{lum}$ of (*R*)- and (*S*)-BINAPO moieties is either identical or opposite to that of the *D*-facam. Thus, one would expect some differences in the magnitude of the overall CPL activity measured for [Eu{(*R*)-BINAPO}(*D*-facam)$_3$] and [Eu{(*S*)-BINAPO}(*D*-facam)$_3$], even if the main contributor to the observed CPL signal is *D*-facam. This is in line with the observation of a relatively small CPL activity for [Eu{(*R*)-BINAPO}$(HFA)_3$] ($g_{lum}$ of 0.03), which contains the chiral (*R*)-BINAPO ligand and three achiral hfa ligand molecules. It is worth noting that a similar CPL activity was reported for the solvated *tris*-complexes of Eu(III) with chiral (*R*)- and (*S*)-1,1′-binaphtyl-2,2′-diyl phosphate (*R* and *S*)-PNB) ligands ($|g_{lum}|$ of 0.04) [109]. In this latter example, observation of almost perfect mirror-image CPL spectra confirmed the influence of the chiral ligand on the chiral configuration of the metal complex. However, it was not possible to determine the absolute configuration around the Eu(III) ion since no suitable crystals for X-ray structure determination were obtained.

### 3.4.3.4 *Chiral Ligands With Pinene-bipyridine Frameworks*

The chiroptical properties of the trinuclear complexes of Eu(III) and Tb(III) with both enantiomeric forms of the pinene-bipyridine carboxylic derivatives (+)- and (−)-$\mathbf{L^9}$, $[Ln_3(\mathbf{L^9})_6(\mu_3\text{-OH})(H_2O)_3](ClO_4)_2$ (tris{Ln[$\mathbf{L^9}$]$_2$}), were investigated by CPL [67]. In addition to the observation of mirror image CPL spectra for these complexes with both enantiomers of $\mathbf{HL^9}$, it was shown that these trinuclear systems exhibited stronger CPL activities ($|g_{lum}|$ of ca. 0.10) than other mononuclear Eu(III) and Tb(III) compounds with bis (pinene terpyridine) or tetraazacyclododecane-*N*,*N*′,*N*,*N*′-tetraacetic acid (DOTA)-type ligands, whereas comparable CPL activities with diethylene triamine pentaacetic acid (DTPA) and DOTA derivatives were reported. The origin of such a CPL activity comes probably from the fact that the trimetallic core is situated in a well-defined chiral environment. The structure of these complex species possesses a crystallographic $C_3$ axis and consists of a trimetallic core and six ligands wrapped around, three by three, in a helical opposite fashion, as shown from the X-ray structure analysis. It must be pointed out that each set of three ligands wraps around the trimetallic core in *P*- and *M*-helical fashions for (+)-$\mathbf{L^9}$, whereas an opposite trend is observed for (−)-$\mathbf{L^9}$. In addition, $^1$H-NMR spectroscopic and electrospray mass spectroscometric studies strongly suggested that these trinuclear species were maintained in $CH_2Cl_2$ solution and, also, stable in the concentration range investigated ($10^{-2}$–$10^{-6}$ M)

### 3.4.3.5 *Chiral Ligands Derived from 2,6-Pyridinedicarboxamide Moieties*

Since the 9-coordinate Ln(III) complexes of ligands derived from DPA exist as racemic mixture of complexes with Δ and Λ helicity in solution, the introduction of chiral substituents may lead to a diastereomeric resolution of these 9-coordinate Ln(III) complexes. This was already evidenced for the substituted pyridine-2,6-dicarboxamide ligand, $\mathbf{L^{10}}$, where a chiral group was grafted onto the *para*-position of the pyridine group [110]. Although this ligand led to the formation of stable triple helical $[Ln(\mathbf{L^{10}})_3]^{3+}$ complexes, there was only a very small excess of one diastereoisomer induced in solution by $\mathbf{L^{10}}$. This was corroborated by a weak CPL activity of the Ln(III)-containing systems ($g_{lum} = 0.02$). Working along these lines, Bonsal et al. reported that the CPL activity for the 9-coordinate Eu(III) complexes with the enantiomers of *N*,*N*′-bis(1-phenylethyl)-2,6-pyridinedicarboxamide (**L**), (*R*,*R*)-**1** and (*S*,*S*)-**1**, was much higher ($|g_{lum}| = 0.19$; see Fig. 3.4 and Table 3.2) [75]. It is worth noting that the diastereomeric nature of $[Eu\mathbf{L}_3]^{3+}$ with (*R*,*R*)-**1** and (*S*,*S*)-**1** prevails in solution as evidenced by the CPL results from the Eu-containing species (observation of mirror-image CPL spectra that are independent of the polarisation of the excitation beam, either right-, left-, or plane-polarised light, and if direct or indirect excitation is used) [86]. This was also supported by the various spectroscopic and X-ray crystallographic studies [111]. In particular, it was shown that all $[Ln\mathbf{L}_3]^{3+}$ complexes

[Ln(III) = Eu, Gd, Tb, Yb and **L** = (*R,R*)-**1** and (*S,S*)-**1**] were isostructural and crystallised in the cubic space group I23. Moreover, it was also shown that the Δ and Λ metal stereochemistry was induced by the chiral nature of the ligand molecules in all of the complexes studied. That is, ligand (*S,S*)-**1** induced a Δ chirality about the Ln(III) ions, whereas ligand (*R,R*)-**1** induced a Λ chirality. These findings are also in agreement with those reported for related highly symmetrical isostructural chiral 1:3 Ln(III) complexes (Ln = Sm, Eu, and Tb) with another chiral pyridyldiamide ligand derivative, pyridine-2,6-dicarboxalic acid bis-[1-naphthalen-1-yl-ethyl)-amide] (**L**′ = (*R,R*)-**1**′ and (*S,S*)-**1**′) [112]. This ligand only differs from **L** by its carboxamide moieties in which the phenyl groups of **L** are replaced by naphthalene substituents. Unlike $[LnL_3]^{3+}$, which is not water-soluble [111], Leonard et al. [112] showed that their related Eu(III) and Tb(III)-containing compounds were relatively water-soluble and, more importantly, that the Ln(III) centre retained its coordination saturation in aqueous solution ($q$ values of ~0, confirming that no water molecules were bound to the Ln(III) ion). It should be mentioned that the CPL sign patterns of the Eu(III) complexes are consistent with what Bonsall et al. reported for their related system with **L**. That is, a sequence of (+, +) followed by (−), and (+, −, +) for the various components (2, 1, and 3) in the spectral range of the $^5D_0 \rightarrow {}^7F_1$, $^5D_0 \rightarrow {}^7F_2$, and $^5D_0 \rightarrow {}^7F_4$ transitions was observed for the (*S,S*)-ligands, respectively (Fig. 3.4 and Table 3.2) [75]. Similarly, an opposite trend was observed for the related Eu(III) compounds with (*R,R*)-ligands.

It is worth pointing out that a CPL study on chiral tris Eu(III) complexes with the asymmetrical pyridyl ligands (*R*)-**3**″ and (*S*)-**4**″ resulted in the same CPL pattern (similar sign sequence for the $^5D_0 \rightarrow {}^7F_1$, $^5D_0 \rightarrow {}^7F_2$, and $^5D_0 \rightarrow {}^7F_4$ transitions and $|g_{lum}|$ values of 0.15) than that observed for the related systems with **L**′ [112,113]. In fact, the ligands (*R*)-**3**″ and (*S*)-**4**″ represent "half the ligands" of (*R,R*)-**1**′ and (*S,S*)-**1**′, since one of the chiral tryptophan-based moieties in **L**′ is substituted by a carboxylic acid function. However, it was shown that (*R*)-**3**″ and (*S*)-**4**″ led to the formation of Eu(III) complexes with a 3:1 stoichiometry and that each "half-helicate" isomer was formed enantiomerically pure and most likely remained homochiral in solution. These latter findings were based on the fact that the CD and CPL spectral features were similar to the ones of the related systems with **L**′ for which the metal helicity was solved using X-ray crystallography. The handedness of these "half-helicates" was thus predicted to be Λ and Δ for [(*R*)-**3**″]$_3$:Eu$_1$ and [(*S*)-**4**″]$_3$:Eu$_1$, respectively. Although no suitable X-ray-quality single crystals for these "half-helicates" **3**″$_3$:Eu$_1$ and **4**″$_3$:Eu$_1$ were obtained, MM2 calculations carried out on these Eu(III)-containing species pointed to the most stable structure as being the one for which the three antennas were located on the same side and directed towards the inside of a "half-helicate". This geometry also favoured the existence of $\pi$-$\pi$ interactions between the antenna and the pyridyl moieties [113], as already observed in the tris Eu(III) compounds with **L**′ [112]. Similarly, the CPL analysis of the 1:3 Eu(III) complexes with the chiral ligand $N_2$-[1-hexadecyl]-$N_6$[1*R* or *S*)-1-(naphthalene-1-yl)ethyl]pyridine-2,6-dicarboxamide, [(*R*)-**1**″]$_3$:Eu$_1$ and [(*S*)-**2**″]$_3$:Eu$_1$, in MeCN led to the same conclusions [114]. The $g_{lum}$ values amounted to −0.177 and +0.106 for the $^5D_0 \rightarrow {}^7F_1$ and $^5D_0 \rightarrow {}^7F_2$ transitions of [(*R*)-**1**″]$_3$:Eu$_1$, whereas opposite values were observed for [(*S*)-**2**″]$_3$:Eu$_1$ (respectively +0.176 and −0.102). In addition to corroborate the CPL results of its parent chiral ligand **L**″ [(*R*)-**3**″ and (*S*)-**4**″], Kitchen et al. investigated the CPL features of the Langmuir-Blodgett films of these [(*R*)-**1**″]$_3$:Eu$_1$ and [(*S*)-**2**″]$_3$:Eu$_1$ systems. It was pointed out that the slight differences between the solution and film results (−0.177 and +0.106 vs. −0.161 and +0.068 for [(*R*)-**1**″]$_3$:Eu$_1$, respectively) were most likely attributed to the influence of the packing of the molecules in the Langmuir–Blodgett films on the local environment of the Eu(III) ion and/or to the effect of the solid surface.

Although the results are of preliminary nature, these findings for the tris Eu(III) complexes with **L** [(*R*,*R*)-**1** and (*S*,*S*)-**1**], **L**′ [(*R*,*R*)-**1**′ and (*S*,*S*)-**1**′], **L**″ [(*R*)-**3**″, and (*S*)-**4**″], and **L**″ [(*R*)-**1**″, and (*S*)-**2**″] seem to support that one may envision using such systems with a trigonal symmetry to develop reliable empirical relationships between chiral structures and CPL sign patterns (Scheme 3.5). Even if a generalisation would be difficult due to the difference in the symmetry of the Ln(III) compounds, we would like to point out that a comparable trend for $C_4$-symmetric Ln(III)-containing complexes was observed (see Section 3.4.1).

### *3.4.3.6 Chiral Macrocycle-type Ligands*

Knowing that a CPL signal is a sensitive "fingerprint" of the local environment of the Ln(III), CPL spectroscopy was used to investigate the helical inversion in Eu(III) and Tb(III) complexes with a chiral nona-aza macrocycle-type ligand [115]. In addition to the observation of almost perfect mirror-image CPL spectra for the Eu(III) and Tb(III) complexes with both enantiomers of this ligand ($\mathbf{L}_{RRRRRR}$ and $\mathbf{L}_{SSSSSS}$), the CPL study confirmed that the polarisation of the emitted light was controlled by the (*M*)/(*P*) helicity at the Ln(III) centre (Fig. 3.8). Indeed, noticeable spectral changes were observed in the Tb (III)-based CPL spectra after the complex solution was heated at 328 K for three weeks. Moreover, the observed CPL changes between the solutions containing the pure (*M*)-isomer and a mixture of the (*M*)- and (*P*)-isomers of the Eu(III) complex with a (*M*)/(*P*) ratio of 3.5:1 were due to the influence of the different local helicities, (*M*) and (*P*), at the Eu(III) centre for the two diastereoisomers, (*M*)-$[\mathrm{Eu}\mathbf{L}_{RRRRRR}]^{3+}$ and (*P*)-$[\mathrm{Eu}\mathbf{L}_{RRRRRR}]^{3+}$, present in solution. Otherwise, mirror-image CPL spectra would have been observed if the polarisation of the emitted light would have been controlled by the absolute configuration at the diaminocyclohexane carbon centres, as shown for (*M*)-$[\mathrm{Ln}\mathbf{L}_{RRRRRR}]^{3+}$ and (*P*)-$[\mathrm{Ln}\mathbf{L}_{SSSSSS}]^{3+}$ in Fig. 3.8. This study also confirmed that the spectral changes observed were more obvious for the Eu(III) systems than for the parent Tb(III) complex solutions since Eu(III), in theory, has a simpler crystal field energy level pattern than Tb(III).

CPL studies have been also used to investigate the solution structure and the species distribution as a function of the pH of a Tb(III) compound with the chiral 3,3′,3″,3‴-[1,4,7,10-tetraazacyclododecane-1,4,7,10-tetrayltetrakis(methylene)]tetrakis[2-hydroxy-*N*-[(1*R*)-1-phenylethyl]carboxamide] hydrogenbromide (1:2) ligand, $H_4L^7{\cdot}2HBr$ [116]. Of special importance it was shown that the chiral structure of this complex was dependent on the nature of the base present in MeOH solution, since differences in the magnitude, sign, and shape of the CPL signals were observed between the complex solutions containing either 1% of $Et_3N$ or pyridine. Additionally, the observation of similar CPL spectra following direct and indirect excitations pointed to the presence of a single species in the solution that contained 1% of $Et_3N$. This was also corroborated by the fact that the $g_{lum}$ values were independent of the polarisation of the excitation beam, either right-, left-, or plane-polarised light. On the other hand, the CPL results following direct and indirect excitations (observation of two different spectra) of the complex solution in the presence of

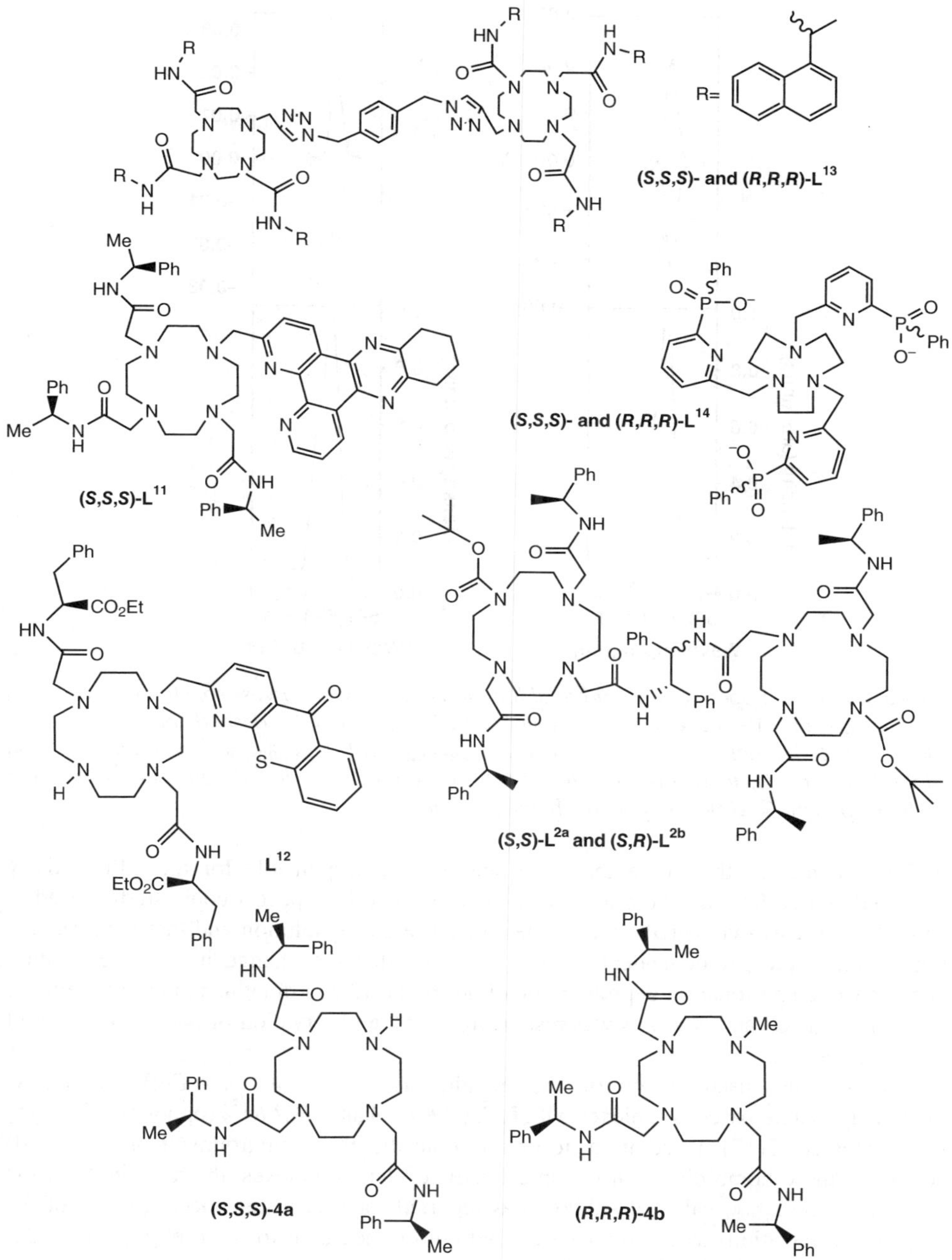


**Scheme 3.5** *Chemical structures of ligands*

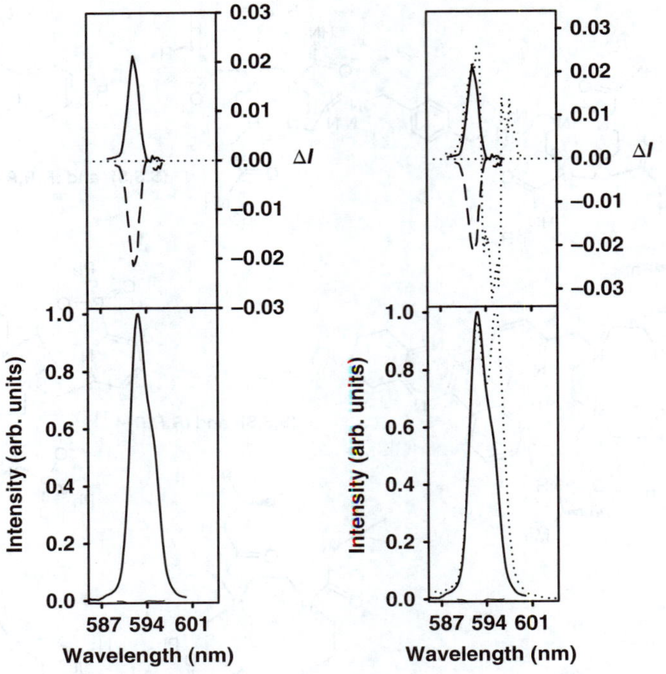


***Figure 3.8*** *CPL (upper curves) and total luminescence (lower curves) spectra for the* $^5D_0 \rightarrow {}^7F_1$ *transition of* (M)-[Eu$\mathbf{L}_{\mathbf{RRRRRR}}$]$^{3+}$ *(solid line) and* (P)-[Ln$\mathbf{L}_{\mathbf{SSSSSS}}$]$^{3+}$ *(dashed line) in* $1 \times 10^{-2}$ *M MeOH at 295 K (left) and a 3.5:1* (M)-[Eu$\mathbf{L}_{\mathbf{RRRRRR}}$]$^{3+}$:(P)-[Eu$\mathbf{L}_{\mathbf{RRRRRR}}$]$^{3+}$ *complex solution (dotted line, right), following excitation at 283–284 nm. Reproduced with permission from [115]. Copyright © 2008, American Chemical Society*

pyridine suggested that more than one species was responsible for the CPL activity observed. These findings were in agreement with the other spectroscopic studies used to show that at least two Tb(III)-like species were present in solution and that their relative populations were pH dependent. It was proposed that the change in the coordination geometry upon protonation/deprotonation of one of the 12 coordinating atoms consisting of eight and four O- and N-atoms was responsible for the observation of these two different complex species.

CPL was also used to confirm the enantiomeric nature of homo-Eu(III) chiral *N*-methylnaphtalene-based complexes with $\mathbf{L}^{13}$, (*R*,*R*,*R*)-**3** and (*R*,*R*,*R*)-**4** (opposite CPL signs were obtained) [117]. In addition, to demonstrate the chiral environment of the Eu(III) centres in these homo-dinuclear cationic Eu(III) cyclen complexes, the magnitude of the $g_{lum}$ values (i.e. $g_{lum}$ value of −0.063 was reported for **4**Eu in the spectral range of the $^5D_0 \rightarrow {}^7F_2$ transition) was also consistent with those measured for $C_3$ (i.e. $g_{lum}$ of −0.06) and $C_4$ symmetrical related phenyl based tri- and tetra-amide complexes [118,119].

Finally, a recent CPL study on chiral $C_3$-symmetric Eu(III) and Tb(III) complexes of a pyridylphenylphosphinate triazacyclononane based ligand ($\mathbf{L}^{14}$) showed that these systems with both enantiomers of $\mathbf{L}^{14}$ exhibited mirror image CPL spectra with strong CPL activities ($|g_{lum}|$ up to 0.4) [120]. It is interesting to note that this study also suggested that there was a correlation between the sign of the torsional angle NCCO or NCCN and the metal stereochemistry. That is, an (*R*) configuration at the P atom resulted in positive and negative signs for the ring NCCN and substituent $NCCN_{py}$ torsion angles, associated with a Λ configuration. This was in line with the findings of previous studies on Ln(III) complexes with $C_4$-symmetrical chiral tetraamide derivative ligands based on 1,4,7,10-tetraazacyclododecane, even though $\mathbf{L^{14}}$ did not exhibit chirality about an amide moiety (121, 122). It was shown that an (*R*) configuration at the C centre resulted in negative and positive signs for the ring NCCO and $C_4$-related NCCN torsion angles, associated with Λ and *δ* configurations, respectively. Although it is premature to generalise such a phenomenon to related chiral Ln(III) compounds, one may wonder if, for example, a negative NCCO or NCCN torsion angle is generally associated with an (*R*)-configuration. However, it is worth pointing out that the same behaviour was observed in each of these Ln(III)-containing systems [120–122]. That is, Δ or Λ stereochemistry about Ln(III) was selectively induced by ligand chirality (*S* or *R*, respectively). It is also worth pointing out that such a pattern may help in the ongoing studies aimed at developing a "helicity rule" that correlates the metal stereochemistry, namely Δ or Λ, to the CPL sign patterns. For instance, it was observed that the CPL sign sequence for the $^5D_0 \rightarrow {}^7F_1$ and $^7F_2$ transitions in the CPL spectra of *SSS*-Δ-[Eu$\mathbf{L^{14}}$] followed a similar behaviour to that observed with enantiopure $C_4$-symmetric Δ Eu(III) tetra-amide complex with an $N_4O_4$ donor set (positive and positive around 590 and 620 nm, respectively) [120–122]. Although the $M^I$[Eu{(+)-hfbc}$_4$] ($M^I$ = Na, K, Rb, and Cs) complexes may not present the same structure than those with the chiral macrocycle-type ligands (e.g. $\mathbf{L^{14}}$), it was recently shown that a positive CPL component at around 595 nm for the $^5D_0 \rightarrow {}^7F_1$ transition of $M^I$[Eu{(+)-hfbc}$_4$] correlated with a Δ-configuration for the local helicity environment around the Eu(III) ion [45]. It is interesting to note that an almost similar pattern was observed in the CPL for the Eu(III) complexes with the chiral 2-hydroxyisophthalamide octadentate ligand [*R*(+) and *S*(−)BnMeH22IAM] [105]. That is, a positive CPL sign was observed for the Δ-Eu(III) compound. To conclude, the combination of these observations may open new perspectives for the development of an empirical spectra:structure correlation based upon chiroptical spectral information (e.g. CPL).

## 3.5 Chiral Ln(III) Complexes to Probe Biologically Relevant Systems

Unlike the Pfeiffer-induced systems discussed in Section 3.4.1, this section details some of the successful recent examples demonstrating the design principles for targeting anions and/or proteins that are of biological relevance [direct coordination to the metal centre of the chiral Ln(III)-based probes], followed by a survey of similar studies for analyte sensing (i.e. cations and/or proteins) in which the coordination occurs between the analytes and some of the antenna/receptor groups of the chiral Ln(III)-based probes.

### 3.5.1 Sensing through Coordination to the Metal Centre

The sensing through coordination to the metal centre approach requires the use of a stable luminescent chiral Ln(III) complex with at least one open coordination site (Fig. 3.9). This site is occupied by a solvent molecule that can be readily displaced when an anion-like molecule is added to the solution of the coordinatively unsaturated chiral Ln(III)-based

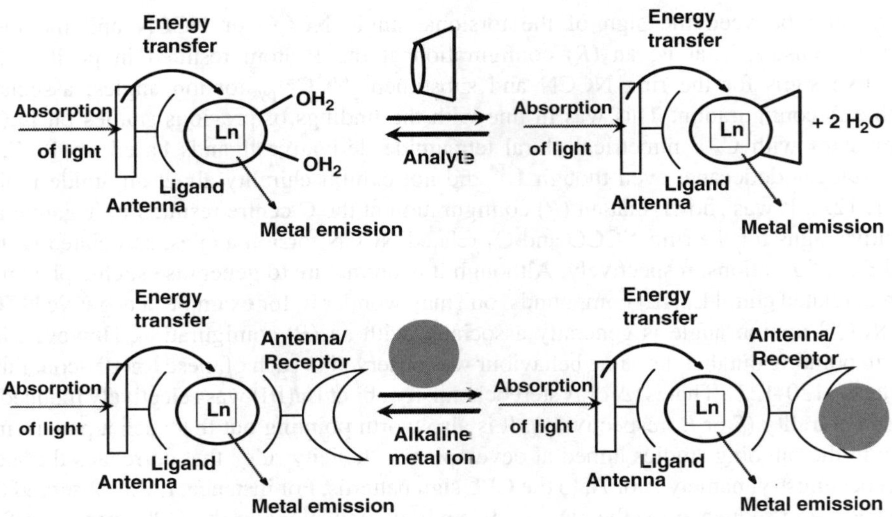


*Figure 3.9* *Schematic illustration on how to design Ln(III)-containing sensors based on the direct coordination of the targeted analyte to the Ln(III) ion (top) and on the covalently or coordinatively binding of the targeted cation to the antenna/receptor group (bottom)*

complex. The key point is that, upon coordination of a given anion to the Ln(III) ion, the structural changes in the new anion-bound complex formed result in a "fingerprint" chiral signal correlated to the added anion. Although the modulation of the luminescent and/or NMR spectral properties of the Ln(III) complex upon the binding of an anionic molecule is commonly used for sensing/recognition purposes, it is important to mention that such studies are not always conclusive. For instance, Atkinson et al. [123] showed that the binding of a series of *O*-phosphorylated amino acids to coordinatively unsaturated chiral DOTA-based macrocyclic, (*S,S,S*)-**4a** and (*S,S,S* and *R,R,R*)-**4b**, complexes of Eu(III) led to the observation of similar $^1$H-NMR spectral profiles, whereas their chiroptical study resulted in the observation of different CPL spectra (Fig. 3.10). This work also indicated that the Ln(III) complexes of interest had a preference for binding *O*-phosphono-L-tyrosine sites. Working along these lines, Murray et al. [63] investigated the anion affinity of ditopic systems based on (*S,S,S*)-**4a** with the expectation that it will allow simultaneous binding to two *O*-phospho-tyrosine sites. Although the results are of preliminary nature, the CPL spectra of the dinuclear Eu(III) aqua complexes with the stereoisomeric ligands (*SS*)-$\mathbf{L^{2a}}$ and (*SR*)-$\mathbf{L^{2b}}$ recorded in the presence and absence of the phosphoryllated glucose species revealed that the change in overall complex chirality was associated with the local helicity at the Eu(III) centre. Both Eu(III) centres in $[Eu_2\mathbf{L^{2b}}]^{6+}$ possess a Δ-configuration and, more importantly, behave independently and identically to the Eu(III) centre in the mononuclear Eu(III) complex of (*S,S,S*)-**4a** (similar CPL spectra were observed). On the other hand, the local helicities at the two Eu(III) centres in $[Eu_2\mathbf{L^{2a}}]^{6+}$ are opposite and, thus, leading to a reduced CPL signal (observation of different sign patterns and very weak CPL signals compared to those of $[Eu_2\mathbf{L^{2b}}]^{6+}$). These findings were attributed to the differing solution conformations of these Ln(III) complexes that were governed by the configuration of the

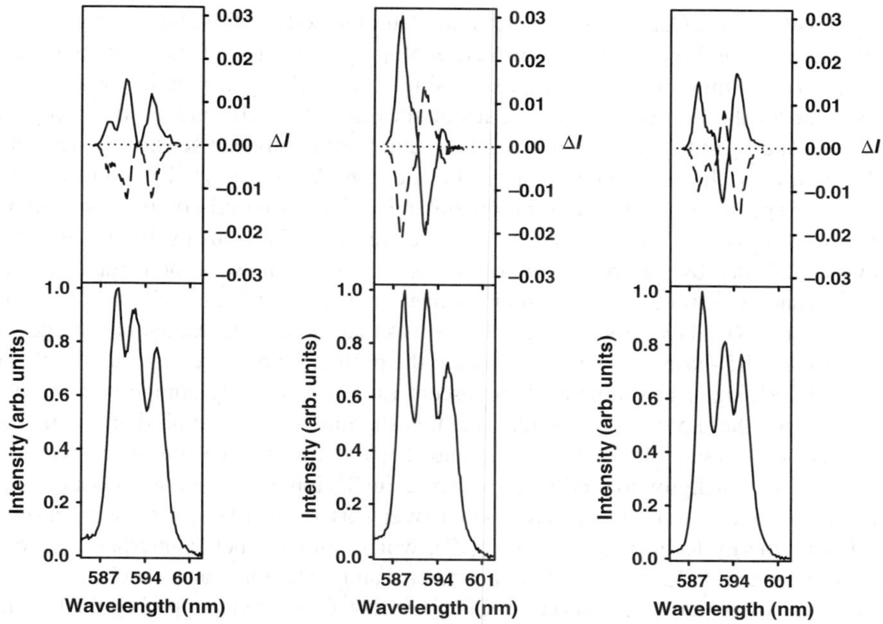


*Figure 3.10* CPL (upper curves) and total luminescence (lower curves) spectra for the $^5D_0 \rightarrow {}^7F_1$ transition of (R,R,R)-Λ-[Eu(**4b**)$(OH_2)_2]^{3+}$ (solid lines, 1 mM) and (S,S,S)-Δ-[Eu(**4b**)$(OH_2)_2]^{3+}$ (dashed lines, 1 mM) in the presence of 10 mM of N-acetyl-O-phosphono-L-tyrosine (left), D-glucose 6-phosphate (middle), and O-phosphono-L-tyrosine (right) anions, following excitation at the maximum of the $^5D_0 \leftarrow {}^7F_0$ transition ($\lambda_{exc}$ = 579.4 nm, pH 7.4, 293 K, 0.1 M MOPS). Reproduced with permission from [123]. Copyright © 2005, John Wiley

linking moiety present in the two diastereomeric ditopic ligands (*S,S*)-$\mathbf{L^{2a}}$ and (*S,R*)-$\mathbf{L^{2b}}$. These two ligands were prepared from the linking of (*S,S,S*)-heptadentate triamide derivative of cyclen with a central (*S,S*)- or (*S,R*)-1,2-diaminoethane derivative.

More recently, work performed by Montgomery et al. [124] resulted in the discovery of a very unique chiroptical probe for serum albumin binding. They showed that the Δ-conformer of the Ln(III) complexes with the enantiopure ligand (*S,S,S*)-$\mathbf{L^{11}}$ bound selectively to "drug site II" in serum albumin. The novelty was that this binding process resulted in an inversion of the complex helicity (Δ to Λ), as signaled by a sign change in the CPL spectra, and was only observed for the Eu(III) and Tb(III) compounds with (*S,S,S*)-$\mathbf{L^{11}}$ and its analogs. No dynamic helicity inversion was observed for the related complexes with (*R,R,R*)-$\mathbf{L^{11}}$, nor in the presence of excess of various substrates such as *B*- or *Z*-DNA, chiral anions (i.e. tartrate, lactate), (*S*)- or (*R*)-*α*-phenylsuccinate, or cyclodextrins. These findings were corroborated by the lack of changes in the sign and form of the measured CPL signals.

This fascinating study by Montgomery et al. [124] clearly opens opportunities for developing chiral luminescent Ln(III) complex probes that can be observed by microscopy and, more importantly, have a cell-penetrating feature in addition to their chiroptical recognition properties. This area of research is currently under investigation, as suggested by a series of articles published in the literature [11,13,125–131]. Most of these articles are mainly devoted to the elaboration of various strategies to address this challenging objective. For instance, the preliminary results of New et al. [128] seem to indicate that the observed properties (i.e. similar toxicity behaviour, similar quenching sensitivity, and common intracellular localisation profile) of a series of six pairs of Eu(III) and Tb(III) compounds of macrocyclic ligands, containing a common tetraazatriphenylene sensitiser, are dictated by the nature of the sensitising moiety. In addition, Yu et al. [131] showed that the intracellular speciation of their europium-containing DOTA-based complex (macrocyclic DOTA-based ligand $\mathbf{L}^{12}$) was consistent with the presence of various Eu(III)-based species. This was confirmed by the observation of distinctive CPL signals for their Eu(III) complex in the presence of citrate, L-Asp, L-Asn, malate, or oxalate and, also, for this compound localised in the NIT 3T3 cells. Of special note was that these CPL results may suggest that some of the Eu(III)-based species are bound to the proteins and, more importantly, that one can record CPL signals from chiral luminescent Ln(III)-containing compounds present in cells. Although these preliminary studies are investigating the most emissive Ln(III)-based probes [typically compounds of Eu(III) and Tb(III)] for the reasons discussed in the introduction and background section, the use of a Eu(III) ion in these systems is a key factor in the understanding steps due to its advantage over Tb(III). Eu(III) has, in theory, a simpler crystal field energy level pattern than Tb(III), which allows a better interpretation of the luminescence and CPL spectra. On the other hand, one may also envisage that the development of a new application of CPL, namely CPL microscopy, could be on its way at some point in the future.

### 3.5.2 Sensing through Coordination to the Antenna/Receptor Groups

As discussed above, the sensing/recognition properties of chiral Ln(III)-based probes operate through the coordination of the molecules of interest to the Ln(III) ions of coordinatively unsaturated Ln(III) systems. In this section, the development of chiral luminescent Ln(III)-based probes of relevant biological molecules uses an approach in which they bind to specific receptor groups of the Ln(III) species (Fig. 3.9). Thus, the changes in the chiroptical properties of the adduct formed are influenced by the recognition processes between the receptor groups and targeted analytes. The approach is that the targeted analyte (often a cation) covalently or coordinatively binds the antenna/receptor groups. This can be assimilated to the “derivatisation” (formation of covalent bonds) and “complexation” (bound via non-covalent interactions) methods, respectively. In addition to the usual excessive synthetic effort to prepare the Ln(III) complexes, the derivatisation technique is also complicated by other factors such as the use of a consequent amount of substrates, the need of coupling reactions and purification steps, the difficulty of substrate recovery, or the possibility of racemisations and/or optical resolutions. One can understand why the complexation method, which is based on non-covalent recognition interactions, is more popular for the sensing of cationic species such as metal ions. However, the result of the recognition (or binding) event is a modulation of the chiroptical properties of the Ln(III) complex-like species that can be detected by CD or CPL.

Since numerous research articles summarise the work of Parker et al. in this area [96,132–140], it is interesting to briefly discuss the recent findings of Kaizaki and his collaborators on the development of chiral Ln(III)-based systems to probe alkali metal ions. Working along these lines, Lunkley et al. [45,46] explained that the constant CPL activity of the tetrakis

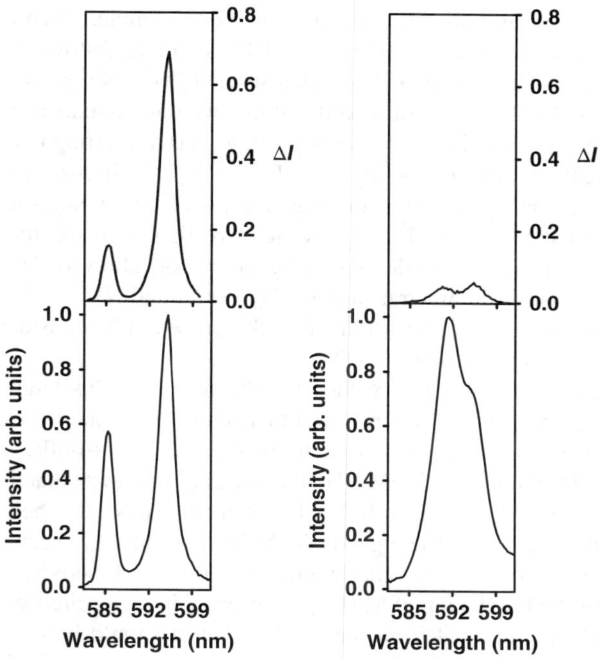


***Figure 3.11*** *CPL (upper curves) and total luminescence (lower curves) spectra for the* $^5D_0 \rightarrow {}^7F_1$ *transition of Cs[Eu{(+)-hfbc}₄] (left) and Na[Eu{(+)-hfbc}₄] (right) in 2 mM* $CHCl_3$ *at 295 K, following excitation at 352 nm. Reproduced with permission from [46]. Copyright © 2008, American Chemical Society and [50] Copyright © 2011, American Chemical Society*

(+)-hfbc Eu(III) complex was due to the helical Δ-SAPR(C4) arrangement of the four (+)-hfbc ligands in the compound of interest. In addition, they also showed that the solvent and alkali metal ion effects were due to the difference in susceptibility to solvation toward the alkali metal ion, but not toward the Eu(III) ion, unlike the case of axially symmetric DOTA-type complexes described by Bruce et al. [38,121]. This conclusion was corroborated by the observation that the helical Δ-SAPR($C_4$) arrangement of the four (+)-hfbc ligands in $CH_3Cl$ and EtOH accounted for the CPL pattern of the $M^I$-Eu complexes ($M^I$ = Na, K, Rb, and Cs), which in turn was controlled by the alkali metal ion sizes. The ninefold decrease in the magnitude of the CPL signal of the Na-Eu compound resulted from a less chiral environment for the Eu(III) ion in the Na-Eu compared to Cs-Eu (Fig. 3.11). In addition, the authors indicated that the constant and large CPL activity ($|g_{lum}|$ values up to 1.38) substantiated the stereospecific formation of chiral Δ-SAPR-($C_4$) configurations with the aid of $M^I \cdots$ FC (fluorocarbon) interactions (an example of the "complexation" approach), as revealed by the X-ray analysis more clearly than the exciton CD spectra [141,142].

In addition to these findings suggesting that such Ln(III)-based compounds can serve as probes for recognition/sensing of $M^I$ cations, the observation of differing CPL activities with variation of alkali metal ions and solvents opens new perspective for the studies aimed at understanding the chiroptical spectra–structure relationships and, in particular, their important roles in configurational chirality for, for example, chemical sensors. The $M^I[Eu\{(+)\text{-hfbc}\}_4]$ complexes showed that the CPL sign was following a parallel pattern amongst the various alkali metals and solvents studied ($M^I$ = Na, K, Rb, Cs, and $CHCl_3$, EtOH, and MeCN) [45,46]. It was observed that the two components at 585 and 595 nm of the magnetic dipole allowed $^5D_0 \rightarrow {}^7F_1$ were positive. It is interesting to note that the single and two components of the $^5D_0 \rightarrow {}^7F_2$ and $^5D_0 \rightarrow {}^7F_3$ transitions were also showing a similar pattern consisting of negative, and negative–positive CPL signs at around 612, and 650 and 653 nm for the various $M^I$ and solvents investigated, respectively. Although the CPL sign pattern result is comparable to the magnetic dipole allowed $^5D_0 \rightarrow {}^7F_1$ transition, one must be aware that these two transitions, $^5D_0 \rightarrow {}^7F_2$ and $^5D_0 \rightarrow {}^7F_3$, are not theoretically the most favourable chiroptical transition-like probes to be considered since they do not obey the magnetic dipole rule.

Since the magnitude of the $g_{lum}$ values for the magnetic dipole-allowed $^5D_0 \rightarrow {}^7F_1$ transition is so large (>1) and considering that this behaviour is parallel to the large exciton CD couplet observed for the systems of interest (i.e. $Cs[Eu\{(+)\text{-hfbc}\}_4]$ exhibits strong CPL and exciton CD activities in 2 mM $CHCl_3$), one may anticipate that the local symmetry of the Eu(III) site in $M^I[Eu\{(+)\text{-hfbc}\}_4]$ is retained. Thus, the helical Δ-SAPR($C_4$) arrangement of the four (+)-hfbc ligands in $M^I[Eu\{(+)\text{-hfbc}\}_4]$ accounts for the CPL pattern of the $M^I$-Eu complexes. In other words, the strong CPL activity observed supports that the chirality of the Eu(III) ion of $M^I[Eu\{(+)\text{-hfbc}\}_4]$ in the excited state corresponds to that in the ground state derived from the exciton CD. It should be mentioned that these findings were also corroborated by the detection of strong CPL ($|g_{lum}| > 1$) and exciton CD activities for the analogous complex with Sm(III) instead of Eu(III).

## 3.6 Concluding Remarks

Although the focus of this review was to exemplify some of the recent successful CPL studies published in the literature, one can envision that a growing interest in using such Ln(III)-based chiral systems for analytical and biomedical applications is on its way. However, the development of new applications taking advantage of the chiroptical properties is probably dependent to the technical advancement in the chiroptical spectroscopic tools such as CD and, more importantly, CPL (i.e. CPL imaging). Although CPL is still an underemployed technique, recent studies and developments suggest that CPL would be an attractive complementary tool to CD, a much less sensitive chiroptical absorption technique [13,36,45,86,124,131].

Another factor to the continued development of CPL is probably a significant advancement in the correlation between the chiral structure of Ln(III)-containing complexes and their CPL spectral features. Although some progress has been made toward this goal [45,46,53,54,121], the main requirements for such analyses are probably the use of enantiopure Ln(III)-containing edifices (chiral structure due to the arrangement of ligand molecules around the luminescent centre) or related compounds for which the presence of at least one asymmetric carbon atom in the ligand molecule(s) surrounding the Ln(III) ion results in a complete diastereomeric resolution of the Ln(III)-based systems. Additionally, these compounds need to possess a defined high-symmetry (i.e. $C_4$-symmetric species or at least a trigonal symmetry) and be conformationally rigid in solution on the time scale of the

CPL measurement. In other words, one may want to consider Ln(III)-containing edifices that possess a limited number of individual crystal field levels. That is the reason why the use of low-symmetric Ln(III) complexes would be more problematic, since they would result in more complex CD and CPL spectra and thus complicate the identification and/or assignment of the various components of the studied transitions. As already mentioned earlier, the choice of studying Eu(III) compounds would also lead to simpler CPL spectra since the splitting of the $^7F_j$ levels is less complicated than that for the other luminescent Ln(III) complexes.

### Acknowledgements

The author thanks all the co-workers and collaborating scientists involved in the various works described in this book chapter and cited in the reference section. Financial support from the National Institute of Health, Minority Biomedical Research Support (1 SC3 GM089589-04 and 3 S06 GM008192-27S1) and the Henry Dreyfus Teacher–Scholar Award is gratefully acknowledged. The author also thanks Mrs Françoise C. Muller for her technical help in the preparation of this book chapter.

## References

1. Maier NM, Franco P, Lindner W. Separation of Enantiomers: Needs, Challenges, Perspectives. J Chromatogr A 2001; 906:3–33.
2. Iisakka K. Single-Enantiomer Drugs: Starting Materials and Intermediates. Chirals 2003 (September):42–43.
3. Shinoda S, Tsukube H. Luminescent Lanthanide Complexes as Analytical Tools in Anion Sensing, pH Indication and Protein Recognition. Analyst 2011; 136:431–435.
4. Ye Z, Chen J, Wang G, Yuan J. Development of a Terbium Complex-Based Luminescent Probe for Imaging Endogenous Hydrogen Peroxide Generation in Plant Tissues. Anal Chem 2011; 83:4163–4169.
5. Hagan AK, Zuchner T. Lanthanide-Based Time-Resolved Luminescence Immunoassays. Anal Bioanal Chem 2011; 400:2847–2864.
6. Wang X, Wang X, Wang Y, Guo Z. Terbium(III) Complex as a Luminescent Sensor for Human Serum Albumin in Aqueous Solution. Chem Commun 2011; 47:8127–8129.
7. Liu X, Ye Z, Wei W, Du Y, Yuan J, Ma D. Artificial Luminescent Protein as a Bioprobe for Time-Gated Luminescence Bioimaging. Chem Commun 2011; 47:8139–8141.
8. Bünzli J-CG. Lanthanide Luminescence for Biomedical Analyses and Imaging. Chem Rev 2010; 110:2729–2755.
9. Eliseeva SV, Bünzli J-CG. Rare Earths: Jewel for Functional Materials of the Future. New J Chem 2011; 35:1165–1176.
10. New EJ, Parker D, Smith DG, Walton JW. Developments of Responsive Lanthanide Probes for Cellular. Curr Opin Chem Bio 2010; 14:238–246.
11. Thibon A, Pierre VC. Principles of Responsive Lanthanide-Based Luminescent Probes for Cellular Imaging. Anal Bioanal Chem 2009; 394:107–120.
12. Moore EG, Samuel APS, Raymond KN. From Antenna to Assay: Lessons Learned in Lanthanide Luminescence. Acc Chem Res 2009; 42(4):542–552.

13. Montgomery CP, Murray BS, New EJ, Pal R, Parker D. Cell-Penetrating Metal Complex Optical Probes: Targeted and Responsive Systems Based on Lanthanide Luminescence. Acc Chem Res 2009; 42:925–937.
14. dos Santos CMG, Harte AJ, Quinn SJ, Gunnlaugsson T. Recent Developments in the Field of Supramolecular Lanthanide Luminescent Sensors and Self-Assemblies. Coord Chem Rev 2008; 252:2512–2527.
15. Muller G. Luminescent Chiral Lanthanide(III) Complexes as Potential Molecular Probes. Dalton Trans 2009; 2009:9692–9707.
16. Stalcup AM. Chiral Separations. Annu Rev Anal Chem 2010; 3:341–363.
17. Berova N, Polavarapu PL, Nakanishi K, Woody RW, editors. Comprehensive Chiroptical Spectroscopy. John Wiley & Sons, Inc., New York; 2011.
18. Bertucci C, Pistolozzi M, De Simone A. Circular Dichroism in Drug Discovery and Development: an Abridged Review. Anal Bioanl Chem 2010; 398:155–166.
19. Wenzel TJ. Discrimination of Chiral Compounds Using NMR Spectroscopy. John Wiley & Sons, Inc., New York; 2007.
20. Polavarapu PL, Nafie LA, Berova N. Special Issue "Advances in Chiroptical Methods". Chirality 2009; 21(1E):E1–E312.
21. Berova N, Nakanishi K, Woody RW. Circular Dichroism: Principles and Applications. 2nd edn. John Wiley & Sons, Inc., New York; 2000.
22. Kusumi T, Ooi T, Ohkubo Y, Yabuuchi T. The Modified Mosher's Method and the Sulfoximine Method. Bull Chem Soc Jpn 2006; 79:965–980.
23. Dale JA, Mosher HS. Nuclear Magnetic Resonance Enantiomer Reagents. Configurational Correlations *via* Nuclear Magnetic Resonance Chemical Shifts of Diastereomeric Mandelate, *O*-Methylmandelate, and $\alpha$-Methoxy-$\alpha$-Trifluoromethylphenylacetate (MTPA) Esters. J Am Chem Soc 1973; 95:512–519.
24. Seco JM, Quiñoá E, Riguera R. The Assignment of Absolute Configuration by NMR. Chem Rev 2004; 104:17–117.
25. Seco JM, Quiñoá E, Riguera R. A Practical Guide for the Assignment of the Absolute Configuration of Alcohols, Amines and Carboxylic Acids by NMR. Tetrahedron: Asymmetry 2001; 12:2915–2925.
26. Marathias VM, Tawa GJ, Goljer I, Bach II AC. Stereochemical Identification of (*R*)- and (*S*)-Ibuprofen Using Residual Dipolar Couplings, NMR, and Modeling. Chirality 2007; 19: 741–750.
27. Schlingmann G, Roll DM. Absolute Stereochemistry of Unusual Biopolymers from *Ascomycete* Culture LL-W1278: Examples that Derivatives of (*S*)-6-Hydroxymellein Are Also Natural Fungal Metabolites. Chirality 2005; 17:S48–S51.
28. Barron LD, Buckingham AD. Vibrational Optical Activity. Chem Phys Lett 2010; 492: 199–213.
29. Polavarapu PL. Renaissance in Chiroptical Spectroscopic Methods for Molecular Structure Determination. Chemical Record 2007; 7:125–136.
30. Bringmann G, Maksimenka K, Mutanyatta-Comar J, Knauer M, Bruhn T. The Absolute Axial Configurations of Knipholone and Knipholone Anthrone by TDDFT and DFT/MRCI CD Calculations: a Revision. Tetrahedron 2007; 63:9810–9824.
31. Corradini R, Sforza S, Tedeschi T, Marchelli R. Chirality as a Tool in Nucleic Acid Recognition: Principles and Relevance in Biotechnology and in Medicinal Chemistry. Chirality 2007; 19:269–294.
32. Sforza S, Haaima G, Marchelli R, Nielsen PE. Chiral Peptide Nucleic Acids (PNAs): Helix Handedness and DNA Recognition. Eur J Org Chem 1999; 1999:197–204.
33. Wittung P, Nielsen PE, Buchardt O, Egholm M, Nordén B. DNA-Like Double Helix Formed by Peptide Nucleic Acid. Nature 1994; 368:561–563.

34. Polavarapu PL, Donahue EA, Shanmugan G, Scalmani G, Hawkins EK, Rizzo C, et al. A Single Chiroptical Spectroscopic Method May Not Be Able to Establish the Absolute Configurations of Diastereomers: Dimethylesters of Hibiscus and Garcinia Acids. J Phys Chem A 2011; 115:5665–5673.
35. Polavarapu PL. Why is it Important to Simultaneously Use More Than One Chiroptical Spectroscopic Method for Determining the Structures of Chiral Molecules? Chirality 2008; 20:664–672.
36. Riehl JP, Muller G. Circularly Polarized Luminescence Spectroscopy and Emission-Detected Circular Dichroism. In: Berova N, Polavarapu PL, Nakanishi K, Woody R, editors. Comprehensive Chiroptical Spectroscopy: John Wiley & Sons, Inc., New York; 2012, pp. 65–90.
37. Hopkins TA, Metcalf DH, Richardson FS. Intermolecular Chiral Recognition Processes Probed by Chiroptical Luminescence. Chirality 2008; 20:511–523.
38. Riehl JP, Muller G. Circularly Polarized Luminescence Spectroscopy from Lanthanide Systems. In: Gschneidner KA, Jr., Bünzli J-CG, Pecharsky VK, editors. Handbook on the Physics and Chemistry of Rare Earths. Amsterdam: North-Holland Publishing Company; 2005, pp. 289–357.
39. Maupin CL, Riehl JP. Circularly Polarized Luminescence and Fluorescence Detected Circular Dichroism. In: Lindon JC, Trantner GE, Holmes JL, editors. Encyclopedia of Spectroscopy and Spectrometry: Academic Press, New York; 2000, pp. 319–326.
40. Riehl JP. Chiroptical Studies of Molecules in Electronically Excited States. In: Purdie N, Brittain HG, editors. Analytical Applications of Circular Dichroism. Amsterdam: Elsevier Science B. V.; 1994, pp. 207–240.
41. Riehl JP, Richardson FS. Circularly Polarized Luminescence Spectroscopy. Chem Rev 1986; 86:1–16.
42. Richardson FS, Riehl JP. Circularly Polarized Luminescence Spectroscopy. Chem Rev 1977; 77:773–792.
43. Richardson FS. Selection Rules for Lanthanide Optical Activity. Inorg Chem 1980; 19:2806–2812.
44. Luk CK, Richardson FS. Circularly Polarized Luminescence and Energy Transfer Studies on Carboxylic Acid Complexes of Europium(III) and Terbium(III) in Solution. J Am Chem Soc 1975; 97:6666–6675.
45. Lunkley JL, Shirotani D, Yamanari K, Kaizaki S, Muller G. Chiroptical Spectra of a Series of Tetrakis((+)-3-heptafluorobutylyrylcamphorato)lanthanide(III) with an Encapsulated Alkali Metal Ion: Circularly Polarized Luminescence and Absolute Chiral Structures for the Eu(III) and Sm(III) Complexes. Inorg Chem 2011; 2011 (50):12724–12732.
46. Lunkley JL, Shirotani D, Yamanari K, Kaizaki S, Muller G. Extraordinary Circularly Polarized Luminescence Activity Exhibited by Cesium Tetrakis(3-heptafluoro-butylryl-(+)-camphorato) Eu(III) Complexes in EtOH and CHCl3 Solutions. J Am Chem Soc 2008; 130:13814–13815.
47. Hilmes GL, Riehl JP. Circularly Polarized Luminescence Spectroscopy of Mixed-Ligand Complexes of Europium(III) with 2,6-Pyridine-Dicarboxylic Acid and *L*-Malic Acid. Inorg Chim Acta 1987; 129:123–125.
48. Kaizaki S. Application in Stereochemical Analysis of Synthetic Compounds, Natural Products and Biomolecules. In: Berova N, Woody RW, Polavarapu PL, Nakanishi K, editors. Advances in Chiroptical Methods: John Wiley & Sons Ltd. Inc., New York; 2012, pp. 449–471.
49. Riehl JP, Kaizaki S. Absolute Chiral Structures of Inorganic Compounds. In: Bakac A, editor. Physical Inorganic Chemistry. John Wiley & Sons. Inc., New York; 2010, pp. 143–197.
50. Kuroda R, Saito Y. Circular Dichroism of Inorganic Complexes: Interpretation and Applications. In: Berova N, Nakanishi K, Woody RW, editors. Circular Dichroism: Principles and Applications. John Wiley & Sons, Inc., New York; 2000, pp. 563–599.
51. Lisowski J. Enantiomeric Self-Recognition in Homo- and Heterodinuclear Macrocyclic Lanthanide(III) Complexes. Inorg Chem 2011; 50:5567–5576

52. Yuasa J, Ohno T, Miyata K, Tsumatori H, Hasegawa Y, Kawai T. Noncovalent Ligand-to-Ligand Interactions Alter Sense of Optical Chirality in Luminescent Tris($\beta$-diketonate) Lanthanide(III) Complexes Containing a Chiral Bis(oxazolinyl) Pyridine Ligand. J Am Chem Soc 2011; 133:9892–9902.
53. Gawryszewska P, Legendziewicz J, Ciunik Z, Esfandiari N, Muller G, Piguet C, et al. On the Determination of Empirical Absolute Chiral Structure: Chiroptical Spectrum Correlations for $D_3$ Lanthanide (III) Complexes. Chirality 2006; 18:406–412.
54. Çoruh N, Hilmes GL, Riehl JP. Use of the Pfeiffer Effect to Probe the Optical Activity of Europium(III) Complexes with 2,6-Pyridinedicarboxylate. Inorg Chem 1988; 27:3647–3651.
55. Morley JP, Saxe JD, Richardson FS. The Chiroptical Luminescence Spectra of $Eu^{3+}$ in Trigonal $Na_3[Eu(oxydiacetate)_3]$-$2NaClO_4$-$6H_2O$. Mol Phys 1982; 47:379–406.
56. Sen AC, Chowdhury M, Schwartz RW. Optical Activity of Rare Earth Crystals. J Chem Soc, Faraday Trans 2 1981; 77:1293–1300.
57. Dickins RS, Parker D, Bruce JI, Tozer DJ. Correlation of Optical and NMR Spectral Information with Coordination Variation for Axially Symmetric Macrocyclic Eu(III) and Yb(III) Complexes: Axial Donor Polarisability Determines Ligand Field and Cation Donor Preference. Dalton Trans 2003; 2003:1264–1271.
58. Di Bari L, Pintacuda G, Salvadori P, Dickins RS, Parker D. Effect of Axial Ligation on the Magnetic and Electronic Properties of Lanthanide Complexes of Octadentate Ligands. J Am Chem Soc 2000; 122:9257–9264.
59. Tsumatori H, Harada T, Yuasa J, Hasegawa Y, Kawai T. Circularly Polarized Light from Chiral Lanthanide(III) Complexes in Single Crystals. Appl Phys Express 2011; 4:011601.
60. Schippers PH. PhD. Dissertation: University of Leiden, The Netherlands; 1982.
61. Steinberg IZ, Gafni A. Sensitive Instrument for the Study of Circular Polarization of Luminescence. Rev Sci Instrum 1972; 43:409–413.
62. Brittain HG. Techniques of Chiroptical Spectroscopy. Appl Spectrosc Rev 2000; 35 (3): 175–201.
63. Murray BS, Parker D, dos Santos CMG, Peacock RD. Synthesis, Chirality and Complexation Phenomena of Two Diastereoisomeric Dinuclear Lanthanide(III) Complexes. Eur J Inorg Chem 2010; 2010:2663–2672.
64. Harada T, Hasegawa Y, Nakano Y, Fujiki M, Naito M, Wada T, et al. Circularly Polarized Luminescence From Chiral Eu(III) Complex with High Emission Quantum Yield. J Alloys Compd 2009; 488:599–602.
65. Harada T, Nakano Y, Fujiki M, Naito M, Kawai T, Hasegawa Y. Circularly Polarized Luminescence of Eu(III) Complexes with Point- and Axis-Chiral Ligands Dependent on Coordination Structures. Inorg Chem 2009; 48:11242–11250.
66. Goto H, Akagi K. Optically Active Conjugated Polymers Prepared from Achiral Monomers by Polycondensation in a Chiral Nematic Solvent. Angew Chem 2005; 117:4396–4402.
67. Lama M, Mamula O, Kottas GS, Rizzo F, De Cola L, Nakamura A, et al. Lanthanide Class of a Trinuclear Enantiopure Helical Architecture Containing Chiral Ligands: Synthesis, Structure, and Properties. Chem Eur J 2007; 13:7358–7373.
68. Riehl JP, Richardson FS. Circularly Polarized Luminescence. Methods in Enzymology 1993; 226:539–553.
69. Anyanwu UK, Luu VQ, Muller G. Unpublished results. 2011.
70. Rexwinkel RB, Schakel P, Meskers SCJ, Dekkers HPJM. Time-Resolved Polarization of Luminescence Spectroscopy: an Accurate and Versatile Digital Instrument for the Sub-$\mu$s Time Domain. Appl Spectrosc 1993; 47 (6):731–740.
71. Metcalf DH, Snyder SW, Wu S, Hilmes GL, Riehl JP, Demas JN, et al. Excited-State Chiral Discrimination Observed by Time-Resolved Circularly Polarized Luminescence Measurements. J Am Chem Soc 1989; 111:3082–3083.

72. Barnett CJ, Drake AF, Mason SF. The Polarized Luminescence and Vibrational Optical Activity of Calycanthine. Bull Soc Chim Belg 1979; 88:853–862.
73. Brittain HG, Richardson FS. Circularly Polarized Emission Studies on the Chiral Nuclear Magnetic Resonance Lanthanide Shift Reagent tris(3-trifluoroacetyl-d-camphonato)Europium(III). J Am Chem Soc 1976; 98:5858–5863.
74. Maupin CL. Polarization-Selective and Wavelength-Selective Luminescence Spectroscopy of Lanthanide (III) Ions [Ph.D. Thesis]. Houghton: Michigan Technological University; 1999.
75. Bonsall SD, Houcheime M, Straus DA, Muller G. Optical Isomers of *N,N'*-bis(1-phenylethyl)-2,6-pyridinedicarboxamide Coordinated to Europium(III) Ions as Reliable Circularly Polarized Luminescence Calibration Standards. Chem Commun 2007; 35:3676–3678.
76. Coughlin FJ, Westrol MS, Oyler KD, Byrne N, Krami C, Zysman-Colman E, et al. Synthesis, Separation, and Circularly Polarized Luminescence Studies of Enantiomers of iridium(III) Luminophores. Inorg Chem 2008; 47:2039–2048.
77. Matsumoto K, Suzuki K, Tsukuda T, Tsubomura T. A Chiral 2,6-Bis(Oxazolinyl)Pyridine Ligand with Amide Groups to Form Isomorphous Complexes Through All the Lanthanoid Series. Inorg Chem 2010; 49:4717–4719.
78. Meskers SCJ, Riehl JP, Dekkers HPJM. Linearly Polarized Luminescence Spectra of Eu(2,6-pyridinedicarboxylate)$_3^{3-}$ in Hydroxylic Solution. Chem Phys Lett 1993; 216:241–246.
79. Dekkers HPJM, Moraal PF, Timper JM, Riehl JP. Optical Artifacts in Circularly Polarized Luminescence Spectroscopy. Appl Spectrosc 1985; 39:818–821.
80. Tinoco J, I., Ehrenberg B, Steinberg IZ. Fluorescence Detected Circular Dichroism and Circular Polarization of Luminescence in Rigid Media: Direction Dependent Optical Activity Obtained by Photoselection. J Chem Phys 1977; 66:916–920.
81. Field JE, Muller G, Riehl JP, Venkataraman D. Circularly Polarized Luminescence from Bridged Triarylamine Helicenes. J Am Chem Soc 2003; 125:11808–11809.
82. Tsumatori H, Nakashima T, Kawai T. Observation of Chiral Aggregate Growth of Perylene Derivative in Opaque Solution by Circularly Polarized Luminescence. Organic Lett 2010; 12:2362–2365.
83. Albertsson J. Structural Studies on the Rare Earth Carboxylates. 10. Crystal and Molecular Structure of Monoclinic Trisodium Tris(pyridine-2,6-dicarboxylato)ytterbate(III) 13-Hydrate. Acta Chem Scand 1972; 26 (3):985–1004.
84. Albertsson J. Structural Studies on the Rare Earth Carboxylates. 12. Crystal and Molecular Structure of Triclinic Trisodium Tris(pyridine-2,6-dicarboxylato)neodymate(III) 15-Hydrate. Acta Chem Scand 1972; 26 (3):1023–1044.
85. Albertsson J. Structural Studies on the Rare Earth Carboxylates. 4. Crystal and Molecular Structure of Orthorhombic Trisodium Tris(pyridine-2,6-dicarboxylato)ytterbiumate(III)-14-Hydrate. Acta Chem Scand 1970; 24 (4):1213–1229.
86. Do K, Muller FC, Muller G. A Promising Change in the Selection of the Circular Polarization Excitation Used in the Measurement of Eu(III) Circularly Polarized Luminescence. J Phys Chem A 2008; 112:6789–6793.
87. Muller FC, Muller G, Riehl JP. Emission Detected Circular Dichroism from Long-Lived Excited States: Application to Chiral Eu(III) Systems. Chirality 2007; 19:826–832.
88. Riehl JP. Excited State Optical Activity. Purdie N, Brittain HG, editors. Amsterdam: Elsevier; 1993.
89. Cantuel M, Bernardinelli G, Muller G, Riehl JP, Piguet C. The First Enantiomerically Pure Helical Noncovalent Tripod for Assembling Nine-Coordinate Lanthanide(III) Podates. Inorg Chem 2004; 43:1840–1849.
90. Gawryszewska P, Sokolnicki J, Dossing A, Riehl JP, Muller G, Legendziewicz J. Characterization of the Electronic Excited-State Energetics and Solution Structure of Lanthanide(III)

Complexes with the Polypyridine Ligand 6,6′-Bis[bis(2-pyridylmethyl)aminomethyl]-2,2′-Bipyridine. J Phys Chem A 2005; 109:3858–3863.
91. Belair SD, Maupin CL, Logue MW, Riehl JP. Analysis of the Temperature Dependence of the Racemization of Eu(III) Complexes Through Measurement of Steady-State Circularly Polarized Luminescence. J Luminesc 2000; 86:61–66.
92. Sokolnicki J, Legendziewicz J, Muller G, Riehl JP. The Luminescence, Molecular and Electronic Structure, and Excited State Energetics of Tris Complexes of 4-Phenylethynyl-2,6-pyridinedicarboxylic Acid with Eu(III) and Tb(III) Prepared in Sol-Gel. Optical Materials 2005; 27:1529–1536.
93. Thompson NC, Parker D, Schmitt-Willich H, Sülzle D, Muller G, Riehl JP. NMR and Chiroptical Examination of the Diastereoisomers of (*S*)-Eu-EOB-DTPA. J Chem Soc, Dalton Trans 2004; 12:1892–1895.
94. George MR, Golden CA, Grossel MC, Curry RJ. Modified Dipicolinic Acid Ligands for Sensitization of Europium(III) Luminescence. Inorg Chem 2006; 45:1739–1744.
95. Tsukube H, Shinoda S. Lanthanide Complexes in Molecular Recognition and Chirality Sensing of Biological Substrates. Chem Rev 2002; 102:2389–2403.
96. Parker D, Dickins RS, Puschmann H, Crossland C, Howard JAK. Being Excited by Lanthanide Coordination Complexes: Aqua Species, Chirality, Excited-State Chemistry, and Exchange Dynamics. Chem Rev 2002; 102:1977–2010.
97. Brittain HG. Circularly Polarized Luminescence Studies of Chiral Lanthanides Complexes. Pract Spectrosc 1991; 12:179–200.
98. Brittain HG. Circularly Polarized Luminescence Studies of Chiral Lanthanide Complexes. J Coord Chem 1989; 20 (4):331–347.
99. Muller G, Muller FC, Maupin CL, Riehl JP. The Measurement of the Fluorescence Detected Circular Dichroism (FDCD) from a Chiral Eu(III) System. Chem Commun 2005; 28:3615–3617.
100. Kirschner S. The Pfeiffer Effect, Outer-Sphere Complexation, and the Absolute Configuration of Coordination Compounds. J Indian Chem Soc 1974; 51:28–31.
101. Moussa A, Pham C, Bommireddy S, Muller G. Importance of Hydrogen-Bonding Sites in the Chiral Recognition Mechanism Between Racemic $D_3$ Terbium(III) Complexes and Amino Acids. Chirality 2009; 21:497–506.
102. Lunkley JL, Nguyen NG, Kosareff MN, Muller G. Unpublished results. 2012.
103. Muller G, Riehl JP. Use of Induced Circularly Polarized Luminescence (CPL) from Racemic $D_3$ Lanthanide Complexes to Determine the Absolute Configuration of Amino Acids. J Fluorescence 2005; 15:553–558.
104. Meskers SCJ, Dekkers HPJM. Enantioselective Quenching of Luminescence: Molecular Recognition of Chiral Lanthanide Complexes by Biomolecules in Solution. J Phys Chem A 2001; 105:4589–4599.
105. Petoud S, Muller G, Moore EG, Xu J, Sokolnicki J, Riehl JP, et al. Brilliant Sm, Eu, Tb, and Dy Chiral Lanthanide Complexes with Strong Circularly Polarized Luminescence. J Am Chem Soc 2007; 129:77–83.
106. Seitz M, Do K, Ingram AJ, Moore EG, Muller G, Raymond KN. Circularly Polarized Luminescence in Enantiopure Europium and Terbium Complexes with Modular, All-Oxygen Donor Ligands. Inorg Chem 2009; 48:8469–8479.
107. Seitz M, Moore EG, Ingram AJ, Muller G, Raymond KN. Enantiopure, Octadentate Ligands as Sensitizers for Europium and Terbium Circularly Polarized Luminescence in Aqueous Solution. J Am Chem Soc 2007; 129:15468–15470.
108. Samuel APS, Lunkley JL, Muller G, Raymond KN. Strong Circularly Polarized Luminescence from Highly Emissive Terbium Complexes in Aqueous Solutions. Eur J Inorg Chem 2010; 2010:3343–3347.

109. Pedziwiatr M, Kosareff NM, Muller G, Koposov AY, Nemykin VN, Riehl JP, et al. Preparation, Characterization, and Circularly Polarized Luminescence of Lanthanum and Europium 1,1′-Binaphthyl-2,2′-Diyl Phosphate Complexes. J Alloys Compnds 2008; 451:251–253.
110. Muller G, Schmidt B, Jiricek J, Hopfgartner G, Riehl JP, Bünzli J-CG, et al. Lanthanide Triple Helical Complexes with a Chiral Ligand Derived from 2,6-Pyridinedicarboxylic Acid. J Chem Soc, Dalton Trans 2001; 2001:2655–2662.
111. Hua KT, Xu J, Quiroz EE, Lopez S, Ingram AJ, Johnson VA, et al. Structural and Photophysical Properties of Visible- and Near-Infrared-Emitting Tris Lanthanide(III) Complexes Formed with the Enantiomers of *N,N′*-Bis(1-phenylethyl)-2,6-pyridinedicarboxamide. Inorg Chem 2012; 2012 (51):647–660.
112. Leonard JP, Jensen P, McCabe T, O'Brien JE, Peacock RD, Kruger PE, et al. Self-Assembly of Chiral Luminescent Lanthanide Coordination Bundles. J Am Chem Soc 2007; 129:10986–10987.
113. Lincheneau C, Destribats C, Barry DE, Kitchen JA, Peacock RD, Gunnlaugsson T. Lanthanide Directed Self-Assembly Synthesis and Photophysical Evaluation of Chiral Eu(III) Luminescent "Half-Helicates". Dalton Trans 2011; 40:12056–12059.
114. Kitchen JA, Barry DE, Mercs L, Albrecht M, Peacock RD, Gunnlaugsson T. Circularly Polarized Lanthanide Luminescence from Langmuir–Blodgett Films Formed from Optically Active and Amphiphilic $Eu^{III}$-Based Self-Assembly Complexes. Angew Chem Int Ed 2012; 51:704–708.
115. Gregoliński J, Starynowicz P, Hua KT, Lunkley JL, Muller G, Lisowski J. Helical Lanthanide(III) Complexes with Chiral Nonaaza Macrocycle. J Am Chem Soc 2008; 130:17761–17773.
116. D'Aléo A, Xu J, Do K, Muller G, Raymond KN. A [Cyclentetrakis(methylene)]tetrakis[2-hydroxybenzamide] Ligand That Complexes and Sensitizes Lanthanide(III) Ions. Helv Chim Acta 2009; 92:2439–2460.
117. Molloy JK, Kotova O, Peacock RD, Gunnlaugsson T. Synthesis of Luminescent Homo-dinuclear Cationic Lanthanide Cyclen Complexes Bearing Amide Pendant Arms Through the Use of Copper Catalysed (1,3-Huisgen, CuAAC) Click Chemistry. Org Biomol Chem 2012; 10:314–322.
118. Bruce JI, Dickins RS, Govenlock LJ, Gunnlaugsson T, Lopinski S, Lowe MP, et al. The Selectivity of Reversible Oxy-anion Binding in Aqueous Solution at a Chiral Europium and Terbium Center: Signaling of Carbonate Chelation by Changes in the Form and Circular Polarization of Luminescence Emission. J Am Chem Soc 2000; 122:9674–9684.
119. Beeby A, Clarkson IM, Dickins RS, Faulkner S, Parker D, Royle L, et al. Non-Radiative Deactivation of the Excited States of Europium, Terbium and Ytterbium Complexes by Proximate Energy-Matched OH, NH and CH Oscillators: an Improved Luminescence Method for Establishing Solution Hydration States. J Chem Soc, Perkin Trans 2 1999; 1999:493–503.
120. Walton JW, Di Bari L, Parker D, Pescitelli G, Puschmann H, Yufit DS. Structure, Resolution and Chiroptical Analysis of Stable Lanthanide Complexes of a Pyridylphenylphosphonate Triaza-cyclononane Ligand. Chem Commun 2011; 47:12289–12291.
121. Bruce JI, Parker D, Lopinski S, Peacock RD. Survey of Factors Determining the Circularly Polarized Luminescence of Macrocyclic Lanthanide Complexes in Solution. Chirality 2002; 14:562–567.
122. Dickins RS, Howard JAK, Maupin CL, Moloney JM, Parker D, Riehl JP, et al. Synthesis, Time-Resolved Luminescence, NMR Spectroscopy, Circular Dichroism and Circularly Polarised Luminescence Studies of Enantiopure Macrocyclic Lanthanide Tetraamide Complexes. Chem Eur J 1999; 5 (3):1095–1105.
123. Atkinson P, Bretonnière Y, Parker D, Muller G. NMR and Luminescence Binding Studies of Ytterbium, Thulium, and Europium Macrocyclic Complexes with Phosphorus(V) Oxy Anions. Helv Chim Acta 2005; 88:391–405.

124. Montgomery CP, New EJ, Parker D, Peacock RD. Enantioselective Regulation of a Metal Complex in Reversible Binding to Serum Albumin: Dynamic Helicity Inversion Signalled by Circularly Polarised Luminescence. Chem Commun 2008; 2008:4261–4263.
125. Bünzli J-CG. Lanthanide Luminescent Bioprobes (LLBs). Chem Letters 2009; 38 (2):104–109.
126. Pandya S, Yu J, Parker D. Engineering Emissive Europium and Terbium Complexes for Molecular Imaging and Sensing. Dalton Trans 2006; 2006:2757–2766.
127. Murray BS, New EJ, Pal R, Parker D. Critical Evaluation of Five Emissive Europium(III) Complexes as Optical Probes: Correlation of Cytotoxicity, Anion and Protein Affinity with Complex Structure, Stability and Intracellular Localisation Profile. Org Biomol Chem 2008; 6:2085–2094.
128. New EJ, Parker D, Peacock RD. Comparative Study of the Constitution and Chiroptical Properties of Emissive Terbium and Europium Complexes with a Common Tetraazatriphenylene Sensitiser; the Nature of the Sensitiser Determines Quenching Sensitivity and Cellular Uptake. Dalton Trans 2009; 2009:672–679.
129. Poole RA, Montgomery CP, New EJ, Congreve A, Parker D, Botta M. Identification of Emissive Lanthanide Complexes Suitable for Cellular Imaging that Resist Quenching by Endogenous Anti-Oxidants. Org Biomol Chem 2007; 5:2055–2062.
130. Montgomery CP, Parker D, Lamarque L. Effective and Efficient Sensitisation of Terbium Luminescence at 355 nm with Cell Permeable Pyrazoyl-1-Azaxanthone Macrocyclic Complexes. Chem Commun 2007; 2007:3841–3843.
131. Yu J, Parker D, Pal R, Poole RA, Cann MJ. A Europium Complex That Selectively Stains Nucleoli of Cells. J Am Chem Soc 2006; 128:2294–2249.
132. Parker D, Bretonnière Y. Luminescent Lanthanide Complexes as Sensors and Imaging Probes. In: Bogdanov AAJ, Licha K, editors. Molecular Imaging: Springer, Heidelberg; 2005. pp. 123–146.
133. Parker D. Luminescent Lanthanide Sensors for pH, $pO_2$ and Selected Anions. Coord Chem Rev 2000; 205:109–130.
134. Gunnlaugsson T, Leonard JP. Responsive Lanthanide Luminescent Cyclen Complexes: from Switching/Sensing to Supramolecular Architectures. Chem Commun 2005; 2005:3114–3131.
135. Bobba G, Bretonnière Y, Frias J-C, Parker D. Enantiopure Lanthanide Complexes Incorporating a Tetraazatriphenylene Sensitiser and Three Naphthyl Groups: Exciton Coupling, Intramolecular Energy Transfer, Efficient Singlet Oxygen Formation and Perturbation by DNA Binding. Org Biomol Chem 2003; 1:1870–1872.
136. Bobba G, Frias J-C, Parker D. Highly Emissive, Nine-Coordinate Enantiopure Lanthanide Complexes Incorporating Tetraazatriphenylenes as Probes for DNA. Chem Commun 2002; 2002:890–891.
137. Bobba G, Dickins RS, Kean SD, Mathieu CE, Parker D, Peacock RD, et al. Chiroptical, ESMS and NMR Spectroscopic Study of the Interaction of Enantiopure Lanthanide Complexes with Selected Self-Complementary Dodecamer Oligonucleotides. J Chem Soc, Perkin Trans 2 2001; 2001:1729–1737.
138. Bobba G, Kean SD, Parker D, Beeby A, Baker G. DNA Binding Studies of Cationic Lanthanide Complexes Bearing a Phenanthridinium Group. J Chem Soc, Perkin Trans 2 2001; 2001:1738–1741.
139. Beeby A, Dickins RS, FitzGerald S, Govenlock LJ, Maupin CL, Parker D, et al. Porphyrin Sensitization of Circularly Polarized Near-IR Lanthanide Luminescence: Enhanced Emission with Nucleic Acid Binding. Chem Commun 2000; 2000:1183–1184.
140. Govenlock LJ, Mathieu CE, Maupin CL, Parker D, Riehl JP, Siligardi G, et al. Stereodifferentiation and Base-Pair Selectivity in the Binding of Δ and Λ Cationic Lanthanide

Complexes to $[(CG)_6]_2$, $[(AT)_6]_2$ and CT-DNA. J Chem Soc, Chem Commun 1999; 1999 (17):1699–1700.
141. Shirotani D, Suzuki T, Yamanari K, Kaizaki S. Crystal Structure and Chiroptical Spectra of Sodium Tetrakis (+)-hfbc Pr(III) Complex. J Alloys Comp 2008; 451:325–328.
142. Shirotani D, Suzuki T, Kaizaki S. Novel Configurational Structures of Sodium Tetrakis(3-heptafluorobutylryl-(+)-camphorato) Ln(III) Complexes with a Trapped $Na^+$ by $Na^+\cdots FC$ Interactions in the Solid State and in Solution. Inorg Chem 2006; 45:6111–6113.

# 4

# Luminescence Bioimaging with Lanthanide Complexes

*Jean-Claude G. Bünzli*

*Swiss Federal Institute of Technology, Switzerland; and Korea University, Republic of Korea*

This chapter is dedicated to Prof. Huang Chunhui for her seminal contribution to the field

## 4.1 Introduction

A key challenge in biology and medicine is to understand the structure and functional properties of cells, organs, and living organisms. Focusing on medicine, imaging techniques are crucial for two aspects of disease treatment: diagnosis and assessment of the treatment efficiency. The development of diagnostic imaging techniques such as positron emission tomography (PET), X-ray computed tomography (CT), or computed axial tomography (CAT), X-ray angiography, magnetic resonance imaging (MRI), or ultrasound imaging and their associated contrast agents has had a tremendous impact on the progress in medicine during the past 30 years, particularly in the case of cancer detection and treatment. However improvements are still needed with respect to sensitivity, spatial resolution, penetration depth, and specificity. Presently, much interest is focused on two fast emerging areas. The first one is *theranostics*, which combines the delivery of a therapeutic drug with a diagnostic agent within the same dose administered to the patient. In this way, the diseased tissue, delivery kinetics, and treatment efficacy are monitored at the same time with the same biodistribution. Ultimately, there is hope to be able to adapt the therapy and the drug dose to the specific case, that is to pave the way for individualised medicine [1]. The second aspect

*Luminescence of Lanthanide Ions in Coordination Compounds and Nanomaterials*, First Edition.
Edited by Ana de Bettencourt-Dias.

deals with *molecular imaging* which aims at gaining *in vivo* target-specific information with high sensitivity; to meet this goal, it seems necessary to design probes mixing different modalities and signal processing [2].

Optical imaging, especially luminescence imaging, emerges as a vital component of the various techniques needed to meet the stringent requirements of modern medicine and biology [3]. Indeed, depending on its wavelength, light can penetrate deeply within biological tissues and its detection is highly sensitive since single-photon detection is achievable. Another advantage is that photons interact with electrons at the molecular level and techniques such as Förster resonant energy transfer (FRET) lead to the detection of molecular interactions, for example DNA hybridisation. Moreover, depending on the lifetime of the excited state generating light emission, time-domain information can be gained thanks to time-gated detection. Finally, since quantum yields are temperature dependent, recording of luminescence intensity variations within cells or tissues leads to thermal imaging. There are several types of optical luminescent probes.

Organic luminophores are well documented and highly emissive; they can also be easily derivatised to make them specific, for example cell penetrating agents localising in given organelles or bioconjugated chromophores for antigen detection. They usually are fluorescent molecules, that is the transition from the excited state to the ground state occurs without spin change ($\Delta S = 0$) and is therefore allowed according to selection rules. This translates into very short excited state lifetimes, on the order of nanoseconds or shorter. Time-resolved detection (TRD), which considerably enhances the signal to noise ratio by getting rid of the probe autofluorescence, therefore necessitates sophisticated and expensive experimental setups. Finally, organic chromophores are often subject to intense photobleaching and some of them are only operative during a few seconds, which is rather short for informative imaging.

Semi-conductor quantum dots (QDs) and their bioconjugates have been tested as alternative probes [4,5]. They consist of nanocrystals with 2–10 nm diameters made of group 12 and 15 elements (e.g. ZnS, CdS, CdSe, InAs or InP). If the size of the nanoparticles is smaller than a critical length known as the exciton Bohr radius (typically 10 nm), electrons are confined into the nanoparticles and their original electronic levels split leading to specific optical properties which can be tuned by varying the particle size. Quantum dots are highly luminescent with fairly sharp (fwhm 25–100 nm) emission bands ranging from 450 to 1250 nm and they display better photostability than organic chromophores. But, they suffer from some flaws such as: (i) the difficulty in getting near-infrared (NIR) emission in the range 0.8–1.0 μm ideal for deep tissue penetration, except with InAs, (ii) flickering of the emission when there are only a small number of QDs in the target material, and (iii) concern about their long-term toxicity, for instance through the release of toxic cadmium, arsenide or selenide ions [6].

Trivalent lanthanide ions, $Ln^{III}$, represent other substitutes to both organic luminophores and quantum dots in view of their remarkable optical properties enabling easy spectral and time discrimination of their emission bands which extend from UV through NIR spectral ranges depending on the ion [7,8]. Another noteworthy and decisive advantage of lanthanide complexes is their low exposition to photobleaching because their luminescence is sensitised through the organic ligands and energy transfer from the chromophore to the $Ln^{III}$ ion is fast enough to avoid photodegradation of the ligand [8]. The use of lanthanide luminescence for bioanalyses was pioneered by I. Hemmilä and co-workers at the end of the 1970s [9–11]. It is now ubiquitous for immunoassays run in every medical laboratory. Examples are testing for early stage of pregnancy by quantifying the human chorionic gonadotropin hormone, detecting the human immunodeficiency virus (HIV), or measuring the prostate-specific antigen (PSA) level in blood or urine. Lanthanide complexes, polyaminocarboxylates or macrocyclic complexes, are attached onto a specific antibody and luminescence from the emitting ion is detected after the biochemical reaction is completed, either in a two-step procedure (heterogeneous assay) or with a one-step, FRET-based protocol (homogeneous assay). The field is the subject of numerous review articles appearing regularly [12–14] and the availability of up-converting nanoparticles (UCNPs) on the one hand [15] and nanoparticles doped with luminescent lanthanide ions or complexes on the other hand [16–18] has added new dimensions to this vivid area of research. As the usefulness of lanthanide luminescent bioprobes and bioconjugates (LLBs) for immunoanalyses unfolded, attempts to apply them to imaging purposes were a logical extension. First experiments were simply taking advantage of either continuous-mode lanthanide emission, for instance for staining bacterial smears of *Escherichia coli* [19], or of time-resolved luminescence, for instance for evidencing fingerprints by laser excitation [20,21]. But since lanthanide complexes and/or their bioconjugates are cell permeable, luminescence microscopy [22] and more recently time-resolved luminescence microscopy (TRLM) images [16] started to be performed with the help of LLBs. While the number of papers dealing with LLB-assisted luminescence imaging remained small between 1988 and 2005 (fewer than 10 per year), it is now on exponential growth, with about 60 papers published in 2011 and more than 100 in 2012.

In this chapter, we review bioimaging experiments based on lanthanide luminescence, essentially restricting ourselves to the cases dealing with lanthanide complexes and single-photon excitation since multiphoton imaging is dealt with more completely in Chapter 5 while bioimaging with UCNPs is reviewed in Chapter 7. Literature is systematically covered up to the end of 2011 while selected newer entries are introduced in Section 4.3.

## 4.2 Luminescence Microscopy

### 4.2.1 Classical Optical Microscopy: a Short Survey

In a classical microscopy setup, images are generated by passing visible light transmitted or reflected by the sample into a magnifying lens system. The image is visualised by human eyes or detected by a charge-couple device (CCD) camera (or recorded on photographic plates or films). One limitation of the technique is intrinsic to visible light: in order to distinguish between two adjacent points in the sample, the illuminating light must have a wavelength on the order of magnitude of the distance between these points, so that resolution is limited to a few hundred Ångströms. More precisely, E. Abbe has demonstrated that monochromatic light converging to a point on a sample with an angle $\theta$ makes a spot with radius $r$:

$$r = \frac{\lambda}{2(n \sin\theta)} \tag{4.1}$$

in which $\lambda$ is the wavelength of the incident light and $n$ the refractive index of the sample. The quantity $n \sin\theta$ is called the numerical aperture (*NA*). In modern microscopes, *NA* can be as large as 1.3–1.4. The resulting resolution limit is called Abbe's diffraction limit. Rayleigh equation is often invoked to calculate the resolution of a microscope in terms of the distance $d$ between adjacent points perceived as separate:

$$d = 1.22\frac{\lambda}{2NA} = 1.22\frac{\lambda}{2(n\sin\theta)} \tag{4.2}$$

For getting higher resolutions, shorter wavelength light has to be used (UV, X-ray, high-energy electrons). Alternatively, special technologies such as interferometric microscopy or near-field techniques can substantially increase resolution, but at the cost of increased complexity.

There are several ways of recording an image. The simplest one is *bright field microscopy* in which the sample is illuminated by white light from below and the image is observed from above. Minimal sample preparation and simplicity are big assets of this technique. Also, low contrast is obtained for biological samples since most of their components are colourless. Staining with organic chromophores which selectively localise into organelles is therefore commonly performed. This sometimes implies fixing the cells; that is killing them. Other ways of recording microscopy images include *oblique illumination* which gives 3D appearance, *cross-polarised light illumination* which generates contrast from differences in light rotation in the sample, *dark-field illumination* which uses light scattered by the sample to produce contrast, *phase contrast microscopy* which translates differences in refractive index into differences in contrast, and various interference techniques [23].

### 4.2.2 Principle of Luminescence Microscopy

Intrinsically, measurement of luminescence intensity is more sensitive than measurement of light absorption, especially when weak signals are concerned. Very sensitive photomultipliers (PMTs) operated in photon counting mode, or CCD cameras detect a few photons per second when the dark noise is reduced by cooling the photocathode. Conversely determining weak absorption means comparing two very large signals, which is obviously less precise. With respect to microscopy, luminescence recording of the image results in enhanced contrast, comparable to the one obtained by the conventional techniques mentioned in Section 4.2.1. In this technology, luminescent stains are introduced into the sample and emission is stimulated by an incident beam of light. These stains are fluorescent organic molecules, phosphorescent transition metal complexes, quantum dots, doped nanoparticles, or their bioconjugates for improved specificity. In life sciences, the experimental setup corresponds usually to an *epifluorescence* microscope in which excitation light is directed onto the sample from above after going through a dichroic mirror and the objective (Fig. 4.1). The excitation wavelength is selected through adequate choices of the dichroic mirror and the excitation filter. The light emitted by the sample is focused onto the detector through the objective used for excitation and the spectral range is selected by the emission filter. A great deal of the excitation light is transmitted by the sample so that only reflected light accompanies sample fluorescence and the signal to noise ratio is therefore large, especially that the emission filter can help get rid of the spurious

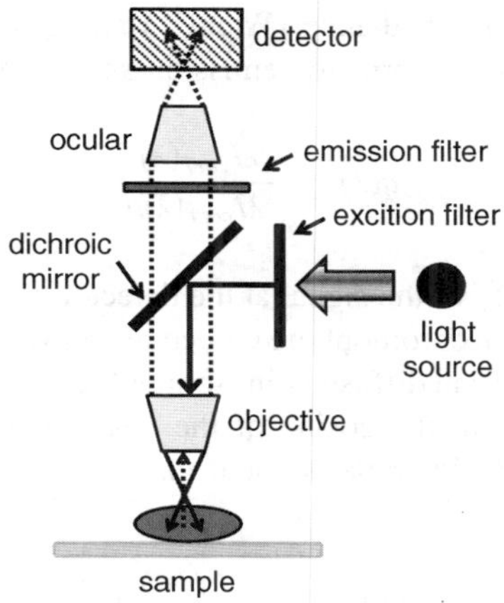


***Figure 4.1*** *Simplified scheme of an epifluorescence microscope*

excitation light. In case discrimination is not large enough, an additional dichroic mirror can be inserted in the light path to filter more efficiently the excitation light. The light source is either non-coherent (e.g. xenon lamp) or coherent (laser, diode laser), polarised or non-polarised depending on the effect sought for.

Luminescence microscopy is ideal when quantitative data are required although several precautions should be exercised. First, saturation of the signal must be avoided: for instance, for a PMT operated in photon-counting mode, linearity between the signal strength and the analyte concentration is lost when the intensity is larger than $10^6$ impulses per second (ips); reducing the illumination intensity usually cures the problem. Second, photobleaching of the luminophore has to be avoided, unless studying translational and rotational diffusion [24]; in this respect, LLBs are ideal. Third, light penetration-dependent effects (inner-filter effect) and light scattering have to be monitored. Fourth, staining by a luminophore may interfere with the cell functions or even may need fixing the cells; in any case, cytotoxicity and morphology tests should be performed.

The photometric throughput of a luminescence microscope is estimated as follows assuming linear and isotropic response and illumination low enough to avoid saturation [25]. Excitation light with wavelength λ is absorbed by the chromophores in the object field with a probability governed by their absorption cross section $\sigma_A(\lambda)$ in $cm^2$ $molecule^{-1}$ defined as:

$$\sigma_A(\lambda) = 10^3 \ln(10)\frac{\varepsilon_\lambda}{N} = 3.82 \times 10^{-21}\,\varepsilon(\lambda) \tag{4.3}$$

where $\varepsilon(\lambda)$ is the molar absorption coefficient ($M^{-1}\,cm^{-1}$) and $N$ is Avogadro's number ($molecule\ mol^{-1}$). The usefulness of the cross section is easy to understand since it avoids defining an absorption pathlength in the sample. In addition to $\varepsilon(\lambda)$, chromophores are characterised by their quantum yield $Q(\lambda_{exc}, \lambda)$ and act as secondary isotropic light sources with spectral radiant intensity $I(\lambda)$:

$$I(\lambda) = \frac{I_{exc}(\lambda)\sigma_A(\lambda)Q(\lambda_{exc}, \lambda)q(\lambda)}{4\pi} \tag{4.4}$$

where $I_{exc}(\lambda)$ is the excitation irradiance (W cm$^{-2}$) and $q(\lambda)$ is the spectral distribution function, which is related to the corrected emission spectrum $I_{corr}(\lambda)$:

$$q(\lambda) = \frac{\lambda I_{corr}(\lambda)}{\int \lambda I_{corr}(\lambda) d\lambda} \tag{4.5}$$

For calculating the intensity of the signal at the detector one has to take into account the solid angle of collection for the chromophore embedded into a medium with refractive index $n$ and acting as a non-Lambertian diffuse point source (i.e. the intensity of which in a given direction is not proportional to the cosine of the angle with respect to the direction of maximum irradiance). This solid angle is equal to:

$$2\pi\left[1 - \sqrt{1 - \left(\frac{NA}{n}\right)^2}\right] \tag{4.6}$$

Further assuming an overall transmittance $\tau_{opt}(\lambda)$ of the entire optical setup, a measuring spectral bandwidth $\Delta\lambda_{em}$, and a spectral sensitivity $\eta_{det}(\lambda)$ of the detector, one finally gets the photoelectron flux $\Phi_{im}$ (in photoelectrons s$^{-1}$ molecule$^{-1}$) in the image plane taking into account that the energy of a photon with wavelength $\lambda$ is $(1.987 \times 10^{-16}/\lambda)$ J $\approx (2 \times 10^{-16}/\lambda)$ J:

$$\Phi_{im} = 10^{16} Q(\lambda_{exc}, \lambda) \cdot [I_{exc}\sigma_A(\lambda)] \cdot [q(\lambda)\Delta\lambda_{em}] \cdot \left[1 - \sqrt{1 - \left(\frac{NA}{n}\right)^2}\right] \cdot \left[\tau_{opt}(\lambda)\eta_{\det}(\lambda)\frac{\lambda}{4}\right] \tag{4.7}$$

with quantities between brackets having to be evaluated by integrating over the entire spectral range. Typically, with a xenon lamp source delivering 25 W cm$^{-2}$ at the specimen and a 63× objective with 1.3 *NA*, this corresponds to about 25 photoelectrons s$^{-1}$ molecule$^{-1}$ if $Q(\lambda_{exc}, \lambda)$ is large.

As for detectors, PMTs have been used, but they are now commonly replaced by CCD cameras which have been introduced in 1969 at AT&T Bell Laboratories. They consist of closely spaced metal oxide/p-doped silicon (MOS) capacitors and an electric field shifts the created photoelectrons between the capacitive stages up to where they can be transformed into a digital signal. The main advantages of CCD detectors over PMTs are low background noise, lack of geometric distortion, and linearity of response over large spectral and dynamic ranges. These are convincing reasons why they are now also proposed as standard detectors for spectrofluorimeters, especially when quantitative measurements such as quantum yield determinations are required [26].

There are various techniques for recording luminescence microscopy images and for improving the contrast and/or the resolution. Their description is out of the scope of this chapter. It is however worth mentioning *confocal microscopy*. In this technique, conventional illumination in which the observed object is completely and evenly irradiated by a light source (wide field illumination) is replaced by point laser illumination. A scanning unit is inserted between the dichroic mirror and the objective and allows the illumination of successive points on the sample. Finally, a pinhole is introduced in front of the detection assembly for eliminating out-of-focus light rays. The name "confocal" stems from the fact that only light emitted very close to the focal plane in the sample can be detected. The setup results in increased optical resolution, though to the cost of signal intensity (mainly because of the pinhole) so that reconstructed 2D or 3D images require long exposure times. Confocal microscopy is now often combined with multi-photon excitation which protects live samples from photodegradation by exciting in the near-infrared instead of in the visible (usually blue) spectral range; this also results in deeper penetration of the excitation light into the sample [27].

### 4.2.3 Principle of Time-resolved Luminescence Microscopy

In the previous section, we have outlined that luminescence microscopy despite its larger sensitivity over conventional microscopy suffers from shortcomings, such as photobleaching of organic chromophores. Another major problem is that both organic and inorganic chromophores are ubiquitous in nature; excitation of the sample therefore provokes an intense luminescence stemming from the sample and its containment. This so-called *autofluorescence* severely obscures the emission from the chromophoric probe. A clever selection of filters and dichroic mirror may reduce the problem, but it never eradicates it. This difficulty can be circumvented if pulsed excitation is used and if images are produced based on the differences in exponential decays of the fluorescent stain and the autofluorescence (fluorescence lifetime imaging; FLIM). This is usually achieved with time-correlated single-photon counting (TCSPC) with resolution down to a few picoseconds or by phase modulation when fluorescence lifetimes are determined in the frequency domain by phase-modulated detection. Fluorescence lifetime imaging is compatible with FRET measurements. However the requirements of the instrumental setup are relatively sophisticated and a way must be found for transferring rapidly all of the data stored in the camera [28], so that another solution is emerging: luminescence imaging with phosphorescent bioprobes.

Autofluorescence occurs in a short span time (typically in the ns range), so that replacement of the fluorescent probes by phosphorescent stains considerably attenuates this problem if time-resolved (or time-gated) detection (TRD) is applied. The microscope is then operated in time domain and detects only the longer time-scale events. The principle of this powerful technique is outlined in Fig. 4.2.

Continuous illumination is replaced by pulsed illumination achieved either by a pulsed light source (flash lamp, laser) or by modulating a continuous excitation light beam with a chopper or a ferroelectric shutter. In addition, the detector is gated, that is turned on or off depending on the advancement of the experiment. The initial pulse defines $t(0)$; at that time, the detector is off. Intense autofluorescence is generated and the detector is maintained off until it disappears (gate or delay time). Phosphorescence from the luminescent probe develops more slowly and the detector captures it until it fades out (acquisition time). At that point the detector is again shut off and a new light pulse is shined onto the sample, restarting the cycle. Referring to lanthanide luminescent probes, the entire experiment lasts about 1–4 ms so that light measurement is repeated about 500 times per second or 3000 times per minute. Since in signal-averaging mode the signal to noise ratio (SNR) is increasing proportionally to the square root of the number of accumulations, it means that it can be improved by a factor of 10 in less than a second and by a factor of 100 in a bit more than

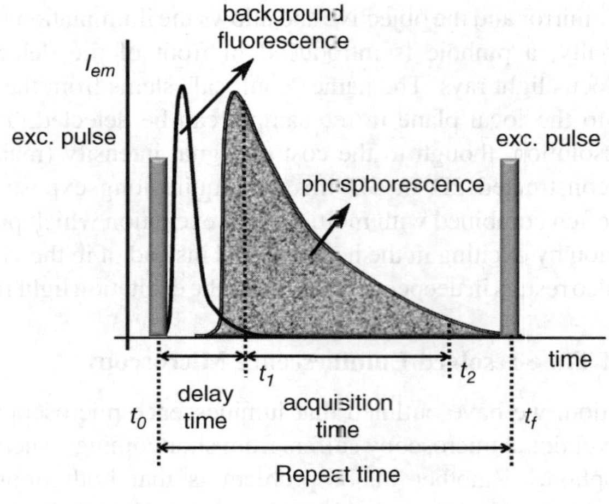


*Figure 4.2 Principle of time-resolved detection*

3 min. The latter time is too long for microscopy experiments, but integration times of this order of magnitude are common for time-resolved immunoassays, considerably boosting their sensitivity.

Both delay and acquisition times have to be chosen carefully in order to maximise SNR. For instance, a too long delay time cuts out the most intense emission from the phosphorescent probes and a too long acquisition time leads to noise integration at the end of the probe decay. If the decay is a single exponential function, the phosphorescence signal intensity decreases as:

$$I_{em} = I_{em}^0 \times e^{-t/\tau} + B \tag{4.8}$$

where $\tau$ is the lifetime of the chromophore and $B$ the background noise. Proportions of the total emission signal measured versus the acquisition time are collected in Table 4.1. From these simulated data it is clear that the delay time should not be longer than 0.1 $\tau$ while detecting light emitted after $3\tau$ does not bring much signal improvement.

***Table 4.1** Rounded proportions (%) of the total emission intensity in function of acquisition time ($t_2 - t_1$) for a normalised single-exponential luminescence decay with lifetime $\tau$ and $B = 0$ (see Equation 4.8)*

| $t_1/\tau \rightarrow t_2/\tau \downarrow$ | 0 | 0.1 | 0.2 | 0.5 | 1.0 | 2.0 |
|---|---|---|---|---|---|---|
| 1 | 63 | 54 | 45 | 24 | 0 | — |
| 2 | 87 | 77 | 68 | 47 | 23 | 0 |
| 3 | 95 | 86 | 77 | 56 | 32 | 9 |
| 4 | 98 | 89 | 80 | 59 | 35 | 12 |
| 5 | 99 | 90 | 81 | 60 | 36 | 13 |

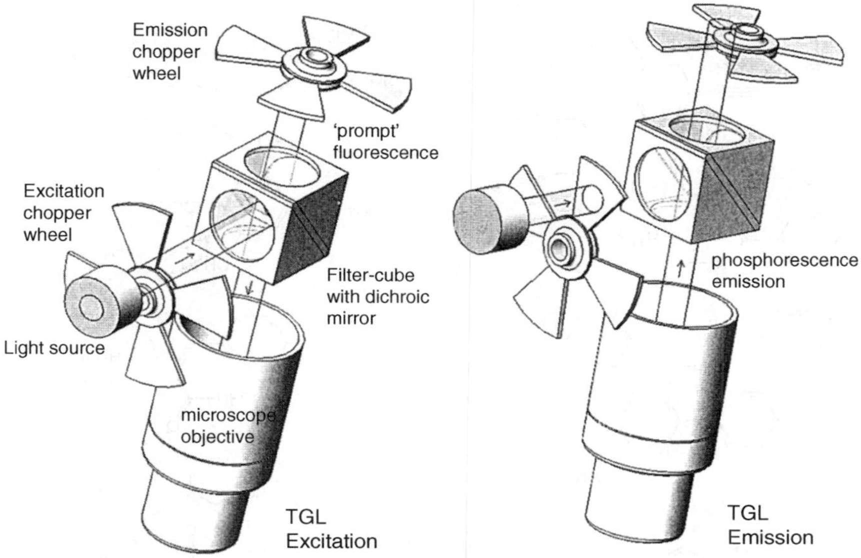


***Figure 4.3*** *Illustration of the isolation of excitation and detection phases in a TRLM setup. Left: Excitation mode with light source open and detector closed. Right: Emission mode with light source closed and detector open (TGL = time-gated luminescence). Reproduced with permission from [30]. Copyright © 2008, John Wiley*

Technically, early time-resolved luminescence microscopes employed two chopper wheels [29] to isolate the excitation and detection phases in a TRLM cycle as demonstrated in Fig. 4.3. The excitation chopper can however be advantageously replaced with a pulsed light source.

Another point stemming out of Table 4.1 is that, ideally, the delay and acquisition times should be adapted to the decay time of the luminescent probe. This is not quite easy with chopper-fitted microscopes since the delay and acquisition times not only depend on the rotation speed of the chopper but, also, on the number of blades. Optimum working conditions would therefore require changing the chopper wheel, depending on the luminescent stain, which is not very practical. Microscopes are therefore usually fitted with chopper wheels adequate for one class of LLBs (e.g. lanthanide polyaminocarboxylates, with lifetimes around 600–700 μs). There are other ways of continuously modifying the time delay, but they also have their drawbacks (see below).

The significant improvement in SNR brought by time-resolved detection is a welcome bonus when trying to identify rare targets in biological or environmental samples. A classic example is the detection of *Giardia lamblia* cysts [31] and *Cryptosporidium* oocytes [32], two important pathogens, in drinking water. The analysis requires filtering large volumes of water which leads to residues containing large amounts of minerals, algae and other plant matter, all highly fluorescent. Time-resolved microscopy considerably improves the sensitivity of the analysis while simplifying it: 30-fold improvements in SNR have been reported. In this context, LLBs with their long lifetimes, in the μs to ms range, are ideal chromophores for these experiments.

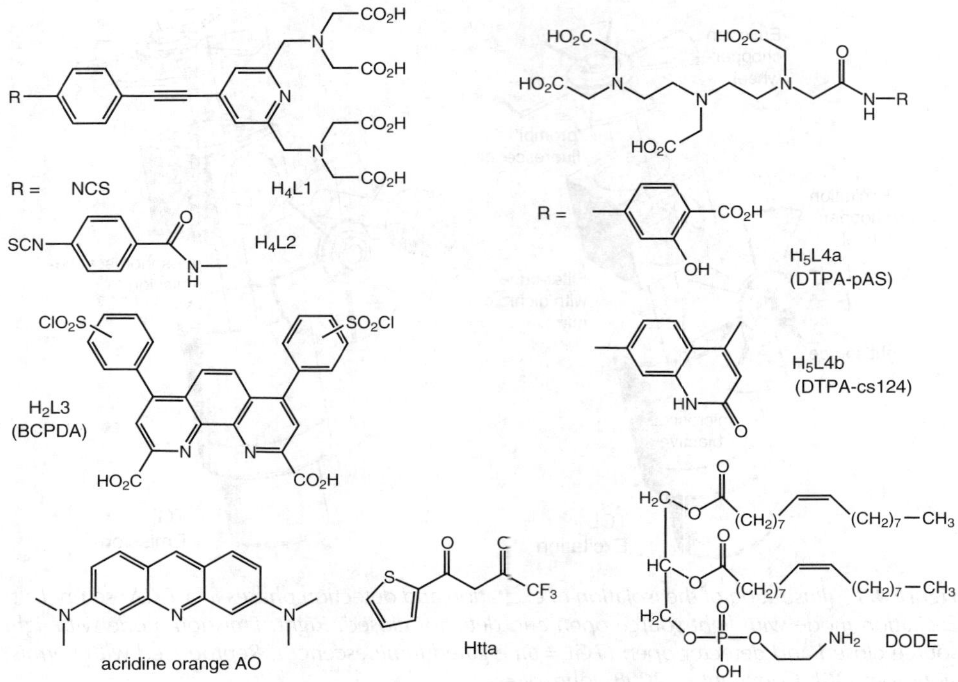


***Scheme 4.1*** *Carboxylate and diketonate ligands used in early imaging experiments with lanthanide luminescent bioprobes*

Given the nature of lanthanide luminescence, with large Stokes's shifts when the chromophores are excited in the organic ligand, simple and clever usage of appropriate filters already leads to background suppression. Further selectivity and discrimination in detection can be gained in combining both spectral and time resolution, as well as spatial and polarisation discrimination. This multiple discrimination technique was demonstrated on crystals of $[EuL4b]^{2-}$ (see Scheme 4.1) [33].

### 4.2.4 Early Instrumental Developments for Time-resolved Microscopy with LLBs

Early experiments on cell and/or organism imaging with the help of lanthanide chelates did not take advantage of time-resolved detection. We however briefly describe these works for historical reasons. To our knowledge, the first successful attempt was the visualisation of *Bacillus mageterium* in 1969. Suspensions of the bacillus were lysed, fixed with formaldehyde and stained with aqueous mM solutions of $Eu^{III}$ and 4,4,4-trifluoro-1-(2-thienyl)-1,3-butanedionate (tta) at pH 7.4. Photomicrographs of whole cells were recorded upon excitation at 365 nm with a xenon lamp and with an exposure time of 8 s (Figure 4.4). Images of protoplasts were also obtained. Interestingly, replacing the tris(β-diketonate) complex with europium chloride produced similar images. In contrast, *Escherichia coli* treated in the same way produced only faint luminescence, which led the authors to conclude that $Eu^{III}$ does not bind to *E. coli* cell walls [19]. Another conclusion of the paper is of general interest in that it is recommended to avoid phosphate buffers such as PBS when

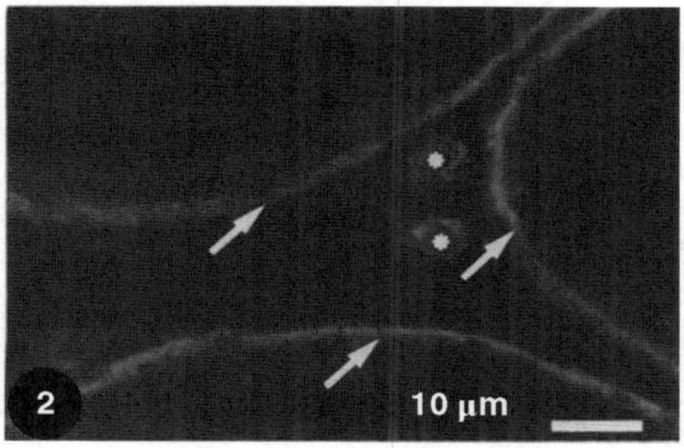


***Figure 4.4*** *Immunocytochemical demonstration of smooth muscle myosin in adult rat epididymis by luminescence of the [Eu(L1)]⁻ chelate. Adapted with permission from [22]. Copyright © 1988, The Histochemical Society*

lanthanide probes are provided because the strong interaction of phosphate with $Ln^{III}$ ions leads to dissociation of the chelates.

This initial work does not seem to have had a follow-up until 1988 when the Finnish group who invented time-resolved luminescence immunoassays at Wallac Oy (Turku, Finland) tested the ability of chelate W1014 = $[Eu(L1)]^-$ (see Scheme 4.1) to enhance contrast in steady-state luminescence microscopy. The chelate was coupled to the IgG fraction of swine anti-rabbit IgG with the help of a bromoacetyl reaction, followed by gel filtration. Immunocytochemical reactions were performed on fixed sections of adult rat epididymis and the sections were analysed with an epifluorescence microscope fitted with a 400-nm short-pass (SP) filter and a 430-nm long-pass (LP) filter. Smooth muscle myosin in the epididymis was illuminated by the europium conjugate (Fig. 4.4) and in view of this achievement, the authors proposed the construction of a time-resolved microscope [22]. Another type of early work with lanthanide probes is exemplified by the investigation of $Eu^{III}$ binding sites on the walls of *Datura innoxia* by means of laser-excited excitation spectra of the $Eu(^5D_0 \leftarrow {}^7F_0)$ transition and luminescence lifetimes [34].

About at the same time, one notes the first use of near-infrared $Yb^{III}$ luminescence for the detection of malignant tumours in C-57 dark-fur mice [35]. Administration of free porphyrin radicals is replaced with $Yb^{III}$ haematoporphyrins to eliminate the masking effect of the background luminescence, since $Yb^{III}$ is emitting at 980 nm, spectrally far from the autofluorescence of the sample. *In vivo* luminescence localisation of tumours could be achieved with tetramethyl-haematoporphyrin $H_2P1$, 2,4-di($\alpha$-methoxyethyl)-deuteroporphyrin $H_4P2$, *meso*-tetra(*p*-carboxyphenyl)porphyrin $H_6P3$ or tetrasulfophenylporphyrin $H_6P4$ (Scheme 4.2). In view of the poor solubility of the complexes, they were introduced into phosphatidylcholine liposomes. A fibre-laser luminescence spectrometer has been especially designed to monitor the $Yb^{III}$ luminescence from sarcoma-implanted mice. Both the central and peripheral regions of the tumours were measured, as well as surrounding healthy skin, which allowed the determination of a contrast ratio $\gamma_S$ from the total emission areas $S$ (emission from tumour) and $S_0$ (emission from healthy skin):

$$\gamma_S = \frac{S - S_0}{S} \tag{4.9}$$

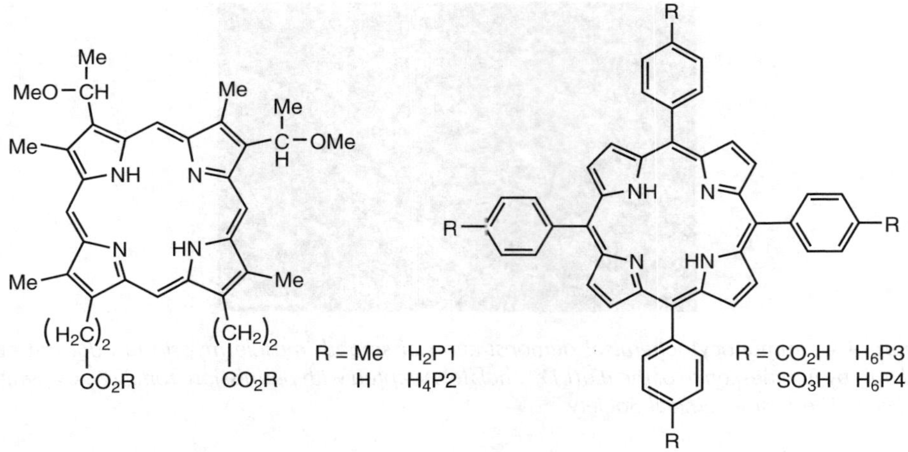


*Scheme 4.2* Porphyrins used for tumour detection by $Yb^{III}$ luminescence

The complexes $[Yb(P1)]^{+}$ and $[Yb(P3)]^{3-}$ yielded a contrast value $\gamma_S = 10 \pm 2$, while the two other complexes were more efficient, with $\gamma_S$ values equal to $19 \pm 4$, and $45 \pm 9$ for $[Yb(P2)]^{-}$ and $[Yb(P4)]^{3-}$, respectively [35,36]. Here again, this initial success was not followed by further experiments until recent times.

It was soon realised that TRD will considerably improve the quality of luminescence microscopy images and the first report describing the design of a time-resolved microscope used in conjunction with $Eu^{III}$ probes stemmed from the group of H. J. Tanke and was published in 1990 [16]. In this work, inorganic phosphors, $Y_2O_2S$:Eu conjugated to various proteins, are proposed as lanthanide luminescent bioprobes. The parent phosphor itself emits at 620 nm upon near-UV excitation (360 nm), with a typical lifetime of 0.7 ms. $Eu^{III}$-doped yttrium oxysulfide nanoparticles were milled to an average size of 100–300 nm in the presence of polycarboxylic acid to avoid aggregation and subsequently conjugated to avidin, protein A, or IgGs by simple adsorption on their surface. Performance of these conjugates was evaluated in three different systems: (i) latex beads with polyclonal goat anti-mouse IgG covalently attached to their surface, (ii) fixed human peripheral blood lymphocytes, and (iii) fixed human erythrocytes. TRLM was performed with a home-modified epifluorescence microscope equipped with a 50 Hz xenon flash lamp synchronised to a mechanical chopper with variable velocity between 1500–6000 rpm so as to generate adjustable delays between 100 and 2000 μs. Position of the blade was detected by an optical sensor and its signal was fed into an electronic delay circuit controlling the flash lamp. Images were visualised with the naked eye and/or a photographic camera. The inorganic-based conjugates performed well as demonstrated by specific immunocytochemical reactions on the three systems mentioned above. In the case of fixed human lymphocytes stained for leu-3a antigen, the conjugated inorganic phosphors were examined under TRD and $Eu^{III}$ luminescence only was observed. It is noteworthy that the delay time ($t_1 = 250\,\mu s$, $t_1/\tau \approx 0.35$) and the acquisition time ($t_2 = 4.71$ ms, $t_2/\tau \approx 6.7$) used ensure sizeable light collection (see Table 4.1). The setup was later fitted with a CCD camera and several other immunophosphors were successfully tested, including $Y_2O_2S$:Tb, ZnS:Ag, ZnS:Cu, and $Zn_2SiO_4$:Mn,As [37].

One year later, G. Marriott et al. [38] also described modification of an epifluorescence microscope, but deviating from the previous setup in three ways. First, the pulsed excitation source was replaced with a continuous source (Hg arc lamp or argon laser) focused on the blades of a rotating chopper phased-locked to the emission chopper (see Fig. 4.3). Second, an acousto-optic modulator (AOM) was incorporated into the excitation beam as a convenient way of easily varying the length of the excitation pulse; the AOM was triggered by the electronics of the excitation chopper and was only actuated during the time excitation light goes through the chopper. Finally, the first-order diffraction exiting the AOM was selected with an iris and directed into a quartz fibre which funnelled light to the sample. In this way, the delay time was electronically continuously adjustable. A filter wheel was implanted in front of the excitation chopper, itself positioned very close to the image plane of the CCD camera (Fig. 4.5). The sensitivity of the setup, determined as the rejection of the prompt image, was one part in $10^7$ and detection could start as close as 50 μs after the end of the excitation pulse without interference from diffuse scattering or prompt fluorescence.

The microscope was tested on solutions of acridine orange (AO) bound to RNA and DNA as well as on 3T3 cells stained with AO. The latter easily permeates into living mammalian

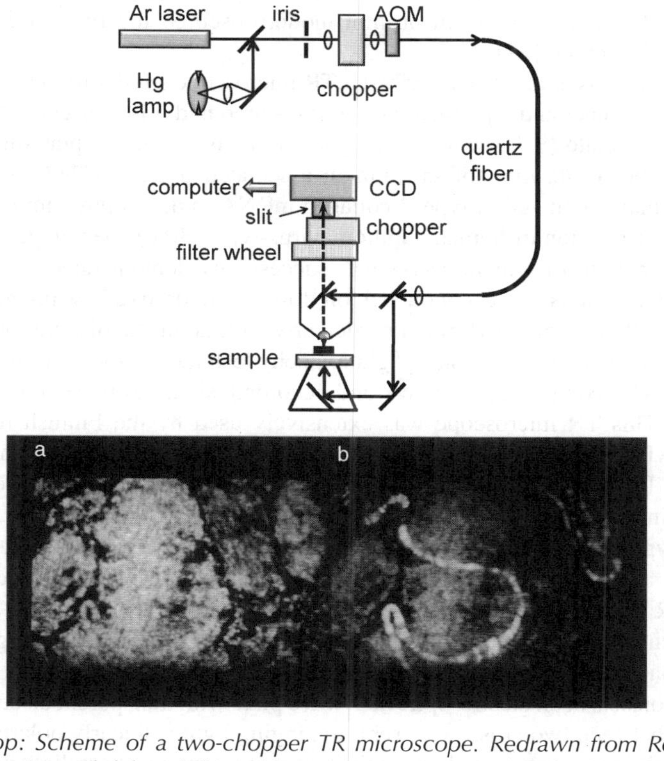


***Figure 4.5*** *Top: Scheme of a two-chopper TR microscope. Redrawn from Reference [38]. Bottom: Prompt (a) and delayed (b) red luminescence (>590 nm) of an AO-stained polytene chromosome covered with tissue (excitation wavelength: Ar laser, 488 nm). Reproduced with permission from [38]. Copyright © 1991, Elsevier*

cells, localising in several organelles. In the presence of oxygen, AO and AO intercalated into DNA or RNA showed predominant green fluorescence (520 nm, $\tau = 1.7$–$1.9$ ns) and at dye concentrations >10 µM (allowing aggregation to take place), weak red fluorescence ($\tau = 5$–$20$ ns). In anoxic solutions, red phosphorescence was detected with a lifetime of 1 ms. For live cells, weak red phosphorescence was detected in the nucleus despite its intensity being about $10^2$–$10^4$ smaller than the prompt fluorescence. TR images revealed interesting details and additional information which could not be obtained in prompt mode. Furthermore, a polytene chromosome covered with tissue could clearly be imaged by AO staining (Fig. 4.5).

The same microscope, operated with a 100 W mercury lamp was used with a lanthanide bioconjugate, named streptavidin-based macromolecular complex (SBMC) featuring streptavidin (SA) cross-linked to thyroglobulin and multiply labelled with the $Eu^{III}$-BCPDA chelate synthesised by R.A. Evangelista et al. in 1988 [39] for immunoassays (see Scheme 4.1). The bioconjugate displayed reasonable photostability, losing 32% of emission intensity after 5 min irradiation with a 340 nm wavelength (10 nm bandpass). Several imaging experiments were conducted in TR mode, in particular, the distribution of biotinylated concanavalin A on the surface of living cells of *Dictyostelium discoidenum* amoebae. In order to demonstrate the power of TRD, the cells were simultaneously incubated with Lucifer yellow, an organic stain displaying a strong fluorescence signal in pinocytosed vesicles which, in prompt mode, obscured the $Eu^{III}$ emission, but was suppressed in TR mode [29].

In 1992, L. Seveus et al. [40] described a TR microscope similar to the one proposed by the group of H. Tanke and equipped with a flash-lamp and a Peltier-cooled CCD camera. The $[Eu(L2)]^-$ chelate (Scheme 4.1) conjugated to antisera and streptavidin proved being adequate for the localisation of the tumour-associated antigen C242 in the malignant mucosa of human colon and of type II collagen mRNA in developing human cartilaginary growth plates. Detection of human papillomavirus-type (HPV) specific gene sequences in squamous epithelium of human cervix was successfully demonstrated as well. Immunohistochemical reactions were conducted on formaldehyde-fixed, wax-embedded tissue sections. The TR Eu-probe performed particularly well upon stabilisation of the cells after the immunocytochemical reaction by glutaraldehyde fixation: the resulting strong autofluorescence was completely avoided and recorded signal-to-noise ratios were up to 2400:1 [41]. This TR microscope was extensively used by the Finnish researchers, for instance for visualising individual PSA molecules covalently linked to one nanoparticle containing a $Eu^{III}$ tris(β-diketonate) [42]. It was the origin for the development of a commercial instrument by Wallac Oy (Turku, Finland; later bought by PerkinElmer). Some prototypes were built with the TR electronics taken from the TR luminescence spectrometer sold by the same company (Victor® series) but for undisclosed reasons, the Signifier® TR microscope (see Fig. 4.6) was never marketed.

Ferro-electric liquid crystals (FLC) represent an alternative to pulsed light sources for eliminating one of the choppers since their closing speed is faster, around 50–80 µs. The FLC device consists in an electro-optic half-wave plate with transparent electrodes mounted at both sides between two cross polarisers. Impinging light is linearly polarised by the first polariser. Applying a negative voltage to the half-wave plate results in light passing through the cell but being blocked by the second polariser, while a positive voltage alters the polarisation angle by 90° so that light can now pass through the second polariser. In a 1994

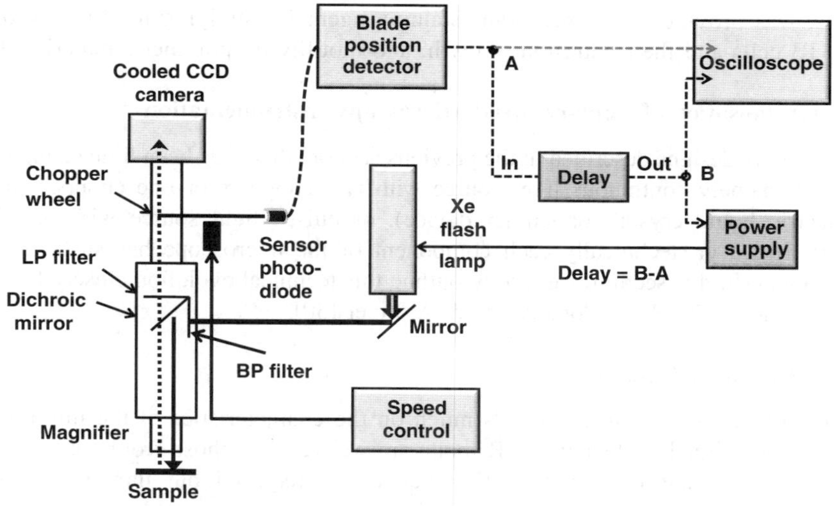


***Figure 4.6*** *Simplified scheme of a flash-lamp operated time-resolved microscope*

microscope design for time-resolved total internal reflection fluorescence microscopy (TR-TIRFM), a FLC shutter was positioned in the emission beam. Its transmission was 28% and 0.04% for open and close situations, respectively, yielding a contrast ratio of about 700. This contrast was increased to about $10^5$ by mounting two FLCs in series, but this resulted in severe reduction in the measured emission intensity, about eight times for one device and 64 times for two. With one device, however, no significant image degradation was observed when inorganic phosphor particles, lanthanide chelates or naturally phosphorescent materials were imaged in TRD mode [43].

TIRFM is a technique designed to probe the surface of living cells labelled with luminescent stains. Excitation of the probe is provided by an evanescent wave generated by a light beam travelling between two media with different refractive indices. Practically, in this case, an incident He-Cd laser beam was reflected at the critical angle for total internal reflection upon encountering the interface between the microscope glass slide and the aqueous medium containing the biological sample. Luminophores within a few nm from the surface are excited by the evanescent wave, but not those lying deeper in the cells or tissues. The technique is particularly well suited to investigate the interaction of molecules with surfaces. Swiss albino mouse 3T3 cells were imaged in this way after being stained with the long-lived ($\tau = 1.5$ ms) terbium chelate $[Tb(L4a)]^-$ (Scheme 4.1) and with 1,1-dioactadecyl-3,3,3′,3′-tetramethylindocarbocyanine perchlorate (DiI) for comparison. One 3T3 cell has autofluorescence equivalent to 33 600 fluorescein molecules so that flow cytometry of these cells is difficult to carry out. In the present case, they were imaged with a TR microscope fitted with a HeCd continuous laser and an FLC modulator. For the organic label, the TR to steady state intensity ratio only reached about 1%, whereas with the Tb chelate this ratio was up to an average 15%, with the intense autofluorescence suppressed [44]. The same chelate was also conjugated to a lipid, dioleoylphosphatidylethanolamine (DOPE), and this conjugate, which has a lifetime of

1.58 ms, was provided as a membrane-staining agent for studying the morphology of fixed 3T3 cells and their adhesion, growth, and motility on polymeric materials [45].

### 4.2.5 Optimisation of Time-resolved Microscopy Instrumentation

Since the early designs described in the previous section, there has been little change in the principles, namely continuous light source with two choppers or one chopper and one ferroelectric liquid crystal (or similar device), or pulsed light source with or without chopper. However, technically each component of the microscope has sustained large improvement. In this section, we briefly outline this technical evolution closely following the review article by R. E. Conally and J. A. Piper [30].

#### *4.2.5.1 Excitation Light Source*

Selection of the light source depends much on the characteristics of the luminophore. Luminescence signals detected in TR mode are weaker than those from a continuously excited stain so that intense excitation beams are required but they must also be compatible with preservation of the sample: strong near UV or blue excitation may lead to photodegradation of the biological sample. A convenient device is a xenon flashlamp which delivers more energetic pulses in a shorter time span (2–10 μs) than a chopper-interrupted continuous light source. As a result inexpensive flashlamps are used in a number of home-made TR luminescence microscopes [31,33,46,47]. A major drawback of flash lamps is the plasma glow that persists for tens of microseconds after the primary discharge. The decay is exponential with reported associated lifetimes between 18 and 19 μs, which sets a lower limit for the delay time at around 50 μs: for this time after the pulse, about 93% of the plasma glow has faded out. This is however not a major problem with lanthanide bioprobes and bioconjugates which usually have lifetimes larger than 500 μs, so that a delay time of 60–100 μs can be conveniently applied (see Table 4.1). Other, smaller deficiencies include some variability in pulse to pulse intensity and a limited repetition rate.

Light-emitting diodes (LEDs) are now available in the whole UV-visible-NIR spectral range. For instance, a UV LED can deliver 500 mW at 365 nm in fast-switched (ns) pulsed mode. The wavelength is somewhat long for optimum excitation of $Eu^{III}$ or $Tb^{III}$ luminophores, but given the power delivered, absorption of light around 340–350 nm remains sufficient to produce bright red or green luminescence. The advantage of such pulsed light sources lies in the low voltage needed to drive them and in their small size which permits mounting of the device in the filter assembly of the microscope. On the other hand, InGaN UV LEDs weakly emit in the visible (550 nm) with a lifetime comparable to that of flash lamps and this emission has to be filtered out [48]. A comparison of the signal-to-noise ratio achieved by a UV LED versus flashlamp excitation has been conducted with *Giardia lamblia* cysts labelled with a $Eu^{III}$ chelate. Because the UV LED can be switched fast, a shorter delay time could be enforced: 5 μs versus 60 μs for the flashlamp, which resulted in an improved SNR, 25-fold for UV LED excitation [32] versus 11-fold with flashlamp excitation [47], compared to continuous excitation.

Lasers achieve maximum excitation thanks to their powerful optical density. A common laser in TR luminescence immunoassays is the pulsed nitrogen laser which delivers 337-nm output, rather convenient for the excitation of lanthanide chelates [49,50]. Curiously, to the best of our knowledge, it has not yet been mounted on a TR luminescence microscope; but one setup described above uses a He-Cd laser with 325-nm output [44]. Visible emitting lasers are less useful for lanthanide probes, but they have their advantage for transition metal stains such as platinum or palladium porphyrins.

#### *4.2.5.2 Chopper Wheels and Opto-electronic Shutters*

The problem of two versus one chopper has been discussed above and it seems that the present tendency is to combine a pulsed light source with one emission chopper for cost and technical reasons. Indeed, optical electric liquid crystals or other opto-electronic shutters have the tendency to attenuate too much the signal, their switching time is still too long, and they are somewhat sensitive to UV light, so that for the time being they seem to be abandoned. The shutters currently used are mechanical choppers because last-generation compact and fast mechanical choppers can be easily inserted into a microscope and, combined with an adequate optical configuration, switching times as low as 10–15 μs can be achieved at a respectable 5 kHz repetition rate. On the other hand, devices such as vibrating choppers have longer switching times (> 50 μs) and their cycling rate is slow (1 kHz). Alternatively, ICCD cameras (see below) can be directly gated fast enough so that their combination with a flashlamp eliminates the need for choppers.

#### *4.2.5.3 Detectors*

CCD cameras are ubiquitous despite their lack of amplification of the number of electrons generated from each photon impact, which results in relatively long integration times (several seconds). A way out of this difficulty is to couple CCD cameras with micro-channel plate (MCP) image intensifiers. These devices were developed at the end of the 1980s and consist of three functional units. The first one is a photocathode which converts incoming photons to electrons. The photoelectrons are subsequently amplified by an array (1024 × 1024) of miniature photomultipliers (10–12 μm in diameter) for which photo-multiplication gains up to $10^5$ have been reported. Finally, a fluorescent screen fitted at the end of the device converts the multiplied electrons back to photons and the resulting intensified image is optically coupled to the CCD camera. Several TR luminescence microscopes have been equipped with such an image-intensifier (ICCD) [31,33,47,51,52] despite a limited dynamic range when the detector is operated in high-gain regime, resulting in somewhat grainy images. At the turn of the century, a landmark was reached by the invention of an electron-multiplying CCD (EMCCD) camera. The device is a solid-state monolithic intensifier produced at low cost on a silicon wafer which reaches gains up to 2000. The detector has sufficient sensitivity to capture the image in real time: a back-illuminated EMCCD chip reaches quantum efficiency of 92% at 600 nm, while a generation III ICCD has 40% quantum efficiency only! When comparing ICCD with EMCCD one has, however, to take into account the amplification of the produced photoelectrons. From this point of view, image intensifiers are much better, by at least a factor of two, which compensates for their reduced quantum efficiency. In addition ICCDs can be operated at room temperature while EMCCDs require cooling to 200 K, usually achieved by Peltier effect. As a result, modern high-speed ICCD cameras now compete efficiently with EMCCD devices.

#### 4.2.5.4 *Recently Described Microscopes*

In the absence of commercially available simple equipment for TRL microscopy, authors have modified existing instruments. For instance, the group of T. Nagano fitted an Olympus fluorescence microscope equipped with a cooled CCD camera with a 60-W xenon flash lamp and an image intensifier unit located in the emission pathway. The controller of the image intensifier provided the delay (52 or 70 μs) and acquisition (808 μs) times (Fig. 4.7). Time-resolved dual-colour images of cells were obtained by simultaneous use of two LLBs ($Eu^{III}$ and $Tb^{III}$) and changes in intracellular $Zn^{II}$ could be monitored with a $Eu^{III}$ chemosensor [53].

In the time-resolved luminescence microscope proposed by R. Connally et al. [54], a commercial Zeiss epifluorescence microscope is adapted by fitting a high-intensity UV LED (100 mW) emitting at 365 nm (full width at half maximum = 10 nm). After exiting the filter cube containing the dichroic mirror, the useful power entering into the objective is reduced to 14 mW $cm^{-2}$. Luminescence detection is achieved with an EMCCD device with shutter speed down to 3 ns. The Peltier-cooled super video graphic array (SVGA) element has a resolution of 1280 × 1024 at 12 bits (4096 counts). The LED filter-box assembly contains a dichroic mirror strongly reflecting light with $\lambda < 400$ nm and provides sufficient wavelength selection to remove both excitation and emission filters, which helps maximising luminescence collection. An improvement in SNR close to 30 is obtained in TR mode compared with conventional epifluorescence mode.

Another design by the same authors combines the excitation and emission choppers as well as the dichroic mirror into a single component; the modified Olympus microscope is fitted with a gated autosynchronous luminescence detector (GALD) and a UV LED as the excitation source. With the $[Eu(BHHCT)_2]$-SA conjugate, which has lifetime of 640 μs, and a time delay of 260 μs imposed by the GALD, 66% of the light emitted by the lanthanide

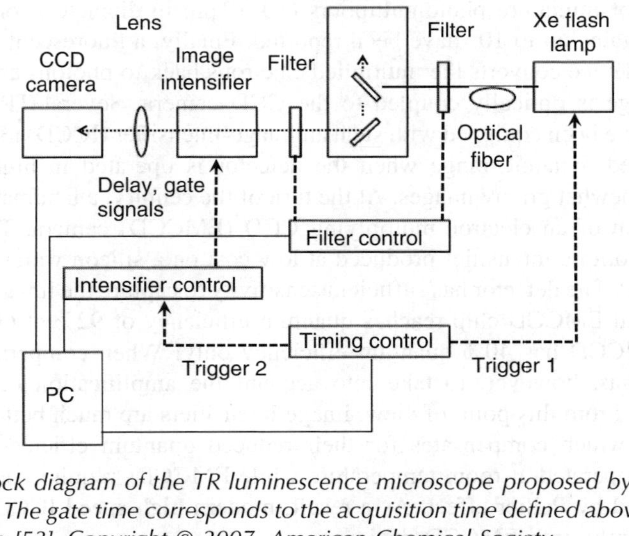


***Figure 4.7*** *Block diagram of the TR luminescence microscope proposed by T. Nagano and coworkers [53]. The gate time corresponds to the acquisition time defined above. Adapted with permission from [53]. Copyright © 2007, American Chemical Society*

***Table 4.2*** *Collection windows and other parameters versus the number of chopper blades at maximum chopper speed (167 rps) [56]*

| Number of blades | Rep. rate (kHz) | On/off duty ratio | Switching time (μs)[a] | Collection window (μs) |
|---|---|---|---|---|
| 30 | 5.0 | 1:1 | 11 | 100 |
| 15 | 2.5 | 3:1 | 11 | 300 |
| 10 | 1.0 | 5:1 | 11 | 500 |

[a] With ×60 objective; 16 μs with ×20 objective

bioprobe can be collected and a ≈ 38-fold improvement in SNR is obtained when *Giardia lamblia* cysts are imaged in TR mode. Replacing the GALD by an EMCCD camera boosts this improvement to 114-fold [55].

More recently, easy and cheap modification of commercial epifluorescence microscopes has been advocated by D. Jin and J.A. Piper [56], leading to a design essentially similar to the one sketched in Fig. 4.6, except for more modern components. The light source is a current-switched UV LED emitting at 350 nm and operating at 2.5 kHz cycle rate resulting in a switch-on-off transition time smaller than 3 μs while an excitation pulse length of 75 μs is chosen. The compact mechanical chopper is positioned next to a 1.1-mm pinhole at the focus of the eyepiece of the microscope and achieves an 11-μs switching time at the maximum 5 kHz repetition rate with a 60× objective (observation spot: 160-μm diameter). The image is reconstructed in the observation plane by a second eyepiece. The pinhole is in fact a near-field spatial filter which determines the effective beam diameter at the chopper wheel. High optical throughput is reached with this arrangement, while minimising the size of the optical beam at the chopper, which results in fast switching times. Different collection windows are obtained by varying the number of blades of the chopper wheel, as shown in Table 4.2. The detector assembly allows for both direct eye examination and recording of images with a conventional CCD camera.

The instrument was again tested on *Giardia lamblia* and on smaller micro-organisms, *Cryptosporidium parvum* oocysts with a diameter of approximately 3 μm. In the first case, an 18-fold improvement in SNR is obtained versus autofluorescence background while a 13-fold improvement is revealed in the second case. These improvements are smaller by a factor of about 2 with respect to the more sophisticated TR microscope fitted with an EMCCD camera and described above [54], but the authors insist on the minimum modification required to the epifluorescence microscope and, also, on the low cost of this modification, about US $2000.

### 4.2.6 Commercial Instruments

Most of the time-resolved microscopy images described in relevant articles published to date rely on home-made modified microscopes. Commercial instruments are becoming available but most of them include the sophisticated technique of TCSPC. Examples are the Microtime 100 and 200 from PicoQuant which ally TR capability and confocal microscopy or the Andor Technology fast gated ICCD camera which can be mounted on microscopes and microspectroscopy devices. Another sophisticated instrument is the Zeiss LSM 710

NLO multiphoton scanning microscope which combines fluorescence and phosphorescence lifetime imaging abilities [57].

## 4.3 Bioimaging with Lanthanide Luminescent Probes and Bioprobes

Time-resolved detection is extensively used in luminescence immunoassays and associated analyses. In some cases, these analyses are also combined with TR imaging of the analyte. Moreover cell and tissue imaging starts to benefit from TR luminescence microscopy based on lanthanide probes. In this section, work published up to the end of 2011 on bioimaging with the help of lanthanide coordination compounds is reviewed and organised according to the chemical class of probes used; selected papers describing new research in the field have been added at the beginning of 2014. Both steady-state and time-resolved experiments are described. Chronological order is followed within the sections unless obvious grouping of works makes reading easier.

### 4.3.1 β-Diketonate Probes

Lanthanide β-diketonates are amongst the best studied rare-earth luminescent complexes [58]. They are brightly luminescent and volatile so that incorporation into various electroluminescent materials is simple. Moreover their photophysical properties are easily tuned by a judicious choice of ancillary ligands. Indeed, conventional synthesis usually yields bis(hydrated) lanthanide tris(β-diketonates), but the two solvent molecules can be substituted by either a fourth diketonate anion or a donor ligand with adequate functionalisation as to provide convenient light harvesting and subsequent energy transfer onto the metal ion. It is noteworthy that not only visible but also near-infrared luminescence [59,60] is efficiently sensitised in lanthanide β-diketonates. In the case of $Eu^{III}$, some ternary complexes have quantum yields up to 85% [8] and the main asset of their luminescent properties is an emission essentially concentrated in the hypersensitive $^5D_0 \rightarrow {}^7F_2$ transition at 615–620 nm. The $Eu(^5D_0)$ lifetimes are in the range 0.3–0.7 ms whereas much longer lifetimes for lanthanide (1.5–2.5 ms) are reported for other classes of compounds with comparable quantum yields. This is due to β-diketonates inducing appreciable admixture of ligand orbitals into 4f wavefunctions, henceforth leading to relatively short radiative lifetimes [61]. Selected β-diketone ligands and associated ancillary ligands used so far in luminescence imaging are sketched in Scheme 4.3. It is noteworthy that early works relied on simple tris(β-diketonates) which are commercially available and this tendency is still present in more recent investigations when these basic chelates are introduced into inorganic–organic hybrid materials or nanoparticles.

In the following, we try to favour the description of quantitative data whenever available. However, dissociation of the chelates is very often not taken into consideration, despite that existing stability constant data on most of the simple lanthanide tris(β-diketonates) demonstrate that these compounds are not that stable. For instance, stability constants for $[Eu(tta)_3]$ in $H_2O$:EtOH, 20:80 v:v are $\log K_1 = 6.44 \pm 0.04$, $\log K_2 = 5.5 \pm 0.04$, and $\log K_3 = 4.36 \pm 0.04$ [62]. For a millimolar solution, this translates into a speciation of about 81% of the 1:3 chelate, 18.5% of the 1:2 complex and 0.5% of the 1:1 species. Diluting at $10^{-6}$ M leads to a speciation of 1, 31, and 56%, respectively, in addition to 11% of uncomplexed $Eu^{III}$.

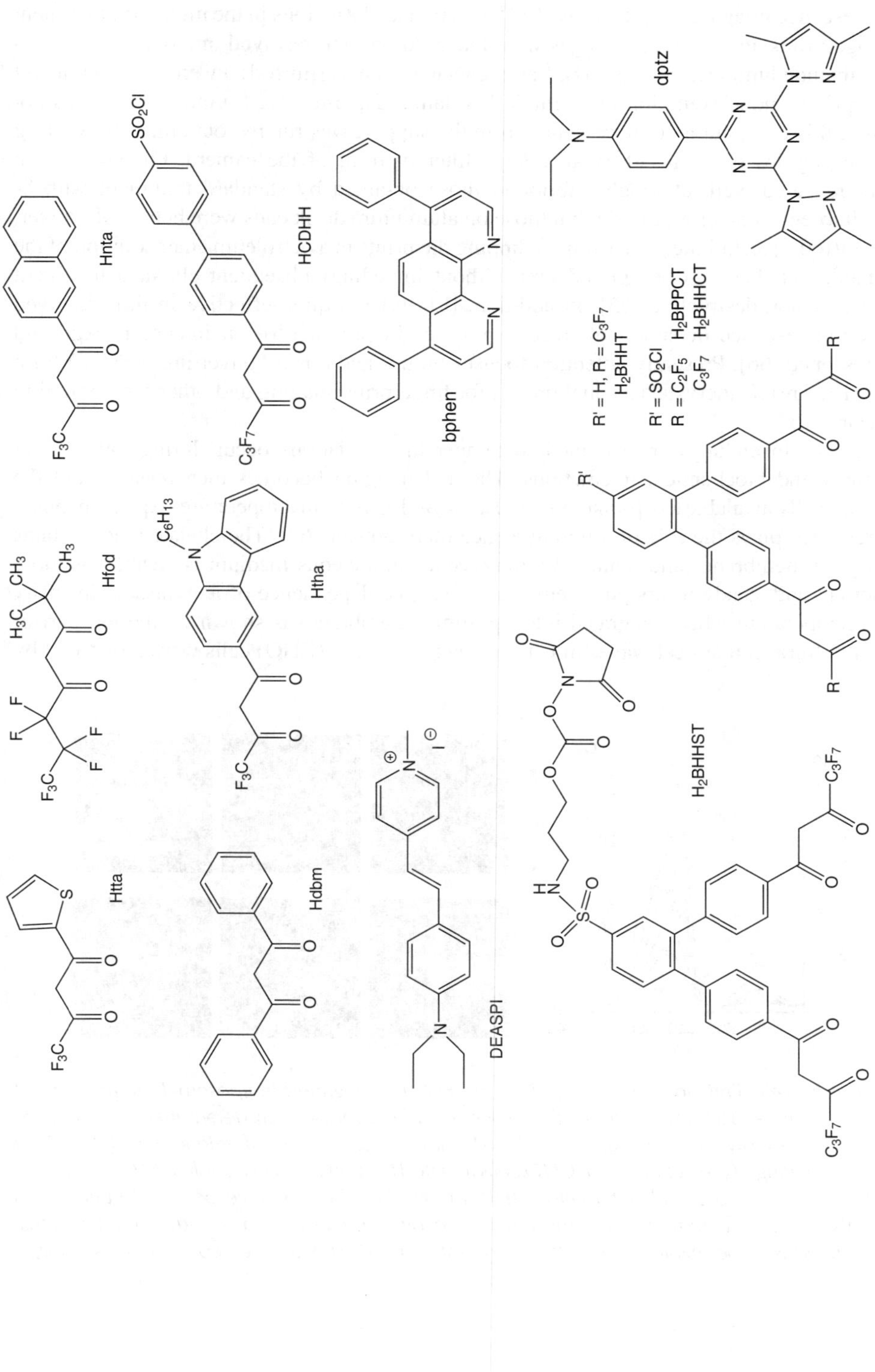


**Scheme 4.3** *β-Diketones and ancillary ligands used in luminescence sensing and imaging*

#### 4.3.1.1 Molecular β-Diketonate Probes and Conjugates

Forensic scientists have been among the first to take advantage of $Eu^{III}$ and $Tb^{III}$ luminescence imaging [63]. Initially, $Ln^{III}$ ions replaced $Zn^{II}$ ions in the ninhydrin treatment of fingerprints and enhanced images were recorded on time-resolved microscopes adapted to lanthanide luminescence [20,64]. Later, chelates such as $[Eu(fod)_3]$ were used to unravel fingerprints under excitation of a simple UV lamp. Time-resolved luminescence was not used in this case, but autofluorescence from the support was filtered out either by wearing orange goggles or by inserting an orange filter in front of the camera. Fingerprints on paper revealed well, although inferior to those obtained by standard treatment with 8-diazafluoren-9-one or ninhydrin, but those on aluminium drink cans were better. Moreover, the $[Eu(fod)_3]$ technique, consisting in dipping the prints in a petroleum ether solution of the lanthanide chelate can be applied first without impeding subsequent classical treatment since it is non-destructive [65]. In addition, $[Eu(tta)_3]$ is quite effective in time-resolved photoluminescence detection of trace explosives in presence of an intense background fluorescence [66]. Presently, attention focuses on up- and down-converting $NaYF_4$:Yb,Ln (Ln = Er, Tm) nanocrystals as dual probes for fingerprint imaging and other counterfeiting measures [67].

Changes in enthalpy translating into temperature variations occur during intracellular chemical and biochemical interactions. Thermal imaging becomes then feasible and the commercially available tris(β-diketonate) $[Eu(tta)_3(H_2O)_2]$ with temperature-dependent phosphorescence plays the role of a luminescence thermometer [68]. The chelate readily stains biological membranes and while its luminescence in aqueous medium is weak, it is more important and steady in lipophilic environments. The dependence of its emission intensity upon temperature when integrated into liposomal membranes is shown on the left part of Fig. 4.8. Intracellular heat waves in Chinese hamster ovary (CHO) cells were monitored by

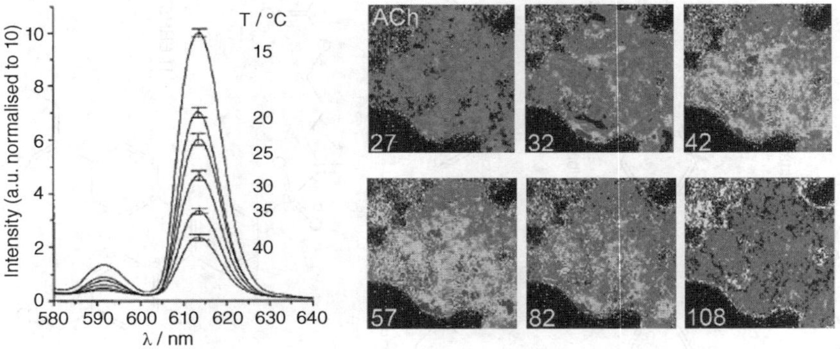


***Figure 4.8*** *Left: Emission intensity of $[Eu(tta)_3(H_2O)_2]$ integrated in liposomal membranes, at pH 7.4 and under 363-nm excitation. Each spectrum is an average of six determinations (standard deviations are shown at the maximum of the 613-nm peak). Redrawn after Reference [68]. Right: Thermal imaging of the heat wave in CHO cells initiated by a puff of acetylcholine (ACh, frame 27, 22 °C) and evidenced with internalised $[Eu(tta)_3(H_2O)_2]$. Images have been recorded at 1-s intervals (lower left numbers give the time in s); blue: coldest regions, red: hottest regions. Reproduced with permission from [68]. Copyright © 1998, Elsevier (See colour plate section)*

recording one image per second after activation of the metabotropic L-muscarinic receptor by application of acetylcholine (ACh, see right part of Fig. 4.8). The ACh puff (at 22 °C) generates a triphasic response. The first phase lasts about 2.5 s and corresponds to a cooling due to endothermic dilution of ACh in the buffer saline solution. Two exothermic phases (rapid and slow) then develop. The first one is short (2.5 s) while the second lasts about 110 s. In parallel, calcium is released from intracellular stores, which suggests that the L-muscarinic signal transduction pathway is involved in the heat production.

The temperature dependence of the luminescence intensity of [Eu(Dk)$_3$(dptz)], where Dk stands for tta, nta, or fod, is the basis for tailoring interesting temperature-sensitive luminescent probes. In particular, simultaneous luminescence sensing of both temperature with these chelates embedded into poly(vinyl methyl ketone) films and oxygen with a palladium porphyrin complex entrenched into a poly(styrene-*co*-acrylonitrile) film can be achieved. The sensors are made of a polyester support and of a hydrogel containing $O_2$- and $T$-sensitive beads including the luminescent metal complexes [69]. The work was extended to the monitoring of the consumption of oxygen in the glucose oxidase-catalysed oxidation of glucose at varying temperatures, with [Eu(tta)$_3$(dpbt)] as lanthanide probe [70]. Potential applications are foreseen in temperature-compensated sensing of oxygen, in contactless sensing of both oxygen and temperature in biology and medicine, as well as in simultaneous imaging of air pressure and temperature in wind tunnels.

When it comes to tissue imaging, [Eu(tta)$_3$($H_2O$)$_2$] has enabled increased differentiation between *n*-butyl-2-cyanoacrylate (NBCA) and an oil, Lipiodol, in embolised vessels, within the frame of an arteriovenous malformation (AVM) model in swine [71]. Arteriovenous malformation consists in an abnormal connection between veins and arteries; the anomaly can occur in any location of the body, but is essentially known because of its incidence in the central nervous system. NBCA polymerisation results in a subocclusive matrix which causes partial or complete thrombosis. Histopathology was performed on both acute and chronic animal models in order to better mimic human AVMs. Embolised networks of blood vessels (retia) from pharyngeal artery were fixed before staining with the europium chelate which localises at the same place as NBCA. The depth of NBCA penetration was determined on paraffin-embedded tissue sections. The photograph reproduced in Fig. 4.9 demonstrates that

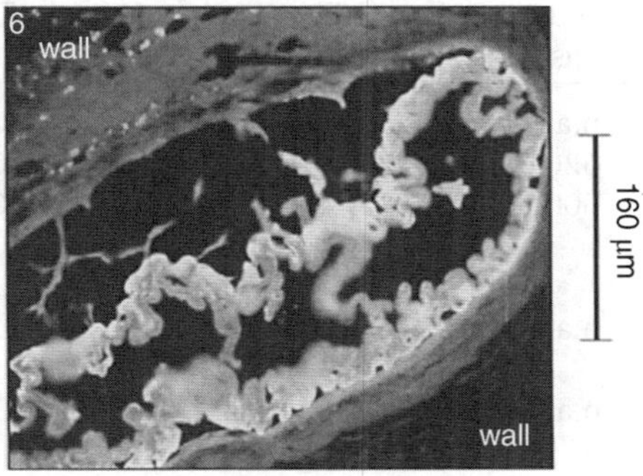


***Figure 4.9*** *Paraffin-embedded excised embolised tissue containing a 50:50 Lipidiol:NBCA ratio, as evidenced by Eu$^{III}$ luminescence. Reproduced with permission from [71]. Copyright © 2001, American Society of Neuroradiology (See colour plate section)*

some regions within the cross section contain an intact endothelium while others show stripping of the endothelium, which may be caused by NBCA. The procedure works well and a patent has been filed [72].

In 1997, J. Yuan, K. Matsumoto et al. [73] proposed a new chlorosulfonated tetradentate bis(diketone) chelating agent BHHCT (Scheme 4.3) tailored for $Eu^{III}$ sensitisation which can easily be covalently bound to bovine serum albumin (BSA) and streptavidin (SA). They used it for highly sensitive time-resolved luminescence immunoassays of immunoglobulin G [74], α-fetoprotein (improving the limit of detection by 4–5 orders of magnitude over conventional assays) [75], and human thyroid-stimulating hormone [76], among others. This and related chelating agents naturally caught the attention of developers of time-resolved luminescence microscopes, especially that they proved more photostable than some of the early polyaminocarboxylate luminescent probes (see next section). Scientists interested in the detection of *Giardia lamblia*, *Cryptosporidium parvum*, *Cryptosporidium muris*, and other microorganisms in drinking water and soft juices made an extensive use of these chelates. The cysts are protozoan parasites that colonise and reproduce in the intestine causing giardiasis in the small intestine and diarrhoea. The analysis protocol calls for the $Eu^{III}$ chelate to be attached onto a specific monoclonal antibody (mAb) before time-resolved imaging. For *Giardia lamblia*, the adequate antibody is the rabbit antimouse IgG antibody (sometimes labelled G203). Analyses and imaging have been carried out with various home-made instrumental setups reflecting the advancement of technology in this field during the past ten years. Published data are summarised in Table 4.3. Comparisons are not always clear-cut since available data are often fragmentary; for instance, to our knowledge, quantum yield has been measured for one conjugate only. The extent of chelate dissociation in solution is also rarely established so that lifetime or quantum yield data may refer to mixtures (1:1 + 1:2 chelates + solvated $Eu^{III}$) rather than to the pure 1:2 chelates.

Close scrutiny of Table 4.3 leads to the following remarks. One point is very clear: time-resolved detection improves the signal-to-noise ratio (SNR) by at least one order of

***Table 4.3*** *Photophysical data for time-resolved imaging of* Giardia lamblia *cysts with europium chelates on various instrumental setups. See Scheme 4.3 for ligand formulae*

| Ligand(s) | $\varepsilon/10^4\,M^{-1}\,cm^{-1}$ | τ(Eu)/ μs[a] | Buffer | τ(Eu)/ μs[b] | Lamp | Delay time/μs | Detector | Δ(SNR) | Ref. |
|---|---|---|---|---|---|---|---|---|---|
| BPPCT | 3.84 | n.a. | PBS | 240 | Flash | 60 | ICCD | 30× | (31) |
| BHHCT | 2.80 | 620 | PBS | 339 | Flash | 50–60 | ICCD | 10× | (47) |
| BHHST | 3.32 | 600 | Bicarb. | >500 | Flash | 60 | ICCD | 11× | (52) |
| | | | Bicarb. | | LED | 5 | ICCD | 12.8× | (32) |
| | | | Bicarb. | | LED | 3.2 | EMCCD | 30×[c] | (54) |
| BHHCT/ dpbt | n.a | n.a. | Bicarb. | 460 | Flash | 400 | CCD | n.a. | (77) |
| BHHCT/ Ag-NPs | n.a. | n.a. | PBS | 380/ 192[d] | LED | 19 | CCD | 2× (w.r. to no Ag)[d] | (78) |

[a] Lifetime of the $Eu(^5D_0)$ level in the chelate
[b] Lifetime of the $Eu(^5D_0)$ level in the conjugate
[c] 7-dimethylaminocoumarin-4-acetic acid added to increase the autofluorescence background
[d] Without/with Ag

magnitude. In our opinion the 2002 report (first entry of the table) might have been somewhat optimistic since the 30-fold improvement has not been reproduced until an EMCCD camera was used.

The second point is that experimental time-resolved conditions used were not always optimum with respect to the ratio lifetime/delay time. One has advantage of dealing with a long-lifetime chelate; in this respect, buffer composition is of course important. For instance, the $Eu^{III}$ lifetime of the BSA-$[Eu(BHHCT)_2]$ conjugate amounts to 380 μs in 0.1 M Tris-HCl (pH 9.1) and to 641 μs in $10^{-5}$ M TOPO – 0.05% sodium dodecylsulfate (SDS) – 0.1 M $NaHCO_3$ (pH 8.4) [75]. TOPO is a strong coordinating group which may remove potential coordinated water molecule(s) on $Eu^{III}$; moreover, SDS helps building a protective micellar environment, preventing second-sphere interactions with water molecules (Triton® 100 also does the same job), so that one understands that the $Eu(^5D_0)$ lifetime is larger in the second medium. For *Giardia lamblia* imaging, PBS was initially chosen as buffer, but was then substituted with a mixture of bicarbonate/TOPO.

Various aspects are important when dealing with a luminescent chelate or conjugate. BPPCT was rapidly substituted with BHHCT because of its short lifetime. With a 60-μs delay, only about 75% of the emission from the BPPCT conjugate can be detected while this proportion increases to 85% for BHHCT. However, BHHCT is hard to prepare in high purity and often contains a side product which actively binds to proteins but which has no influence on luminescence enhancement. Purification by silica-gel chromatography is not feasible because diketonate groups of terphenyl derivatives are labile and do not survive the passage through the chromatography column [79]. Moreover, conjugates with BHHCT are quite hydrophobic and consequently display poorer stability at high luminophore/protein ratio. For these reasons, the chlorosulfonyl group on BHHCT was substituted with a sulfonylaminopropyl-ester-*N*-succinimide moiety, leading to BHHST. Not only did the storage stability of the immune conjugate increase substantially (from 1 to 6 months), but a conjugate against *Cryptosporidium parvum* could be prepared with properties comparable to those achieved for *Giardia lamblia*, whereas conjugates with BHHCT all proved inactive. Furthermore, the processed images can be analysed by machine-vision systems to automate the analysis of samples [79].

Different research groups have also different approaches to bioconjugation. R. Connally's team couples the chelate directly onto the specific mAb, with a luminophore-to-protein ratio of about six, while J. Yuan's group goes through a BSA or BSA-SA conjugate first [77]. This is exemplified in their efforts to obtain a luminescent bioprobe with a better quantum yield. To this end, one BHHCT molecule is replaced with the triazine derivative dpbt, which is known as being an excellent antenna for $Eu^{III}$. The ternary $[Eu(BHHCT)(dpbt)]^+$ complex is then coupled to BSA to yield the BSA-EuBd conjugate (Fig. 4.10, top). The latter displays a wide excitation range up to 387 nm, a relatively long lifetime, 460 μs, and a quantum yield of 27% in $NaHCO_3$ 0.05 M, pH 8.5. The remaining primary amino acids of BSA are then cross-linked to SA with glutaraldehyde to produce an active BSA-EuBd-SA bioconjugate able to bind various microorganisms. To this end, *Giardia lamblia* cysts, *Cryptosporidium parvum* and *Cryptosporidium muris* oocysts are immuno-stained with their respective antibody and with a secondary biotinylated antibody which reacts with the lanthanide bioconjugate through the remarkably strong interaction between streptavidin and biotin ($\log K = 15$ [80]) as sketched in the middle panel of Fig. 4.10. The three microorganisms detected by this indirect immunochemical analysis and imaged in time-resolved mode are seen with good contrast. Control experiments with no primary antibody confirm the specificity of the

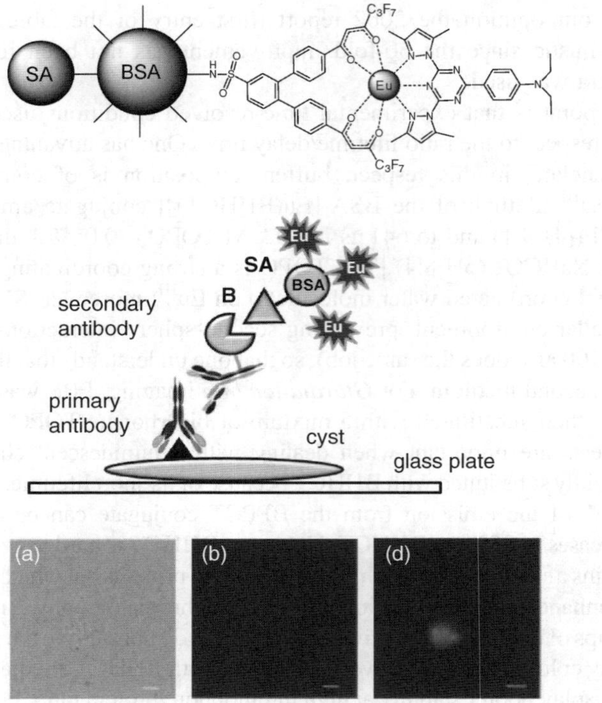


*Figure 4.10 Top: Sketch of the SA-BSA-[Eu(BHHCT)(dpbt)] (SA-BSA-EuBd) conjugate. Middle: Principle of the indirect immunochemical analysis of* Giardia gardia *cysts. Bottom: Bright field (a), TR luminescence without (b) and with (d) antibody images of* Cryptosporidium parvum *in a water sample immuno-stained as described in the text. Adapted with permission from [77]. Copyright © 2010, American Chemical Society (See colour plate section)*

detection (Figure 4.10, bottom panel). Unfortunately, SNR data for prompt luminescence and time-resolved luminescence are not reported [77]. In the same article, the authors propose to use an aqueous micellar solution of 0.1 M phthalate-HCl (pH 3.2) containing 60 μM BHHT, 60 μM dpbt and 0.1% Triton X100 as a dissociation-enhancement solution for heterogeneous luminescent immunoassays; this solution has a Eu($^5D_0$) lifetime of 520 μs and a test with a μM solution of $[Eu(EDTA)]^-$ shows the aminocarboxylate being totally replaced by BHHT/dpbt in about 10 min.

The use of the BHHCT complexes is of course not restricted to micro-organism imaging. For instance, the BHHCT chelate has been successfully applied in the investigation of the interaction between oligochitosan and tobacco cells. Chitosan (350 kDa) and oligochitosan (6 kDa, prepared by enzymatic hydrolysis of de-acetylated chitosan) are natural polysaccharides (poly β-*N*-acetyl-D-glucosamines) derived from chitin. They are presently considered for biomedical applications such as wound healing, artificial organ membranes, or drug carriers, in view of their biocompatibility and non-toxicity. But their main applications lie in agriculture. Indeed, they have broad-spectrum antibacterial activity in addition to being effective fungicides, inhibiting spore germination and mycelial growth, so that they are interesting alternatives to chemicals for controlling post-harvest diseases and prolonging fruit and vegetable storage life. Finally, they also promote the growth of plants according to a mechanism which is still not completely elucidated.

Tobacco is a plant benefitting from the growth-promoting effect of oligochitosan, which prompted this investigation. The experiment required derivatisation of BHHCT with an amine group (-$SO_2$-NH-$CH_2$-$CH_2$-$NH_2$), followed by Schiff reaction of the end-aldehyde group of oligochitosan with the primary amine function before final reduction with $NaBH_4$. The resulting conjugate was simply added to suspensions of tobacco cells. Luminescence microscopy images allowed determining the binding kinetics of oligochitosan on the cell walls. Furthermore, another $Eu^{III}$ luminescent probe for the detection of singlet oxygen developed by the same group [81] was used for quantifying the amount of peroxidase induced by the interaction of oligochitosan in tobacco cells. This experiment gave reliable evidence for explaining the growth-promoting mechanism of oligochitosan in tobacco plants [82].

#### *4.3.1.2 Nanoscopic β-Diketonate Probes*

As seen from Table 4.3 the various procedures and luminescent chelates used to image microorganisms seem to have reached an optimum, with improved SNR between 10 and 30 over prompt fluorescence detection. For this reason, researchers have been turning to more exotic materials such as nanoparticles, both regular nanoparticles and silver-coated ones, to induce plasmonic effects, or hybrid materials in which the luminescent chelate is adsorbed on an inorganic surface, for increasing the sensitivity of bio-analyses and bio-imaging. For instance, incorporation of europium tris(β-diketonates) into polystyrene nanoparticles with 107-nm size effectively leads to ultra-sensitive analyses. The NPs contain more than 30 000 chelates and are covalently coated with streptavidin. The detection limit of biotinylated prostate specific antigen immunoassay is lowered to $10^{-14}$ M, corresponding to $6 \times 10^{-19}$ mol in a 60 μl microtiter well. Single nanoparticles, representing individual PSA molecules, are visualised in the same microtiter well in time-resolved mode by the microscope sketched in Fig. 4.6. The image superimposes with that obtained by scanning electron microscopy [42]. It is noteworthy that despite the large concentration of $Eu^{III}$ chelate in the NPs ($\approx$80 mM), no inner-filter effect is operating, as observed with organic chromophores, because the emitted light cannot be re-absorbed by $Eu^{III}$. As a consequence, the loss of intensity due to TR operations is compensated by the absence of inner-filter effect.

The nano- and micro-structures into which lanthanide chelates are embedded often play a protective role, restraining non-radiative deactivation processes. For instance, the Eu($^5D_0$) lifetime of the [Eu(nta)$_3$(bphen)] chelate (Scheme 4.3) introduced into 40-nm NeutrAvidin® beads (each bead contains $\approx$ 370 chelates) is lengthened from 450 μs for the molecular complex in toluene to 590 μs for the beads in water [51].

A proof of concept has been brought for the imaging of *Giardia lamblia* by luminescent conjugated nanoparticles. In this assay, BHHCT is substituted by CDHH (Scheme 4.3) and the [Eu(CDHH)(dpbt)]$^+$ complex is linked to 3-aminopropyl(triethoxy)silane (APTES) by formation of an amide bridge between the chlorosulfonyl group of CDHH and the primary amine of APTES. The APTES-[Eu(CDHH)(dpbt)] entity is then hydrolysed and copolymerised. Nanoparticles are formed in presence of tetraethylorthosilicate (TEOS) and free

APTES in water/oil emulsion so that they bear free amino groups on their surface, an indispensable requirement for bioconjugation. The luminescent properties of the NPs are very sensitive to the composition of the water/oil and for optimum performances the micro-emulsion composition is a mixture of dioctyl sulfosuccinate, heptane, toluene and a small amount of water. The NPs obtained in this way are spherical with 10-nm diameter; they have an excitation maximum at 406 nm and their quantum yield in borate buffer amounts to 21% with the Eu($^5D_0$) lifetime equal to 397 μs. Conjugation of the nanoparticles is conducted as described above: NPs are added to BSA in PBS buffer and treated with glutaraldehyde followed by sodium boron tetrahydride. The treatment is repeated in presence of SA and, finally, luminescent nanoparticle-conjugated SA is used for the immunochemical detection of the microorganisms. Sharp images are obtained but again, no quantitative SNR data are available [83]. In fact, the replacement of BHHCT by CDHH did not prove to be very successful. Indeed, in a subsequent paper, the same authors reported a large improvement in quantum yield with BHHCT-based nanoparticles: 66% in Tris-HCl buffer, as compared to 21% for CDHH in borate buffer with the same excitation wavelength. In contrast, the Eu($^5D_0$) lifetime remains the same within experimental error, 389 μs [84].

#### *4.3.1.3 Metal-enhanced Luminescent Probes (Plasmonics)*

A way of enhancing luminescence is to bring the emitting moiety in close contact with a metallic surface or with small colloidal silver or gold islands. The physics behind the phenomenon is rather complex and, depending on the size of the metallic colloidal particles, either luminescence quenching or enhancement can be generated [85]. Small colloids have the tendency to quench luminescence because light absorption is dominant over scattering while larger colloids are expected to enhance luminescence. The incident electric field from the electromagnetic wave is modified by interacting with the metal surface, resulting in strong coupling between the surface plasmons of the metal particles and the transition dipoles of the chromophore. The radiative lifetime is substantially decreased and, in parallel, luminescence intensity is enhanced. In practice, nanoparticles can be used for this purpose, for instance Au-core Ag-shell nanoparticles with size >50 nm [86]. Since not only metal-enhanced luminescence is achieved but also reduction in photobleaching [87], the technique is finding applications in bioanalyses and bioimaging.

In a first trial, a twofold improvement was obtained with the SA-BSA[Eu(BHHCT)$_2$] conjugate and a detection protocol similar to the one sketched in Fig. 4.10 (middle panel) except that the glass plate was decorated with silver nanostructures or Ag-coated silica beads used [78]. A first step in producing the Ag coating on glass slides or silica beads is to synthesise a highly uniform monolayer of gold nanoparticles with 8 nm diameter, which enables controlling the growth of silver nanoparticles. Note that the substrates have to be preliminary thoroughly cleaned and surface treated with (3-aminopropyl)trimethoxysilane (APTMS). The glass slides or silica beads are then immersed in a commercially available silver enhancing solution during different times (10 s, 1 min, 3 min). This procedure is carried out in the dark to avoid photodecomposition. Luminescence enhancement depends on the incubation time in the Ag-containing solution and for the longest time, enhancement factors of 9.1 and 11.3 were obtained for glass slides and silica beads, respectively. Concomitantly, the Eu($^5D_0$) lifetime is reduced from 380 μs to 227 and 192 μs, respectively.

The recorded enhancement of luminescence results from two effects: (i) the increase in local electromagnetic field near silver nanoparticles causes an increase in the excitation rate of the chromophoric ligand as well as an increase in its decay rate (enhancement of antenna effect) and (ii) simultaneously the radiative lifetime of the $Eu^{III}$ ion is shortened, leading to an increase in quantum yield of the metal-centred luminescence according to the following equation:

$$Q_L^{Ln} = \eta_{sens} Q_{Ln}^{Ln} = \eta_{sens} \frac{\tau_{obs}}{\tau_{rad}} \tag{4.10}$$

in which $Q_L^{Ln}$ is the overall quantum yield of the metal-centred luminescence upon ligand excitation, $\eta_{sens}$ is the efficacy of the luminescence sensitisation by the ligand(s), $Q_{Ln}^{Ln}$ is the intrinsic quantum yield upon direct excitation into the *f*-levels, $\tau_{obs}$ is the observed lifetime of the $Eu(^5D_0)$ emissive level and $\tau_{rad}$ the radiative lifetime of this level in absence of non-radiative deactivation processes. The radiative lifetime depends on the metal ion environment, that is on the amount of mixing between the *f*-orbitals and the ligand orbitals, as well as on the refractive index [88]. Using collected data on emission intensities, lifetimes, and quantum yields, the enhancement factors mentioned above can be split according to the various contributions. Roughly speaking, excitation improvement contributes a factor of 4–5 while emission enhancement accounts for a factor slightly larger than 2, of which 81–87% is due to the decrease in radiative lifetime (Fig. 4.11).

Despite these very encouraging results, the SNR enhancement for *Giardia lamblia* imaging amounts to twofold only. This is traced back to the fact that enhancement of luminescence by metal islands or surfaces is most pronounced when the luminophore is close to the metal surface (usually 5–20 nm), which is not the case in the bioassay in which the distances between the chromophore and the metal structure span a much larger range. Enhancement is thus averaged over the cell thickness.

A similar enhancement in luminescence intensity is obtained with the ternary APTES-[Eu(BHHCT)(dpbt)] conjugate covalently bound to $Ag/SiO_2$ nanocomposites. With a silver core diameter of 57 nm and a 12-nm silica shell, intensity enhancement of one order of

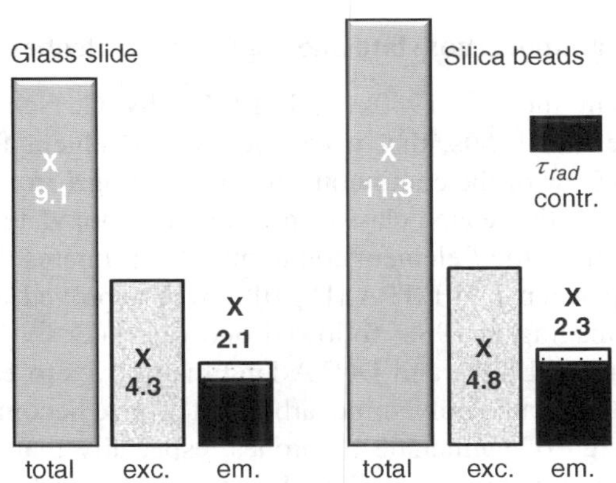


*Figure 4.11 Factors contributing to the luminescence enhancement of the SA-BSA-[Eu(BHHCT)$_2$] bioconjugate by silver islands. From data reported in Reference [78]*

magnitude are reported while the Eu($^5D_0$) lifetime is reduced from 300 to 60 μs [89]. These microspheres were then coupled to a biotinylated rabbit antimouse IgG antibody and the bioconjugate was tested by fixing it on glass plates coated with streptavidin. Single nanoparticles could be imaged as confirmed by TEM measurements. This demonstrates the sensitivity of the method, but no microorganism was investigated at that stage [90].

#### *4.3.1.4 Inorganic–Organic Hybrid Probes*

The simple [Eu(dbm)$_3$(Phen)] chelate has been coated on the surface of single-walled carbon nanotubes (SWNT), rendering them highly luminescent, but no imaging has been performed yet with this material [91]. The same chelate has also been covalently linked to attapulgite, a very stable fibrillar natural mineral. Modifications by coupling agents containing alkoxide groups such as APTES allow one to change the grafting ratio. The attapulgite-based lanthanide composite linked by APTES features a wide excitation range from UV to visible, up to 398 nm, a luminescence lifetime close to 0.5 ms, a high quantum yield (48%), high photostability, and low cytotoxicity. When incubated with human cervix cancer HeLa cells it localises on the cell membranes, brightly illuminating them [92].

#### *4.3.1.5 Imaging with Multi-photon Excitation*

Deep penetration examination of biological tissues by optical methods require excitation light with wavelength in their transparency window (ideally 650–1000 nm), therefore the recourse to multiphoton excitation with either a Ti:sapphire (780 nm) or a Nd:YAG laser (1.06 μm). There are not many reports on such experiments with β-diketonates to date, although the tetrakis complex DEASPI[Eu(tta)$_4$] (Scheme 4.3) with one-photon quantum yield equal to 28% and a sizeable two-photon cross section of 112 GM at 1.06 μm could be a potential TPA bioprobe [93], considering that the less luminescent [Eu(tha)$_3$(Phen)] (1-photon quantum yield = 8%, 2-photon cross section at 720 nm = 80 GM) has been fruitfully used for staining DNA in the nuclei of human breast cancer MCF-7 cells by confocal luminescence microscopy [94].

### 4.3.2 Aliphatic Polyaminocarboxylate and Carboxylate Probes

Since the discovery in the mid-1940s of $H_4$EDTA by G. Schwarzenbach and the demonstration, in the mid-1950s, that it can be used as eluent for the separation of the lanthanide ions in view of the continuous increase of Log*K* from approximately 15.5 for $La^{III}$ to 19.8 for $Lu^{III}$, the related class of polyaminocarboxylate chelating agents has drawn hefty attention from the *f*-element community [95]. Interest for these compounds made a quantum leap when $[Gd(DTPA)(H_2O)]^{2-}$ was approved as contrast agent for magnetic resonance imaging in 1988, followed one year later by $[Gd(DOTA)(H_2O)]^{-}$. The ligand frameworks of DTPA and DOTA lend themselves to easy derivatisation so that these molecules and related polyaminocarboxylates have become common complexation units in the design of lanthanide bioprobes, especially that the thermodynamics and kinetics of complex formation are well understood [96]. Hundreds of these molecules have been synthesised and tested for various bio-applications, in particular for luminescent immunoassays [97]. In this section, we describe bioimaging applications of aliphatic polyaminocarboxylates and their bioconjugates, while those with macrocyclic representatives of this class of compounds (mainly cyclen derivatives) are dealt with in Section 4.3.3.

#### 4.3.2.1 Molecular Probes and Conjugates

We have described early time-resolved imaging experiments based on $Eu^{III}$ and $Tb^{III}$ polyaminocarboxylates with L1–L4 (Scheme 4.1) for visualising rat epididymis [22], malignant mucosa of human colon [40], *D. discoidenum* amoebae [29], and Swiss albino mouse 3T3 cells [44,45] in Section 4.2.4, so that no further details will be presented here. To overcome the somewhat low quantum yields of these first luminescent lanthanide bio-probes, the group of H. Tanke has developed a peroxidase-mediated amplification system. The technique produces large amounts of biotin-tyramide subsequently detected by the streptavidin conjugate of $[EuL1]^-$. Strong signals from $Eu^{III}$ luminescence were observed even by the naked eye for vimentin in cryo-sections of rat liver, as well as for DNA and RNA *in situ* hybridisation [98].

An interesting approach to multiparameter immunocytochemistry is the use of four different metal-based and one organic luminescent tag for the identification of peripheral blood leukocytes [46]. Leukocytes or white blood cells defend the body against infectious diseases. There exist five types of these cells: (i) neutrophils, which are filled with microscopic granules containing enzymes digesting microorganisms, (ii) eosinophils, which are responsible for fighting multicellular parasites, (iii) basophils, which release histamine, causing vasodilatation, (iv) leukocytes, which bind antigens present on virus-infected or cancerous cells and kill them, and (v) monocytes, which are long-lived cells phagocyting foreign organisms. Leukocytes make up about 1% of blood in a healthy adult and their number is a good indication of the health status of the individual. The proposed assay aims at identifying neutrophils, eosinophils and T-lymphocytes by time-resolved microscopy with the help of the metal chelates $[Eu(L5)]^-$ (Eu-W8044), $[Tb(L6)]^-$ (Tb-16034), $[Pd(L7a)]^{3-}$, and $[Pt(H_3L7b)]^{3-}$ (Scheme 4.4), while Syto 25$^{TM}$ is chosen as a green nucleus stain which works with all cells. The photophysical properties of these stains are summarised in Table 4.4.

The leucocyte mixtures are incubated in a one-step procedure in PBS buffer with a combination of luminophore-conjugated antibodies recognising neutrophil- and

**Table 4.4** *Photophysical properties of the luminescent bioprobes used for multiparameter detection of leucocytes [46]*

| Label | $\lambda_{max}$ (exc), nm | Excitation λ/nm (bandpass) | $\lambda_{max}$ (em), nm | Detection λ/nm (bandpass) | Decay time (μs) |
|---|---|---|---|---|---|
| $[Eu(L5)]^-$ | 292 | 340 (25) | 613 | 615 (4) | 1600 |
| $[Tb(L6)]^-$ | 335 | 340 (25) | 545 | 545 (4) | 1100 |
| $[Pd(L7a)]^{3-}$ | 390 | 545 (10) | 667 | 673 (10) | 700 |
| $[Pt(H_3L7b)]^{3-}$ | 380 | 530 (10) | 647 | 642 (10) | 80 |
| Syto 25$^{TM}$ | 521 | 470 (20) | 556 | 525 (25) | [a] |

[a] Prompt fluorescence

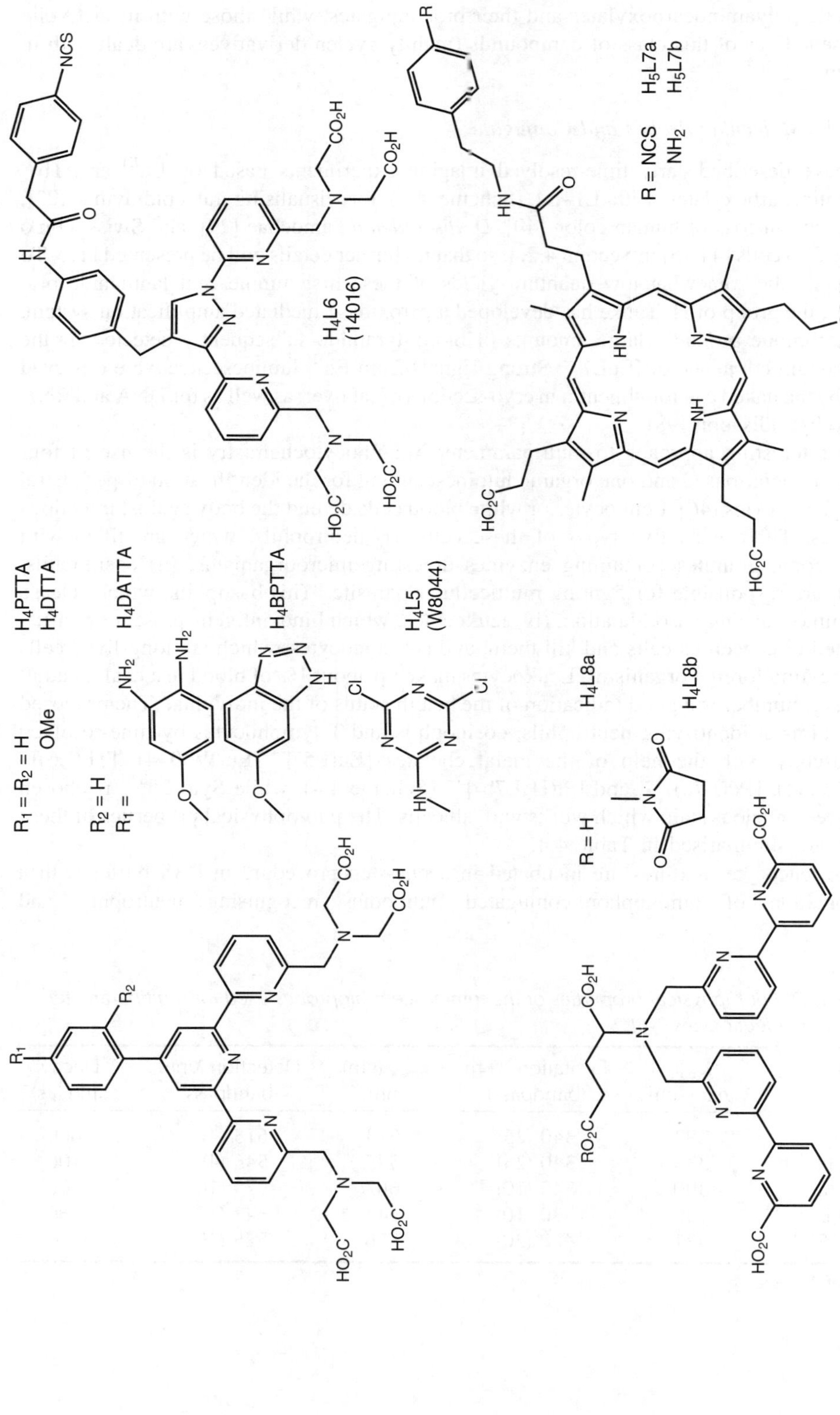


***Scheme 4.4*** *Some terpyridine-, bipyridine- and porphyrin-based ligands bearing carboxylic acid functions*

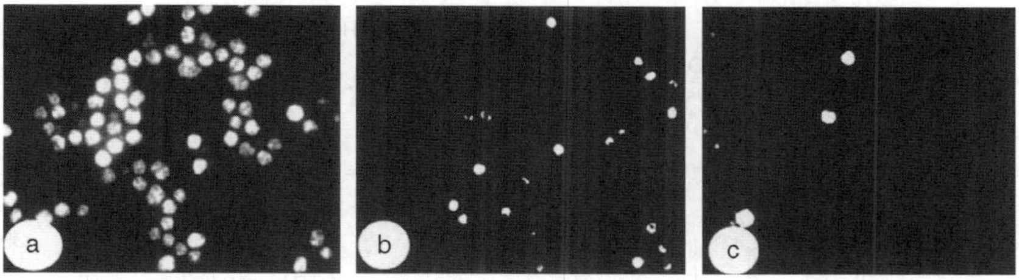


*Figure 4.12* Identification of different cell types in a mixed population of human peripheral blood leukocytes: (a) neutrophils identified with [Eu(L5)]-anti HNL, (b) T-lymphocytes illuminated by $[Tb(L6)]^{-}$-CD3, and (c) eosinophils imaged by $[Pt(L7a)]^{3-}$. Reproduced with permission from [46]. Copyright © 2003, John Wiley

lymphocyte-specific markers. The $Pd^{II}$ porphyrin is conjugated with a monoclonal antibody to human myeloperoxidase present in neutrophils and monocytes, the $Eu^{III}$ chelate is attached to an antibody recognising the neutrophil-specific marker human neutrophil lipocalin (HNL), while the $Tb^{III}$ complex is conjugated to anti CD3, an antigen on the cell surface of T-lymphocytes.

Identification of eosinophils takes advantage of electrostatic binding between the triply negatively charged $Pt^{II}$ porphyrin complex and the cationic proteins of eosinophils. Fig. 4.12 gives an example of the differentiation of three cell types with $Eu^{III}$, $Tb^{III}$, and $Pt^{II}$ markers. Superposition of the three images (not shown) clearly reveals juxtaposition of the various cells.

Several other ligand systems have been tested for potential TR imaging. For instance, $[Ln(L8b)(H_2O)]^{-}$ complexes (Scheme 4.4) were conjugated to BSA through the *N*-succinimide group (complex to BSA ratios of 8 and 7 for $Eu^{III}$ and $Tb^{III}$, respectively) and their luminescence was compared to fluorescein-labelled rabbit immunoglobulin, TR detection giving much higher contrast with the $Ln^{III}$ probes [99]. Moreover, the $Tb^{III}$ complex $[Tb(L8a)(H_2O)]^{-}$ is a reporter of local hydrophilicity thanks to its high polarity: rat brain slices incubated with this marker exhibit highly luminescent zones in the choroid plexus, a domain of the blood brain barrier containing hydrophilic epithelial cells [100].

Another, simpler ligand is HNTTA (Scheme 4.5): $[Eu(NTTA)]^{2+}$ features a reasonably large molar absorption coefficient at 330 nm ($1.9 \times 10^4\ M^{-1}\ cm^{-1}$), a quantum yield of 15.3% in aqueous solution and a $Eu(^5D_0)$ lifetime of 1.33 ms, ideal for TR experiments. In view of its low cell permeability, the complex was encapsulated into the cavity of apoferritin (AFt), a ubiquitous protein with iron-storage capability and which can be easily internalised by endocytosis. About 14 complex molecules are trapped into the cavity of AFt and the luminescence lifetime of $Eu(^5D_0)$ remains unchanged. Incubation of human hepatoma liver cancer cells HepG2 with Eu-NTTA@AFt results in the internalisation of the lanthanide bioprobe and recording of bright confocal microscopy images of these cells under TR mode [101].

Many luminescent tags are designed with the aim of imaging various analytes and proteins within living cells. For instance, $Tb^{III}$ luminophores have been elaborated by linking trimethoprim (TMP) with cs124-DTPA to yield $H_3L9a$, with cs-124-triethylenetetraaminohexaacetic acid to produce $H_3L9b$, or with a proprietary $Tb^{III}$ macrocyclic complex (Lumi4; see Section 4.3.3). All these complexes display bright luminescence:

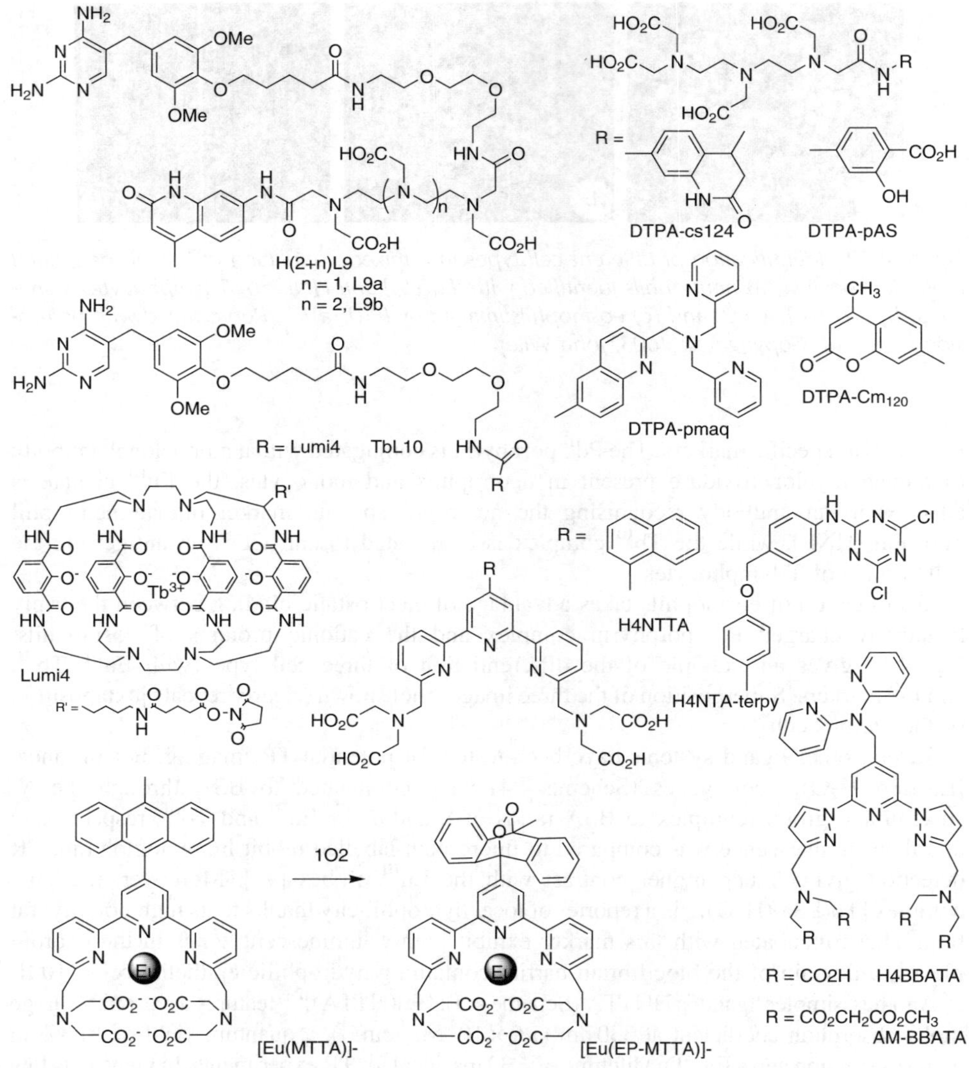


***Scheme 4.5*** *Polyaminocarboxylate ligands and Eu$^{III}$, Tb$^{III}$ complexes for molecular bioprobes*

quantum yields in water are as high as 0.32, 0.40, and 0.60 for [Tb(L9a)], [Tb(L9b)], and TbL10, respectively. They bind *Escherichia coli* dihydrofolate reductase (eDHFR) fusion proteins with high affinity (Log$K \approx 7.7$–$8.7$, as compared to 9.0 for free TMP). As a consequence, eDHFR can be tagged on the surface of living mammalian cells (e.g. NIH3T3 cells) with TbL10. In addition all three complexes show sensitising ability for GFP, the green fluorescent protein, through Förster resonant energy transfer [102]. This FRET system is able to image protein–protein interactions such as the one between the first PDZ domain of *Zonula occludens-1* fused to eDHFR and the C-terminal YV motif of claudin-1 fused to

GFP [103]. PDZ is a structural domain found in signalling proteins of living organisms and plants (PDZ stands for *P*ost synaptic density protein, *D*rosophilia disc large tumour suppressor, and *Z*onula occludens-1) which is a multidomain protein localised at intercellular junctions. Claudin-1 is a human protein which regulates cellular transformation and plays a role in metastatic colon cancer [104].

Circulating islet cell and β-cell specific autoantibodies are characteristic of prodromal periods of insulin-dependent diabetes mellitus. Conventional quantitation of these is often prone to errors because specific and unspecific fluorescence signals are difficult to separate. The situation can be dramatically improved if the $Eu^{III}$ chelate $[Eu(NTA\text{-}terpy)]^-$ conjugated to anti-human polyclonal IgG is used as the secondary antibody: in a test of 57 patients, 96.5% were detected as being positive for islet cell autoantibodies compared to only 89.5% with the conventional test [105].

Zinc is a ubiquitous cation in biological systems and several luminescent responsive probes have been tailored for its quantification and imaging in living cells. One is $[Tb(DTPA\text{-}pmaq]^-$ (Scheme 4.5), the luminescence of which is switched on upon $Zn^{II}$ complexation [106] and which helps visualising this cation into HeLa cells, even in presence of rhodamine 6G, thanks to TR detection [53]. Di(2-picolyl)amine is another specific chelating agent for $Zn^{II}$ and the bifunctional ligand $H_4BBATA$ (Scheme 4.5) has therefore been engineered. In absence of zinc, the di(2-picolyl)amine function of $[Tb(BBATA)]^-$ quenches the excited state of the pyrazolyl-pyridine complexing moiety by photoinduced electron transfer, therefore $Tb^{III}$ luminescence is turned off. This transfer is suppressed upon $Zn^{II}$ complexation, which restores the green $Ln^{III}$ emission. The $Tb^{III}$ complex can be introduced into cells by incubating them into a mixture of the cell-permeable form of $H_4BBATA$, AM-BBATA, and $Tb^{III}$ ions. Stable $[Tb(BBATA)]^-$ forms *in situ* after hydrolysis of the acetoxymethyl ester groups by intracellular esterases. The system is not only a zinc-sensitive but also a zinc-specific probe. Indeed, there is no interference from a wealth of cations such as $Na^I$, $K^I$, $Ag^I$, $Mg^{II}$, $Ca^{II}$, $Mn^{II}$, $Fe^{II}$, $Co^{II}$, $Ni^{II}$, $Cu^{II}$, $Fe^{III}$; only $Cd^{II}$ interferes somewhat. As a result, time-resolved luminescence imaging of $Zn^{II}$ in living HeLa cells is very sensitive [107]. In a similar approach, the same authors developed a ratiometric method for imaging and quantifying peroxynitrite in HeLa cells. For this assay, $H_4DTTA$ (Scheme 4.4) was chosen and fitted with acetoxymethyl ester groups. Only $Tb^{III}$ luminescence is quenched by addition of $ONOO^-$ ions, so that cells were incubated both with $Tb^{III}$ and $Eu^{III}$ ions, the complex $[Eu(DTTA)]^-$ playing the role of reference [108]. Further modulation of photoinduced electron transfer led to the use of $[Eu(DATTA)]^-$ as probe for imaging nitrogen monoxide in biological samples. The complex is almost non luminescent due to photoinduced electron transfer from the *o*-diaminophenyl group. Under aerobic conditions in an aqueous buffer at pH 7.4, NO reacts with the complex and transforms it into the highly luminescent $[Eu(BPTTA)]^-$ chelate (Scheme 4.4). Time-gated images of $[Eu(BPTTA)]^-$ in onion epidermal peels serve as proof of concept for this protocol [109].

The europium chelate $[Eu(MTTA)]^-$ acts as a highly sensitive and selected time-gated luminescence stain for probing singlet oxygen, as demonstrated by the group of J. L. Yuan [81]. The probe has large stability in the pH range 3–10 ($\text{Log}K \approx 21$). It reacts in a fast way ($k = 2.8 \times 10^{10}\,M^{-1}\,s^{-1}$) with $^1O_2$ to yield the endoperoxide shown on Scheme 4.5. In fact, $[Eu(MTTA)]^-$ is almost non-luminescent with $Q_{Ln}^L = 0.9\%$ at pH 9.1 (borate buffer) owing to photoinduced electron transfer. After the anthracene unit is cut out upon reaction with singlet oxygen, the resulting EP-MTTA complex is luminescent ($Q_{Ln}^{L}$ = 13.8%) with lifetime $\tau(^5D_0)$ = 1.3 ms and $\varepsilon_{max}$(335 nm) = 16 400 $M^{-1}$ $cm^{-1}$. Incubation of HeLa cells with $[Eu(MTTA)]^-$ and the photosensitiser 5,10,15,20-tetrakis(1-methyl-4-pyridinio)-porphyrin tetrakis(*p*-toluenesulfonate) followed by irradiation at 450–490 nm generates $^1O_2$, which can be visualised *in cellulo* in TR mode via its reaction with $[Eu(MTTA)]^-$ triggering the luminescent switch [81].

#### 4.3.2.2 Nanoparticle Probes

Nanoparticles are ideal substrates for designing various functional materials for bioanalyses [110–112], because their size and structure (core–shell) can be controlled at will and because they lend themselves to easy surface derivatisation [113]. Several types of NPs are available, with either organic substrates, usually based on polymers [114], or inorganic matrices, based on lanthanide salts [110] or silica [18]. The latter are quite popular and many of them are prepared by the now classical reverse microemulsion method: an oil consisting in a cyclohexane/*n*-octanol mixture is added to water and Triton X-100 as surfactant and vigorously stirred for about 1 h, which generates nano-sized droplets of water surrounded by the surfactant and dispersed into the oil; these droplets serve as confined medium for the formation of nanoparticles upon addition of tetraethylorthosilicate (TEOS); the size of the nanoparticles is determined by the size of the nanodroplets itself controlled by the water to surfactant ratio and by the concentration of ammonium hydroxide used for the hydrolysis of TEOS. Since silica is easily derivatised, inorganic nanoparticles, including UCNPs, are often coated with this material.

Several hundreds of lanthanide complexes can be inserted into a single nanoparticle, so that substantial improvements in detection limits for bioanalyses and imaging are documented, particularly if sensitisers are coated on the surface. For instance, increase in $Eu^{III}$-centred emission up to 100-fold upon doping this ion into $LaF_3$ NPs coated with 6-carboxy-5′-methyl-2,2′-bipyridine has been reported [115]. Both fully inorganic, for example $YVO_4$:Eu [116] or UCNPs [117,118], and complex-containing NPs have been proposed for bioimaging. An example of surface derivatisation is sketched in Scheme 4.6, in which a $SiO_2$ nanoparticle is fitted with 3-aminopropyl trimethoxysilane (APTS). The primary amino groups can then be easily conjugated to various biological compounds and/or lanthanide complexes. For instance, $[Ln(L8b)]^-$ (Ln = Eu, Tb; see Scheme 4.4) inserted into derivatised $SiO_2$ NPs yield bright luminescence under TR detection [119], mixed $SiO_2$/$[Eu(DTPA\text{-}Cm120)]^-$ (Scheme 4.5) NPs (1300 complexes/NP) provide a sensitive analytical procedure for DNA with detection limit down to $1.6 \times 10^{-10}$ M, 100-fold better than with fluorescein isothiocyanate (FITC) [120], and core-shell $SiO_2$ NPs consisting of a $[Ru(bpy)_3]^{2+}$-doped core and $[Ln(DTPA\text{-}pAs)]^-$ (Scheme 4.5) derivatised shell (Ln = Eu, Tb, Gd) are proposed for multiplex analyses [121].

Introduction of luminophores in NPs is often performed during NP synthesis as sketched in Scheme 4.6. In this example, the $[Eu(PTTA)]^-$ complex is covalently linked to APTS first, introduced into the water in oil emulsion and after adequate treatment, NPs with about 40-nm size are obtained. Further surface treatment includes derivatisation by SA with a procedure similar to the one adopted by the authors for molecular probes and mentioned above (Section 4.3.1). Such NPs, co-doped with the $Eu^{III}$ and $Tb^{III}$ complexes to achieve multimodal capability, are uptaken by mouse LAF1 pituitary gland tumour cells and serve as stains to

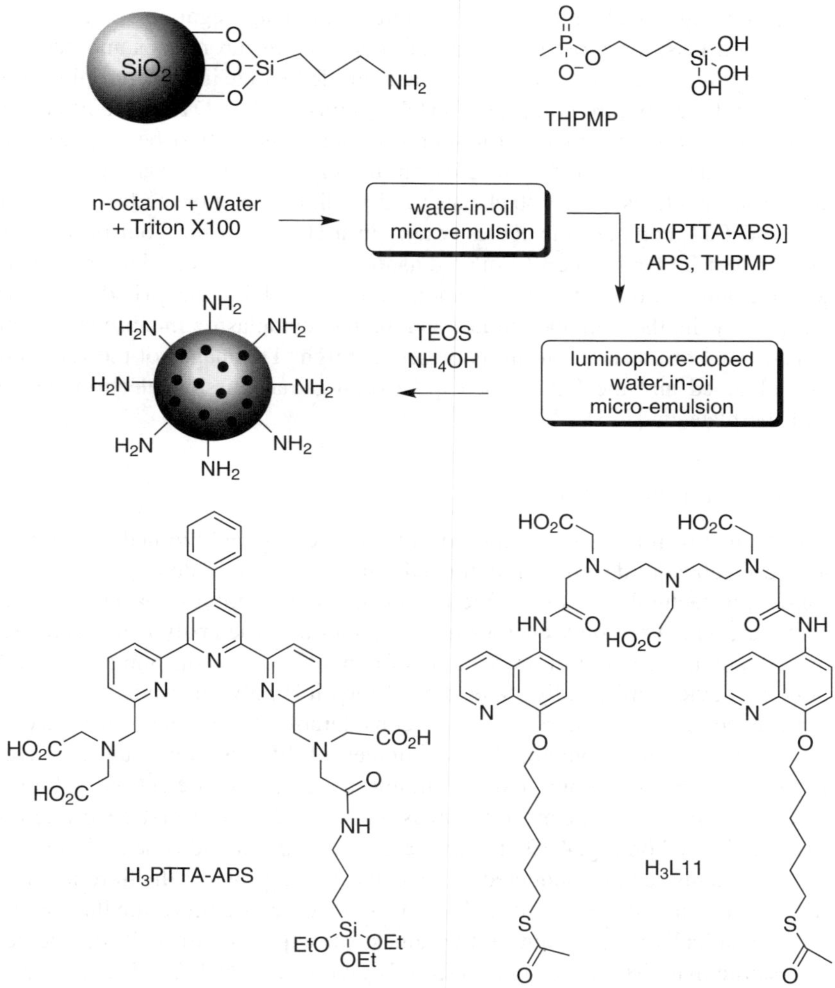


**Scheme 4.6** *Top and middle: Derivatised $SiO_2$ NP with 3-aminopropyl triethoxysilane (APS) and synthetic path for obtaining [Ln(PTTA-APS)]-doped nanoparticles via reverse emulsion procedure [122]. Bottom: Ligand $H_3$PTTA-APS [122] and ligand used for evidencing human platelets [123]*

image them [122]. They also act as efficient labels in PSA immunoassay and in the imaging of water-borne *Giardia lamblia* [124]. Another benefit of these NPs is their photobleaching resistance, which is larger compared to that of the corresponding molecular complexes: after 1 h exposure to a 30-W deuterium lamp, they only lose 11% of their emissive properties, compared to 24% for the molecular complex (and 98% for rhodamine 6G!).

Another noteworthy contribution stems for the group of Z. Pikramenou. In this investigation, human platelets are imaged with internalised gold nanoparticles co-coated with a europium complex and a pH low insertion peptide (pHLIP). Platelets enable fast activation following vessel wall damage and initiate rapid aggregation and clot formation to impede blood loss; therefore they play a critical role in thrombosis and haemostasis. The pHLIP peptide is attached to the NPs by a disulfide linkage in the C-terminus domain. The ligand for the europium complex [Eu(L11)] consists in a DTPA-modified skeleton fitted with two quinoline moieties for luminescence sensitisation bearing each a long substituent decorated with a surface-reactive sulfur-containing anchor group (Scheme 4.6). One nanoparticle contains about 640 chelates and modification of the surface by attachment of pHLIP and $Eu^{III}$ complex molecules results in a change in the gold plasmon visible absorption at about 500 nm. The 13-nm gold nanoparticles are delivered in vitro to platelets upon a pH change stimulus; internalisation is facilitated by the pHLIP and localisation is primarily in the cell membrane and in the cytoplasm. In absence of pHLIP, nanoparticles are not uptaken, but in its presence, bright TR images of the $Eu^{III}$ luminescence are obtained and the NPs also help recording transmission electron microscopy images of the platelets [123].

### *4.3.2.3 Lab-on-a-Chip Technology*

The large volume of information required in bio-screening and biomedical analyses has prompted researchers to develop miniaturised sensors in a methodology often referred to as lab-on-a-chip technology [125]. One aspect of this technology makes use of microfluidic devices, which not only enables miniaturisation of the analytical procedures, but leads to substantial reduction in the volume of expensive reactants and solvents. As the corresponding devices are usually made of biocompatible polydimethylsiloxane (PDMS), autofluorescence of this material usually considerably limits the sensitivity of the analyses. To avoid this problem, the advantages of PDMS microfluidics have been combined with LLB technology and TR luminescence for the detection of biomarkers expressed by MCF-7 cells in cancerous breast tissues [126]. The first-generation microfluidic device showed on Fig. 4.13a features a zigzag 100-μm wide channel in which two types of experiments can be conducted. First, cells can be grown in the fibronectin-coated microchannel in a much shorter time than in the usual cell culture medium and to the benefit of dramatically reduced nutrient volumes; subsequently, the cells can be detected via their specific markers. A proof of concept is given with MCF-7 cells, spotted via the 5D10 antigens they express. To this end, $[Eu(L5)]^-$ (see Scheme 4.4; quantum yield 23.6%, $Eu(^5D_0)$ lifetime 1.42 ms) is conjugated to avidin and reacted with a biotinylated 5D10 antibody. The sensitivity of the TR detection assay (Fig. 4.13c) is almost six times larger than a comparable assay conducted with FITC-labelled streptavidin. Dual immunocytochemical detection of two separate antigens on the same cells (5D10 and substance P peptide) can also be performed with a mixture of $[Eu(L5)]^-$ and $[Tb(L6)]^-$ (Scheme 4.4).

The second type of experiments involves the detection of biomarkers in cancerous tissues. A thin section of the tissue is pressed onto the microchannel and reactants are sequentially flown. For this experiment, two types of receptors expressed by MCF-7 cells have been chosen: the human epidermal growth factor receptor 2 (Her2/*neu*) expressed on the cell surface and the oestrogen receptors (ER) expressed on the nucleus, because their relative proportions determine the therapeutic outcome for the patient. The former receptors are detected indirectly, Her2/*neu* – anti Her2/*neu* IgG – biotinylated anti IgG – $Eu^{III}$-labelled avidin,

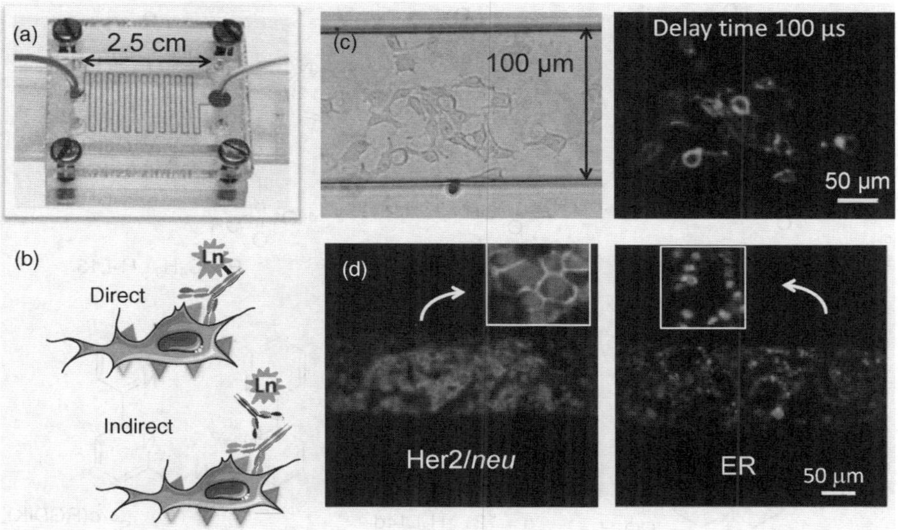


*Figure 4.13* *(a) First-generation microfluidic device. (b) Principle of direct and indirect histochemical assays. (c) Detection of MCF-7 breast cancer cells through the 5D10 antigen they expressed itself recognised by its antibody decorated with [Eu(L5)]$^-$: bright-field image (left) and TR detected Eu$^{III}$ luminescence (right). (d) TR luminescence detection of Her2/*neu *and ER receptors on human breast cancer tissue sections; see text for details. Reproduced with permission from [126]. Royal Society of Chemistry (See colour plate section)*

while the latter are imaged directly by Eu$^{III}$-labelled anti-lgG recognising specifically the anti-ER IgG (Fig. 4.13d).

### 4.3.3 Macrocyclic Probes

This class of compounds has generated a wealth of studies, particularly those derived from the cyclen framework since DOTA-class complexes are amongst the most stable lanthanide chelates (Log*K* between 22.9 for La$^{III}$ and 25.4 for Lu$^{III}$). Moreover the chemical, physico-chemical, and biochemical properties of these complexes have been thoroughly investigated during the development of contrast agents for magnetic resonance imaging.

Suggestion to use luminescent macrocyclic probes for cell imaging, namely lanthanide cryptates, was made in 1995 [127], but the first images were obtained with Tb$^{III}$ macrocyclic complexes with aza crown ether derivatives featuring fused or substituted pyridines as sensitising group [128]. The chelates [Tb(L12a)]$^{3-}$, [Tb(L12b)], and [Tb(L12c)]$^{3-}$ (Scheme 4.7) have sizeable quantum yields in water: 21, 48, and 40%, respectively, upon excitation in the range 260–270 nm, and [Tb(L12a)] has a Tb($^5D_4$) lifetime of 2.7 ms. Biodistribution experiments conducted with [$^{153}$Sm(L12a)] and [$^{153}$Sm(L12c)]$^{3-}$ evidence a preferential localisation in the bone tissue of Sprague–Dawley rats after venous tail injection, which allows luminescence imaging of the rat femur (Fig. 4.14) [128]. This prompted the construction of an endoscopic spectrometer for this type of imaging. On the other hand, the more lipophilic [Tb(L12b)] preferentially localises in the epithelial layer of the colon.

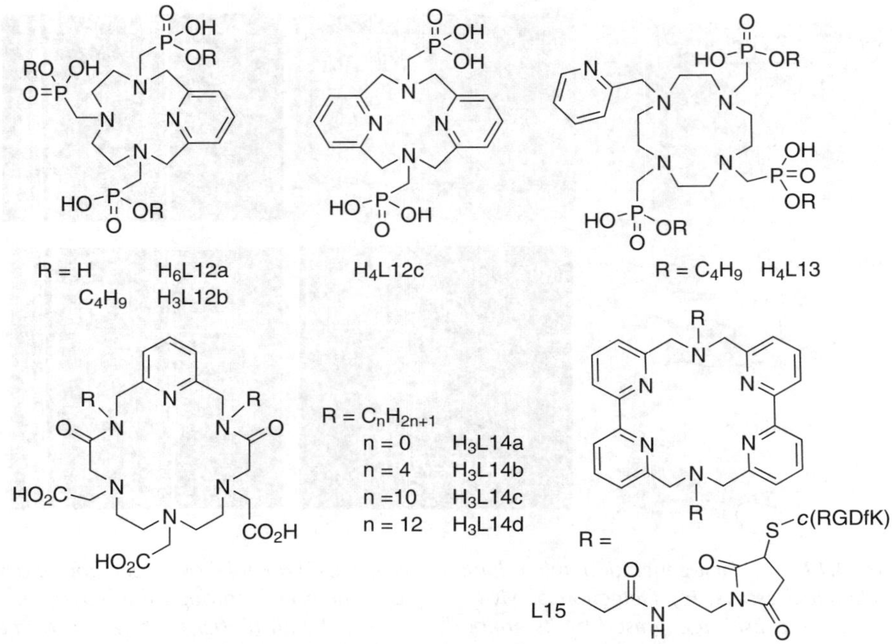


***Scheme 4.7*** *Aza macrocyclic ligands.*

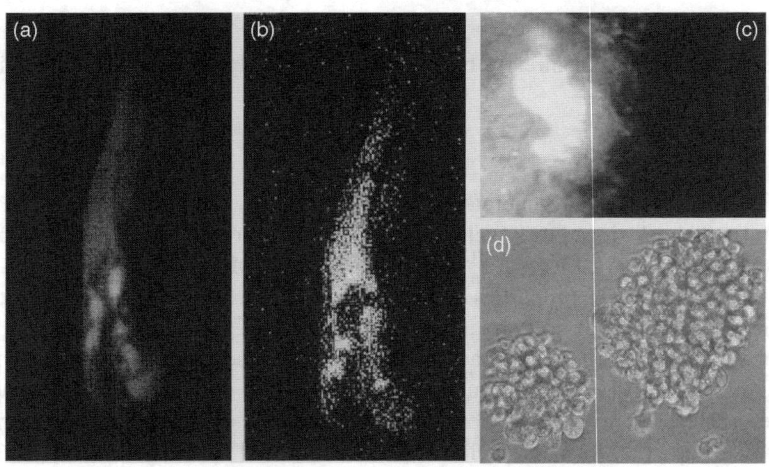


***Figure 4.14*** *Rat femur images: (a) white light reflectance image, (b) luminescence image recorded under monochromatic excitation at 270 nm and with a 550-nm band pass filter after tail-injection of the $Tb^{III}$ chelate $[Tb(L12c)]^-$. Reproduced with permission from [136]. Copyright © John Wiley (c) Colon adenocarcinoma suspect mass illuminated by [Tb(L12b)]; Reproduced with permission from [129]. Copyright © 1999, American Chemical Society. (d) U87-MG cells incubated with $[Eu(L15)]^{3+}$ (TR mode); Reproduced with permission from [133]. Copyright © 2011, Elsevier*

Moreover, carcinogenic cells have greater affinity for the chelate compared to normal cells. The luminescent [Tb(L12b)] chelate was therefore introduced into the intestine of rats having colon cancer through a microendoscope, which allowed to clearly evidence a suspect mass in the colon [129]. Later, the authors switched to the cyclen framework [130], grafting it with a pyridine antenna to give the luminescent contrast agent [Tb(L13)] with a long Tb($^5D_4$) lifetime of 3.5 ms. It was used for the detection of early-stage malignant lesions in Syrian hamster cheek pouch [131]. In subsequent work with similar probes, [Ln(L14)], the $Gd^{III}$ lipophilic complexes with $H_3$L14c,d, were internalised into cultured mammalian Jurkat cells and the chelates displayed increases in T1-weighted MRI images. The $Tb^{III}$ analogues were then used to visualise the uptake mechanism: Förster resonant energy transfer occurs between the $Tb^{III}$ chelate and extracellular, but not intracellular, calcein, which emits its characteristic green fluorescence. Therefore the complexes label the cells by inserting the two hydrophobic alkyl chains into the cell membrane [132]. The disadvantage of these aza-crown probes though is their need for UV excitation and their low molar absorption coefficients (2500–3500 $M^{-1}$ $cm^{-1}$).

Complexes of macrocyclic bipyridine ligand L15 (Scheme 4.7) bearing two cyclic peptides *c*(RGDfK) are well suited for the visualisation of tumour cells expressing integrins on their surface. The cytotoxicity of both $Eu^{III}$ and $Tb^{III}$ complexes is low ($IC_{50} > 100$ μM) and they are sufficiently stable to illuminate $\alpha_v\beta_3$-integrin expressing human glioblastoma U87-MG cells in TR mode, as shown on Fig. 4.14d. The integrin $\alpha_v\beta_3$ plays apparently an important role in tumour growth, metastasis, and angiogenesis and therefore expression of this glycoprotein is in direct correlation with the tumour aggressiveness [133].

A large body of very fine work is being performed by the group of D. Parker at Durham University on luminescent cyclen derivatives with two major aims. One is the design of responsive ratiometric probes for the determination of several analytes, including pH, $O_2$ and various anions within the frame of luminescent time-resolved immunoassays [134]. In this context, a rapid and reliable analytical method enabling the measurement of citrate in seminal fluids has been validated; citrate is a key metabolite the concentration of which falls by up to a factor of ten in individuals having prostate cancer [135].

The second goal is the selective staining of cells as well as luminescent imaging of analytes within cells. Indeed, spatial and temporal variations of analytes in live cells are vital information for the biochemist. In the course of an extensive and seminal investigation of over 60 different cyclen-based $Eu^{III}$ and $Tb^{III}$ LLBs fitted with various chromophores featuring tetraazatriphenylene, acridone, azaxanthone, azathiaxanthone or pyrazolyl azaxanthone sensitising units [137–144], Parker's group has established their localisation and the mechanism of cellular uptake (Scheme 4.8). The cell lines studied include mouse skin fibroblasts (NIH-T3), Chinese hamster ovarian (CHO), or cervix carcinoma HeLa cells. These investigations are thorough and photophysical properties of the luminescent probes such as quantum yield, lifetimes, and quenching rate constants are established, in addition to the uptake mechanism. They are summarised in two recent reviews [145,146] so that only salient aspects are presented here.

Quantum yields of the luminescent chelates published to date are on the order of 8–10% and 40–60% for $Eu^{III}$ and $Tb^{III}$, respectively. Some representative examples are given in Table 4.5 for $Tb^{III}$ chelates. It is noteworthy to point out that the triplet state energy, as determined at 77 K on the $Gd^{III}$ complexes, is about the same for all complexes, 23 450–24 800 $cm^{-1}$ and therefore there is no correlation between this energy and the quantum yields

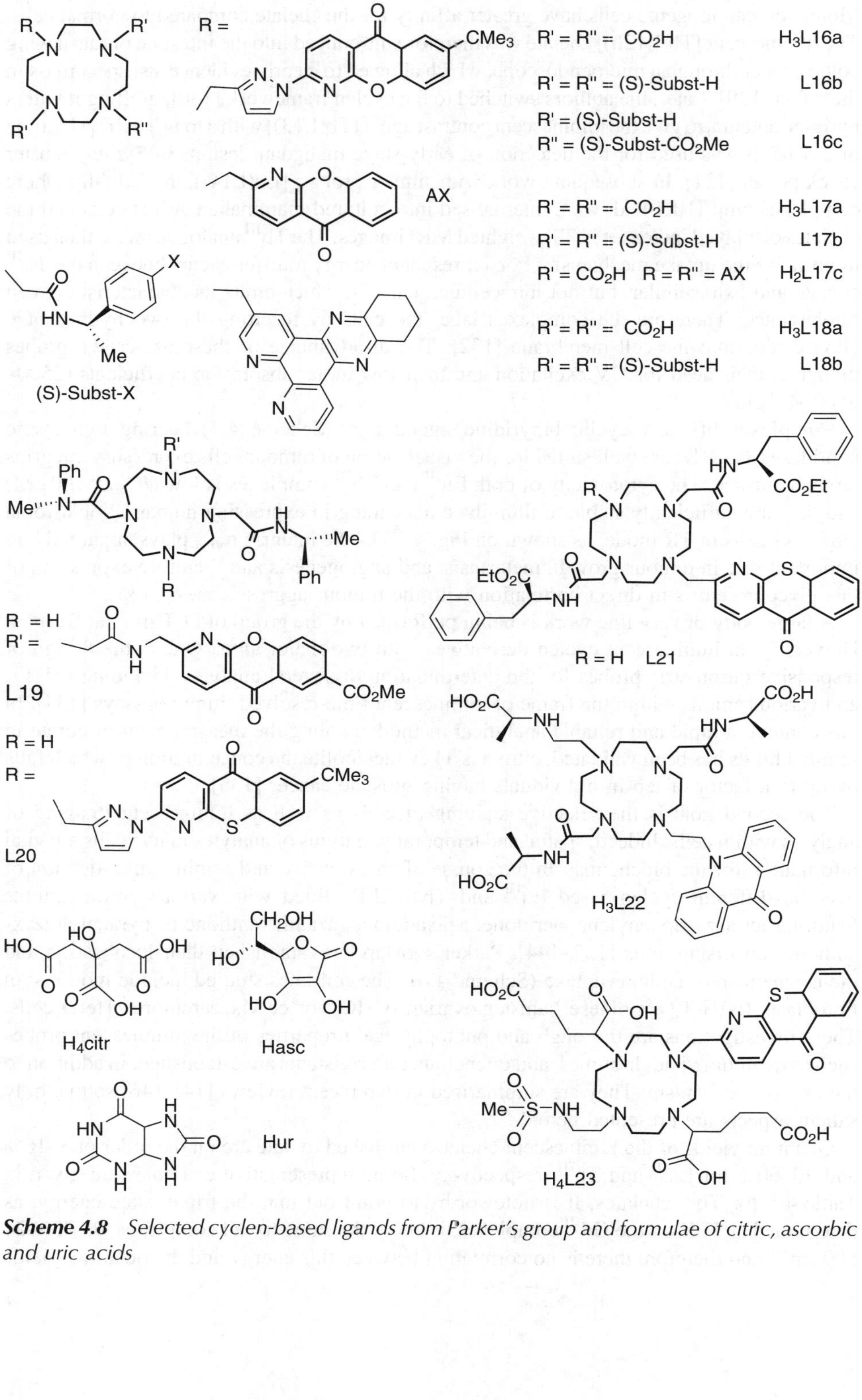


*Scheme 4.8* Selected cyclen-based ligands from Parker's group and formulae of citric, ascorbic and uric acids

*Table 4.5* *Photophysical properties of selected cyclen-based terbium chelates [138]. See Scheme 4.8 for ligand formulae*

| Chelate | $\varepsilon$, $M^{-1}$ $cm^{-1}$ | $\tau(^5D_4)$, ms | $q^a$ | $Q^L_{Ln}$, % | $K^{-1}_{SV}(u)$, $mM^b$ | $K^{-1}_{SV}(a)$, $mM^b$ |
|---|---|---|---|---|---|---|
| [Tb(L16a)] | 15 050 | 2.27 | 0.05 | 15 | 0.03 | 1.39 |
| $[Tb(L16b)]^{3+}$ | 15 050 | 2.00 | 0.1 | 18 | 0.055 | 0.36 |
| $[Tb(L16c)]^{3+}$ | 15 050 | 2.22 | 0.15 | 61 | n.a. | n.a. |
| [Tb(L17a)] | 6900 | 1.81 | 0.6 | 24 | 0.012 | 0.57 |
| $[Tb(L17b)]^{3+}$ | 6900 | 1.64 | 1.0 | 37 | 0.04 | 0.37 |
| [Tb(L18a)] | 8300 | 1.47 | 0.1 | 33 | 0.006 | 0.38 |
| $[Tb(L18b)]^{3+}$ | 8300 | 1.56 | 0 | 40 | 0.025 | 0.18 |

[a] Hydration number
[b] $K_{SV}$ is the Stern–Volmer quenching constant for urate (u) and ascorbate (a) in water, at pH 7.4

of the $Tb^{III}$-centred luminescence. An additional remark is that $Q^L_{Ln}$ values also do not directly correlate with the hydration number of the complex.

Data in Table 4.5 include Stern–Volmer constants for dynamic quenching by urate and ascorbate. These endogenous electron-rich species present in the cells indeed may transfer electrons on the free or protein-bound chelate resulting in a loss in luminescence intensity. Globally, $Tb^{III}$ macrocyclic complexes resist quenching well, meaning that they are suitable for intracellular applications [138]. $Eu^{III}$ chelates are more sensitive to quenching, because it is the excited state of the lanthanide ion which drives the quenching process, and $^5D_0$ (17 300 $cm^{-1}$) lies lower in energy than $^5D_4$ (21 500 $cm^{-1}$) [140]. Urate appears to be much more effective (10–60 times) than ascorbate in quenching the complexes, despite its larger one-electron oxidation potential (0.59 V vs. NHS) compared to ascorbate (0.30 V).

With respect to cellular uptake (Fig. 4.15, top), the nature of the chromophore and its attachment mode to the macrocycle primarily determine the localisation and the kinetics of the cell internalisation, but not the charge of the complex or its lipophilicity. The reason invoked is that polycyclic sensitiser units are recognised by protein association. Therefore, the nature of the substituent on the sensitiser units is a key parameter for protein affinity [147].

Complexes staying in mitochondria have usually low $IC_{50}$ values and cause cell apoptosis while those internalised in lysosomes are non-toxic and therefore can act as responsive probes. About 80% of the macrocyclic complexes have endosomal/lysosomal localisation; in these cases the rates of uptake and egress are fast. Half a dozen complexes also display fast uptake, but slower egress and are shuttling between mitochondria and endosomal/lysosomal compartments; $[Eu(L19)(H_2O)]^{3+}$ is such a chelate, with initial mitochondrial internalisation (<4 h), final lysosomal localisation (>14 h), and $IC_{50} > 175$ μM. In contrast, $[Eu(L16c)]^{3+}$ with $IC_{50} = 58$ μM stains both mitochondria and lysosomes and its localisation is independent of time. Finally, a few complexes display slow uptake and egress and enter the ribosomes and the nucleoli (Fig. 4.15, bottom). Their $IC_{50}$ values are in the range 40–90 μM.

The mechanism with which lanthanide cyclen chelates penetrate into live cells has been established since it is critical to the development of imaging probes and therapeutic vectors. One of the major pathways is endocytosis (either clathrin- or caveolae-mediated) involving

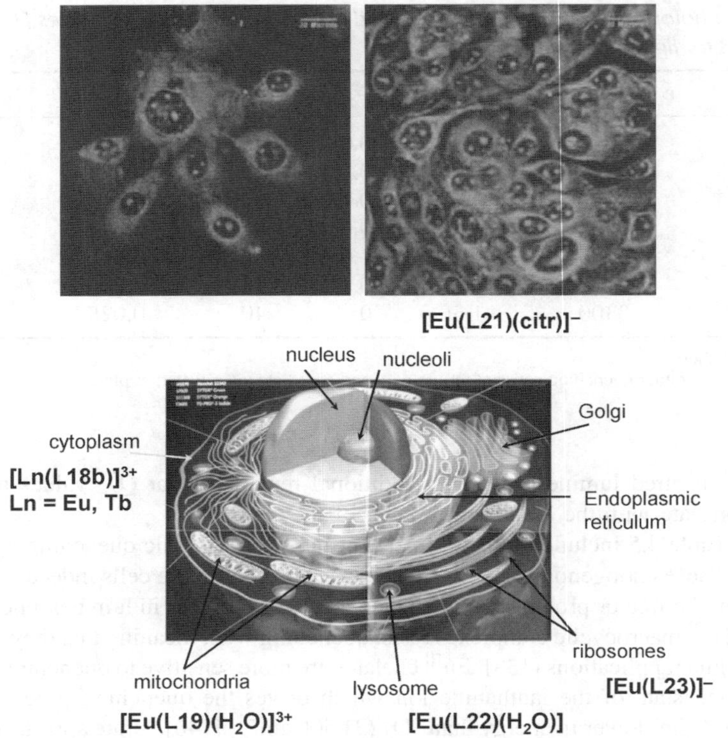


***Figure 4.15*** *Top: Confocal microscopy images ($\lambda_{exc}$ = 405 nm) showing the intracellular localisation of [Eu(L20)($H_2O$)]$^{3+}$ in NIH 3T3 cells. Adapted with permission from [146]. Copyright © 2009, American Chemical Society. Bottom: Dominant intracellular localisation of some cyclen complexes (See colour plate section)*

a vesicular uptake of the incoming chelate followed by invagination of the cell membrane. Other mechanisms are macropinocytosis and clathrin/caveolae-independent cytosis. Distinction between these mechanisms is made by monitoring the uptake in presence of inhibitors/promoters of specific pathways. The dominant mechanism for cyclen Eu$^{III}$ and Tb$^{III}$ complexes seems to be macropinocytosis [148]. In a recent development, [Ln(17c)]$^+$ (Ln = Eu, Tb) cationic chelates have been shown to have enhanced emission intensity in presence of serum albumin, which permitted staining dividing HeLa cells and visualising mitotic chromosomes [149]. Binding to HSA leading to higher luminescence intensity has been confirmed with dinuclear complexes with ligands featuring two DO3A units linked by bridging molecules such as 4,4′-diaminophenylamine or *o*-dianisidine. An apparent binding constant of Log*K* = 4.8 has been obtained in a simulated cellular medium and HeLa cells could be imaged in TR mode. Molar absorption coefficients (5 000–5 500 $M^{-1}$ $cm^{-1}$) and quantum yields (4–7% for Eu$^{III}$ and 8–13% for Tb$^{III}$) for these complexes remain however modest [150].

Another, simpler, series of cyclen derivatives is fitted with 6-derivatised quinoline moieties as sensitisers, yielding neutral chelates [Eu(L24)] (Scheme 4.9); the quantum

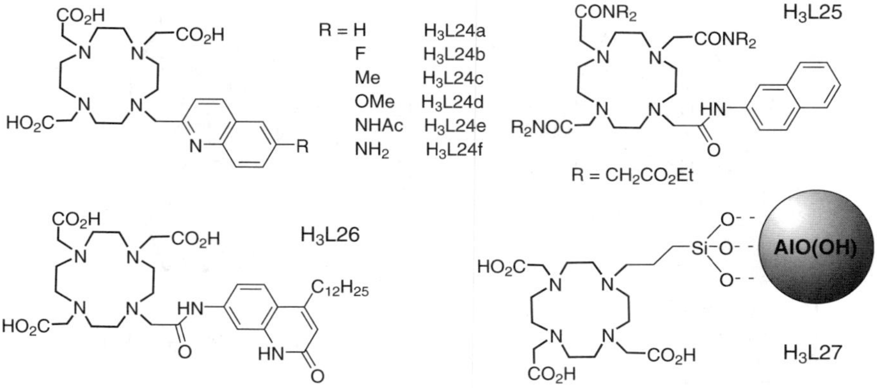


***Scheme 4.9*** *Other cyclen-based ligands and cryptands*

yields are smaller than those exhibited by Parker's compounds (2.6–5.9%) but HeLa cells can be conveniently imaged [53]. In addition, emission from mixtures of [Eu(L24d-e)] and $[Tb(DTPA\text{-}cs124)]^-$ (Scheme 4.5) injected into the same cells can easily be spectrally separated by means of filters.

Further applications of cyclen-derived chelates include imaging of bone cracks with $[Eu(L25)]^{3+}$ (Scheme 4.9) after hydrolysis in $MeOH/H_2O$. A sample of bovine tibia is immersed in a solution containing the $Eu^{III}$ chelate for 24 h and imaged by confocal luminescence microscopy: a healthy bone surface does not show emission while a scratched one is highly luminescent, with an intensity ratio of 1:28. The explanation is that scratched surfaces expose $Ca^{II}$ sites which bind to the $Eu^{III}$ chelate [151].

Energy transfer between $Tb^{III}$ and $Eu^{III}$ is known to be efficient. The carbostyril Cs124 derivatised $H_3DOA$ ligand $H_3L26$ (Scheme 4.9) is fitted with a long hydrophobic $C_{12}$ chain which helps the corresponding $Eu^{III}$ and $Tb^{III}$ chelates localise on cell membranes. The antenna sensitises $Tb^{III}$ much more efficiently than $Eu^{III}$ as demonstrated by images of HEK-293T cells incubated with: (i) the $Eu^{III}$, or (ii) $Tb^{III}$ chelate, and (iii) with a cocktail of both. No luminescence is seen for the former, while $Tb^{III}$ emission clearly stains the cell membrane in the second case and $Eu^{III}$ emission only in the third case, owing to almost quantitative Tb-to-Eu energy transfer [152].

Polyaminocarboxylate chelates with $H_3L27$ (Scheme 4.9) and DTPA have been used to modify boehmite nanoparticles with sizes $18 \pm 1$ nm for the former and $9 \pm 1$ nm for the latter. In aqueous solution, both types of $Gd^{III}$-containing NPs have relaxivity of about $35\ mM^{-1}\ s^{-1}$ at 37 °C and 20 MHz while $Eu^{III}$- and $Tb^{III}$-containing NPs are luminescent, as demonstrated by confocal luminescence microscopy of HeLa cells incubated with the NPs, which localise in the cytoplasm. These results open the way for dual MRI/luminescence imaging [153].

As mentioned at the beginning of this section, lanthanide cryptates are also suitable for TR luminescence imaging. Few of them, however, have been tested for this purpose, with the exception of the highly luminescent commercial stain Tb-Lumi4 (Scheme 4.5) the use of which is ubiquitous [51] and the cryptates marketed by Cisbio-Bioassays for

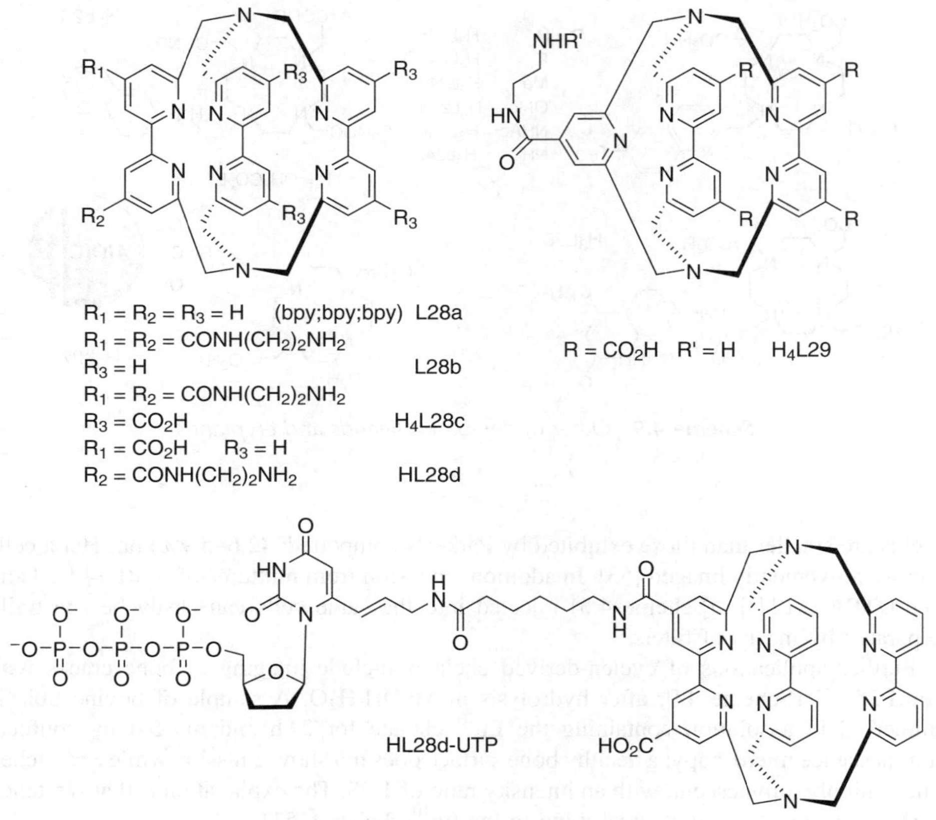


***Scheme 4.10*** *Cryptands for LLBs*

time-resolved immunoassays with TRACE® (time-resolved amplified cryptate emission) technology. For instance, bioconjugated $[Eu(L28b)]^{3+}$ (Scheme 4.10) detects polymerase chain reaction products [154] and $[Eu(L28d\text{-}UTP)]^{2+}$ acts as luminescent label for RNA analysis [155]. Optimisation of the presently used cryptate, [Eu(L29)], is an interesting case study for the design of lanthanide luminescent bioprobes. Indeed, $[Eu(L28a)]^{3+}$ has a very low 2% quantum yield [156] because of two reasons: (i) the cryptate is hydrated with at least two water molecules coordinated onto the metal ion and (ii) a low-lying ligand-to-metal charge transfer state is present. The resulting photoinduced electron transfer leads to the formation of $Eu^{II}$, which has an ionic radius of 1.25 Å, fitting very well into the cryptate cavity. Introducing fluoride ions into the solution expels water molecules from the inner coordination sphere and the quantum yield goes up to about 7%. Reducing the cavity size by cropping one pyridine off a bpy unit gives cryptand $H_4L29$, which displays larger quantum yield, therefore avoiding the use of cytotoxic fluoride ions and opening the way for *in vivo* imaging experiments [157]. This is exemplified by the imaging of γ-aminobutyrate B ($GABA_B$) receptors on the surface of transfected HEK cells (Fig. 4.16).

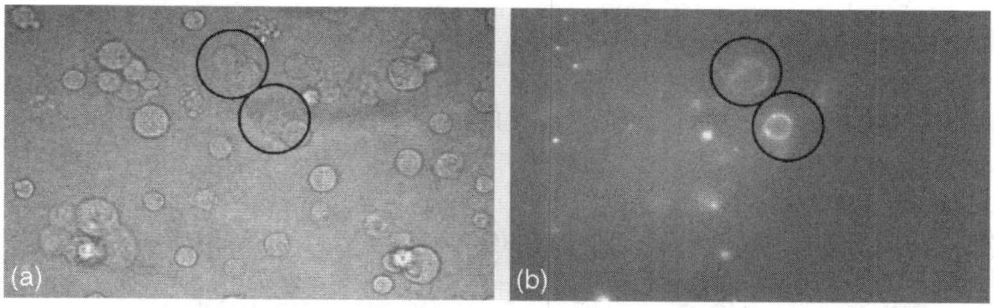


***Figure 4.16*** *Bright field (a) and time-resolved luminescence (b; delay 100 µs) of transfected living HEK cells expressing the tagged membrane receptor GABAB2 as revealed by [Eu(L29)]$^{2+}$-conjugated antibodies. Reproduced with permission from [158]. Copyright © 2007, Elsevier*

### 4.3.4 Self-assembled Triple Helical Bioprobes

Tridentate ligands, such as bis(benzimidazole)pyridines [159] or dipicolinates (DPA) [160], wrap around $Ln^{III}$ ions to yield saturated 9-coordinate complexes in which the ligand strands are hold together by weak intramolecular π-π interactions. Lanthanide tris(dipicolinates) are fairly stable in water at physiological pH, a reason why they have been tested as bioprobes, in particular, after suitable derivatisation, as multi-photon-excited stains [161], see Chapter 5. The use of tris(dipicolinates) as luminescent analytical probes is well established, for instance for analysing triphosphate nucleotides [162]. Remarkably enough, dipicolinic acid is a constituent of *Bacillus anthraces* spores, which have been the biological vectors in anthrax terrorist attacks. These spores may therefore be detected through complexation with a non-luminescent $Tb^{III}$ complex with a cyclen derivative, which becomes luminescent upon formation of a ternary complex with DPA [163].

Double lanthanide tags taking advantage of two lanthanide ions in one molecule have been recently advocated for protein structure investigation [164]. For testing the utility of dinuclear LLBs for bioimaging, a library of water-soluble hexadentate ditopic ligands with DPA core fitted with two benzimidazole chromophoric units has been developed (see Scheme 4.11). These ligands self-assemble in water, at room temperature and physiological pH 7.4, to give the corresponding homodinuclear helicates in large yields. Several criteria have to be met for a complex to be considered as a biological probe: thermodynamic stability, kinetic inertness, photophysical properties, cytotoxicity, and ability to conjugate with biomolecules. All these criteria have been tested for the helicates.

#### *4.3.4.1 Structure, Thermodynamic Stability, and Kinetic Inertness*

Crystal structures are only available for $[Ln_2(L30a)_3]$ (Ln = Eu, Tb). They show the two metal ions lying on a pseudo-$C_3$ axis, at a distance of about 9 Å, and the three ligand strands helically wrapped around them, forming protective $N_6O_3$ nine-coordinate environments (Fig. 4.17, left).

This structure is maintained in solution, as demonstrated by lanthanide-induced shift analysis on eight different helicates [165]. Although quite precise, the procedure is somewhat long and involved so that the structure of the other dinuclear bioprobes has simply been ascertained by the NMR spectra of the chelates with diamagnetic $Ln^{III}$ ions, which point to

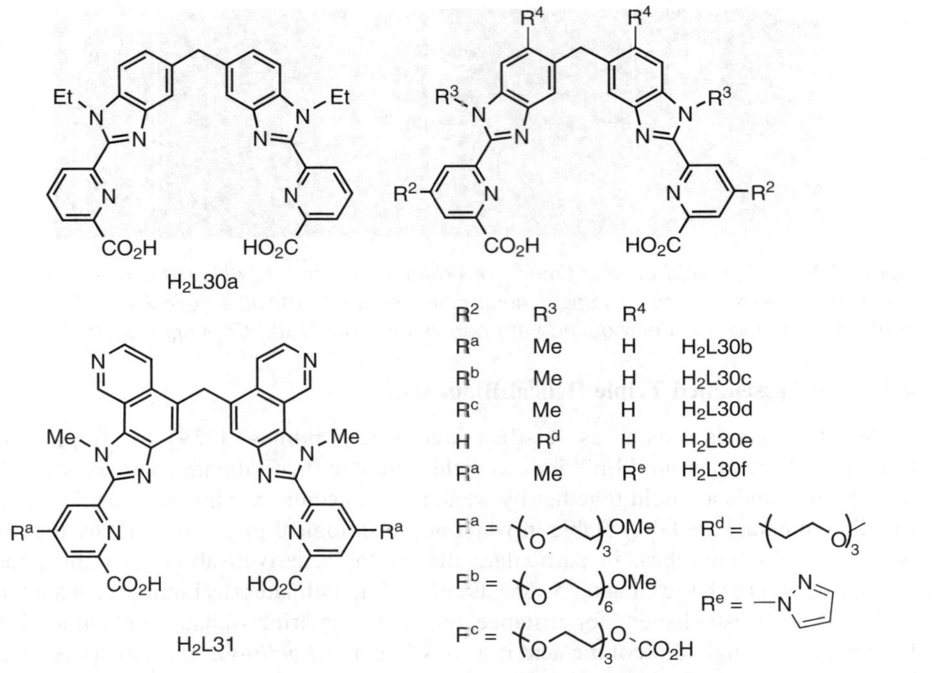


***Scheme 4.11*** *Hexadentate ditopic ligands for the self-assembly of lanthanide dinuclear helicates*

averaged $D_3$ symmetry around the metal ions, and by luminescence spectra. Indeed the ligand-field splitting of the Eu($^5D_0$) emission bands is a signature for the symmetry around the metal ion. In particular, all the helicates with the ligands sketched in Scheme 4.11 display emission spectra characteristic of a pseudo-trigonal symmetry, the $^7F_1$ level being split into two A and E

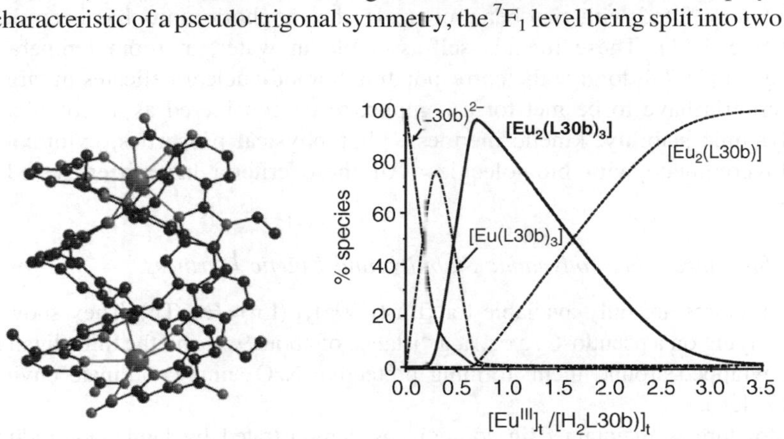


***Figure 4.17*** *Left: Crystal structure of [$Tb_2(L30a)_3$]. redrawn from Ref. [165]. Right: Speciation diagram for $Eu^{III}$ in presence of [$H_2L30b$)]$_t$ = $4.5 \times 10^{-4}$ M in Tris-HCl buffer (pH 7.4); Reproduced with permission from [166]. Royal Society of Chemistry*

*Table 4.6 Photophysical properties of solutions of the helicates $[Eu_2(L30i)_3]$ (i = a-f)$^-$ and $[Eu_2(L31)_3]^{6-}$ in aqueous solution (Tris-HCl 0.1 M) at pH 7.4 and 295 K. Data from references [88,169,170]*

| Ligand | % 2:3$^a$ | Log$\varepsilon^b$ | (A-E)$^c$, cm$^{-1}$ | (E-E)$^c$, cm$^{-1}$ | $Q^L_{Eu}$, ±10% | $\tau_{obs}$, ms ±2% | $\tau_{rad}$,$^d$ ms ±12% | $Q^{Eu}_{Eu}$,$^e$ ±12% | $\eta_{sens}$,$^f$ ±16% |
|---|---|---|---|---|---|---|---|---|---|
| $H_2L30a$ | 99.5 | n.a. | 156 | 28 | 0.24 | 2.4 | 6.9 | 0.37 | 0.67 |
| $H_2L30b$ | 97.1 | 4.93 | 161 | 31 | 0.21 | 2.4 | 6.9 | 0.36 | 0.58 |
| $H_2L30c$ | 92.9 | n.a. | 146 | 31 | 0.19 | 2.4 | 6.3 | 0.37 | 0.52 |
| $H_4L30d$ | n.a. | n.a. | 180 | n.a. | 0.15 | 2.4 | 7.4 | 0.33 | 0.45 |
| $H_2L30e$ | 89.2 | n.a. | 162 | 38 | 0.11 | 2.2 | 6.2 | 0.36 | 0.30 |
| $H_2L30f$ | 92.5 | 4.97 | 168 | 26 | 0.15 | 2.5 | 6.4 | 0.40 | 0.38 |
| $H_2L31$ | 94.6 | 4.90 | 177 | 34 | 0.09 | 2.3 | 6.7 | 0.35 | 0.26 |

$^a$ Percentage of the $[Eu_2L_3]$ helicate for $[L]_t = 4.5 \times 10^{-4}$ M
$^b$ Molar absorption coefficient of the helicate ($M^{-1}$ $cm^{-1}$)
$^c$ Splitting of the $^5D_0 \rightarrow {}^7F_1$ transition at 10 K in water/glycerol 9/1, v/v (see text)
$^d$ $Eu(^5D_0)$ radiative lifetime (see Equation 4.10)
$^e$ Intrinsic quantum yield (see Equation 4.10)
$^f$ Ligand sensitisation efficiency (see Equation 4.10)

components. The latter is further split into a tight doublet. According to ligand-field considerations [167 and Chapter 1], the A–E energy difference is proportional to the $B^2_0$ crystal-field parameter while the E–E energy difference gives the extent of the distortion from the idealised $D_3$ symmetry. These energy differences are reported in Table 4.6. The dinuclear helicates are quite stable, with Log$\beta_{23}$ values ranging between 23 and 28. As a consequence, the percentage of 2:3 species at concentrations used in bioanalyses is larger than 95% for both systems with $H_2L30b$ and $H_2L31$ (Table 4.6, Fig. 4.17, right). Moreover, the next more abundant species (1:2 or 1:3) accounts for most of the remaining speciation, that is, stoichiometric 2:3 solutions contain only minute concentrations of non-complexed $Ln^{III}$, an important feature when it comes to put them into contact with living materials. Another finding is that the stability does not vary much along the lanthanide series, so that one lanthanide ion may be replaced by another one with different photophysical properties without significantly affecting the speciation, another useful feature for multiplex experiments. Biological media contain a wealth of substances potentially able to interfere with LLBs, either chemically by ligand exchange or transmetallation, or photophysically by quenching of the metal-centred luminescence by endogenous antioxidants. In order to test the action of potential interfering substances, several experiments were conducted on $[Eu_2(L30b)_3]$ to which were added 10–100 equivalents of EDTA, DTPA, citrate, L-ascorbate, or zinc. The luminescence of the resulting solutions did not decrease substantially (0–10%) 24 h after the addition, meaning that the dinuclear LLB is well suited for bio-analyses. Moreover, no ligand exchange occurs with polyaminocarboxylates and transmetallation with zinc starts only to be problematic when 100 equivalents are added, corresponding to a concentration far more important than the largest one found in bio-systems [168].

### 4.3.4.2 *Photophysical Properties*

Relevant photophysical parameters for the $[Eu_2(L)_3]$ helicates are reported in Table 4.6. The most striking characteristics are the large molar absorption coefficients, around

85 000 $M^{-1}$ $cm^{-1}$, that are far larger than those exhibited by most lanthanide bioprobes, long lifetimes, and sizeable quantum yields in aqueous solution. Among the helicates tested, the one with $H_2L30b$ proved to be the best, especially that the ligand sensitises also $Tb^{III}$ ($Q^L_{Tb} = 0.11$), $Sm^{III}$ ($Q^L_{Sm} = 0.004$), and $Yb^{III}$ ($Q^L_{Yb} = 0.0015$).

#### *4.3.4.3 Cytotoxicity*

Absence of or small cytotoxicity is an important prerequisite for LLBs intended for the sensing and imaging of live biomaterials. WST-1 cell viability and proliferation tests have been conducted for $[Eu_2(L30i)_3]$ and $[Eu_2(L31)_3]$ helicates on various cancerous cell lines, 5D10 (mouse hybridoma), Jurkat (human T leukaemia), MCF-7 (human breast carcinoma), HeLa (human cervical adenocarcinoma), as well as a non-cancerous line, HaCat (human keratinocyte). The resulting data clearly demonstrate that the cytotoxicity of all the tested helicates is very small, with half inhibitory concentration $IC_{50} > 500\,\mu M$. Some of these results were also confirmed by the lactase dehydrogenase (LDH) test, which allows an assessment of damages caused to the cell membrane by determining the LDH leakage out of the cell. In the case of $[Eu_2(L30e)_3]$, this leakage was found to be less than 4% for the three tested cell lines: Jurkat, 5D10, and MCF-7 [171]. Not only cell viability and proliferation are not affected by the presence of the dinuclear LLBs in the cell culture medium but, also, their morphology remains unchanged.

#### *4.3.4.4 Cell Penetration and Imaging*

When incubated with the $[Ln_2(L30i)_3]$ helicates, the live cells described above uptake the LLBs in a concentration-dependent way [166,168,169,171–174]. In view of their large molecular weight (2.6–3.1 kDa) the most likely operating mechanism is receptor-mediated endocytosis. In this process the cell membrane engulfs the incoming molecule to form a pocket which escapes into the cell to produce a vesicle filled with extracellular fluid, including the exogenous substance. The vesicle moves into the cytosol and then merges with endosomes and liposomes. An easy way to prove this mechanism is to conduct the uptake experiments at 4 °C because endocytosis is not occurring at low temperature. Additional proofs for this receptor-mediated uptake mechanism come from the use of endocytosis inhibitors. When the HeLa cells are incubated with $[Eu_2(L30b)_3]$ in a hypertonic medium containing 0.45 M of sucrose, a 92% decrease in the uptake of the LLB is observed because hypertonicity prevents the association between the receptor-LLB complex and the clathrin lattice to form the coated pit; it also blocks the receptor clustering necessary to the formation of the endosomes. Similarly, potassium depletion blocks the receptor clustering and indeed, low concentration of this ion in the culture medium results in a decrease of 83% of the LLB uptake in the cells [174].

Another important information to unravel is where the luminescent helicates are located within the cytosol of the live cells. To this end, co-localisation experiments have been conducted with organic stains known to localise in specific compartments of the cells, such as acridine orange (AO), which stains the cell nuclei. In the top part of Fig. 4.18, the helicate is shown to be entrapped into isolated vesicles which diffuse into the cytoplasm and the size of which is usually around 0.5–1.5 μm, with shapes ranging from flecks, spheres to ellipsoids. After 1 h incubation, vesicles have concentrated around the nucleus, as ascertained by the counter-staining experiment with AO (Fig. 4.18). In fact, other counter-staining experiments

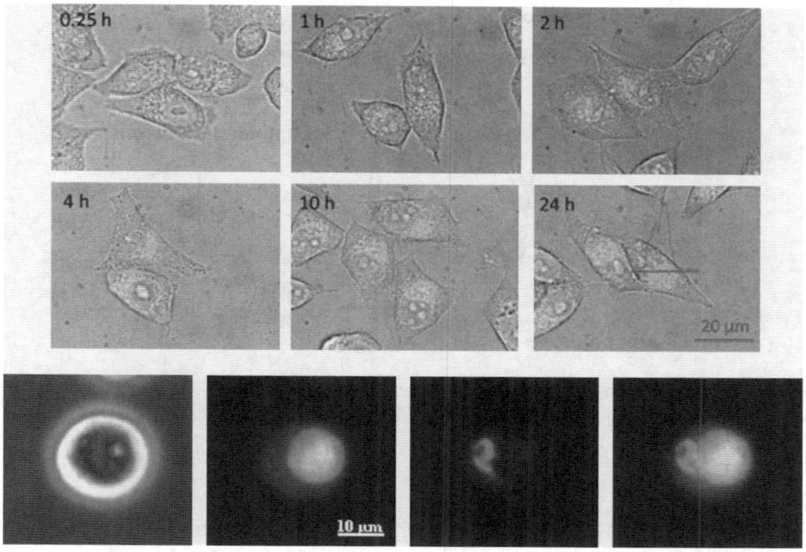


***Figure 4.18*** *HeLa cell imaging. Top: TR luminescence microscopy of the penetration of* $[Eu_2(L30b)_3]$ *in cells incubated at 37 °C for various times; Reproduced with permission from [174]. Royal Society of Chemistry. Bottom: Co-localisation experiment with AO and* $[Eu_2(L30b)_3]$ *showing penetration of the lanthanide probe outside the nucleus; Reproduced with permission from [172]. Copyright © 2008, John Wiley (See colour plate section)*

with Lyso-Tracker blue (localising in the lysosomes), Golgi-Tracker (Golgi apparatus), and ER-Tracker (endoplasmic reticulum) demonstrate localisation in the endoplasmic reticulum [174]. As said before, $(L30b)^{2-}$ sensitises the luminescence of several $Ln^{III}$ ions, and similar images have been obtained with $Tb^{III}$ and $Sm^{III}$.

#### *4.3.4.5 Specific Targeting with Bioconjugated Luminescent Probes*

The TR images of live cells are interesting, but the uptake is not specific, as the helicates penetrate all cell lines, cancerous or non-cancerous. To detect cancerous cells specifically, one has to conduct histochemical analyses aiming at targeting distinctive receptors expressed by the cells. To this end, the helicates are coupled to antibodies either directly, that is the lanthanide probe is linked to a monoclonal antibody mAb, or indirectly with the chelate being covalently bound first to avidin (or biotin); the resulting duplex is then fixed onto a biotinylated (or avidin-derivatised) monoclonal antibody, B-mAb. The procedure is schematically illustrated in Fig. 4.19. A relatively easy way to achieve bioconjugation is to graft a carboxylic acid group onto the ligand, which will then react with amine groups of the biochemical molecule. This has been done with $H_4L30d$, yielding $[Ln(H_2L30d)_3]$-avidin conjugates (LnL30d-A). MALDI-TOF spectra and a quantitative UV-vis analysis point to the binding being covalent and to an average of 3.2 helicates bound per avidin molecule. Except for the quantum yield which decreases by about half, all photophysical properties are maintained in the conjugate and, most importantly, the biochemical activity of avidin is unchanged. Finally, bio-affinity assays

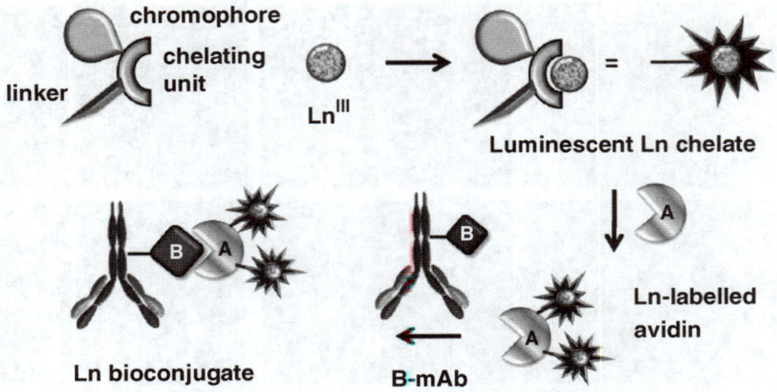


*Figure 4.19 Top: Synthesis of a monoclonal antibody decorated with luminescent lanthanide chelates. Bottom: Specific detection of a biomarker expressed by a cell. Adapted with permission from [12]. Copyright © 2010, American Chemical Society*

demonstrate that LnL30d-A linked to a specific monoclonal antibody can target the 5D10 antigen expressed by MCF-7 human breast cancer cells and that it performs as well as a commercially available lanthanoid tag and much better than an all-organic conjugate. The method is particularly effective in the detection of cultured cells grown in microchannels, a way of achieving screening tests more rapidly (cells grow more rapidly in fibronectin-coated microchannels than in a bulk culture medium) and in a more cost-effective way. The same detection scheme has been applied to visualise MCF-7 cells on magnetic beads after their separation from Jurkat cells on the microfluidic device [175].

An interesting development of this technology is the simultaneous immunohistochemical detection of two biomarkers expressed by cells in human breast cancer tissue sections. These markers are oestrogen receptors (ER) expressed by the nucleus and human epidermal growth factor receptors (Her2/*neu*) expressed by the cell membrane. The relative concentrations of the two receptors are relevant for deciding which therapy has to be applied to the patient. For the analysis, 4-μm sections of formalin-fixed breast carcinomas are pressed against the microchannel and the two markers are visualised by a dual indirect immunohistochemical luminescent assay under time-resolved conditions, as illustrated in Fig. 4.20. The two bioconjugates used are a goat anti-mouse IgG antibody labelled with $[Eu_2(H_2L30d)_3]$ (EuB1) and a goat anti-rabbit IgG antibody labelled with $[Tb_2(H_2L30d)_3]$ (TbB2). The first and simplistic microfluidic device designed allowed the test to be completed in 20 min versus more than 2 h in the hospital and using only one-fifth of the expensive reactants [170,176]. Currently, a second-generation microfluidic device reduces the analysis time down to 3–4 min [177].

### 4.3.4.6 Further Developments

In an attempt to improve the sensitivity of the $[Eu_2(L30b)_3]$ dinuclear helicates, they have been embedded into bare and $NH_2$-functionalised monodispersed silica nanoparticles with average sizes of 55 and 90 nm, respectively. On average, 2380 helicates are inserted into one derivatised NP and their quantum yield amounts to 0.28, a 30% increase with respect to the aqueous solution, due to the protective effect of NPs. The $Ln^{III}$-loaded NPs

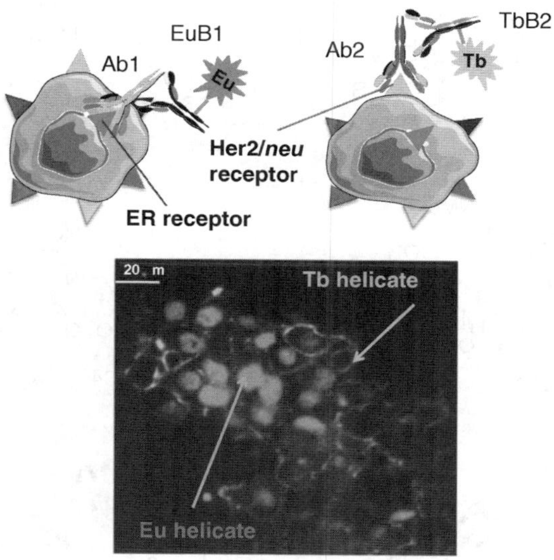


***Figure 4.20*** *Principle of the dual immunohistochemical detection of biomarkers expressed by human breast cancer cells. Ab1 = anti-human ER mouse mAb; Ab2 = polyclonal rabbit anti-human c-erbB-2 oncoprotein Ab. EuB1 and TbB2: see text. Reproduced with permission from [170]. Royal Society of Chemistry (See colour plate section)*

are amenable to bioconjugation to IgG antibodies and were provided for the detection of the 5D10 antigen on MCF-7 cells. The signal to noise ratio obtained is 25% better compared to the molecular probe [178].

A potential drawback of most LLBs described in this chapter, including the helicates, is their excitation wavelength, which usually lies between 320 and 350 nm. Shifting it toward less phototoxic wavelengths can only be done to the detriment of the quantum yield and not further than 400–410 nm in the case of $Eu^{III}$. An alternative is to turn to multiphoton excitation (MPE). Although the hexadentate ligands drawn in Scheme 4.11 are not at all optimised for MPE, they display two- or three-photon excited luminescence. The three-photon cross sections are quite modest for $[Ln_2(L30b)_3]$, 0.26 and 0.27 GM for Ln = Eu and Tb, respectively, while the two-photon cross section for $[Eu_2(L31)_3]$ is equal to ≈ 9 GM, which allows confocal microscopy images of HeLa cells to be recorded under 750-nm excitation [179]. We note that optimised ligands for two-photon excitation have cross sections up to 775 GM [180] so that multi-photon bioimaging could become very sensitive and practical.

### 4.3.5 Other Bioprobes

In addition to the ligands described above, several other classes of compounds have been tested as efficient chromophores for lanthanide luminescent probes. For instance, the cationic chelates $[Eu(L32)_3]^{3+}$ (Scheme 4.12) and their $Zn^{II}$ analogues are effective luminescent markers for the diagnosis of oesophageal squamous carcinoma cells at low concentrations, 0.7–1.6 μM [181].

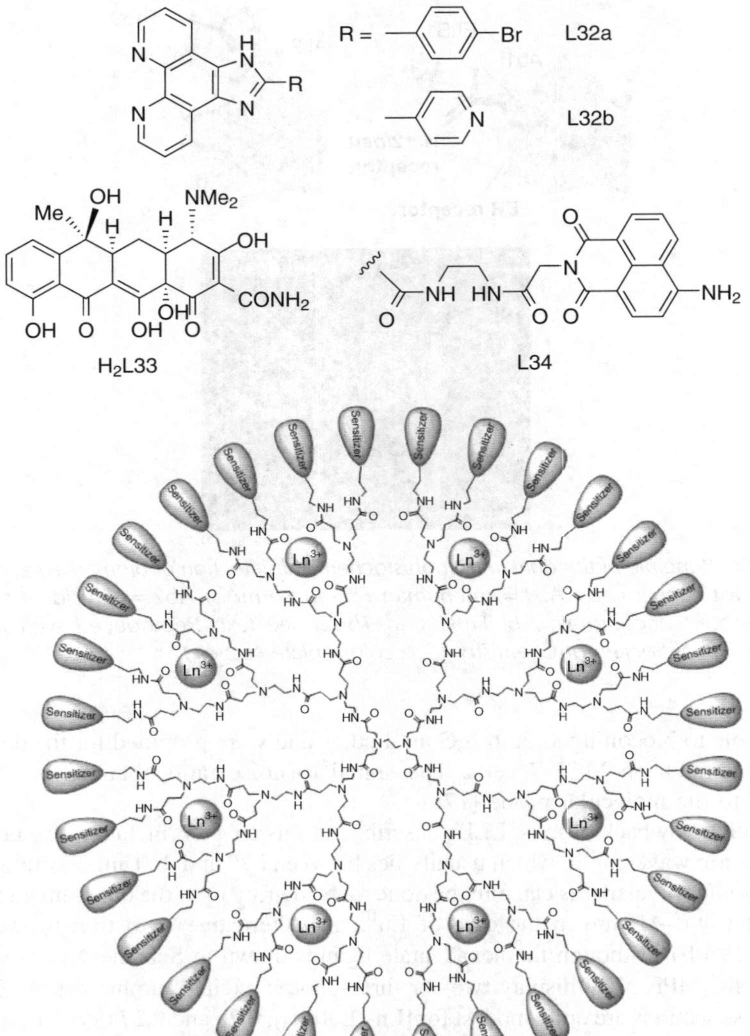


***Scheme 4.12*** *Top: Terpyridine, tetracycline derivatives and sensitiser for dendrimer. Bottom: G3 PAMAM dendrimer $Ln_8[G3\text{-}PAMAM\text{-}(L34)_{32}]$; sensitiser = L34. Redrawn from Reference [190]*

Hydrogen peroxide is produced by the activity of oxidases and determining its intracellular concentration provides assays of the activity of the corresponding enzyme or of substrates such as glucose. It can be detected with high sensitivity by chemiluminescence or electroluminescence, however only at relatively high pH values. A lanthanide-based analytical procedure was proposed in 2002 by O. Wolfbeis and collaborators, relying on the finding that the europium complex with tetracycline $H_2L33$ (Scheme 4.12) binds $H_2O_2$, forming a luminescent ternary complex with a quantum yield of 4% at pH 6.9, representing a 15-fold increase with respect to the initial chelate [182]. The stoichiometry of this complex is not well established but in the analytical procedure, a 3:1 (Eu:L) stoichiometric ratio gives the highest sensitivity. Addition of an enzyme to catalyse the reaction and of a surfactant to enhance the luminescence intensity is not necessary and the reaction is complete in 10 min. The probe is excited at 405 nm by a diode laser and is effective in the range 2–400 μM (limit of detection = 1.8 μM, linear range 2–200 μM). Building on this technology, the authors have engineered an imaging device in which a light-guiding adapter consisting of 96 optical fibres with 3 mm O.D. is inserted between the CCD camera and the 96-well microtitre plate. Lifetime determination imaging of the activity of glucose oxidase in the wells yields a linear range from 0.3 to 2.7 mUnits $ml^{-1}$ with a detection limit corresponding to 1.7 ng $ml^{-1}$ [183]. The method has been applied to an enzyme-linked immunosorbent assay [184].

Dendrimers are very efficient light-harvesting molecules [185,186] besides being excellent energy donors; they are easily functionalised and can encapsulate several lanthanide ions, which may compensate for low quantum yields, particularly with NIR-emitting $Ln^{III}$ ions [187,188]. They are built from successive replication (generation) of an initial dendron so that their radius grows linearly with the generation number while the number of coordinating groups grows geometrically. Some dendrimers such as polyamidoamine (PAMAM) are biocompatible and have demonstrated great potential in cancer therapy. To provide an imaging stain, a third-generation PAMAM dendrimer has been linked through a glycine bridge with 32 sensitising 1,8-naphthalimide groups bearing an amino function in position 4 for further bioconjugation [189,190]. Reaction with $Eu^{III}$ leads to the coordination of eight cations (Scheme 4.12), as confirmed by NMR and elemental analyses. The absorption spectrum of $Eu_8$[G3-PAMAM-$(L34)_{32}$] reveals a maximum at 440 nm but absorption at 630 nm, an attractive wavelength for bioimaging, is still sizeable, with a molar absorption coefficient of 5000 $M^{-1}$ $cm^{-1}$. Upon excitation at 450 nm, the quantum yield of the metalated dendrimer is only $2.9 \pm 0.1\%$, but it is fairly insensitive to photobleaching, the integrated absorbance decreasing only by 4% after 10 h of white-light irradiation, as compared to 30% for the uncomplexed dendrimer. In fact, the dendrimeric complex is tailored for the detection of liver cancers, which often develop as a metastasis of colorectal cancers, and indeed it accumulates in the liver after hepatic arterial infusion in six-week-old male WAG/RijHsd rats. Luminescence from the tumour is visible a few seconds after the injection and blank experiments with $Eu^{III}$ or [G3-PAMAM-$(L34)_{32}$] alone show less good signal-to-noise ratio. This makes rapid visualisation of the tumour feasible, even without TR detection (Fig. 4.21). The $Eu^{III}$ luminescence persists for up to 72 h. Two-photon confocal luminescence microscopy with excitation at 820 nm confirms the presence of the dendrimeric complex outside of the vessels in the perivascular space due to its reduced size (about 3 nm). Other applications of this versatile dendrimer platform are foreseen: the amino group on the 1,8-isophthalamide sensitiser can be conjugated to peptides and antibodies with specific tumour targeting, which would improve accumulation into the cancerous tissue [189]. Moreover, the dendrimers are good conductors of electrical and thermal energy and can serve as target for non-invasive radiofrequency fields releasing heat for generating localised hyperthermia [190].

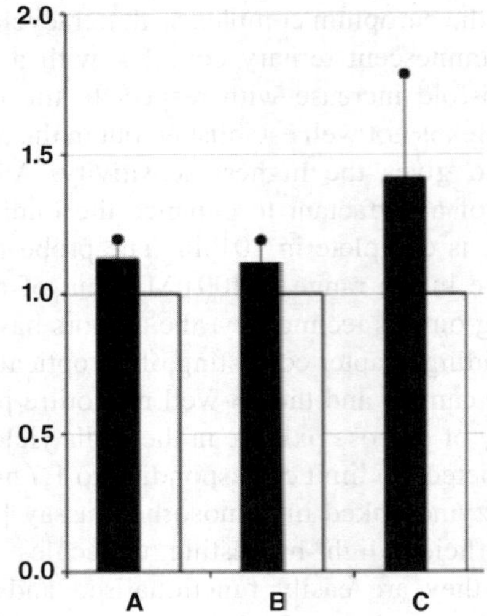


*Figure 4.21 Average relative luminescence intensity of background fluorescence (white rectangles) and tumour luminescence (black rectangles) of rat liver after arterial infusion of $Eu^{III}$ alone (A), G3-PMAML34 alone (B), and $Eu8[G3\text{-}PAMAM\text{-}(L34)_{32}]$ (C). Reproduced with permission from [189]. Copyright © 2011, Elsevier*

## 4.4 Conclusions and Perspectives

After a slow start due to the lack of suitable instrumentation and of highly luminescent, photostable and thermodynamically robust lanthanide bioconjugates, biological imaging with the help of lanthanide luminescent probes and time-resolved detection is now a burgeoning field, as a quick glance through the previous sections demonstrates. With respect to coordination compounds, several classes of highly performing ligands are now at hand and synthetic chemists can modulate their functionalities almost at will. This review has only covered the use of lanthanide coordination compounds but a number of other classes of lanthanide chromophores, mostly inorganic, such as hydroxyapatites [191], doped lanthanide fluoride [192], or oxysulfide [193] nanoparticles, or up-converting nanoparticles, UCNPs, [118,194] are in the race too, particularly when it comes to NIR imaging [113,195].

One developing strategy is indeed to take advantage of the transparency of biological tissues to NIR light (650–950 nm – NIR I window; and 1100–1400 nm – NIR II window) in order to increase the examination depth and to decrease the autofluorescence of the sample. This is by no means easy for two reasons. First, NIR-emitting ions are extremely sensitive to de-activation by collisions with solvent molecules and by high energy vibrations (OH, NH, CH, C=O), henceforth the tendency to protect these emitters into nanostructures with low frequency phonon distribution (ideally < 600 $cm^{-1}$). Second, even if excitation is performed with NIR light, upconverting luminescent probes often emit in the visible which offsets part of the NIR advantages, especially that multiphoton excitation or up-conversion processes have lower probability than one-photon absorption. The real way out is to combine NIR excitation with NIR emission (NIR-NIR imaging). This can be achieved with UCNPs provided the red emission of $Er^{III}$ (640–670 nm) is favoured over the green one (520–550 nm), for instance by controlling the particle shape [196]. Another interesting dopant is $Tm^{III}$ which emits also around 650 nm, but most of its upconverted light is given off in the blue and yellow. A small animal up-conversion imager has nevertheless been developed and *in vivo* imaging performed on female nude mice after injection of $Tm^{III}$-activated UCNPs into the tail vein shows UCNPs localising in the liver [197]. Moreover, another NIR-NIR imaging system has been proposed for imaging mice *in vivo*, which relies on the 1.55 μm emission of $Er^{III}$ [198–200] and similar systems are being developed for various applications, including photodynamic therapy of cancer [201].

Therefore an excruciating question for coordination chemists is: "What is the future of lanthanide chelates as LLBs?" In my opinion, the answer is rather positive and is shared by others [202]. Indeed, chelates will certainly continue to constitute the heart of immunoassays and be quite helpful in immunocyto- and immunohisto-chemical analyses and bioimaging of either whole cells or intracellular analytes. In this respect, the cyclen framework continues to generate an amazing number of new bioprobes [203], with another cyclic platform, triazacyclononane, also currently being tested [204]. Owing to their bright luminescence, diketonates remain strong candidates too [205] but many other ligands, for example polyaminocarboxylates [206], substituted pyridines [207], or task-specific ionic liquids [208], are also demonstrating interesting ability for sensitising lanthanide luminescence. There is certainly a lot of space for imaginative solutions in this field, as demonstrated by pentadecanuclear hydroxo clusters supported on peptides and accumulating in the endosomal/lysosomal system of HeLa cells [209].

When it comes to NIR bioimaging with coordination compounds, $Yb^{III}$ appears to be rather appropriate with emission in the range 980–1020 nm. The energy gap between the emitting level and the ground state being approximately 10 000 $cm^{-1}$, this ion is less sensitive to vibrational de-activation than $Nd^{III}$ or $Er^{III}$ [60] and molecular coordination compounds with organic ligands featuring quantum yields in the range 3–4% are now available [210,211]. As a comparison, the yield of upconversion in $NaYF_4$:Yb(18%), Er(2%) nanoparticles is only $3 \times 10^{-4}$ for each W $cm^{-2}$ [212], that is a monochromatic illuminating power of 100 W $cm^{-2}$ is required to get a yield of 3%. It turns out that NIR imaging is often combined with other functionalities, for instance, photodynamic treatment of cancer [213] or, more frequently, magnetic resonance imaging [207,214–216]. Referring to the latter, combining MRI capability with optical imaging, either in the visible or in the NIR, seems to be ideal in that the good spatial resolution of MRI can be allied with the high sensitivity of optical probes [217,218].

Encapsulation of the chelates into nanoparticles often results in further improvements of their photophysical properties [219,220]. We have also seen above that relatively modest quantum yields can be compensated by introducing several emitting ions in one probe. This is achieved in luminescent metal-organic frameworks [221] or dendrimeric complexes [214] which then present attractive properties, including multifunctionality and nanoscale processability, for chemical sensors and bioprobes.

Another reason to be optimistic with respect to luminescent bio-chelates is the present massive need for more molecular information at the cell level aimed at both diagnosis and treatment (theranostics) in view of setting up personalised medicine. Therefore molecular frameworks combining drug transport and imaging ability [1] or enabling two imaging modes (bimodal probes) are in high demand. Specificity is a key concept here and the almost unlimited versatility of organic ligands allied with new developments in time-resolved luminescence instrumentation [222] are invaluable assets for bioprobes based on coordination compounds.

## Acknowledgements

The author's research was funded by the Swiss National Science Foundation (grant 20020_119866) and the Swiss Office for Science and Education (grant C07.0116) within the frame of the European Union Cost Action D38. The author thanks the World Class University Program from the Ministry of Education, Science and Technology of the Republic of Korea (grant R31-2012-000-10035-0) and Dr Svetlana V. Eliseeva for careful reading of the manuscript.

## Acronyms and Abbreviations

| | |
|---|---|
| 3D | 3-dimensional |
| ACh | acetylcholine |
| AFt | apoferritin |
| AO | acridine orange |
| AOM | acousto-optic modulator |
| APTES | 3-aminopropyl(triethoxy)silane (also referred to as APS, APT or APTS) |
| APTMS | (3-aminopropyl)trimethoxysilane |
| AVM | arteriovenous malformation |
| BHHCT | 4,4′-bis(4,4,5,5,6,6,6-heptafluoro-1,3-hexanedionate-1-yl)-[1,1′:2′,1″-terphenyl]-4′-sulfonyl chloride |
| BPPCT | 4,4′-bis(4,4,5,5,5-pentafluoro-1,3-pentanedionate-1-yl)-[1,1′:2′,1″-terphenyl]-4′-sulfonyl chloride |
| BSA | bovine serum albumin |
| CAT | computed axial tomography |
| CCD | charge-coupled device |
| CDHH | 4′-(4,4,6,6,7,7,7-heptafluoro-1,3-hexanedionate-1-yl)-[1,1′-biphenyl]-4-sulfonylchloride |
| CHO | Chinese hamster ovarian (cells) |
| CT | computed tomography |
| cyclen | 1,4,7,10-tetraazacyclododecane |
| dbm | 1,3-diphenyl-1,3-propanedionate (dibenzoylmethanate) |
| DiI | 1,1-dioactadecyl-3,3,3′,3′-tetramethylindocarbocyanine perchlorate |
| DNA | deoxyribonucleic acid |
| DO3A | 1,4,7,10-tetraazacyclododecane-1,4,7-triacetate amide |
| DOPE | dioleoyl-phosphatidyl-ethanolamine |
| DOTA | 1,4,7,10-tetraazacyclododecane-1,4,7,10-tetraacetate |
| DPA | dipicolinate (2,6-pyridine-dicarboxylate) |
| DTPA | diethylenetriaminepentaacetate |
| eDHFR | *Escherichia coli* dihydrofolate reductase |

| | |
|---|---|
| EDTA | ethylenediamine tetraacetate |
| EMCCD | electron-multiplying charge-coupled device |
| ER | oestrogen receptor |
| FITC | fluorescein isothiocyanate |
| FLC | ferro-electric liquid crystal |
| FLIM | fluorescence lifetime imaging |
| fod | 6,6,7,7,8,8,8-heptafluoro-2,2-dimethyl-3,5-octanedionate |
| FRET | Förster resonant energy transfer |
| fwhh | full width at half height |
| GALD | gated autosynchronous luminescence detector |
| GFP | green fluorescent protein |
| GM | Göppert-Mayer, unit of two-photon cross sections ($1\,GM = 10^{-50}\,cm^4\,phot^{-1}\,s^{-1}$) |
| Her2/*neu* | human epidermal growth factor receptor |
| hfa | 1,1,1,5,5,5-hexafluoro-2,4-pentanedionate (or hexafluoroacetylacetonate) |
| HIV | human immunodeficiency virus |
| HNL | human neutrophil lipocalin |
| HPV | human papillomavirus |
| $IC_{50}$ | half inhibitory concentration |
| IDA | iminodiacetate |
| IgG | immunoglobulin G |
| ICCD | image intensified charged coupled device |
| ips | impulse per second |
| LDH | lactase dehydrogenase |
| LED | light-emitting diode |
| LLB | lanthanide luminescent bioprobe or bioconjugate |
| LP | long pass |
| mAb | monoclonal antibody |
| MCP | micro-channel plate |
| MOS | metal oxide *p*-doped silicon |
| MPE | multiphoton excitation |
| MRI | magnetic resonance imaging |
| *NA* | numerical aperture |
| NBCA | *n*-butyl-2-cyanoacrylate |
| NHE | normal hydrogen electrode |
| NIR | near infrared |
| nta | 1-(2-naphtoyl)- 4,4,4 trifluoro-1,3-butanedionate |
| NTA | nitrilotriacetate |
| O.D. | outer diameter |
| PAMAM | polyamidoamine |
| PBS | phosphate buffered saline (pH 7.4 buffer) |
| PDMS | polydimethylsiloxane |
| PET | positron emission tomography |
| Phen | *o*-phenanthroline |
| PMT | photomultiplier tube |
| pLHIP | pH low insertion peptide |

| | |
|---|---|
| PSA | prostate specific antigen |
| QD | quantum dot |
| RNA | ribonucleic acid |
| rpm | revolution per minute |
| SA | streptavidin |
| SDS | sodium dodecylsulfate |
| SNR | signal to noise ratio |
| SP | short pass |
| SVGA | super video graphic array |
| SWNT | single-walled carbon nanotube |
| TCSPC | time-correlated single-photon counting |
| THPMP | 3-(trihydroxysilyl)propyl methylphosphonate |
| TEM | transmission electron microscopy |
| TEOS | tetraethylorthosilicate |
| tfa | 1,1,1-trifluoro-2,4-pentanedionate (or trifluoroacetylacetonate) |
| tha | 4,4,4-trifluoro-1-(9-hexylcarbazole-3-yl)-1,3-butanedionate |
| TIRFM | total internal reflection fluorescence microscopy |
| TMP | trimethoprim |
| TOPO | trioctylphosphine oxide |
| TPA | two-photon absorption |
| TRACE | time-resolved amplified cryptate emission |
| Tris | tris(hydroxymethyl)aminomethane |
| TR | time-resolved |
| TRD | time-resolved detection |
| TRLM | time-resolved luminescence microscopy or microscope |
| tta | 4,4,4-trifluoro-1-(2-thienyl)-1,3-butanedionate (or thenoyltrifluoro acetonate) |
| UCNP | up-converting nanoparticle |
| UV | ultraviolet |
| $\varepsilon(\lambda)$ | molar absorption coefficient (in $M^{-1}\,cm^{-1}$) |
| $\sigma_A(\lambda)$ | absorption cross section (in $cm^2$ $molecule^{-1}$) |

## References


1. Kelkar S. S., T. M. Reineke. Theranostics: Combining Imaging and Therapy, *Bioconjugate Chem.*, **22**, 1879–1903 (2011).
2. Kobayashi H., M. R. Longmire, M. Ogawa, P. L. Choyke. Rational chemical design of the next generation of molecular imaging probes based on physics and biology: mixing modalities, colors and signals, *Chem. Soc. Rev.*, **40**, 4626–4648 (2011).
3. Kobayashi H., P. L. Choyke. Target-Cancer-Cell-Specific Activatable Fluorescence Imaging Probes: Rational Design and in Vivo Applications, *Acc. Chem. Res.*, **44**, 83–90 (2011).
4. Michalet X., F. F. Pinaud, L. A. Bentolila, J. M. Tsay, S. Doose, J. J. Li, G. Sundaresan, A. M. Wu, S. S. Gambhir, S. Weiss. Quantum dots for live cells, in vivo imaging, and diagnostics, *Science*, **307**, 538–544 (2005).
5. Zrazhevskiy P., M. Sena, X. Gao. Designing multifunctional quantum dots for bioimaging, detection, and drug delivery, *Chem. Soc. Rev.*, **39**, 4326–4354 (2010).
6. Gao X., L. Yang, J. A. Petros, F. F. Marshall, J. W. Simons, S. Nie. In vivo molecular and cellular imaging with quantum dots, *Curr. Op. Biotechnol.*, **16**, 63–72 (2005).

7. Bünzli J.-C. G., Eliseeva S. V. Basics of Lanthanide Photophysics. In Hänninen P., Härmä H., editors. *Springer Series on Fluorescence. Lanthanide Luminescence: Photophysical, Analytical and Biological Aspects*. Ch. 1, 1–45. 2011. Berlin, Springer Verlag.
8. Bünzli J.-C. G., Eliseeva S. V. Photophysics of Lanthanoid Coordination Compounds. In Yam V. W.-W., editor. *Comprehensive Inorganic Chemistry II*. Ch. 8. 03, in press. 2012. Amsterdam, Elsevier B.V.
9. Hemmilä I. A. Time-resolved Fluorimetric Immunoassays; Instrumentation, Application, Unresolved Issues, Future Trends. In Resch-Genger U., editor. *Springer Series on Fluorescence. Standardization and Quality Assurance in Fluorescence Measurements II*. 429–447. 2008. Berlin, Springer Verlag.
10. Hemmilä I., V. M. Mukkala. Time-resolution in fluorometry technologies, labels, and applications in bioanalytical assays, *Crit. Rev. Clin. Lab. Sci.*, **38**, 441–519 (2001).
11. Soini E., I. Hemmilä. Fluoroimmunoassay: Present Status and Key Problems, *Clin. Chem.*, **25**, 353–361 (1979).
12. Bünzli J.-C. G. Lanthanide Luminescence for Biomedical Analyses and Imaging, *Chem. Rev.*, **110**, 2729–2755 (2010).
13. Borisov S. M., O. S. Wolfbeis. Optical biosensors, *Chem. Rev.*, **108**, 423–461 (2008).
14. Hagan A. K., T. Zuchner. Lanthanide-based Time-resolved Luminescence Immunoassays, *Anal. Bioanal. Chem.*, **400**, 2847–2864 (2011).
15. Cooper D. E., D'Andrea A., Faris G. W., MacQueen B., Wright W. H. Up-converting phosphors for detection and identification using antibodies. In Van Emon J. M., editor. *Immunoassay and Other Bioanalytical Techniques*. Ch. 9, 217–247. 2007. Boca Raton, CRC Press, Taylor & Francis group.
16. Beverloo H. B., A. Van Schadewijk, S. Van Gelderenboele, H. J. Tanke. Inorganic Phosphors As New Luminescent Labels for Immunocytochemistry and Time-Resolved Microscopy, *Cytometry*, **11**, 784–792 (1990).
17. Ye Z. Q., M. Q. Tan, G. L. Wang, J. G. Yuan. Preparation, Characterization, and Time-Resolved Fluorometric Application of Silica-Coated Terbium(III) Fluorescent Nanoparticles, *Anal. Chem.*, **76**, 513–518 (2004).
18. Bae S. W., W. Tan, J. I. Hong. Fluorescent dye-doped silica nanoparticles: new tools for bioapplications, *Chem. Commun.*, **48**, 2270–2282 (2012).
19. Scaff W. L., D. L. Dyer, K. Mori. Fluorescent Europium Chelate Stain, *J. Bacter.*, **98**, 246–248 (1969).
20. Murdock R. H., E. R. Menzel. A computer interfaced time-resolved luminescence imaging system, *J. Forens. Sci.*, **38**, 521–529 (1993).
21. Menzel E. R. Detection of latent fingerprints by laser-excited luminescence, *Anal. Chem.*, **61**, 557A–561A (1989).
22. Soini E. J., L. J. Pelliniemi, I. A. Hemmilä, V. M. Mukkala, J. J. Kankare, K. Frojdman. Lanthanide Chelates As New Fluorochrome Labels for Cyto-Chemistry, *J. Histochem. Cytochem.*, **36**, 1449–1451 (1988).
23. Davidson MW, Abromowicz M. Optical Microscopy. 3rd edn. Tallahassee: Florida State University and Olympus America Inc.; 2001.
24. Jovin T. M., W. L. C. Vaz. Rotational and Translational Diffusion in Membranes Measured by Fluorescence and Phosphorescence Methods, *Meth. Enzym.*, **172**, 471–513 (1989).
25. Jovin T. M., D. J. Arndt-Jovin. Luminescence digital imaging microscopy, *An. Rev. Biophys. Biophys. Chem.*, **18**, 271–308 (1989).
26. Ishida H., S. Tobita, Y. Hasegawa, R. Katoh, K. Nozaki. Recent advances in instrumentation for absolute emission quantum yield measurements, *Coord. Chem. Rev.*, **254**, 2449–2458 (2010).
27. Weissleder R., V. Ntziachristos. Shedding light onto live molecular targets, *Nature Med.*, **9**, 123–128 (2003).

28. Shia V., D. Watt, G. W. Faris. High-speed camera with real time processing for frequency domain imaging, *Biomed. Opt. Express*, **2**, 1931–1945 (2011).
29. Marriott G., M. Heidecker, E. P. Diamandis, Y. Yanmarriott. Time-Resolved Delayed Luminescence Image Microscopy Using An Europium Ion Chelate Complex, *Biophys. J.*, **67**, 957–965 (1994).
30. Connally R. E., J. A. Piper. Time-gated luminescence microscopy, *An. N. Y. Acad. Sci.*, **1130**, 106–116 (2008).
31. Connally R., D. Veal, J. Piper. High resolution detection of fluorescently labeled micro-organisms in environmental samples using time-resolved fluorescence microscopy, *Fems Microbiol. Ecol.*, **41**, 239–245 (2002).
32. Connally R., D. Y. Jin, J. Piper. High intensity solid-state UV source for time-gated luminescence microscopy, *Cytometry Part A*, **69A**, 1020–1027 (2006).
33. Vereb G., E. Jares-Erijman, P. R. Selvin, T. M. Jovin. Temporally and spectrally resolved imaging microscopy of lanthanide chelates, *Biophys. J.*, **74**, 2210–2222 (1998).
34. Ke H. Y. D., E. R. Birnbaum, D. W. Darnall, P. J. Jackson, G. D. Rayson. Investigation of Eu(III) Binding Sites on Datura-Innoxia Using Eu(III) Luminescence, *Appl. Spectrosc.*, **46**, 479–488 (1992).
35. Gaiduck M. I., V. V. Grigoryants, A. F. Mironov, V. D. Rumyantseva, V. I. Chissov, G. M. Sukhin. Fiber-laser IR luminescence diagnostics of malignant tumors using rare-earth porphyrins, *J. Photochem. Photobiol. B Biol.*, **7**, 15–20 (1990).
36. Gaiduck M. I., V. V. Grigoryants, A. F. Mironov, L. D. Roitman, V. I. Chissov, V. D. Rumiantseva, G. M. Sukhin. IR Luminescence Diagnostics of Malignant-Tumors with the Help of Metallocomplexes of Porphyrins, *Dokl. Acad. Nauk. SSSR*, **309**, 980–983 (1989).
37. Beverloo H. B., A. Vanschadewijk, J. Bonnet, R. Vandergeest, R. Runia, N. P. Verwoerd, J. Vrolijk, J. S. Ploem, H. J. Tanke. Preparation and Microscopic Visualization of Multicolor Luminescent Immunophosphors, *Cytometry*, **13**, 561–570 (1992).
38. Marriott G., R. M. Clegg, D. J. Arndtjovin, T. M. Jovin. Time Resolved Imaging Microscopy – Phosphorescence and Delayed Fluorescence Imaging, *Biophys. J.*, **60**, 1374–1387 (1991).
39. Evangelista R. A., A. Pollak, B. D. Allore, A. F. Templeton, R. C. Morton, E. P. Diamandis. A New Europium Chelate for Protein Labeling and Time-resolved Fluorometric Applications, *Clin. Biochem.*, **21**, 173–178 (1988).
40. Seveus L., M. Vaisala, S. Syrjanen, M. Sandberg, A. Kuusisto, R. Harju, J. Salo, I. Hemmilä, H. Kojola, E. Soini. Time-Resolved Fluorescence Imaging of Europium Chelate Label in Immunohistochemistry and In situ Hybridization, *Cytometry*, **13**, 329–338 (1992).
41. Seveus L., M. Vaisala, I. Hemmilä, H. Kojola, G. M. Roomans, E. Soini. Use of Fluorescent Europium Chelates As Labels in Microscopy Allows Glutaraldehyde Fixation and Permanent Mounting and Leads to Reduced Autofluorescence and Good Long-Term Stability, *Microsc. Res. Techn.*, **28**, 149–154 (1994).
42. Harma H., T. Soukka, T. Lovgren. Europium nanoparticles and time-resolved fluorescence for ultrasensitive detection of prostate-specific antigen, *Clin. Chem.*, **47**, 561–568 (2001).
43. Verwoerd N. P., E. J. Hennink, J. Bonnet, C. R. G. Vandergeest, H. J. Tanke. Use of Ferroelectric Liquid-Crystal Shutters for Time-Resolved Fluorescence Microscopy, *Cytometry*, **16**, 113–117 (1994).
44. Phimphivong S., S. Kolchens, P. L. Edmiston, S. S. Saavedra. Time-Resolved, Total Internal-Reflection Fluorescence Microscopy of Cultured-Cells Using A Tb Chelate Label, *Anal. Chim. Acta*, **307**, 403–417 (1995).
45. Phimphivong S., S. S. Saavedra. Terbium chelate membrane label for time-resolved, total internal reflection fluorescence microscopy of substrate-adherent cells, *Bioconjugate Chem.*, **9**, 350–357 (1998).
46. Soini A. E., A. Kuusisto, N. J. Meltola, E. Soini, L. Seveus. A New Technique for Multiparameter Imaging Microscopy: Use of Long Decay Time Photoluminescent Labels Enables

Multiple Color Immunocytochemistry with Low Channel-to-Channel Crosstalk, *Microsc. Res. Techn.*, **62**, 396–407 (2003).

47. Connally R., D. Veal, J. Piper. Flash lamp-excited time-resolved fluorescence microscope suppresses autofluorescence in water concentrates to deliver an 11-fold increase in signal to noise ratio, *J. Biomed. Opt.*, **9**, 725–734 (2004).
48. Jin D., R. Connally, J. Piper. Long-lived visible luminescence of UV LEDs and impact on LED excited time-resolved fluorescence applications, *J. Phys. D–Appl. Phys.*, **39**, 461–465 (2006).
49. Diamandis E. P., R. C. Morton, E. Reichstein, M. J. Khosravi. Multiple fluorescence labeling with europium chelators. Application to time-resolved fluoroimmunoassays, *Anal. Chem.*, **61**, 48–53 (1989).
50. Reichstein E., Y. Shami, M. Ramjeesingh, E. P. Diamandis. Laser-Excited Time-Resolved Solid-Phase Fluoroimmunoassays with the New Europium Chelate 4,7-Bis(chlorosulfophenyl)-1,10-phenanthroline-2,9-dicarboxylic Acid as Label, *Anal. Chem.*, **60**, 1069–1074 (1988).
51. Gahlaut N., L. W. Miller. Time-Resolved Microscopy for Imaging Lanthanide Luminescence in Living Cells, *Cytometry Part A*, **77A**, 1113–1125 (2010).
52. Connally R., D. Veal, J. Piper. Time-resolved fluorescence microscopy using an improved europium chelate BHHST for the in situ detection of Cryptosporidium and Giardia, *Microsc. Res. Techn.*, **64**, 312–322 (2004).
53. Hanaoka K., K. Kikuchi, S. Kobayashi, T. Nagano. Time-Resolved Long-Lived Luminescence Imaging Method Employing Luminescent Lanthanide Probes with a New Microscopy System, *J. Am. Chem. Soc.*, **129**, 13502–13509 (2007).
54. Connally R., J. Piper. Solid-state time-gated luminescence microscope with ultraviolet light-emitting diode excitation and electron-multiplying charge-coupled device detection, *J. Biomed. Opt.*, **13**, 034022-1 to 034022-6 (2008).
55. Connally R. A Device for Gated Autosynchronous Luminescence Detection, *Anal. Chem.*, **83**, 4782–4787 (2011).
56. Jin D. Y., J. A. Piper. Time-Gated Luminescence Microscopy Allowing Direct Visual Inspection of Lanthanide-Stained Microorganisms in Background-Free Condition, *Anal. Chem.*, **83**, 2294–2300 (2011).
57. Becker W., B. Su, A. Bergmann, K. Weisshart, O. Holub. Simultaneous Fluorescence and Phosphorescence Lifetime Imaging, *SPIE Proceedings*, **7903**, Art. Nr. 790320 (2011).
58. Binnemans K. Rare earth β-diketonate complexes: functionalities and applications. In *Handbook on the Physics and Chemistry of Rare Earths*, K. A. Gschneidner Jr., J.-C. G. Bünzli, V. K. Pecharsky, editors. Amsterdam, Elsevier Science B.V. Vol. 35, Ch. 225, 107–272 (2005).
59. Tsvirko M. P., S. B. Meshkova, V. Y. Venchikov, Z. M. Topilova, D. V. Bol'shoi. Determination of contributions of various molecular groups to nonradiative deactivation of electronic excitation energy in β-diketonate complexes of ytterbium(III), *Opt. Spectrosc. (Engl. Transl.)*, **90**, 669–673 (2001).
60. Comby S., J.-C. G. Bünzli. Lanthanide Near-Infrared Luminescence in Molecular Probes and Devices. In *Handbook on the Physics and Chemistry of Rare Earths*, K. A. Gschneidner Jr., J.-C. G. Bünzli, V. K. Pecharsky, editors. Amsterdam, Elsevier Science B.V. Vol. 37, Ch. 235, 217–470 (2007).
61. Eliseeva S. V., D. N. Pleshkov, K. A. Lyssenko, L. S. Lepnev, J.-C. G. Bünzli, N. P. Kuzmina. Deciphering three beneficial effects of 2,2'-bipyridine,*N,N'*-dioxide on the luminescence sensitization of lanthanide(III) hexafluoroacetylacetonate ternary complexes, *Inorg. Chem.*, **50**, 5137–5144 (2011).
62. Meshkova S. B., Z. M. Topilova, M. O. Lozinskii, D. V. Bolshoi. Periodicity of Variation of Stability-Constants of Lanthanide(III) Complexes with Fluorine Derivatives of Acetylacetone, *Russ. J. Inorg. Chem.*, **40**, 1296–1301 (1995).

63. Menzel E. R. Recent advances in photoluminescence detection of fingerprints, *The Scientific World Journal*, **1**, 498–509 (2001).
64. Mekkaoui A. I., E. R. Menzel. Spectroscopy of Rare Earth Ruhemanns Purple Complexes, *J. Forens. Sci.*, **38**, 506–520 (1993).
65. Caldwell J. P., W. Henderson, N. D. Kim. Luminescent visualization of latent fingerprints by direct reaction with a lanthanide shift reagent, *J. Forens. Sci.*, **46**, 1332–1341 (2001).
66. Menzel E. R., K. K. Bouldin, R. H. Murdock. Trace explosives detection by photoluminescence, *The Scientific World Journal*, **4**, 55–66 (2004).
67. Liu Y., K. Ai, L. Lu. Designing lanthanide-doped nanocrystals with both up- and down-conversion luminescence for anti-counterfeiting, *Nanoscale*, **3**, 4804–4810 (2011).
68. Zohar O., M. Ikeda, H. Shinagawa, H. Inoue, H. Nakamura, D. Elbaum, D. L. Alkon, T. Yoshioka. Thermal imaging of receptor-activated heat production in single cells, *Biophys. J.*, **74**, 82–89 (1998).
69. Borisov S. M., O. S. Wolfbeis. Temperature-sensitive europium(III) probes and their use for simultaneous luminescent sensing of temperature and oxygen, *Anal. Chem.*, **78**, 5094–5101 (2006).
70. Nagl S., M. I. J. Stich, M. Schaferling, O. S. Wolfbeis. Method for simultaneous luminescence sensing of two species using optical probes of different decay time, and its application to an enzymatic reaction at varying temperature, *Anal. Bioanal. Chem.*, **393**, 1199–1207 (2009).
71. Calvo W. J., B. B. Lieber, N. Hopkins, A. K. Wakhloo. Europium fluorescence to visualize *N*-butyl 2-cyanoacrylate in embolized vessels of an arteriovenous malformation swine model, *Am. J. Neurorad.*, **22**, 691–697 (2001).
72. Calvo WJ, Lieber BB, inventors; Histological *visualization* of cyanoacrylate embolization. United States patent US 20020031443 A1 20020314. 2002, March 14.
73. Yuan J. L., K. Matsumoto. Synthesis of a new tetradentate beta-diketonate-europium chelate and its application for time-resolved fluorimetry of albumin, *J. Pharm. Biomed. Anal.*, **15**, 1397–1403 (1997).
74. Yuan J. L., G. L. Wang, H. Kimura, K. Matsumoto. Highly sensitive time-resolved fluoroimmunoassay of human immunoglobulin E by using a new europium fluorescent chelate as a label, *Anal. Biochem.*, **254**, 283–287 (1997).
75. Yuan J. L., K. Matsumoto, H. Kimura. A new tetradentate beta-diketonate-Europium chelate that can be covalently bound to proteins for time-resolved fluoroimmunoasaay, *Anal. Chem.*, **70**, 596–601 (1998).
76. Yuan J. L., G. Wang, H. Kimura, K. Matsumoto. Sensitive time-resolved fluoroimmunoassay of human thyroid- stimulating hormone by using a new europium fluorescent chelate as a label, *Anal. Sci.*, **14**, 421–423 (1998).
77. Jiang L. N., J. Wu, G. L. Wang, Z. Q. Ye, W. Z. Zhang, D. Y. Jin, J. L. Yuan, J. Piper. Development of a Visible-Light-Sensitized Europium Complex for Time-Resolved Fluorometric Application, *Anal. Chem.*, **82**, 2529–2535 (2010).
78. Deng W., D. Y. Jin, K. Drozdowicz-Tomsia, J. L. Yuan, E. M. Goldys. Europium Chelate (BHHCT-$Eu^{3+}$) and Its Metal Nanostructure Enhanced Luminescence Applied to Bioassays and Time-Gated Bioimaging, *Langmuir*, **26**, 10036–10043 (2010).
79. Connally R., D. Jin, J. Piper. BHHST: An improved lanthanide chelate for time-resolved fluorescence applications, *SPIE Proceedings*, **5704**, 93–104 (2005).
80. Green N. M. Avidin. 1. Use of [14C]Biotin for Kinetic Studies and for Assay, *Biochem. J.*, **89**, 585–591 (1963).
81. Song B., G. Wang, M. Tan, J. L. Yuan. A Europium(III) Complex as an Efficient Singlet Oxygen Luminescence Probe, *J. Am. Chem. Soc.*, **128**, 13442–13450 (2006).
82. Guo W. H., Z. Q. Ye, G. L. Wang, X. M. Zhao, J. L. Yuan, Y. G. Du. Measurement of oligochitosan-tobacco cell interaction by fluorometric method using europium complexes as fluorescence probes, *Talanta*, **78**, 977–982 (2009).

83. Wu J., G. Wang, D. Y. Jin, J. G. Yuan, Y. F. Guan, J. Piper. Luminescent europium nanoparticles with a wide excitation range from UV to visible light for biolabeling and time-gated luminescence bioimaging, *Chem. Commun.*, 365–367 (2008).
84. Wu J., Z. Ye, G. Wang, D. Jin, J. Yuan, Y. Guang, J. Piper. Visible-light-sensitized highly luminescent europium nanoparticles: preparation and application for time-gated luminescence bioimaging, *J. Mater. Chem.*, **19**, 1258–1264 (2009).
85. Lakowicz J. R. Radiative decay engineering 5: metal-enhanced fluorescence and plasmon emission, *Anal. Biochem.*, **337**, 171–194 (2005).
86. Xie F., M. S. Baker, E. M. Goldys. Homogeneous silver-coated nanoparticle substrates for enhanced fluorescence detection, *J. Phys. Chem. B*, **110**, 23085–23091 (2006).
87. Muthu P., N. Calander, I. Gryczynski, Z. Gryczynski, J. M. Talent, T. Shtoyko, I. Akopova, J. Borejdo. Monolayers of silver nanoparticles decrease photobleaching: Application to muscle myofibrils, *Biophys. J.*, **95**, 3429–3438 (2008).
88. Bünzli J.-C. G., A.-S. Chauvin, H. K. Kim, E. Deiters, S. V. Eliseeva. Lanthanide Luminescence Efficiency in Eight- and Nine-coordinate Complexes: Role of the Radiative Lifetime, *Coord. Chem. Rev.*, **254**, 2623–2633 (2010).
89. Deng W., D. Y. Jin, K. Drozdowicz-Tomsia, J. L. Yuan, J. Wu, E. M. Goldys. Plasmonic Ag/SiO(2) composite nanoparticles doped with europium chelate and their metal enhanced fluorescence, *SPIE Proceedings*, **7909**, 79090Z-1-79090Z-11 (2011).
90. Deng W., D. Y. Jin, K. Drozdowicz-Tomsia, J. L. Yuan, J. Wu, E. M. Goldys. Ultrabright Eu-Doped Plasmonic Ag@SiO(2) Nanostructures: Time-gated Bioprobes with Single Particle Sensitivity and Negligible Background, *Adv. Mater.*, **23**, 4649–4654 (2011).
91. Accorsi G., N. Armaroli, A. Parisini, M. Meneghetti, R. Marega, M. Prato, D. Bonifazi. Wet adsorption of a luminescent Eu-III complex on carbon nanotubes sidewalls, *Adv. Funct. Mater.*, **17**, 2975–2982 (2007).
92. Xu J., Z. Sun, L. Jia, B. Li, L. Zhao, X. Liu, Y. Ma, H. Tian, Q. Wang, W. Liu, Y. Tang. Visible light sensitized attapulgite-based lanthanide composites: microstructure, photophysical behaviour and biological application, *Dalton Trans.*, **40**, 12909–12916 (2011).
93. Shi M., C. R. Ding, J. W. Dong, H. Z. Wang, Y. P. Tian, Z. J. Hu. A novel europium(III) complex with versatility in excitation ranging from infrared to ultraviolet, *Phys. Chem. Chem. Phys.*, **11**, 5119–5123 (2009).
94. Hu Z. J., X. H. Tian, X. H. Zhao, P. Wang, Q. Zhang, P. P. Sun, J. Y. Wu, J. X. Yang, Y. P. Tian. Efficient two-photon-sensitized luminescence of a novel europium(III) β-diketonate complex and application in biological imaging, *Chem. Commun.*, **47**, 12467–12469 (2011).
95. Choppin G. R. A Half-Century of Lanthanide Aminopolycarboxylates, *J. Alloys Compd.*, **192**, 256–261 (1993).
96. Bianchi A., L. Calabi, F. Corana, S. Fontana, P. Losi, A. Maiocchi, L. Paleari, B. Valtancoli. Thermodynamic and structural properties of Gd(III) complexes with polyamino-polycarboxylic ligands: basic compounds for the development of MRI contrast agents, *Coord. Chem. Rev.*, **204**, 309–393 (2000).
97. Yang Y. S., B. L. An, M.-L. Gong, H. H. Shi, H. Y. Lei, J. X. Meng. Progress on study of luminescence of rare earth organic chelates, *J. Rare Earths*, **20**, 161–166 (2002).
98. de Haas R. R., N. P. Verwoerd, M. P. vanderCorput, R. P. vanGijlswijk, H. Siitari, H. J. Tanke. The use of peroxidase-mediated deposition of biotin-tyramide in combination with time-resolved fluorescence imaging of europium chelate label in immunohistochemistry and in situ hybridization, *J. Histochem. Cytochem.*, **44**, 1091–1099 (1996).
99. Weibel N., L. J. Charbonnière, M. Guardigli, A. Roda, R. F. Ziessel. Engineering of highly luminescent lanthanide tags suitable for protein labelling and time-resolved luminescence imaging, *J. Am. Chem. Soc.*, **26**, 4888–4896 (2004).

100. Claudel-Gillet S. P., J. Steibel, N. Weibel, T. Chauvin, M. Port, I. Raynal, E. Toth, R. F. Ziessel, L. J. Charbonnière. Lanthanide-based Conjugates as Polyvalent Probes for Biological Labeling, *Eur. J. Inorg. Chem.*, 2856–2862 (2008).
101. Liu X. Y., Z. Q. Ye, W. Wei, Y. G. Du, J. L. Yuan, D. Ma. Artificial luminescent protein as a bioprobe for time-gated luminescence bioimaging, *Chem. Commun.*, **47**, 8139–8141 (2011).
102. Rajapakse H. E., D. R. Reddy, S. Mohandessi, N. G. Butlin, L. W. Miller. Luminescent Terbium Protein Labels for Time-Resolved Microscopy and Screening, *Angew. Chem., Int. Ed. Engl.*, **48**, 4990–4992 (2009).
103. Rajapakse H. E., N. Gahlaut, S. Mohandessi, D. Yu, J. R. Turner, L. W. Miller. Time-resolved luminescence resonance energy transfer imaging of protein–protein interactions in living cells, *Proc. Natl Acad. Sci. USA*, **107**, 13582–13587 (2010).
104. Dhawan P., A. B. Singh, N. G. Deane, Y. No, S. R. Shiou, C. Schmidt, J. Neff, M. K. Washington, R. D. Beauchamp. Claudin-1 regulates cellular transformation and metastatic behavior in colon cancer, *J Clin Invest*, **115**, 1765–1776 (2005).
105. Rulli M., A. Kuusisto, J. Salo, H. Kojola, O. Simell. Time-resolved fluorescence imaging in islet cell autoantibody quantitation, *J. Immunol. Methods*, **208**, 169–179 (1997).
106. Hanaoka K., K. Kikuchi, H. Kojima, Y. Urano, T. Nagano. Development of a Zinc Ion-Selective Luminescent Lanthanide Chemosensor for Biological Applications, *J. Am. Chem. Soc.*, **126**, 12470–12476 (2004).
107. Ye Z. Q., G. L. Wang, J. X. Chen, X. Y. Fu, W. Z. Zhang, J. L. Yuan. Development of a novel terbium chelate-based luminescent chemosensor for time-resolved luminescence detection of intracellular $Zn^{2+}$ ions, *Biosens. Bioelectr.*, **26**, 1043–1048 (2010).
108. Song C. H., Z. Q. Ye, G. L. Wang, J. L. Yuan, Y. F. Guan. A Lanthanide-Complex-Based Ratiometric Luminescent Probe Specific for Peroxynitrite, *Chem. Eur. J.*, **16**, 6464–6472 (2010).
109. Chen Y., W. Guo, Z. Ye, G. Wang, J. Yuan. A europium(III) chelate as an efficient time-gated luminescent probe for nitric oxide, *Chem. Commun.*, **47**, 6266–6268 (2011).
110. Yan C., Z. Yan, Y. P. Du, J. Shen, C. Zhang. Controlled Synthesis and Properties of Rare Earth Nanomaterials. In *Handbook on the Physics and Chemistry of Rare Earths*, K. A. Gschneidner Jr., J.-C. G. Bünzli, V. K. Pecharsky, editors. Amsterdam, Elsevier Science B.V. Vol. 41, Ch. 251, 275–472 (2011).
111. Feng W., L. D. Sun, Y. W. Zhang, C. H. Yan. Synthesis and assembly of rare earth nanostructures directed by the principle of coordination chemistry in solution-based process, *Coord. Chem. Rev.*, **254**, 1038–1053 (2010).
112. Shen J., L. D. Sun, C. H. Yan. Luminescent rare earth nanomaterials for bioprobe applications, *Dalton Trans.*, 5687–5697 (2008).
113. Wang M., G. Abbineni, A. Clevenger, C. B. Mao, S. K. Xu. Upconversion nanoparticles: synthesis, surface modification and biological applications, *Nanomedicine: Nanotech. Biol. Med.*, **7**, 710–729 (2011).
114. Klapper M., S. Nenov, R. Haschick, K. Müller, K. Müllen. Oil-in-Oil Emulsions: A Unique Tool for the Formation of Polymer Nanoparticles, *Acc. Chem. Res.*, **41**, 1190–1201 (2008).
115. Charbonnière L. J., J.-L. Rehspringer, Y. Zimmermann. Highly luminescent water-soluble lanthanide nanoparticles through surface coating sensitization, *New J. Chem.*, **32**, 1055–1059 (2008).
116. Shen J., L. D. Sun, J. D. Zhu, L. H. Wei, H. F. Sun, C. H. Yan. Biocompatible Bright YVO4:Eu Nanoparticles as Versatile Optical Bioprobes, *Adv. Funct. Mater.*, **20**, 3708–3714 (2010).
117. Zhou J., Z. Liu, F. Li. Upconversion nanophosphors for small-animal imaging, *Chem. Soc. Rev.*, **41**, 1323–1349 (2012).
118. Wang H.-Q., Nann T. Upconverting nanoparticles. In Hänninen P., Härmä H., editors. *Springer Series on Fluorescence. Lanthanide Luminescence: Photophysical, Analytical and Biological Aspects*. Ch. 4, 115–132 (2011) Berlin, Springer Verlag.

119. Charbonnière L. J., N. Weibel, C. Estournes, C. Leuvrey, R. F. Ziessel. Spatial and temporal discrimination of silica particles functionalised with luminescent lanthanide markers using time-resolved luminescence microscopyy, *New J. Chem.*, **28**, 777–781 (2004).
120. Chen Y., Z. H. Lu. Dye sensitized luminescent europium nanoparticles and its time-resolved fluorometric assay for DNA, *Anal. Chim. Acta*, **587**, 180–186 (2007).
121. Wu C. L., J. Q. Hong, X. Q. Guo, C. B. Huang, J. P. Lai, J. S. Zheng, J. B. Chen, X. Mu, Y. B. Zhao. Fluorescent core-shell silica nanoparticles as tunable precursors: towards encoding and multifunctional nano-probes, *Chem. Commun.*, 750–752 (2008).
122. Jin D. Y., J. Piper, J. L. Yuan, R. Leif. Time-gated real-time bioimaging system using multicolor microsecond-lifetime silica nanoparticles, *SPIE Proceedings*, **7568**, Art. Nr. 1756819 (2010).
123. Davies A., D. J. Lewis, S. P. Watson, S. G. Thomas, Z. Pikramenou. pH-controlled delivery of luminescent europium coated nanoparticles into platelets, *Proc. Natl Acad. Sci. USA*, **109**, 1862–1867 (2012).
124. Jiang H. F., G. L. Wang, W. Z. Zhang, X. Y. Liu, Z. Q. Ye, D. Y. Jin, J. L. Yuan, Z. G. Liu. Preparation and Time-Resolved Luminescence Bioassay Application of Multicolor Luminescent Lanthanide Nanoparticles, *J. Fluoresc.*, **20**, 321–328 (2010).
125. Dupont E. P., E. Labonne, C. Vandevyver, U. Lehmann, E. Charbon, M. A. M. Gijs. Monolithic Silicon Chip for Immunofluorescence Detection on Single Magnetic Beads, *Anal. Chem.*, **82**, 49–52 (2010).
126. Song B., V. Sivagnanam, C. D. B. Vandevyver, I. A. Hemmilä, H.-A. Lehr, M. A. M. Gijs, J.-C. G. Bünzli. Time-Resolved Lanthanide Luminescence for Lab-on-a-Chip Detection of Biomarkers on Cancerous Tissues, *Analyst*, **134**, 1991–1993 (2009).
127. Delain E., A. Barbinarbogast, C. A. Bourgeois, G. Mathis, C. Mory, C. Favard, P. Vigny, A. Niveleau. The limits in electron microscopy of macromolecular interactions. The use of new labels based on lanthanide cryptates. An interdisciplinary approach, *J. Trace Microprobe Tech.*, **13**, 371–381 (1995).
128. Houlne M. P., T. S. Agent, G. E. Kiefer, K. Mcmilian, D. J. Bornhop. Spectroscopic characterization and tissue imaging using site-selective polyazacyclic terbium(III) chelates, *Appl. Spectrosc.*, **50**, 1221–1228 (1996).
129. Bornhop D. J., D. S. Hubbard, M. P. Houlne, C. Adair, G. E. Kiefer, B. C. Pence, D. L. Morgan. Fluorescent tissue site-selective lanthanide chelate, Tb-PCTMB for enhanced imaging of cancer, *Anal. Chem.*, **71**, 2607–2615 (1999).
130. Griffin J. M. M., A. M. Skwierawska, H. C. Manning, J. N. Marx, D. J. Bornhop. Simple, high yielding synthesis of trifunctional fluorescent lanthanide chelates, *Tetrahedron Lett.*, **42**, 3823–3825 (2001).
131. Bornhop D. J., J. M. M. Griffin, T. S. Goebel, M. R. Sudduth, B. Bell, M. Motamedi. Luminescent lanthanide chelate contrast agents and detection of lesions in the hamster oral cancer model, *Appl. Spectrosc.*, **57**, 1216–1222 (2003).
132. Zheng Q., H. Dai, M. E. Merritt, C. Malloy, C. Y. Pan, W. H. Li. A New Class of Macrocyclic Lanthanide Complexes for Cell Labeling and Magnetic Resonance Imaging Applications, *J. Am. Chem. Soc.*, **127**, 16178–16188 (2005).
133. Ito T., M. Inoue, K. Akamatsu, E. Kusaka, K. Tanabe, S. Nishimoto. alpha(v)beta(3)-Integrin-targeting lanthanide complex: Synthesis and evaluation as a tumor-homing luminescent probe, *Bioorg. & Med. Chem. Lett.*, **21**, 3515–3518 (2011).
134. Parker D., J. A. G. Williams. Responsive luminescent lanthanide complexes. In *Metal Ions in Biological Systems*, A. Sigel, H. Sigel, editors. New York, Marcel Dekker Inc. Vol. 40, 233–280 (2003).
135. Pal R., A. Beeby, D. Parker. Analysis of citrate in low-volume seminal fluid samples using a time-gated measurement of europium luminescence, *J. Pharm. Biomed. Anal.*, **56**, 352–358 (2011).

136. Hubbard D. S., M. P. Houlne, G. E. Kiefer, K. McMillan, D. J. Bornhop. Endoscopic fluorescence imaging of tissue selective lanthanide chelates, *Bioimaging*, **6**, 63–70 (1998).
137. Montgomery C. P., E. J. New, L. O. Palsson, D. Parker, A. S. Batsanov, L. Lamarque. Emissive and Cell-Permeable 3-Pyridyl- and 3-Pyrazolyl-4-azaxanthone Lanthanide Complexes and Their Behaviour in cellulo, *Helv. Chim. Acta*, **92**, 2186–2213 (2009).
138. Montgomery C. P., D. Parker, L. Lamarque. Effective and efficient sensitisation of terbium luminescence at 355 nm with cell permeable pyrazoyl-1-azaxanthone macrocyclic complexes, *Chem. Commun.*, 3841–3843 (2007).
139. Pal R., D. Parker. A single component ratiometric pH probe with long wavelength excitation of europium emission, *Chem. Commun.*, 474–476 (2007).
140. Poole R. A., C. P. Montgomery, E. J. New, A. Congreve, D. Parker, M. Botta. Identification of emissive lanthanide complexes suitable for cellular imaging that resist quenching by endogenous anti-oxidants, *Org. Biomol. Chem.*, **5**, 2055–2062 (2007).
141. Yu J. H., D. Parker, R. Pal, R. A. Poole, M. J. Cann. A Europium Complex That Selectively Stains Nucleoli of Cells, *J. Am. Chem. Soc.*, **128**, 2294–2299 (2006).
142. Poole R. A., G. Bobba, M. J. Cann, J. C. Frias, D. Parker, R. D. Peacock. Synthesis and characterisation of highly emissive and kinetically stable lanthanide complexes suitable for usage 'in cellulo', *Org. Biomol. Chem.*, **3**, 1013–1024 (2005).
143. Frias J. C., G. Bobba, M. J. Cann, C. J. Hutchison, D. Parker. Luminescent nonacoordinate cationic lanthanide complexes as potential cellular imaging and reactive probes, *Org. Biomol. Chem.*, **1**, 905–907 (2003).
144. Beeby A., S. W. Botchway, I. M. Clarkson, S. Faulkner, A. W. Parker, D. Parker, J. A. G. Williams. Luminescence imaging microscopy and lifetime mapping using kinetically stable lanthanide(III) complexes, *J. Photochem. Photobiol. B Biol.*, **57**, 83–89 (2000).
145. New E. J., D. Parker, D. G. Smith, J. W. Walton. Development of responsive lanthanide probes for cellular applications, *Curr. Opin. Chem. Biol.*, **14**, 238–246 (2010).
146. Montgomery C. P., B. S. Murray, E. J. New, R. Pal, D. Parker. Cell-Penetrating Metal Complex Optical Probes: Targeted and Responsive Systems Based on Lanthanide Luminescence, *Acc. Chem. Res.*, **42**, 925–937 (2009).
147. Kielar F., G. L. Law, E. J. New, D. Parker. The nature of the sensitiser substituent determines quenching sensitivity and protein affinity and influences the design of emissive lanthanide complexes as optical probes for intracellular use, *Org. Biomol. Chem.*, **6**, 2256–2258 (2008).
148. New E. J., D. Parker. The mechanism of cell uptake for luminescent lanthanide optical probes: the role of macropinocytosis and the effect of enhanced membrane permeability on compartmentalisation, *Org. Biomol. Chem.*, **7**, 851–855 (2009).
149. Law G. L., C. W.-Y. Man, D. Parker, J. W. Walton. Observation of the selective staining of chromosomal DNA in dividing cells using a luminescent terbium(III) complex, *Chem. Commun.*, **46**, 2391–2393 (2010).
150. Fung Y. O., W. Wu, C.T. Yeung, H. K. Kong, K. K.-C. Wong, W. S. Lo, G. L. Law, K. L. Wong, C. K. Lau, C. S. Lee, W. T. Wong. In Vitro Imaging and Human Serum Albumin Responsive Dimeric Lanthanide DO3A Complex, *Inorg. Chem.*, **50**, 5517–5525 (2011).
151. McMahon B., P. Mauer, C. P. McCoy, T. C. Lee, T. Gunnlaugsson. Selective Imaging of Damaged Bone Structure (Microcracks) Using a Targeting Supramolecular Eu(III) Complex As a Lanthanide Luminescent Contrast Agent, *J. Am. Chem. Soc.*, **131**, 17542–17543 (2009).
152. Lee M., M. S. Tremblay, S. Jockusch, N. J. Turro, D. Sames. Intermolecular Energy Transfer from $Tb^{3+}$ to $Eu^{3+}$ in Aqueous Aggregates and on the Surface of Human Cells, *Org. Lett.*, **13**, 2802–2805 (2011).
153. Delgado-Pinar E., M. T. Albelda, J. C. Frias, O. Barreiro, E. Tejera, V. Kubicek, L. J. Jimenez-Borreguero, F. Sanchez-Madrid, E. Toth, J. Alarcon, E. Garcia-Espana. Lanthanide complexes

as imaging agents anchored on nano-sized particles of boehmite, *Dalton Trans.*, **40**, 6451–6457 (2011).

154. Lopez E., C. Chypre, B. Alpha, G. Mathis. Europium tris(bipyridyne) cryptate label for time-resolved luminescence detection of polymerase chain reaction products fixed on a solid support, *Clin. Chem.*, **39**, 196–201 (1993).
155. Alpha-Bazin B., H. Bazin, L. Boissy, G. Mathis. Europium cryptate-tethered ribonucleotide for the labeling of RNA and its detection by time-resolved amplification of cryptate emission, *Anal. Biochem.*, **286**, 17–25 (2000).
156. Sabbatini N., A. Mecati, M. Guardigli, V. Balzani, J.-M. Lehn, R. Zeissel, R. Ungaro. Lanthanide luminescence in supramolecular species, *J. Lumin.*, **48/49**, 463–468 (1991).
157. Bazin H., E. Trinquet, G. Mathis. Time resolved amplification of cryptate emission: a versatile technology to trace biomolecular interactions, *Rev. Mol. Biotechnol.*, **82**, 233–250 (2002).
158. Ghose S., E. Trinquet, M. Laget, H. Bazin, G. Mathis. Rare earth cryptates for the investigation of molecular interactions in vitro and in living cells, *J. Alloys Compd.*, **451**, 35–37 (2008).
159. Piguet C., A. F. Williams, G. Bernardinelli, J.-C. G. Bünzli. Structural and Photophysical Properties of Lanthanide Complexes with Planar Aromatic Tridentate Nitrogen Ligands as Luminescent Building Blocks for Triple-Helical Structures, *Inorg. Chem.*, **32**, 4139–4149 (1993).
160. Gassner A.-L., C. Duhot, J.-C. G. Bünzli, A.-S. Chauvin. Remarkable Tuning of the Photophysical Properties of Bifunctional Lanthanide tris(Dipicolinates) and its Consequence on the Design of Bio-probes, *Inorg. Chem.*, **47**, 7802–7812 (2008).
161. Picot A., A. D'Aléo, P. L. Baldeck, A. Grichine, A. Duperray, C. Andraud, O. Maury. Long-Lived Two-Photon Excited Luminescence of Water-Soluble Europium Complex: Applications in Biological Imaging Using Two-Photon Scanning Microscopy, *J. Am. Chem. Soc.*, **130**, 1532–1533 (2008).
162. Lamture J. B., T. G. Wensel. Intensively Luminescent Immunoreactive Conjugates of Proteins and Dipicolinate-based Polymeric Tb(III) Chelates., *Bioconjugate Chem.*, **6**, 88–92 (1995).
163. Cable M. L., J. P. Kirby, D. J. Levine, M. J. Manary, H. B. Gray, A. Ponce. Detection of Bacterial Spores with Lanthanide–Macrocycle Binary Complexes, *J. Am. Chem. Soc.*, **131**, 9562–9570 (2009).
164. Barthelmes K., A. M. Reynolds, E. Peisach, H. R. A. Jonker, N. J. DeNunzio, K. N. Allen, B. Imperiali, H. Schwalbe. Engineering Encodable Lanthanide-Binding Tags into Loop Regions of Proteins, *J. Am. Chem. Soc.*, **133**, 808–819 (2011).
165. Elhabiri M., R. Scopelliti, J.-C. G. Bünzli, C. Piguet. Lanthanide Helicates Self-assembled in Water: a New Class of Highly Stable and Luminescent Dimetallic Carboxylates., *J. Am. Chem. Soc.*, **121**, 10747–10762 (1999).
166. Vandevyver C. D. B., A.-S. Chauvin, S. Comby, J.-C. G. Bünzli. Luminescent lanthanide bimetallic triple-stranded helicates as potential cellular imaging probes, *Chem. Commun.*, 1716–1718 (2007).
167. Görller-Walrand C., K. Binnemans. Rationalization of Crystal Field Parameterization. In *Handbook on the Physics and Chemistry of Rare Earths*, K. A. Gschneidner Jr., L. Eyring, editors. Amsterdam, Elsevier Science B.V. Vol. 23, Ch. 155, 121–283 (1996).
168. Bünzli J.-C. G., A.-S. Chauvin, C. D. B. Vandevyver, B. Song, S. Comby. Lanthanide Bimetallic Helicates for *in vitro* Imaging and Sensing, *An. N. Y. Acad. Sci.*, **1130**, 97–105 (2008).
169. Deiters E., B. Song, A.-S. Chauvin, C. Vandevyver, J.-C. G. Bünzli. Luminescent Bimetallic Lanthanide Bioprobes for Cellular Imaging with Excitation into the Visible, *Chem. Eur. J.*, **15**, 885–900 (2009).
170. Fernandez-Moreira V., B. Song, V. Sivagnanam, A.-S. Chauvin, C. D. B. Vandevyver, M. A. M. Gijs, I. A. Hemmilä, H.-A. Lehr, J.-C. G. Bünzli. Bioconjugated Lanthanide Luminescent Helicates as Multilabels for Lab-on-a-Chip Detection of Cancer Biomarkers, *Analyst*, **135**, 42–52 (2010).

171. Chauvin A.-S., S. Comby, B. Song, C. D. B. Vandevyver, J.-C. G. Bünzli. A Polyoxyethylene-Substituted Bimetallic Europium Helicate for Luminescent Staining of Living Cells, *Chem. Eur. J.*, **13**, 9515–9526 (2007).
172. Chauvin A.-S., S. Comby, B. Song, C. D. B. Vandevyver, J.-C. G. Bünzli. A Versatile Ditopic Ligand System for Sensitizing the Luminescence of Bimetallic Lanthanide Bio-imaging Probes, *Chem. Eur. J.*, **14**, 1726–1739 (2008).
173. Deiters E., B. Song, A.-S. Chauvin, C. D. B. Vandevyver, J.-C. G. Bünzli. Testing the Length of the Pyridine Substituent on Cellular Uptake in a Bimetallic Europium Luminescent Probe., *New J. Chem.*, **32**, 1140–1152 (2008).
174. Song B., C. D. B. Vandevyver, A.-S. Chauvin, J.-C. G. Bünzli. Time-resolved Luminescence Microscopy of Bimetallic Lanthanide Helicates in Living Cells, *Org. Biomol. Chem.*, **6**, 4125–4133 (2008).
175. Sivagnanam V., B. Song, C. D. B. Vandevyver, J.-C. G. Bünzli, M. A. M. Gijs. Selective breast cancer cells capture, culture, and immunocytochemical analysis using self-assembled magnetic bead patterns in a microfluidic chip, *Langmuir*, **26**, 6091–6096 (2010).
176. Bünzli J.-C. G., C. D. B. Vandevyver, A.-S. Chauvin, M. A. M. Gijs, H.-A. Lehr. Lighting up Cancerous Cells with Lanthanide Luminescence, *Chimia*, **65**, 361 (2011).
177. Ciftlik A.T., B. Song, C. D. B. Vandevyver, J.-C. G. Bünzli, H.-A. Lehr, M. A. M. Gijs. Fast Immunohistochemical BioMarker Detection Device for Cancer Tissue Slices. Proceedings of 14th International Conference on Miniaturized Systems for Chemistry and Life Sciences; Groningen, The Netherlands, October 3–7: Royal Chemical Society (2010) pp. 699–701.
178. Eliseeva S. V., B. Song, C. D. B. Vandevyver, A.-S. Chauvin, J. B. Wacker, J.-C. G. Bünzli. Increasing the Efficiency of Lanthanide Luminescent Bioprobes: Bioconjugated Silica Nanoparticles as Markers for Cancerous Cells, *New J. Chem.*, **34**, 2915–2921 (2010).
179. Eliseeva S. V., G. Auböck, F. van Mourik, A. Cannizo, B. Song, E. Deiters, A.-S. Chauvin, M. Chergui, J.-C. G. Bünzli. Multiphoton-excited Luminescent Lanthanide Bioprobes: Two- and Three-photon Cross Sections of Dipicolinates Derivatives and Binuclear Helicates, *J. Phys. Chem. B*, **114**, 2932–2937 (2010).
180. Andraud C., O. Maury. Lanthanide Complexes for Nonlinear Optics: From Fundamental Aspects to Applications, *Eur. J. Inorg. Chem.*, **29/30**, 4357–4371 (2009).
181. Sergeeva N. N., M. Donnier-Marechal, G. Vaz A. M. Davies, M. O. Senge. Synthesis and evaluation of the europium(III) and zinc(II) complexes as luminescent bioprobes in high content cell-imaging analysis, *J. Inorg. Biochem.*, **105**, 1589–1595 (2011).
182. Wolfbeis O. S., A. Durkop, M. Wu, Z. H. Lin. A europium-ion-based luminescent sensing probe for hydrogen peroxide, *Angew. Chem., Int. Ed. Engl.*, **41**, 4495–4498 (2002).
183. Wu M., Z. H. Lin, M. Schaferling, A. Durkop, O. S. Wolfbeis. Fluorescence imaging of the activity of glucose oxidase using a hydrogen-peroxide-sensitive europium probe, *Anal. Biochem.*, **340**, 66–73 (2005).
184. Lin Z., M. Wu, O. S. Wolfbeis, M. Ferling. A novel method for time-resolved fluorimetric determination and imaging of the activity of peroxidase, and its application to an enzyme-linked immunosorbent assay, *Chem. Eur. J.*, **12**, 2730–2738 (2006).
185. Balzani V., G. Bergamini, P. Ceroni, E. Marchi. Designing light harvesting antennas by luminescent dendrimers, *New J. Chem.*, **35**, 1944–1954 (2011).
186. Astruc D., E. Boisselier, C. Ornelas. Dendrimers Designed for Functions: From Physical, Photophysical, and Supramolecular Properties to Applications in Sensing, Catalysis, Molecular Electronics, Photonics, and Nanomedicine, *Chem. Rev.*, **110**, 1857–1959 (2010).
187. Bergamini G., E. Marchi, P. Ceroni. Metal ion complexes of cyclam-cored dendrimers for molecular photonics, *Coord. Chem. Rev.*, **255**, 2458–2468 (2011).
188. Branchi B., P. Ceroni, V. Balzani, F. G. Klärner, F. Vögtle. A light-harvesting antenna resulting from the self-assembly of five luminescent components: a dendrimer, two clips, and two lanthanide ions, *Chem. Eur. J.*, **16**, 6048–6055 (2010).

189. Alcala M. A., C. M. Shade, H. Uh, S. Y. Kwan, M. Bischof, Z. P. Thompson, K. A. Gogick, A. R. Meier, T. G. Strein, D. L. Bartlett, R. A. Modzelewski, Y. J. Lee, S. Petoud, C. K. Brown. Preferential accumulation within tumors and in vivo imaging by functionalized luminescent dendrimer lanthanide complexes, *Biomaterials*, **32**, 9343–9352 (2011).
190. Alcala M. A., S. Y. Kwan, C. M. Shade, M. Lang, H. Uh, M. Wang, S. G. Weber, D. L. Bartlett, S. Petoud, Y. J. Lee. Luminescence targeting and imaging using a nanoscale generation 3 dendrimer in an in vivo colorectal metastatic rat model, *Nanomedicine: Nanotech. Biol. Med.*, **7**, 249–258 (2011).
191. Li L., Y. Liu, J. Tao, M. Zhang, H. Pan, X. Xu, R. Tang. Surface Modification of Hydroxyapatite Nanocrystallite by a Small Amount of Terbium Provides a Biocompatible Fluorescent Probe, *J. Phys. Chem. C*, **112**, 12219–12224 (2008).
192. Li X., S. Gai, C. Li, D. Wang, N. Niu, F. He, P. Yang. Monodisperse Lanthanide Fluoride Nanocrystals: Synthesis and Luminescent Properties, *Inorg. Chem.*, **51**, 3963–3971 (2012).
193. Osseni S. A., S. Lechevallier, M. Verelst, C. Dujardin, J. Dexpert-Ghys, D. Neumeyer, M. Leclercq, H. Baaziz, D. Cussac, V. Santran, R. Mauricot. New nanoplatform based on $Gd_2O_2S$: $Eu^{3+}$ core: synthesis, characterization and use for in vitro bio-labelling, *J. Mater. Chem.*, **21**, 18365–18372 (2011).
194. Hahn M. A., A. K. Singh, P. Sharma, B. Parvesh, C. Scott, B. M. Moudgil. Nanoparticles as contrast agents for in-vivo bioimaging: current status and future perspectives, *Anal. Bioanal. Chem.*, **399**, 3–27 (2011).
195. Liu Z., L. N. Sun, L. Y. Shi, D. S. Zhang. Near-Infrared Lanthanide Luminescence for Functional Materials, *Progress in Chemistry*, **23**, 153–164 (2011).
196. Gao D. L., X. Y. Zhang, W. Gao. Tuning upconversion emission by controlling particle shape in $NaYF_4$:$Yb^{3+}Er^{3+}$ nanocrystals, *J. Appl. Phys.*, **111**, Art. Nr. 033505 (2012).
197. Salthouse C., S. Hildebrand, R. Weissleder, U. Mahmood. Design and demonstration of a small-animal up-conversion imager, *Optics Express*, **16**, 21731–21737 (2008).
198. Soga K. Development of NIR Fluorescence Bioimaging System through Polyscale Technology, *Bunseki Kagaku*, **58**, 461–471 (2009).
199. Kamimura M., N. Kanayama, K. Tokuzen, K. Soga, Y. Nagasaki. Near-infrared (1550 nm) in vivo bioimaging based on rare-earth doped ceramic nanophosphors modified with PEG-b-poly(4-vinylbenzylphosphonate), *Nanoscale*, **3**, 3705–3713 (2011).
200. Xiao Q., H. Zhu, D. Tu, E. Ma, X. Chen. Near-Infrared-to-Near-Infrared Downshifting and Near-Infrared-to-Visible Upconverting Luminescence of $Er^{3+}$-Doped $In_2O_3$ Nanocrystals, *J. Phys. Chem. C*, **117**, 10834–10841 (2013).
201. Chen G., C. H. Yang, P. N. Prasad. Nanophotonics and Nanochemistry: Controlling the Excitation Dynamics for Frequency Up- and Down-Conversion in Lanthanide-Doped Nanoparticles, *Acc. Chem. Res.* **46**, 1474–1486 (2013).
202. Coogan M. P., V. Fernandez-Moreira. Progress with, and prospects for, metal complexes in cell imaging, *Chem. Commun.*, **50**, 384–399 (2014).
203. McMahon B. K., R. Pal, D. Parker. A bright and responsive europium probe for determination of pH change within the endoplasmic reticulum of living cells. *Chem. Commun.*, **49**, 5363–5365 (2013).
204. Butler S. J., B. K. McMahon, R. Pal, D. Parker, J. W. Walton. Bright Mono-aqua Europium Complexes Based on Triazacyclononane That Bind Anions Reversibly and Permeate Cells Efficiently, *Chem. Eur. J.*, **19**, 9511–9517 (2013).
205. Zhang L., L. Tian, Z. Ye, B. Song, J. L. Yuan. Preparation of visible-light-excited europium biolabels for time-resolved luminescence cell imaging application, *Talanta*, **108**, 143–149 (2013).
206. Liu M. J., Z. Q. Ye, C. L. Xin, J. L. Yuan. Development of a ratiometric time-resolved luminescence sensor for pH based on lanthanide complexes, *Anal. Chim. Acta* **761**, 149–156 (2013).

207. Bonnet C. S., F. Buron, F. Caillé, C. M. Shade, B. Drahos, L. Pellegatti, J. Zhang, S. Villette, L. Helm, C. Pichon, F. Suzenet, S. Petoud, E. Toth. Pyridine-based lanthanide complexes combining MRI and NIR luminescence activities, *Chem. Eur. J.*, **18**, 1419–1431 (2012).
208. Diniz J. R., J. R. Correa, D. de Moreira, R. S. Fontanele, A. L. de Oliveira, P. V. Abdelnur, J. D. L. Dutra, R. O. Freire, M. O. Rodrigues, B. A. D. Neto, Water-Soluble $Tb^{3+}$ and $Eu^{3+}$ Complexes with Ionophilic (Ionically Tagged) Ligands as Fluorescence Imaging Probes, *Inorg. Chem.*, **52**, 10199–10205 (2013).
209. Thielemann D. T., A. T. Wagner, E. Rösch, D. K. Kölmel, J. G. Heck, B. Rudat, M. Neumaier, C. Feldmann, U. Schepers, S. Bräse, P. W. Roesky. Luminescent Cell-penetrating Pentadecanuclear Lanthanide Clusters, *J. Am. Chem. Soc.*, **135**, 7454–7457 (2013).
210. Zhang J., S. Petoud. Azulene-moiety-based ligand for the efficient Sensitization of four near-infrared luminescent lanthanide cations: $Nd^{3+}$, $Er^{3+}$, $Tm^{3+}$, and $Yb^{3+}$, *Chem. Eur. J.*, **14**, 1264–1272 (2008).
211. Shavaleev N. M., R. Scopelliti, F. Gumy, J.-C. G. Bünzli. Surprisingly Bright Near-Infrared Luminescence and Short Radiative Lifetimes of Ytterbium in Hetero-Binuclear Yb-Na Chelates, *Inorg. Chem.*, **48**, 7937–7946 (2009).
212. Hehlen M. P., Phillips M. L. F., Cockroft N. J., Güdel H. U. Up-conversion phosphors. In Buschow K. H. J., Cahn R. W., Flemings M. C., Ilschner B., Kramer E. J., Mahajan S., editors. *Encyclopedia of Materials: Science and Technology*. 9456–9458 (2001) Oxford, Elsevier Science Ltd.
213. Zhang J. X., H. G. Li, C. F. Chan, R. F. Lan, W. L. Chan, G. L. Law, W. K. Wong, K. L. Wong, A potential water-soluble ytterbium-based porphyrin-cyclen dual bioprobe for Golgi apparatus imaging and photodynamic therapy, *Chem. Commun.*, **48**, 9646–9648 (2012).
214. Andolina C. M., P. J. Klemm, W. C. Floyd, J. M. J. Frechet, K. N. Raymond. Analysis of Lanthanide Complex Dendrimer Conjugates for Bimodal NIR and MRI Imaging, *Macromolecules*, **45**, 8982–8990 (2012).
215. Caillé F., C. S. Bonnet, F. Buron, S. Villette, L. Helm, S. Petoud, F. Suzenet, E. Toth. Isoquinoline-Based Lanthanide Complexes: Bright NIR Optical Probes and Efficient MRI Agents, *Inorg. Chem.* **51**, 2522–2532 (2012).
216. Liu Y., G. Yu, M. Tian, H. Zhang. Optical probes and the applications in multimodality imaging, *Contr. Med. Mol. Im.*, **6**, 169–177 (2011).
217. Bonnet C. S., E. Toth. Towards highly efficient, intelligent and bimodal imaging probes: Novel approaches provided by lanthanide coordination chemistry, *C. R. Chimie*, **13**, 700–714 (2010).
218. Janczewski D., Y. Zhang, G. K. Das, D. K. Yi, P. Padmanabhan, K. K. Bhakoo, T. T. Y. Tan, S. T. Selvan. Bimodal Magnetic-Fluorescent Probes for Bioimaging, *Microsc. Res. Techn.*, **74**, 563–576 (2011).
219. Maldonado C. R., L. Salassa, N. Gomez-Blanco, J. C. Mareque-Rivas. Nano-functionalization of metal complexes for molecular imaging and anticancer therapy, *Coord. Chem. Rev.* **257**, 2668–2688 (2013).
220. Gomez M. C., R. Fernandes, A. Cunha, J. P. Tome, T. Trindade. Fluorescence biolabeling using methylated silica nanoparticles containing a lanthanide complex, *J. Mater. Chem. B*, **1**, 5429–5435 (2013).
221. Cui Y., Y. Yue, G. Qian, B. Chen. Luminescent Functional Metal–Organic Frameworks, *Chem. Rev.*, **111**, 1126–1162 (2012).
222. Lu Y. Q., P. Xi, J. A. Piper, Y. J. Huo, D. Y. Jin. Time-Gated Orthogonal Scanning Automated Microscopy (OSAM) for High-Speed Cell Detection and Analysis, *Sci. Rep.*, **2**, Rp. Nr. 837 (2012).

# 5

# Two-photon Absorption of Lanthanide Complexes: from Fundamental Aspects to Biphotonic Imaging Applications

*Anthony D'Aléo*[1], *Chantal Andraud*[2] *and Olivier Maury*[2]

[1]*CINaM, UMR 7325 CNRS-Aix Marseille Université, France*

[2]*Laboratoire de chimie, UMR 5281 ENS Lyon-CNRS-Université de Lyon, France*

## 5.1 Introduction

The gradual filling of the *4f* shell, shielded from the surrounding by the filled $5s^2 5p^6$ orbitals, confers on the *f*-block elements unique luminescence properties arising from *f-f* transitions, that is sharp emission bands characteristic of the metal and ranging from the visible to the near infra-red (NIR) with a long excited state lifetime (μs to ms). As illustrated in this book and in numerous reviews, these emission properties have found very useful applications in material science for lighting or light-emitting devices [1–3] or in biology for (time-resolved) imaging, sensing [4–9] or time-gated fluoroassays [10].

Whereas the optical or chiroptical [11 and Chapter 3] properties of lanthanide have been extensively studied, less interest was devoted to their potential in nonlinear optics (NLO) until this last decade, when it became an emerging and also challenging field of research [12,13]. As explained in the first part of this chapter, NLO phenomena occur when a molecule or a material is irradiated with a high-power light beam, typically a laser. Depending on the incident laser wavelength, two types of NLO phenomena can be distinguished: nonresonant (e.g., harmonic generation) and resonant (e.g., multi-photon

*Luminescence of Lanthanide Ions in Coordination Compounds and Nanomaterials*, First Edition.
Edited by Ana de Bettencourt-Dias.

absorption) [12]. On the one hand, only a few studies describing the second-harmonic generation properties of lanthanide complexes are present in the literature [14–18] and will be beyond the scope of the present review. On the other hand, the major part of the research endeavour was devoted to the multi-photon absorption properties of lanthanide complexes, to combine the advantages of this nonlinear excitation process and that of the lanthanide luminescence. In this chapter, after an introduction on the basics of multi-photon absorption, we will describe, in an exhaustive way, the results of the last decade devoted to the optimisation of the two-photon absorption efficiency of lanthanide complexes, and we will illustrate their stepwise application to nonlinear microscopy for bio-imaging purposes.

## 5.2 Two-photon Absorption, a Third Nonlinear Optical Phenomenon

### 5.2.1 Theoretical and Historical Background

The two-photon absorption (2PA) process was predicted by M. Göppert-Mayer in 1931 during her PhD thesis [19]. Based on the quantum theory of radiation, M. Göppert-Mayer introduced the concept of "virtual" or "intermediate" states and provided the theoretical expression the 2PA probability. However, due to the very low value of this feature, the first experimental observation of this phenomenon could be only obtained for the first time at the advent of power lasers in 1960s, by the two-photon excited (2PE) up-converted fluorescence of a $CaF_2$:Eu(II) crystal using 694.3 nm ruby laser excitation, as described in detail below [20].

As a general point of view, the 2PA process corresponds to the simultaneous interaction of two photons with matter, leading to the absorption transition from a lower level to a higher excited one, through a virtual level. It is worth noting that multi-higher order (>2) absorption processes can be also observed (multiphoton absorption, MPA, e.g. 3PA, 4PA, etc.); 2PA and 3PA processes are displayed in Fig. 5.1, as examples.

The attenuation of the light beam intensity, which propagates in the $z$ direction through the absorbing medium, is described by Equation 5.1 in the case of a 2PA process with no linear absorption ($\alpha = 0$); in this equation, $\alpha'$ is the 2PA coefficient. A relevant parameter characterising 2PA properties of molecular materials is the 2PA cross-section $\sigma_{2PA}$, generally expressed in Göppert-Mayer (GM), defined as $1\,\mathrm{GM} = 10^{-50}\,\mathrm{cm^4.s.photon^{-1}}$. $\sigma_{2PA}$ is expressed as a function of $\alpha'$ following Equation 5.2, in which $h\upsilon$ is the energy of the incident photon, $N_0$ the molecular concentration, $N_a$ the Avogadro number, $c$ the

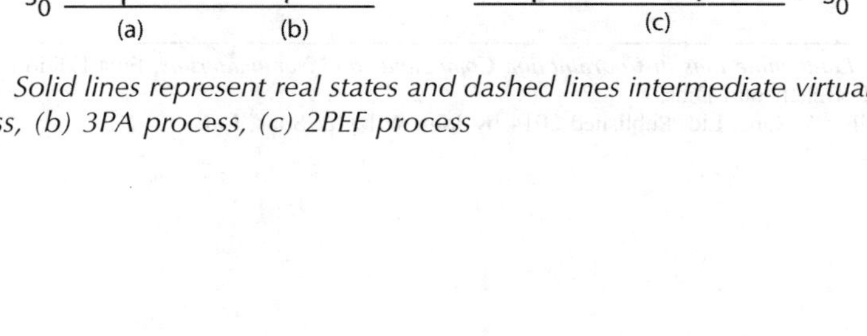


***Figure 5.1*** *Solid lines represent real states and dashed lines intermediate virtual levels in: (a) 2PA process, (b) 3PA process, (c) 2PEF process*

concentration, $\hbar$ the Planck constant and $f$ the local field factor [21]:

$$\frac{dI}{dz} = -\alpha' I^2 + \ldots \tag{5.1}$$

$$\sigma_{2PA} = \frac{h\upsilon}{N_0}\alpha' = \frac{h\upsilon}{N_a c 10^{-3}}\alpha' \tag{5.2}$$

### 5.2.2 Experimental Determination of the 2PA Efficiency of Molecules

Two main methods are generally used for the determination of $\sigma_{2PA}$ values.

#### 5.2.2.1 Nonlinear Transmission Method and Z-Scan Method

The Z-scan method is a single beam technique allowing the measurements of the nonlinear refractive index and the nonlinear absorption. This technique, introduced in 1990 by Van Stryland et al. [22], is based on the measurement of the transmission of a laser through the sample moving around the focal point of the beam. Variations of the beam radius at the position $z$ with respect to the focal $z = 0$ are expressed following Equation 5.3. In this equation, $z_R$ is the Rayleigh length with $z_R = n\pi w_0^2/\lambda$.

$$w(z) = w_0\left[1 + \left(\frac{z}{z_R}\right)^2\right]^{1/2} \tag{5.3}$$

Two different situations are possible: (1) the closed-aperture case, which allows the determination of the nonlinear refractive index of the sample, and (2) the open-aperture situation, in which the light intensity variations are obtained. The nonlinear transmittance $T_{nl}$ can be expressed by Equation 5.4, where $L$ represents the length of the sample [23]:

$$T_{nl} = \frac{1}{\sqrt{\pi\alpha'}\,LI}\int_{-\infty}^{\infty} \ln\left(1 + \alpha' LIe^{-u^2}\right)du \tag{5.4}$$

The 2PA coefficient $\alpha'$ and then the $\sigma_{2PA}$ values using Equation 5.2 are deduced from open-aperture Z-scan data.

#### 5.2.2.2 Two-photon Excited Fluorescence (TPEF) Procedure

An alternative method, used in the case of highly fluorescent samples, is based on the measurements of the 2PA induced luminescence intensity [24]. This method requires a reference molecule, for which the value of the 2PA cross-section $\sigma^r_{2PA}$ is known in the studied spectral range, and consists in the calibration on the studied spectral range of the two-photon induced excitation spectra (Fig. 5.1c) from reference molecule data, following Equation 5.5.

$$\sigma_{2PA} = \frac{F}{F^r}\frac{\varphi^r c^r}{\varphi\, c}\,\sigma^r_{2PA} \tag{5.5}$$

In this relationship, $F$ represents the fluorescence intensities and $\varphi$ the luminescence quantum yields, while the superscript $r$ refers to the reference molecule. The comparison between the reference and the molecule under study leads to the determination of the biphotonic brightness $\sigma_{2PA}.\varphi$, from which the value of $\sigma_{2PA}$ is extracted after division by the value of the luminescence quantum yield.

### 5.2.3 Two-photon Fluorescence Microscopy for Biological Imaging

The 2PA effect has been widely involved in numerous applications as reviewed by Prasad et al. [25], such as frequency-up conversion lasing [26], optical power limiting [27,28], microfabrication [29,30], optical data storage and processing [31] and biological or medical applications [32]. As this chapter, which addresses 2PA of lanthanide complexes, will concern mainly biological applications, we will focus in this part on two-photon excited fluorescence microcopy in biology.

Numerous advantages of biphotonic microscopy can be noted:

1. The use of a Ti:sapphire laser, which produces a light in the range between 700 and 1200 nm, in the so-called "transparency window" of biological tissues [33], leads to decreased linear absorption (Fig. 5.2a) and scattering along the beam [34] (Fig. 5.2b). This allows a deeper penetration of the beam within tissues up to 600–800 mm in the neocortex for instance [35], and is illustrated by *in vivo* two-photon microscopy blood vessels in the mouse brain [36] (Fig. 5.3a).
2. The use of NIR incident beams induces a decrease of auto-fluorescence background compared to excitation localised in UV or blue range of the visible. Furthermore NIR excitation improves the viability of biological media and allows more scans to optimise the 3D image, as obtained in the case of chromosomes during the cell division by multiphonic excitation of 4′,6′-diamidino-2-phenylindol (DAPI) [37] (Fig. 5.3b).

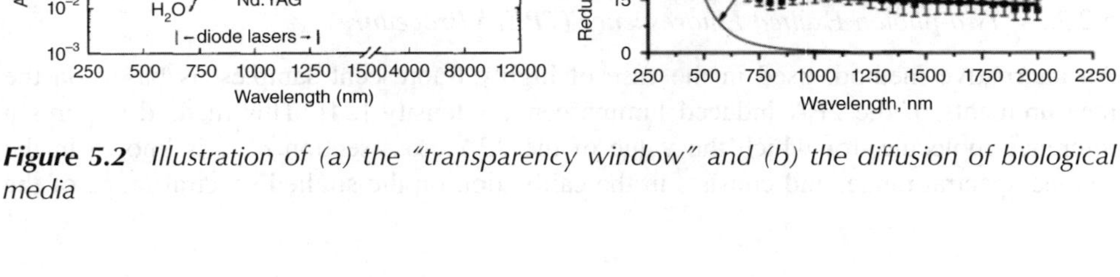


*Figure 5.2* Illustration of (a) the "transparency window" and (b) the diffusion of biological media

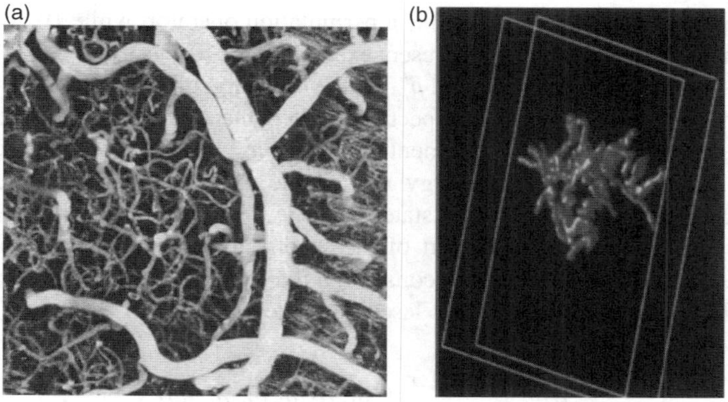


***Figure 5.3*** *(a)* In vivo *blood vessel biphotonic microscopy imaging of brain mouse. (b) 3D image of chromosomes during cell division by multiphotonic excitation of DAPI (See colour plate section)*

3. A reduced photobleaching is obtained, since only the region at the focal point is excited, as 2PA varies quadratically with the incident laser intensity [38] (Equation 5.1).

### 5.2.4 Molecular Engineering for Multiphonic Imaging

The 2PA nonlinear absorption process can be described from the semi-classical theory, which leads to express the nonlinear electric polarisation $\vec{P}$ of an optical medium and the electronic dipole moment $\mu$ induced at molecular scale following Equations 5.6 and 5.7, respectively, as a function of odd-order susceptibilities and hyperpolarisabilities, respectively [39]:

$$\vec{P}(\omega) = \varepsilon_0 \left[\chi^{(1)}(\omega)\vec{E}(\omega) + \chi^{(3)}(\omega,\omega,-\omega)\vec{E}(\omega)\vec{E}(\omega)\vec{E}^*(\omega) + \ldots\right] \tag{5.6}$$

$$\vec{\mu} = \vec{\mu_0} + \alpha\vec{E}(\omega) + \gamma(\omega,\omega,-\omega)\vec{E}(\omega)\vec{E}(\omega)\vec{E}^*(\omega) + \ldots \tag{5.7}$$

In these equations, $\varepsilon_0$ and $\alpha$ are respectively the vacuum permittivity and the linear absorption coefficient. $\vec{E}(\omega)$ represents the applied electric field at the frequency $\omega$ [$\vec{E}^*(\omega)$ is the complex conjugate of $\vec{E}(\omega)$], while $\chi^{(3)}(\omega,\omega,-\omega)$ and $\gamma(\omega,\omega,-\omega)$, which are fourth-order tensors, are respectively the third-order susceptibility and hyperpolarisability. The real part of these tensors represents the induced refractive index change and is a function of the laser intensity $(\omega) \propto \left|\vec{E}(\omega)\vec{E}^*(\omega)\right|$, while the imaginary part describes the 2PA process [40].

The second hyperpolarisability $\gamma$, is expressed by the sum-over states (SOS), Equation 5.8 [41].

$$\gamma_{ijkl}(-\omega;\omega,\omega,-\omega) = \frac{1}{6\hbar^3} P(i,j,k,l;-\omega;\omega,\omega,-\omega)$$
$$\left[\sum_{m\neq 0}\sum_{n\neq 0}\sum_{p\neq 0} \frac{\langle 0|\mu_i|m\rangle\langle m|\bar{\mu}_j|n\rangle\langle n|\bar{\mu}_k|p\rangle\langle p|\mu_l|0\rangle}{(\omega_{m0}-\omega-i\Gamma_{m0})(\omega_{n0}-2\omega-i\Gamma_{n0})(\omega_{p0}-\omega-i\Gamma_{p0})} - \sum_{m\neq 0}\sum_{n\neq 0} \frac{\langle 0|\mu_i|m\rangle\langle m|\mu_j|0\rangle\langle 0|\mu_k|n\rangle\langle n|\mu_l|0\rangle}{(\omega_{m0}-\omega-i\Gamma_{m0})(\omega_{n0}-\omega-i\Gamma_{n0})(\omega_{n0}+\omega-i\Gamma_{n0})}\right] \tag{5.8}$$

In this expression, $P(i,j,k,l;\omega,\omega,-\omega)$ is a permutation operator, while $|0\rangle$ corresponds to the ground state and $|m\rangle$, $|n\rangle$ and $|p\rangle$ represent excited states. $\langle 0|\mu_i|m\rangle$ and $\langle m|\overline{\mu}_i|n\rangle$ correspond respectively to the components $\mu^i_{0m}$ and $\mu^i_{mn}$ of the transition dipole moment between levels $|0\rangle$ and $|m\rangle$ and between $|m\rangle$ and $|n\rangle$, respectively, along the axis $i$ of the molecule; $\langle n|\overline{\mu}_i|n\rangle$ is the component of the static dipole moment difference $\mu^i_{nn} = \Delta\mu^i_{0n}$ along the axis $i$ between $|0\rangle$ and $|n\rangle$. $\hbar\omega_{m0}$ represents the energy $E_{0m}$ of the excited sate $|m\rangle$, while $\Gamma_{m0}$ is the homogeneous width associated to the state $|m\rangle$.

$\sigma_{2PA}$ is related to the imaginary part of the second hyperpolarisability $\gamma$ according to Equation 5.9, in which $\hbar$ is the Planck constant, $n$ the refraction index of the medium and $f$ the local field factor; $\omega$ is the incident laser frequency [21].

$$\sigma_{2PA} = \frac{8\pi^2\hbar\omega^2}{n^2c^2} f^4 \mathrm{Im}\,\gamma(-\omega;\omega,\omega,-\omega) \tag{5.9}$$

The design of molecules for 2PA is closely related to Equation 5.8. This expression can be considerably simplified by using the three-level Equation 5.10.

$$\gamma_{xxxx} \propto \left[ \begin{array}{l} -\dfrac{\mu_{01}^4}{(E_{01}-\hbar\omega-i\Gamma)^2(E_{01}+\hbar\omega-i\Gamma)}|\boldsymbol{N} \\ +\dfrac{\mu_{01}^2\Delta\mu_{01}^2}{(E_{01}-\hbar\omega-i\Gamma)^2(E_{01}-2\hbar\omega-i\Gamma)}|\boldsymbol{D} \\ +\dfrac{\mu_{01}^2\mu_{12}^2}{(E_{01}-\hbar\omega-i\Gamma)^2(E_{02}-2\hbar\omega-i\Gamma)}|\boldsymbol{2P} \end{array} \right] \tag{5.10}$$

This relationship contains three terms: (i) the negative term $\boldsymbol{N}$ corresponds only to a one-photon resonance and can be neglected in the case of 2PA; (ii) the dipolar term $\boldsymbol{D}$ takes into account a two-photon resonance with the low-lying excited state $S_1$ and is only present in the case of dipolar molecules; (iii) the two-photon term ***2P*** is related to a two-photon resonance with the excited state $S_2$ and is strongly dependent on the coupling between excited states.

This simplified model is illustrated by Fig. 5.4 in both cases of centrosymmetric and noncentrosymmetric molecules. In the first case, the dipolar term of Equation 5.10 cancels and only the 2P part must be taken into account; Equation 5.10 becomes Equation 5.11 [21]:

$$\sigma_{2PA} \propto \frac{\mu_{01}^2\mu_{12}^2}{\Delta^2\Gamma} \quad \text{with} \quad \Delta = \left(E_{01} - \frac{E_{02}}{2}\right) < \hbar\omega \tag{5.11}$$

The three-level model Equation 5.10 shows that the 2PA efficiency is strongly correlated to the molecular charge transfer. Several reviews describe different molecular engineering approaches for the optimisation of $\sigma_{2PA}$ values [25,42,43]. Several classes of molecules were studied for their 2PA properties, as displayed in Fig. 5.5: noncentrosymmetric molecules such as dipoles and octupoles, and centrosymmetric systems such as quadrupoles. In the last case, the dipolar term $D$ of Equation 5.10 cancels, but the possibility of a 2PA resonance with the lowest excited state $S_1$ (Fig. 5.4 and Equation 5.11) leads to enhanced 2PA properties [44].

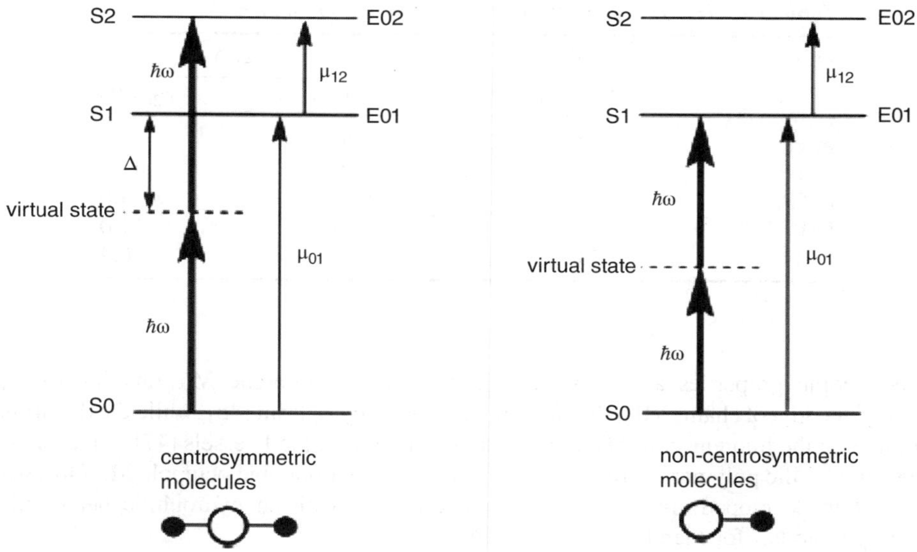


*Figure 5.4* Schematic representation of 2PA three-level model for centrosymmetric and noncentrosymmetric molecules

There is room for molecular engineering in the case of molecules for biological imaging, since $\sigma_{2PA}$ values of proteins used as biological labels are weak, as illustrated in Table 5.1 [45].

To be used in 2PA imaging, labels must combine high hydrosolubility and biphotonic brightness defined as the product of $\sigma_{2PA}$ and the luminescence quantum yield $\varphi$. Figure 5.6 shows several systems specifically designed for biphotonic bioimaging, while their

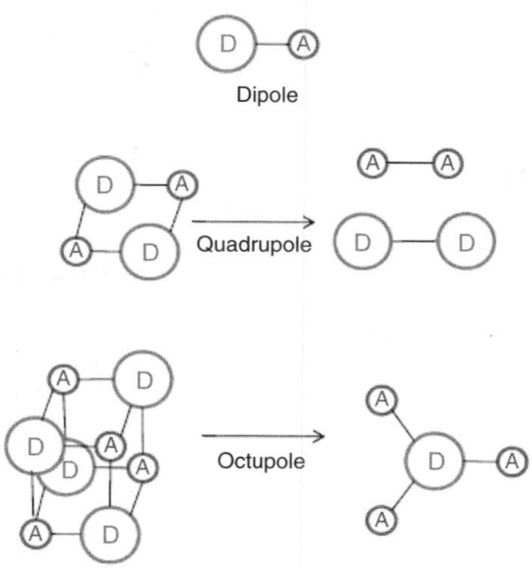


*Figure 5.5* Different classes of molecules for 2PA

***Table 5.1*** *The 1PA and 2PA properties of some proteins*

| Proteins | 1PA | 2PA | |
|---|---|---|---|
| | $\lambda^{(1)}$ (nm) | $\lambda^{(2)}$ (nm) | $\sigma_{2PA}$ (GM) |
| eCFP | 434 | 860 | 7.9 |
| eGFP | 489 | 920 | 41.2 |
| eYFP | 514 | 960 | 25.0 |
| DsRed | 588 | 960 | 11.0 |
| FMN | 373 | 760 | 0.4 |

spectrocopic properties are reported in Table 5.2: the fluorene **M1** functionalised by triethylene glycol chains led to biphotonic imaging of cytoplasm [46], while the dendrimers containing the fluorene core **M2** were used as a probe for blood vessels [47]. Jen et al. used the amphiphilic polymer approach to encapsulate the hydrophobic octupole **M3** [48], while our group developed amphiphilic chromophores **M4** with an hydrophilic head and an hydrophobic tail for membrane imaging [49].

The last example is an octupole with three pyridinium as hydrosolubilising groups. This molecule, weakly fluorescent in water, presents a strong increase of its fluorescence quantum yield after incorporation in the double helix of DNA. This property was

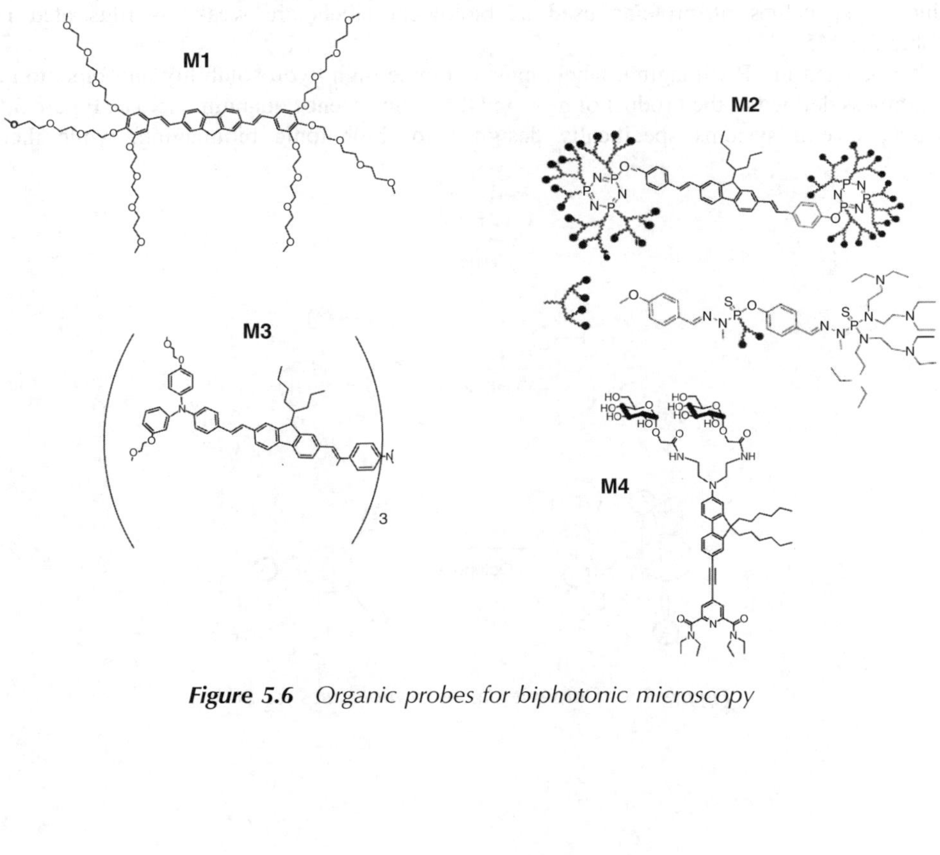


***Figure 5.6*** *Organic probes for biphotonic microscopy*

*Table 5.2* Spectrocopic properties in water of some organic probes

| Chromophore | 1PA | 2PA | |
|---|---|---|---|
| | $\varphi$ | $\lambda^{(2)}$ (nm) | $\sigma_{2PA}$ (GM) |
| M1 | 0.72 | 705 | 119 |
| M2 | 0.71 | 750 | 2790 |
| M3 | 0.56 | 720 | 160 |

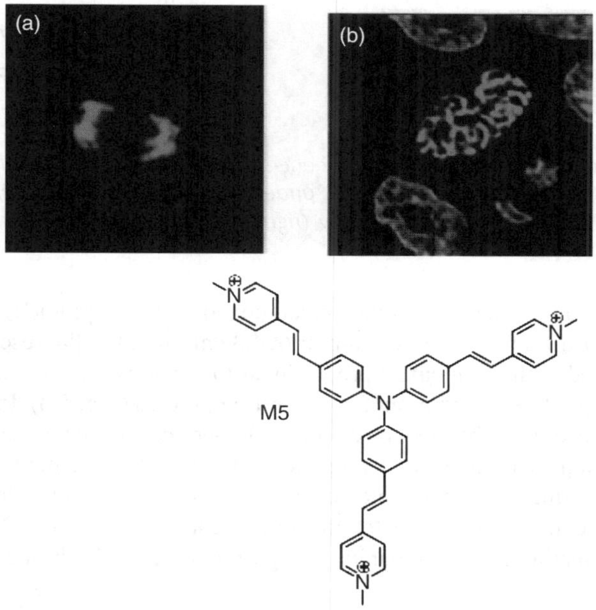


*Figure 5.7* Imaging of DNA by (a) mono and (b) biphotonic fluorescence microscopy of M5

used for selective imaging of DNA by mono and biphotonic fluorescence microscopy (Fig. 5.7) [50].

## 5.3 Spectroscopic Evidence for the Two-photon Sensitisation of Lanthanide Luminescence

### 5.3.1 1961: The Breakthrough Experiments

As mentioned in Section 5.2.1, the two-photon absorption phenomenon was theoretically predicted in 1932 by M. Göppert-Mayer [19] but was demonstrated experimentally only 30 years later thanks to the development of laser sources. Interestingly, the first observation of two-photon absorption induced fluorescence phenomenon was reported in 1961 by

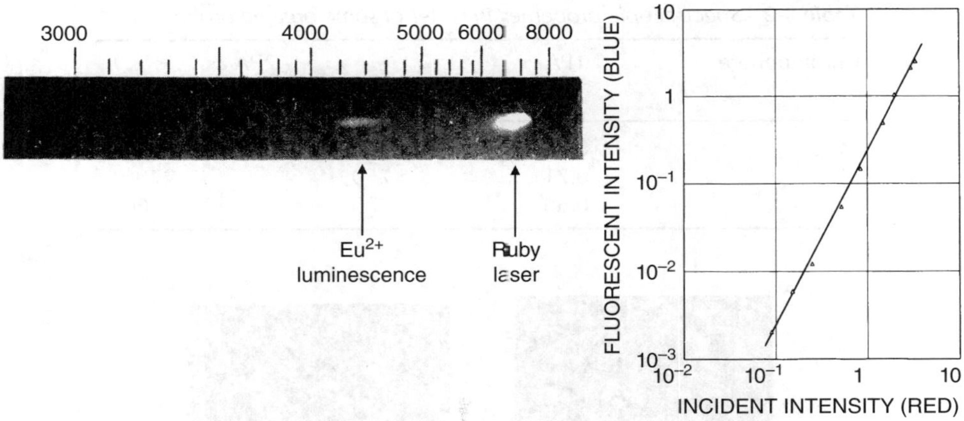


*Figure 5.8 Seminal report on 2PA luminescence: positive photographic plate indicating the blue Eu(II) emission in $CaF_2$:Eu(II) crystal under ruby laser irradiation (left) and quadric dependence with the incident laser intensity (right). Inspired from Reference [20]*

Kaiser and Garrett in a calcium fluoride crystal doped with europium(II) ions [20]. In this experiment, under a ruby laser irradiation at 694.3 nm, the blue fluorescence of Eu(II) at 425 nm is observed with an intensity proportional to the square of the incident laser one, which is a clear signature of a two-photon absorption process (Fig. 5.8). Properly speaking, Eu(II) luminescence is a band-shaped emission and cannot be considered as a real lanthanide emission, a term generally restricted to pure *f-f* transition. Amusingly, this breakthrough experiment was performed using an *f*-block element but triggered the design of mainly organic or transition metal-containing chromophores for 2PA applications, neglecting the potentialities of lanthanide complexes in this field during almost 40 years.

### 5.3.2 Two-photon Excitation of *f-f* Transitions

As mentioned earlier in this book and in several important reviews [51], the *Laporte parity selection rules* make the intra-configurational *f-f* transitions forbidden according to the electric dipole mechanism. Owing to the influence of the crystal- (or ligand-) field, these transitions are in fact partially allowed but exhibit very low molar absorption coefficients ($<10\,L mol^{-1} cm^{-1}$). By contrast, since two-photon selection rules are different, the parity-conserving electronic transitions become allowed. As a consequence, *f-f* transitions can be easily observed in inorganic materials upon monitoring a two-photon excitation spectrum [52,53]. As an example, Denning and Morrison took benefit from this property to establish experimentally the symmetry of crystal field effect on Tb(III) ion in elpsasolites ($Cs_2NaTbCl_6$) [54,55]. In 2001, the group of Lakowicz expanded the scope of the photon-excitation of *f-f* transitions to the Eu(III) or Tb(III) ions solved in (deuterated) water solution [56]. Finally, very recently, Shi and co-workers generalised this phenomenon to crystals of coordination complexes [57]. The bright red europium luminescence is observed upon efficient ultra-narrow two-photon excitation of the $^7F_0 \rightarrow {}^5D_2$ or $^7F_0 \rightarrow {}^5D_1$ transitions. It is important to note that this two-photon excited luminescence is very different from up-conversion processes frequently encountered in lanthanides spectroscopy. Up-conversion processes involve either multi-step excitations due to excited state absorption (ESA) or sequential energy transfer up-conversion (ETU) evidenced by Auzel in the 1960s in Yb-Er systems [58]. Up-conversion is a succession of linear excitations between real excited states from one or several entities, whereas two-photon absorption involves a simultaneous absorption of two photons via a virtual excited state. The frequent confusion between the two processes comes from the fact that up-conversion effects can be considered as nonlinear, due to a quadratic dependence of their efficiency with the laser incident intensity.

### 5.3.3 The Two-photon Antenna Effect

In the early 2000s, Lakowitz et al. reported qualitatively the feasibility of the multi-photon antenna effect of lanthanide ions [56], using *in situ* chelated organic ligand (dipicolinic acid - DPA) or biomolecule (phospholipid, or protein, e.g., transferrin [59]) bonded to the lanthanides. This seminal report was completed a few years later using isolated tris-dipicolinate lanthanide complexes (Ln = Eu, Tb ligand = **DPA**, $\mathbf{DPA^{1-2}}$; Fig. 5.9), well known for their high luminescence properties. Identical Tb(III) emission spectra were obtained using either a direct one-photon excitation (1PE) in the absorption band of the DPA antenna ($\lambda = 275$ nm), or two- or three-photon excitation ($\lambda^{(2)} = 532$ nm [60], or $\lambda^{(3)} = 790$–$800$ nm [61,62], respectively). The quadratic or cubic dependence of the emission intensity versus the incident laser power clearly illustrated the two- (respectively, three-) photon antenna sensitisation process.

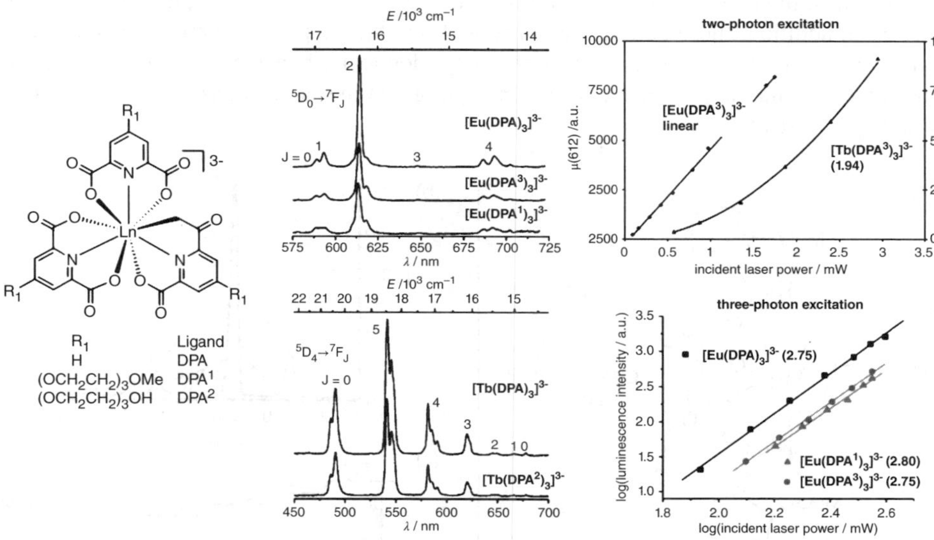


***Figure 5.9*** *From left to right: structure of water soluble tris-dipicolinate complexes (Ln = Tb or Eu), three photon excited ($\lambda_{ex} = 800$ nm) luminescence, variation of the luminescence intensity with the incident laser power for a two (right, top) or three (right, bottom) photon excitation ($\lambda_{ex} = 532$ or $800$ nm respectively) in linear or log scale, respectively (between parenthesis is indicated the experimental determination of the multi-photon absorption)*

The case of Eu(III) complexes is quite different: whereas a 800 nm excitation clearly induced a three-photon antenna effect, the 532 nm excitation gives a luminescence signal whose intensity is two orders of magnitude higher than in the terbium case and is directly proportional to the incident laser power. This striking difference is due to the competitive direct 1PA of the parity forbidden $^7F_1 \rightarrow {}^5D_4$ transition featuring a very small molar absorption coefficient at 532 nm ($\varepsilon_{(532)} = 0.015\ \mathrm{L mol^{-1} cm^{-1}}$). This linear absorption, even negligible, completely masked the 2PA phenomenon, giving a qualitative indication on the relative intensity of the linear and nonlinear excitation.

It was not possible to determine quantitatively the two-photon cross-section of the $[Tb(DPA)_3]^{3-}$ because of its low efficiency and because the use of ns-pulsed laser induced important error in the $\sigma_{2PA}$ determination. Using a fs Ti:Sa laser as source and an integration sphere, Bünzli et al. [62] determined the three-photon cross-section by comparison of the 3PA and 1PA luminescence. This absolute method does not require any standard and the $\sigma_{3PA}$ values of 0.011 and 0.014 GM for $Cs_3[Eu(DPA)_3]$ and $Cs_3[Tb(DPA)_3]$, respectively, can be considered as references and further used as standards for forthcoming determinations.

Following the pioneering works of Lakowicz, Wong and co-workers strongly contributed to the generalisation of the multi-photon sensitisation of lanthanide luminescence. These authors qualitatively reported the two-, three-, and even four-photon absorption sensitised luminescence of terbium or europium polymeric complexes in single crystals [63–66]. Figure 5.10 describes one representative example: the tripodal ligand featuring a methoxy-benzamide antenna (Fig. 5.10a) is able to form nice single crystals after coordination with $Ln(NO_3)_3$ (Ln = Eu, Tb). After excitation at 845 nm (fs – Ti:sapphire source), the characteristic red or green luminescence was detected (Fig. 5.10b, c), and its intensity varies cubically with the incident laser source power clearly indicating a three-photon absorption process. In addition, these authors demonstrated that the shape of the emission spectra as well as the excited state lifetime are rigorously identical when measured from linear or multiphoton excitation processes, clearly indicating that the same emissive state is reached by both processes [67].

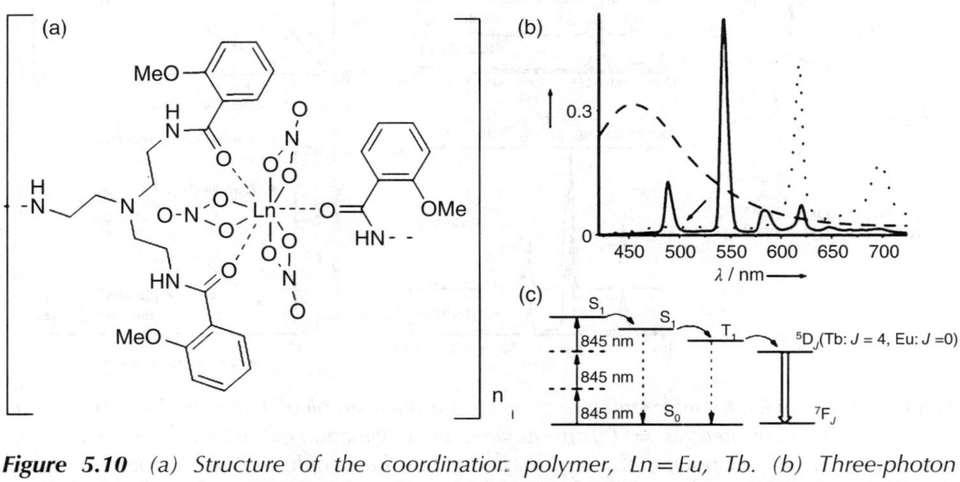


***Figure 5.10*** *(a) Structure of the coordination polymer, Ln = Eu, Tb. (b) Three-photon excited emission of the Eu(III) (dotted) and Tb(III) (straight) complexes (RT, $\lambda_{ex}$ = 845 nm). (c) Proposed sensitisation mechanism (from Reference [65])*

### 5.3.4 The Charge Transfer State Mediated Sensitisation Process

In the first part, we highlighted that significant nonlinear activities are generally obtained using highly polarisable systems generally containing delocalised π-electrons in (CT) derivatives. In order to combine lanthanide luminescence and high nonlinearity, it is therefore necessary to design complexes featuring a CT-mediated sensitisation process. In this context, the first charge transfer excited state sensitising Ln(III) emission involved metal-to-ligand charge transfer transitions (MLCT). The induced triplet MLCT transition associated to such transition metals (belonging to the second or third row), has been extensively studied for sensitising lanthanide ions [68–71]. Indeed, the *d*-block metal complexes sensitisers afford several advantages as compared to organic chromophores such as low energy triplet excited states with relatively important intersystem crossing quantum yield because of strong spin-orbit coupling and rather long triplet excited state lifetimes enabling efficient energy transfer to the lanthanide ions. Numerous examples have been published showing the possibility to sensitise Yb(III), Nd(III) and Er(III) using MLCT associated to ruthenium, osmium, rhenium, chromium or iridium, and exhaustive reviews were recently published on that topic [68–71]. However, such CT states are generally localised and therefore present quite low NLO efficiency.

In contrast, the sensitisation of Ln(III) through singlet excited states has been debated since 1969 and a paper from Kleinerman [72]. For a long time, it was believed that this type of energy transfer was not possible and this hypothesis was supported by the characteristics of this short-lived state making the energy transfer unlikely to occur. However, even to date, no real proof of singlet local excited state has been shown. Only recently, the singlet delocalised intra-ligand charge transfer state (ILCT) was clearly demonstrated to be able to sensitise *f-f* emission, as highlighted by a recent review [73]. As for classical sensitisation, two types of sensitiser using an antenna possessing a CT excited state can be used. The antenna can be: (i) pending close to the chelator unit or (ii) conjugated with the chelating moiety generally featuring an electron withdrawing character. In the latter case, the lanthanide can have a dual function: (i) to be an emitter and (ii) to act as a Lewis acid enforcing the acceptor strength of the push-pull and therefore lowering the energy of the relaxed ILCT excited state. It is important to note that the presence of such low lying CT transitions induces a complete modification of the sensitisation process. Generally, sensitisation of lanthanide luminescence proceeds in three steps: (i) singlet–singlet absorption of the antenna ligand, (ii) intersystem crossing (ISC) to the triplet lowest excited state ($T_1$) and (iii) energy transfer (ET) to the metal accepting state. If the antenna ligand presents a low energy CT transition, the sensitisation can occur directly from the relaxed CT state without any participation of the triplet excited state.

Sensitisation through ILCT was only recently demonstrated to be active. However, a few examples were published before those demonstrations that led one to think that the sensitisation can occur from the ILCT and not from a classical triplet excited state. In the early 2000s, Lakowicz et al. [61] showed that a pending coumarin (an antenna that is known to present charge transfer character) covalently bound to a DTPA chelator (Fig. 5.11) was able to efficiently sensitise Eu(III) luminescence.

In the same context, Parker et al. and Sammes et al. published independently interesting Eu(III)-DO3A complexes containing a pendant acridone chromophore. In those studies, the authors noticed that at 77 K using the Gd(III) complex, most of the energy was driven on the

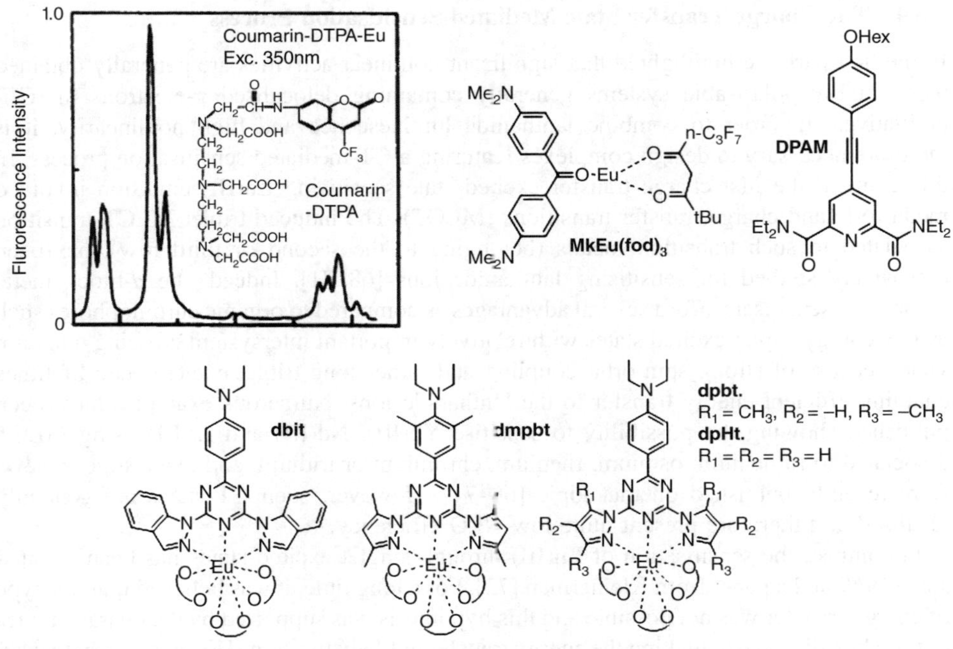


***Figure 5.11** Structure and luminescence spectrum of coumarin-DTPA Eu complex [61]; structure of Michler ketone, functionalised dipicamide [74] and bis-pyrazolyl-triazine ligands [75–77]*

triplet excited state lying at 21 500 $cm^{-1}$ [78]. However, they also indicated a dependence of the luminescence wavelength maximum of the acridone moiety with Reichardt's empirical solvent polarity parameter $E_T(30)$, a characteristic behaviour of CT transition [79]. In such a case, a definitive conclusion about the sensitisation process is difficult, and the above mentioned results suggest that the Eu(III) sensitisation might partially proceed through a CT excited state, although the classical triplet excited state mediated process cannot be ruled out. The first example that strongly suggests the involvement of singlet ILCT excited state in europium sensitisation was published in 1999 by Verhoeven et al. [80]. In this work, the authors noticed the push–pull character of the antenna (a Michler's ketone featuring two strong electro-donating moieties, **Mk**Eu(fod)$_3$; Fig. 5.11). Furthermore, they also succeeded in shifting the absorption upon complexation with the Lewis acid (the lanthanide) to 450 nm with an overall luminescence quantum yield of 17%. Finally, the singlet energy transfer (transferring from the excited ILCT state) was first demonstrated by Wang and co-worker in 2004 [81]. In this interesting luminescence lifetime experiment, the authors demonstrated that, in a **dpbt**Eu(tta)$_3$ complex (Fig. 5.11), the europium *f-f* transitions are sensitised through energy transfer via a singlet ILCT excited state. In addition, it must be noted that the luminescence quantum yield of this latter complex was excellent, about 52% in methanol, highlighting the very high efficiency of this singlet CT sensitisation process. In 2008, our group described a complete family of functionalised tris-dipicolinate europium complexes **DPA**$^{3\text{-}8}$ (Scheme 5.1) [82], featuring various electro-donating groups and allowing the fine tuning of the ILCT excited state over a large energy scale. The charge transfer character of

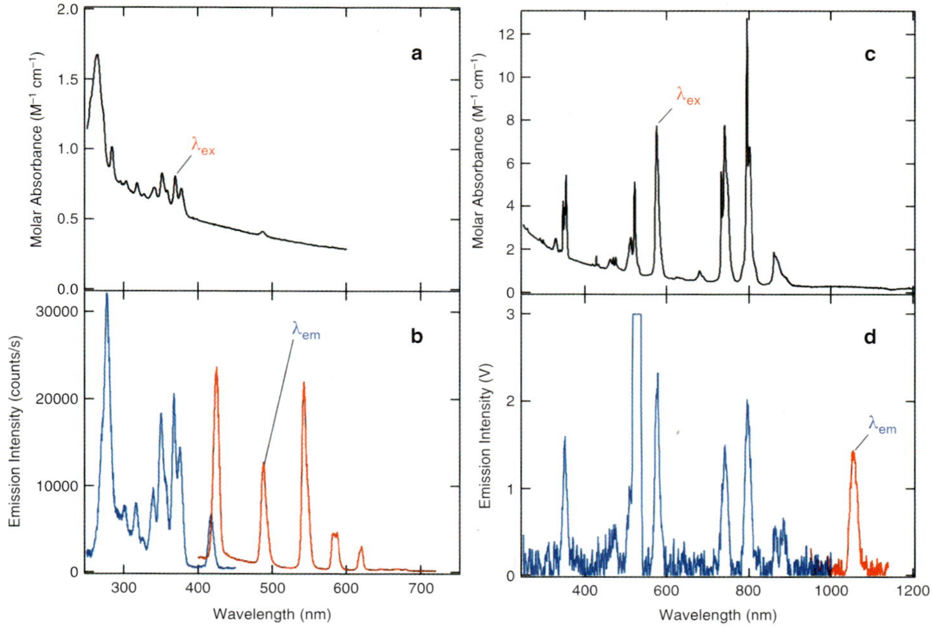


***Figure 2.1*** *Typical absorption and emission spectra for $Ln^{3+}$ ions in dilute aqueous solution: (black lines) absorption data; (red lines) emission data; (blue lines) emission excitation data. (a, b) $Tb^{3+}$. (c, d) $Nd^{3+}$. $[Tb^{3+}] = 1.3$ mM for (b); $[Nd^{3+}] = 27.6$ mM for (d). Emission and emission excitation data were collected using continuous wave excitation from a Xe arc lamp. Wavelengths for exciting the emission ($\lambda_{ex}$) and monitoring the emission excitation ($\lambda_{em}$) spectra are indicated. The detector in B was a photomultiplier tube. The detector in d was a single-element InGaAs device. © David E. Morris, Los Alamos National Laboratory*

*Luminescence of Lanthanide Ions in Coordination Compounds and Nanomaterials*, First Edition.
Edited by Ana de Bettencourt-Dias.

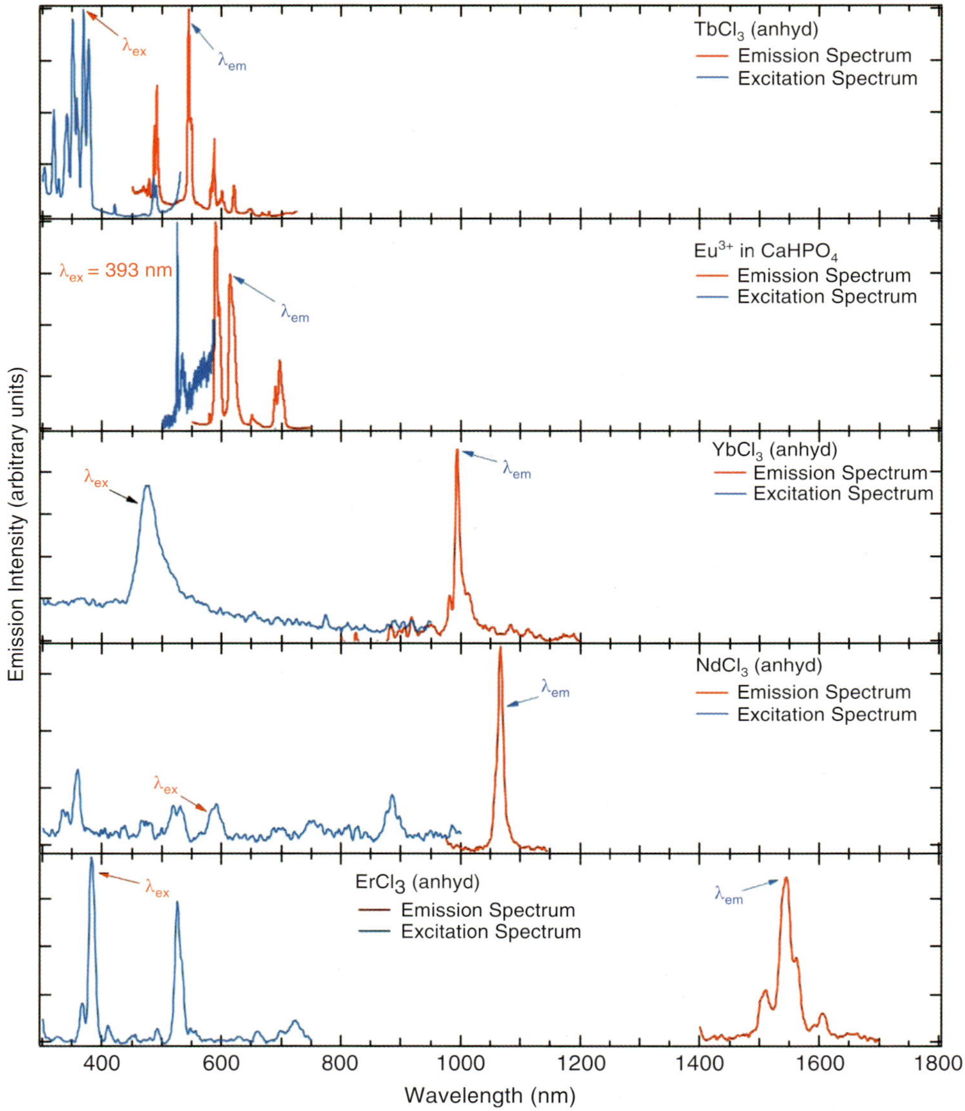


***Figure 2.2*** *Emission (red) and emission excitation (blue) spectral data for a series of $Ln^{3+}$ ion in solid-state matrices illustrating the range of excitation and emission wavelengths over which data need to be collected to characterise these materials. All data were collected using continuous wave excitation from a Xe arc lamp. Wavelengths for exciting the emission ($\lambda_{ex}$) and monitoring the emission excitation ($\lambda_{em}$) spectra are indicated. The detector for the $TbCl_3$ and $Eu^{3+}$ data was a photomultiplier tube. The detector for the $YbCl_3$, $NdCl_3$, and $ErCl_3$ data was a single-element InGaAs device. © David E. Morris, Los Alamos National Laboratory*

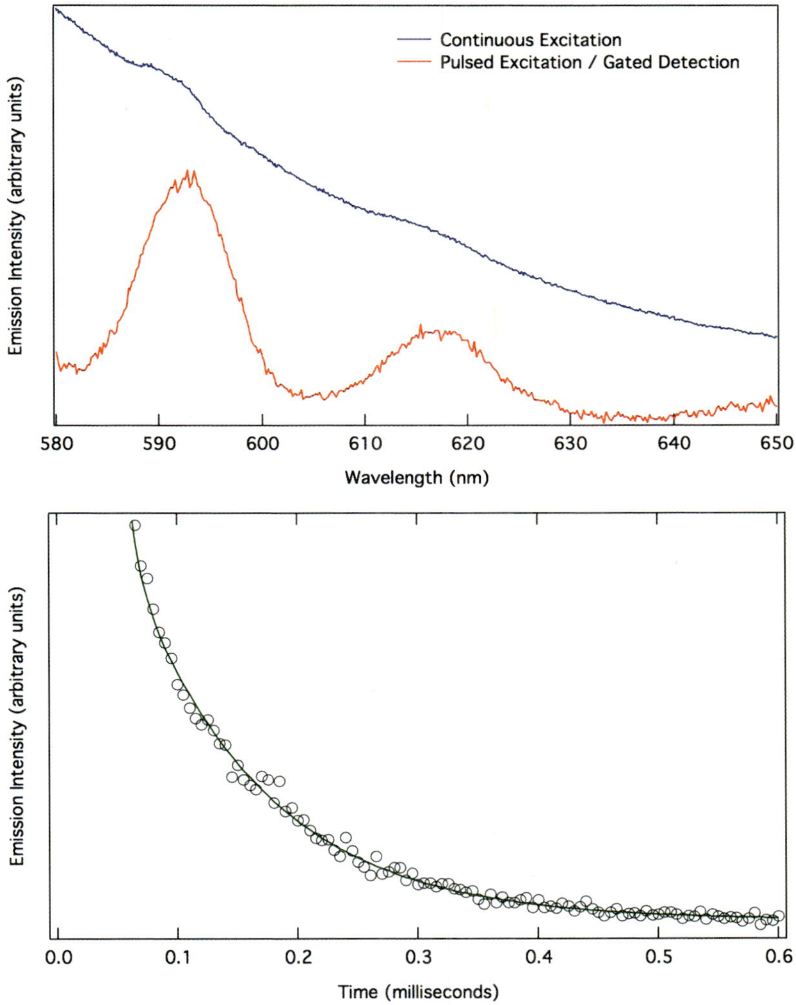


***Figure 2.3*** *Top: Comparison of continuous-wave (cw) excitation versus pulsed excitation/time-delayed gated detection in the emission spectrum of $Eu^{3+}$ adsorbed on Whatman CF-1 cellulose. The excitation source was a Xe arc lamp for the cw spectrum and a Xe flash lamp for the pulsed excitation spectrum. The excitation wavelength for both spectra was 393 nm. The detection in the pulsed spectrum entailed a 50 μs delay and a 10 ms gate width. The strong background emission tail seen in the cw spectrum originates from the cellulose matrix. This signal has a nanosecond lifetime and as illustrated the contribution from this background can be discriminated from the signal for the $Eu^{3+}$ species using gated detection with a delay time longer than the background emission lifetime. Bottom: Emission decay curve for this $Eu^{3+}$/cellulose sample using 393 nm pulsed excitation. Open circles are the experimental data. The green line is an unconstrained double-exponential fit to the data with lifetime values of 7 μs and 110 μs. The short lifetime component required for this fit is an experimental artifact from the residual intensity in the flashlamp pulse. © David E. Morris, Los Alamos National Laboratory*

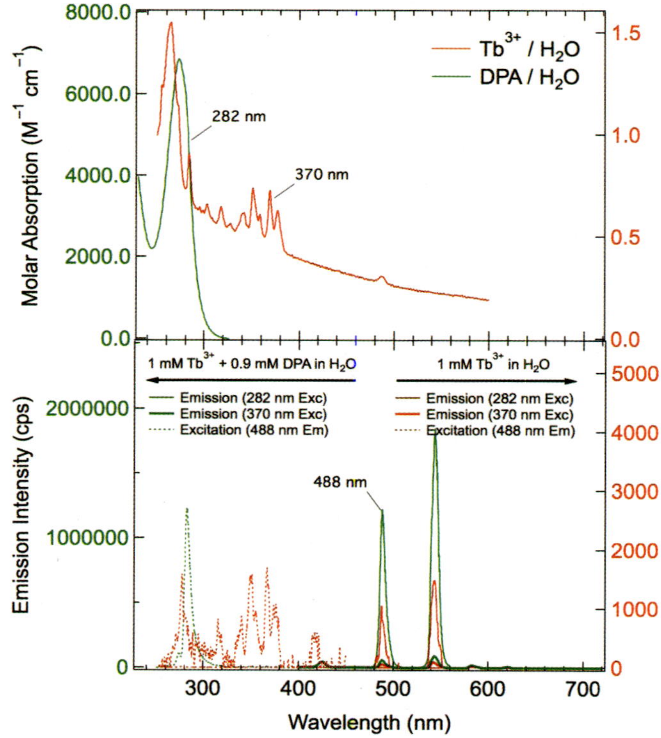


***Figure 2.4*** *Example of the "antenna" (sensitiser) effect on the emission properties of $Tb^{3+}$.* Top*: UV-visible absorption spectra of $Tb^{3+}$ (red) and pyridine-2,6-dicarboxylic acid (dipicolinic acid, DPA, green) in $H_2O$ illustrating the much greater absorption cross-section for the DPA sensitiser relative to that of the fully aquated $Tb^{3+}$ ion.* Bottom*: Continuous-wave emission [solid lines] and emission excitation [dashed lines] spectra for the aquated $Tb^{3+}$ ion (red; right hand intensity scale) and the $Tb^{3+}$/DPA complex (green; left hand intensity scale). Experimental parameters (slit widths and integration times) are identical for both samples. Excitation in resonance with the strong DPA absorption band (282 nm) leads to an enormous increase (~×1250) in the $Tb^{3+}$ emission, and the correspondence between the excitation spectrum for the $Tb^{3+}$/DPA sample and the DPA absorption spectrum clearly reflects that the DPA excited state is strongly coupled to emissive $Tb^{3+}$ states. Tuning off this ligand absorption band (370 nm) leads to a dramatic decrease in emission intensity. © David E. Morris, Los Alamos National Laboratory*

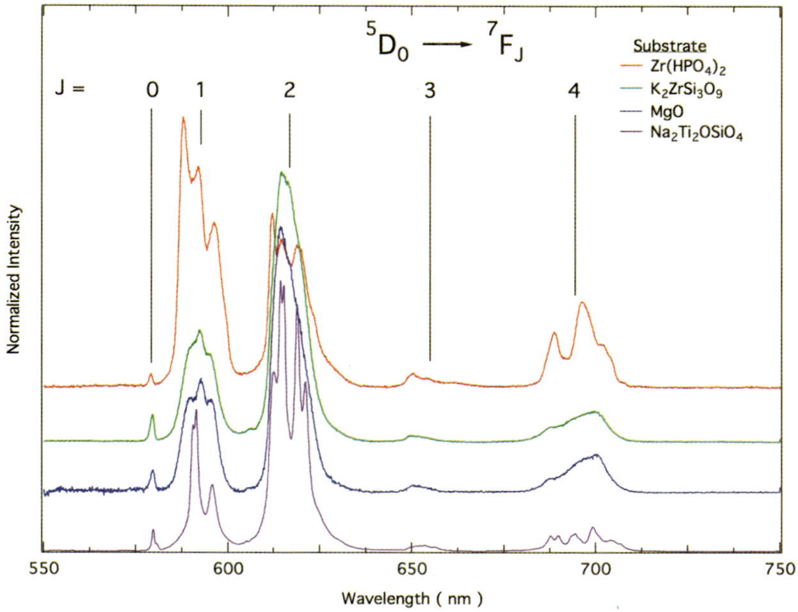


***Figure 2.5*** *Emission spectra for $Eu^{3+}$ ions in a series of solid-state inorganic ion exchange matrices illustrating the often high density of observable spectral transitions from the $^5D_0$ excited state into the manifold of lower-lying $^7F_J$ spin–orbit states and the variation in the relative intensities and degree of resolution in these bands. All spectra were acquired using 393 nm continuous-wave excitation. © David E. Morris, Los Alamos National Laboratory*

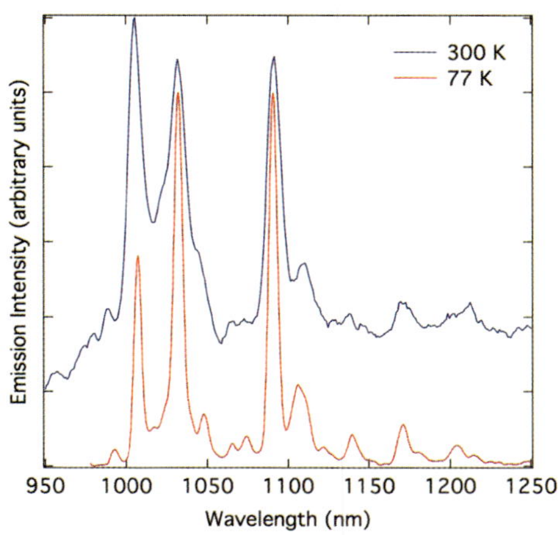


***Figure 2.6*** *Comparison of room-temperature versus liquid-nitrogen temperature emission spectra for $[(C_5Me_5)_2Yb(2,2'\text{-bipyridine})]^+$ in $CH_2Cl_2$ solution illustrating the substantial improvement in band resolution at cryogenic temperature. These vibronically resolved emission bands for the $4f^{13}$ $Yb^{3+}$ ion arise from transitions between the $^2F_{7/2}$ excited state and the $^2F_{5/2}$ ground state manifolds. The excitation in both spectra occurred at 440 nm from a continuous wave Xe arc lamp. © Ryan E. Da Re and David E. Morris, Los Alamos National Laboratory*

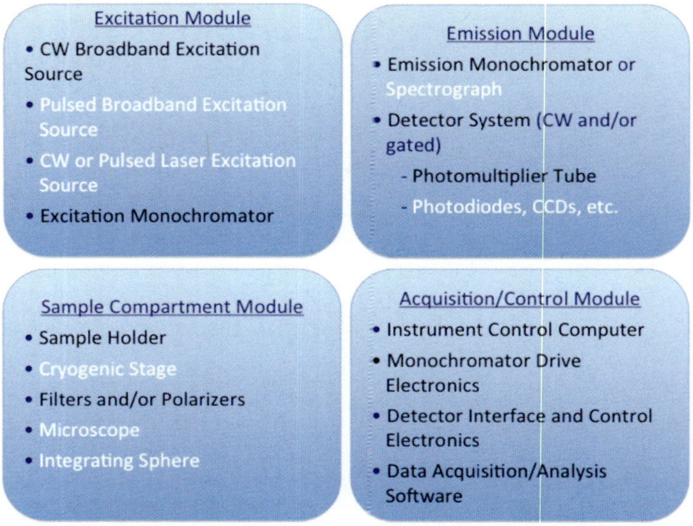


***Figure 2.7*** *The four instrument modules common to most optical emission spectrometer systems. Components in black are the standard elements in a UV-visible emission system. Components in white are optional accessories that enhance the versatility of the system and extend the wavelength range for excitation and/or emission*

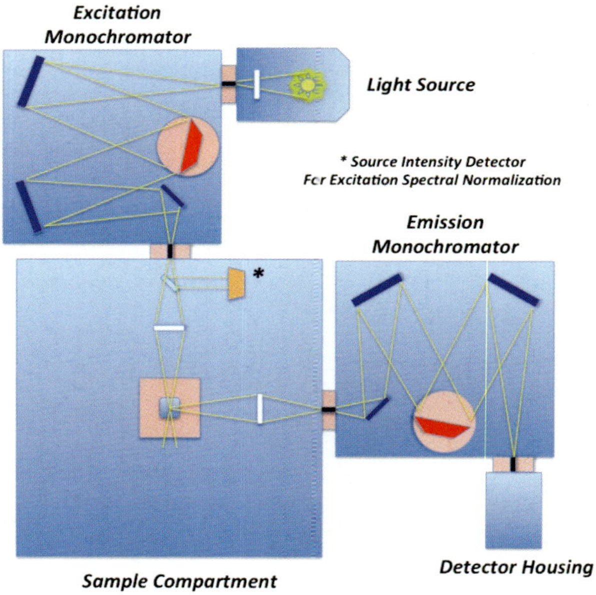


***Figure 2.8*** *A common configuration of the modules in an optical emission spectrometer system. This representation illustrates single-stage monochromators for both excitation and emission. The optical elements illustrated here are the diffraction gratings (red), steering and focusing mirrors (blue), focusing lenses (white), and adjustable entrance and exit slits (black)*

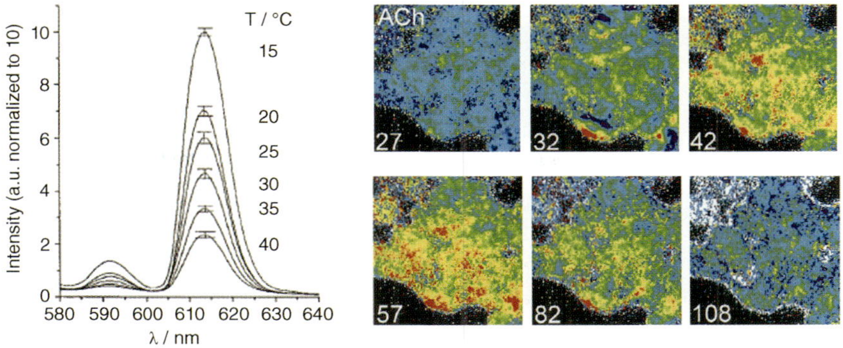


***Figure 4.8*** *Left: Emission intensity of [Eu(tta)$_3$(H$_2$O)$_2$] integrated in liposomal membranes, at pH 7.4 and under 363-nm excitation. Each spectrum is an average of six determinations (standard deviations are shown at the maximum of the 613-nm peak). Redrawn after Reference [68]. Right: Thermal imaging of the heat wave in CHO cells initiated by a puff of acetylcholine (ACh, frame 27, 22 °C) and evidenced with internalised [Eu(tta)$_3$(H$_2$O)$_2$]. Images have been recorded at 1-s intervals (lower left numbers give the time in s); blue: coldest regions, red: hottest regions. Reproduced with permission from [68]. Copyright © 1998, Elsevier*

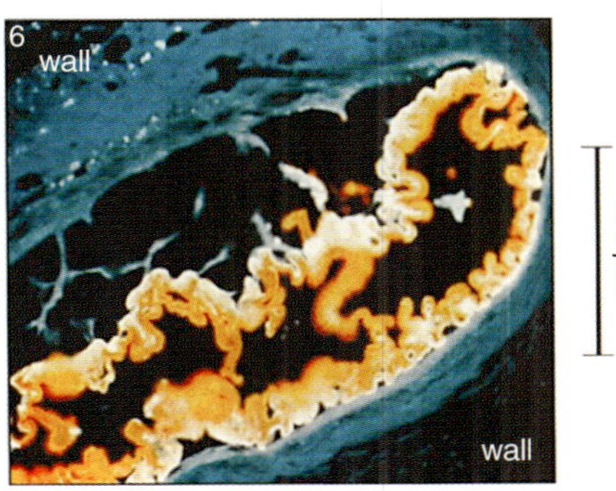


***Figure 4.9*** *Paraffin-embedded excised embolised tissue containing a 50:50 Lipidiol:NBCA ratio, as evidenced by Eu$^{III}$ luminescence. Reproduced with permission from [71]. Copyright © 2001, American Society of Neuroradiology*

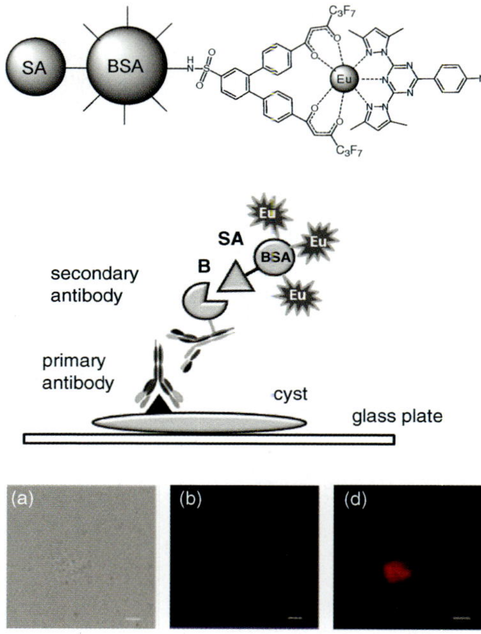


***Figure 4.10*** *Top: Sketch of the SA-BSA-[Eu(BHHCT)(dpbt)] (SA-BSA-EuBd) conjugate. Middle: Principle of the indirect immunochemical analysis of* Giardia gardia *cysts. Bottom: Bright field (a), TR luminescence without (b) and with (d) antibody images of* Cryptosporidium parvum *in a water sample immuno-stained as described in the text. Adapted with permission from [77]. Copyright © 2010, American Chemical Society*

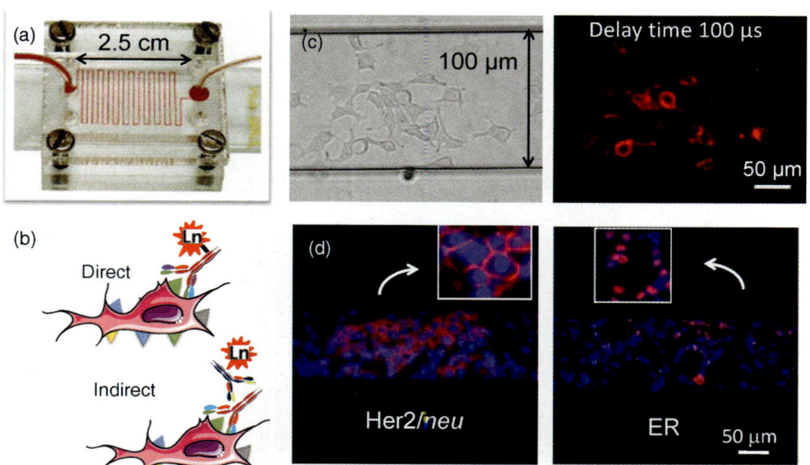


***Figure 4.13*** *(a) First-generation microfluidic device. (b) Principle of direct and indirect histochemical assays. (c) Detection of MCF-7 breast cancer cells through the 5D10 antigen they expressed itself recognised by its antibody decorated with [Eu(L5)]$^{-}$: bright-field image (left) and TR detected Eu$^{III}$ luminescence (right). (d) TR luminescence detection of Her2/*neu *and ER receptors on human breast cancer tissue sections; see text for details. Reproduced with permission from [126]. Royal Society of Chemistry*

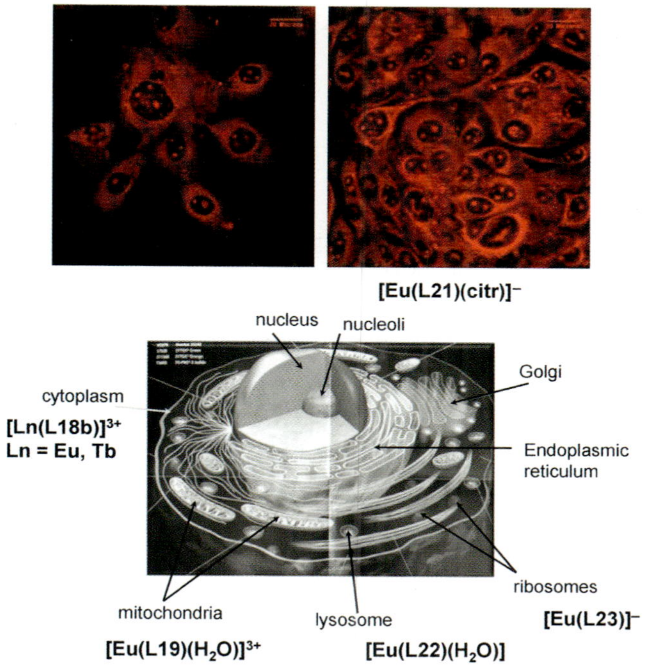


***Figure 4.15*** *Top: Confocal microscopy images ($\lambda_{exc} = 405$ nm) showing the intracellular localisation of $[Eu(L20)(H_2O)]^{3+}$ in NIH 3T3 cells. Adapted with permission from [146]. Copyright © 2009, American Chemical Society. Bottom: Dominant intracellular localisation of some cyclen complexes*

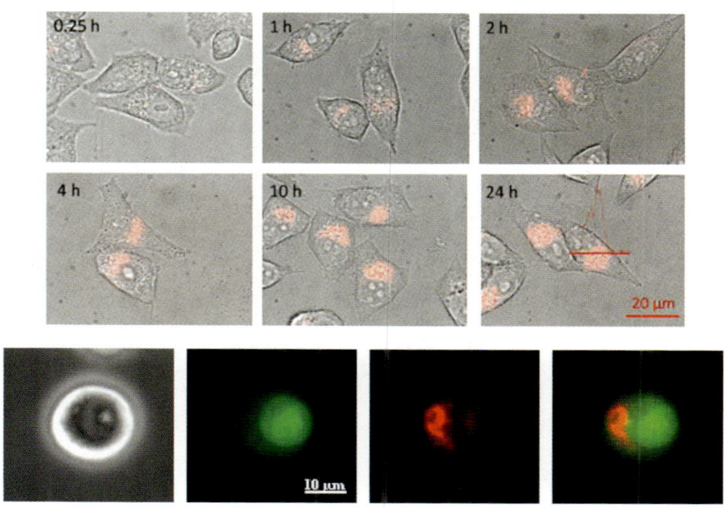


***Figure 4.18*** *HeLa cell imaging. Top: TR luminescence microscopy of the penetration of $[Eu_2(L30b)_3]$ in cells incubated at 37 °C for various times; Reproduced with permission from [174]. Royal Society of Chemistry. Bottom: Co-localisation experiment with AO and $[Eu_2(L30b)_3]$ showing penetration of the lanthanide probe outside the nucleus; Reproduced with permission from [172]. Copyright © 2008, John Wiley*

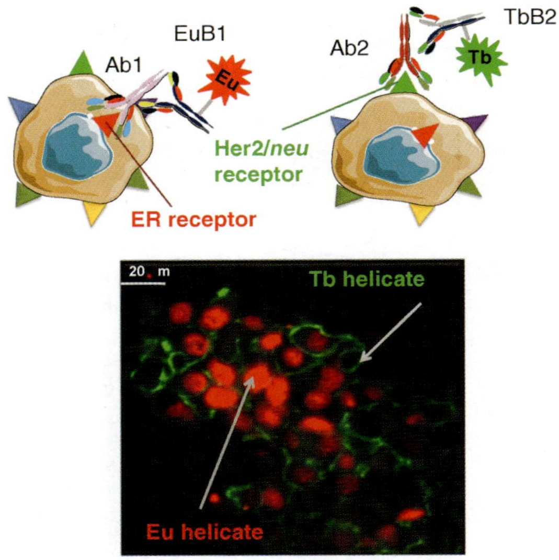


***Figure 4.20*** *Principle of the dual immunohistochemical detection of biomarkers expressed by human breast cancer cells. Ab1 = anti-human ER mouse mAb; Ab2 = polyclonal rabbit anti-human c-erbB-2 oncoprotein Ab. EuB1 and TbB2: see text. Reproduced with permission from [170]. Royal Society of Chemistry*

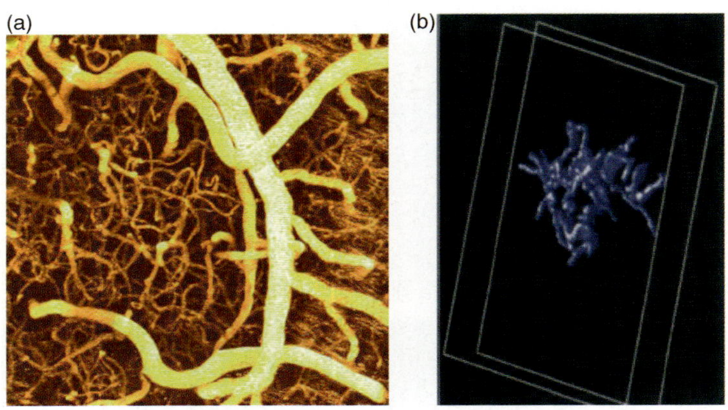


***Figure 5.3*** *(a)* In vivo *blood vessel biphotonic microscopy imaging of brain mouse. (b) 3D image of chromosomes during cell division by multiphotonic excitation of DAPI*

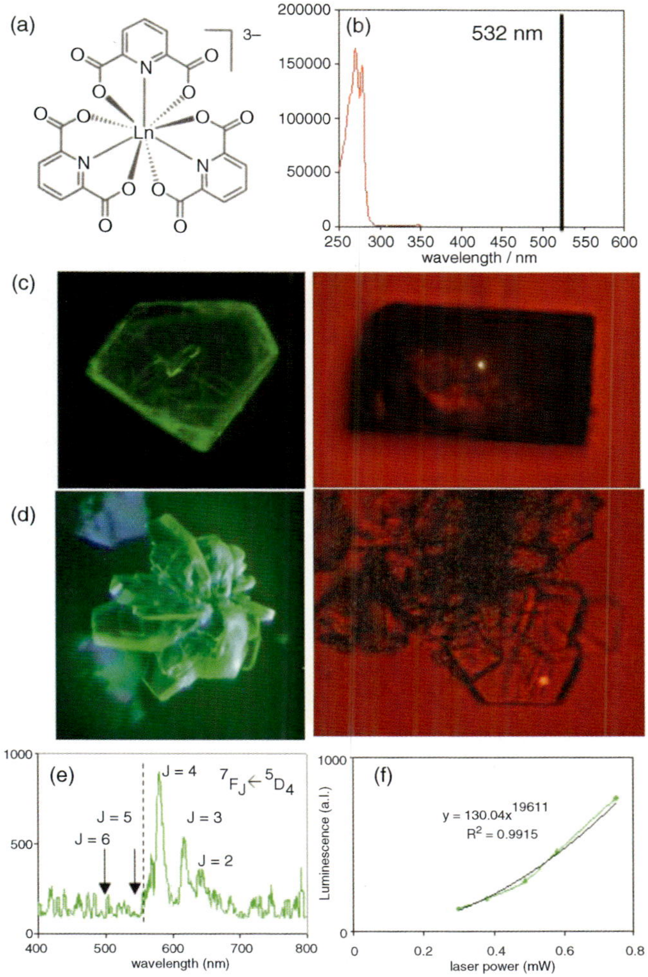


***Figure 5.14*** *Linear microscopy imaging (UV irradiation at 365 nm) of a lysosyme-derivative crystal containing terbium tris dipicolinate (left), nonlinear two-photon microscopy under 532 nm laser irradiation (right). Reproduced from [60]. Copyright © 2007, John Wiley*

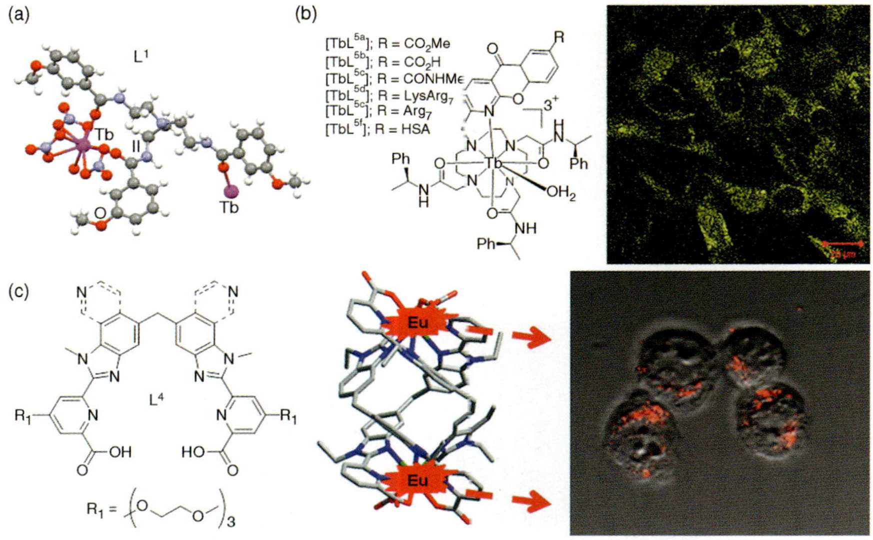


*Figure 5.15* Water soluble complex used for biphotonic cell imaging. Adapted with permission from [88]. Copyright © 2008, American Chemical Society, [103] Copyright © 2008, Royal Society of Chemistry, and Copyright © 2010, American Chemical Society

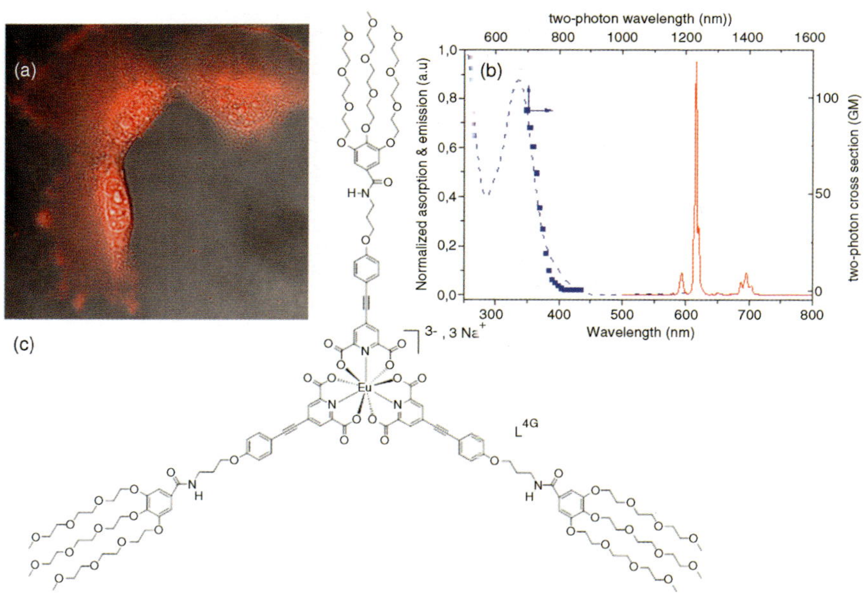


*Figure 5.17* (a) Two-photon scanning image of T24 fixed cells using a 760 nm incident wavelength. (b) Spectroscopic properties in water (blue dashed line, absorption; blue squares two-photon absorption, red solid line, emission). (c) Structure of the probe. Adapted with permission from [92]. Copyright © 2008, American Chemical Society

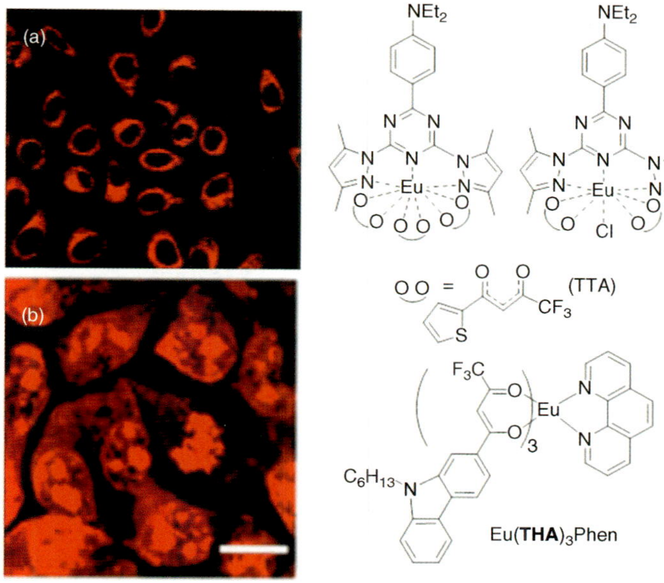


*Figure 5.18* *Two-photon microscopy images of (a) HeLa and (b) MCF-7 cell, using Eu(III) probes (right). Adapted with permission from [77]. Copyright © 2011, American Chemical Society and [91] Copyright © 2008, Royal Society of Chemistry*

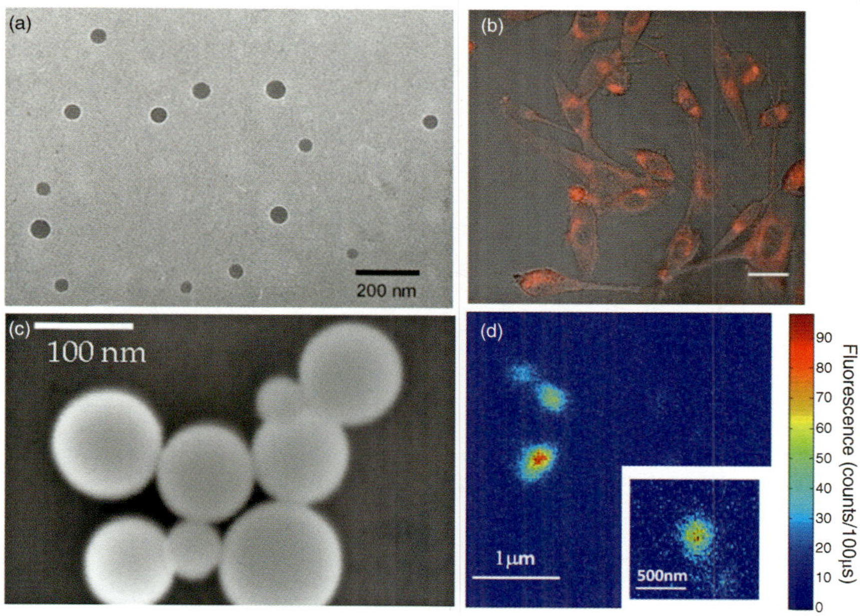


*Figure 5.19* *TEM image of nano-spheres prepared by: (a) micro-emulsion or (c) sol-gel techniques. (b) Two-photon scanning microscopy of MDA-MB-231 cancer cells stained with anti-GFR conjugated nanoparticle. (d) Two-photon luminescence images of single silica nanoparticles. Adapted with permission from [124]. Copyright © 2010, John Wiley and [125] Copyright © 2008, Royal Society of Chemistry*

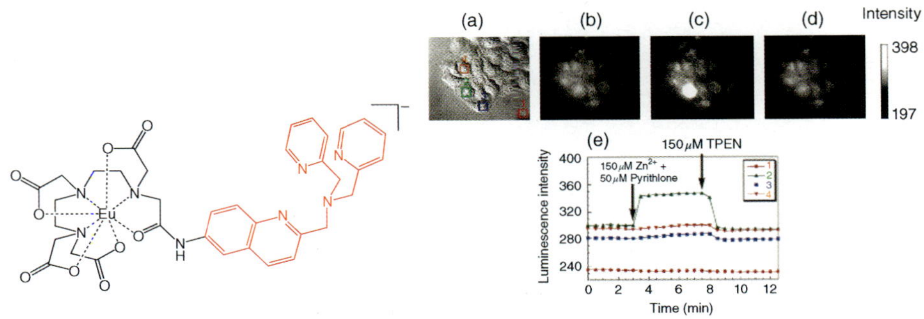


**Figure 6.6** *Eu$^{III}$ complex structure (left) of the zinc-responsive probe showing the zinc binding site in red. Bright-field transmission and luminescence images of $Zn^{2+}$ in HeLa cells loaded with Eu complex in HBSS buffer. The luminescence at 580–654 nm, excited at 320–400 nm, was measured at 30-s intervals. (a) Bright-field transmission image (0 min). (b) Luminescence image of (a) (0 min). (c) Luminescence image (5 min) following an addition of 50 μM pyrithione (zinc ionophore) and 150 μM ZnSO4 to the medium at 3 min. (d) Luminescence image (10 min) following an addition of 150 μM TPEN to the medium at 7.5 min. Luminescence images (b–d) correspond to the luminescence intensity data in (e), which shows the average intensity of the corresponding area or cell area (1, extracellular region; 2, intracellular region of the injected cell; 3, 4, intracellular regions of non-injected cells). Reproduced with permission from [22]. Copyright © 2004, American Chemical Society*

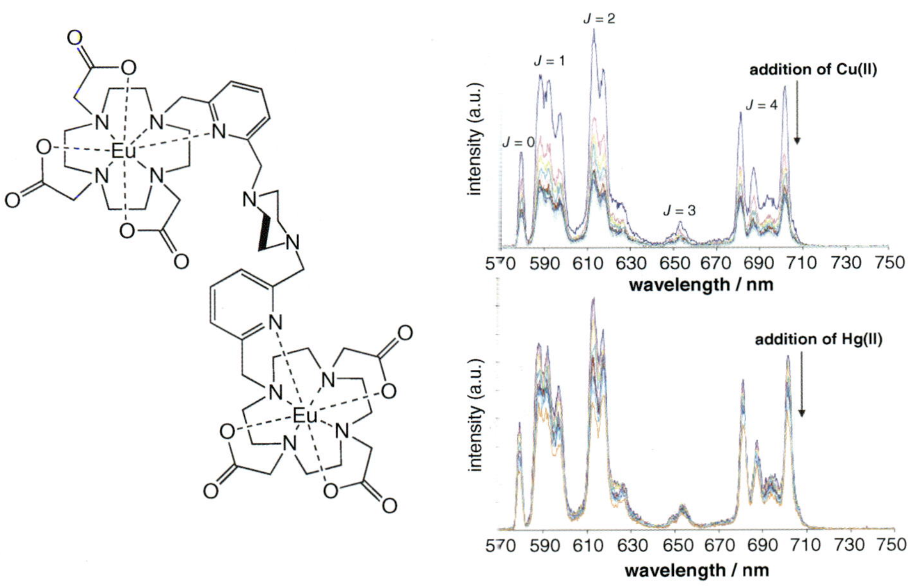


**Figure 6.10** *A dimetallic Eu$^{III}$ complex that responds to Cu$^{II}$ and Hg$^{II}$*

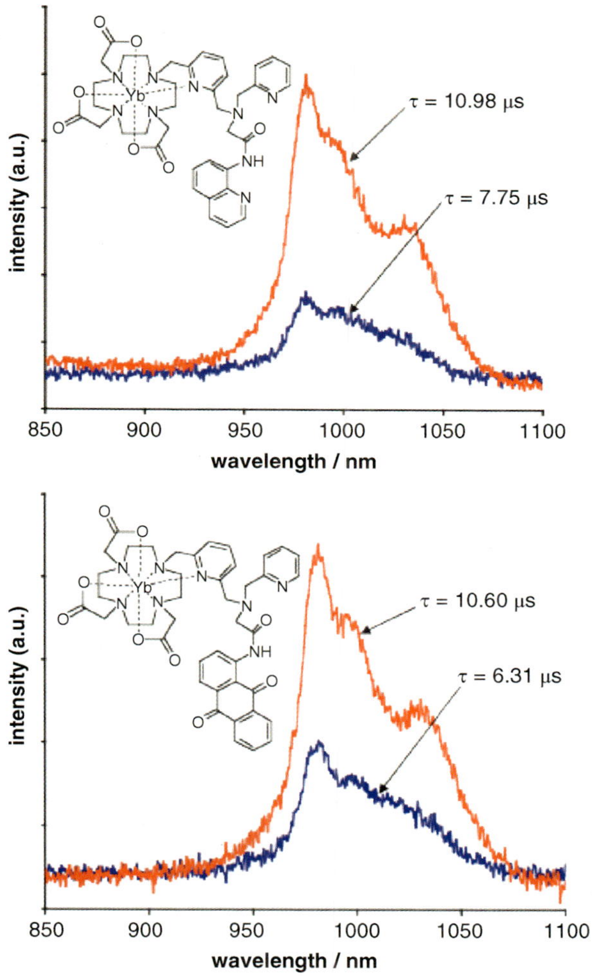


***Figure 6.11*** *Modulation of $Yb^{III}$ near-IR emission after addition of $Hg^{II}$ (red traces)*

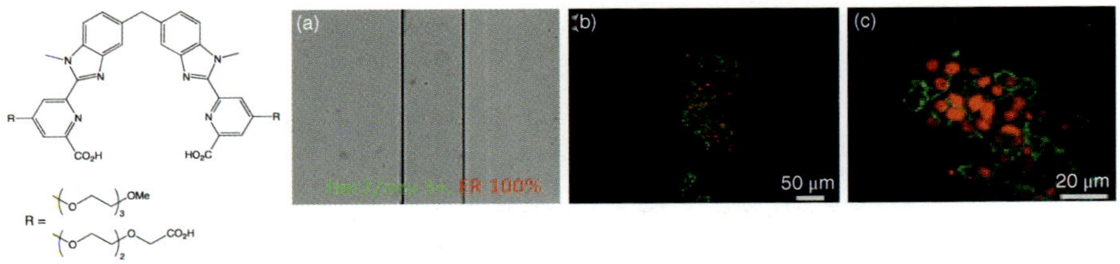


***Figure 6.13*** *An example of a Bünzli ligand fo- the bimetallic helical complexes (left), and 'on-chip' immunohistochemical detection of Her2/neu and ER in a breast cancer tissue sample. (a) Bright field image; (b) merged luminescent image, Her2/neu detected by a green-emitting* $Tb^{III}$ *complex and ER stained with a red-emitting* $Eu^{III}$ *complex; (c) magnified image. Reprinted with permission from [52]. Royal Society of Chemistry.*

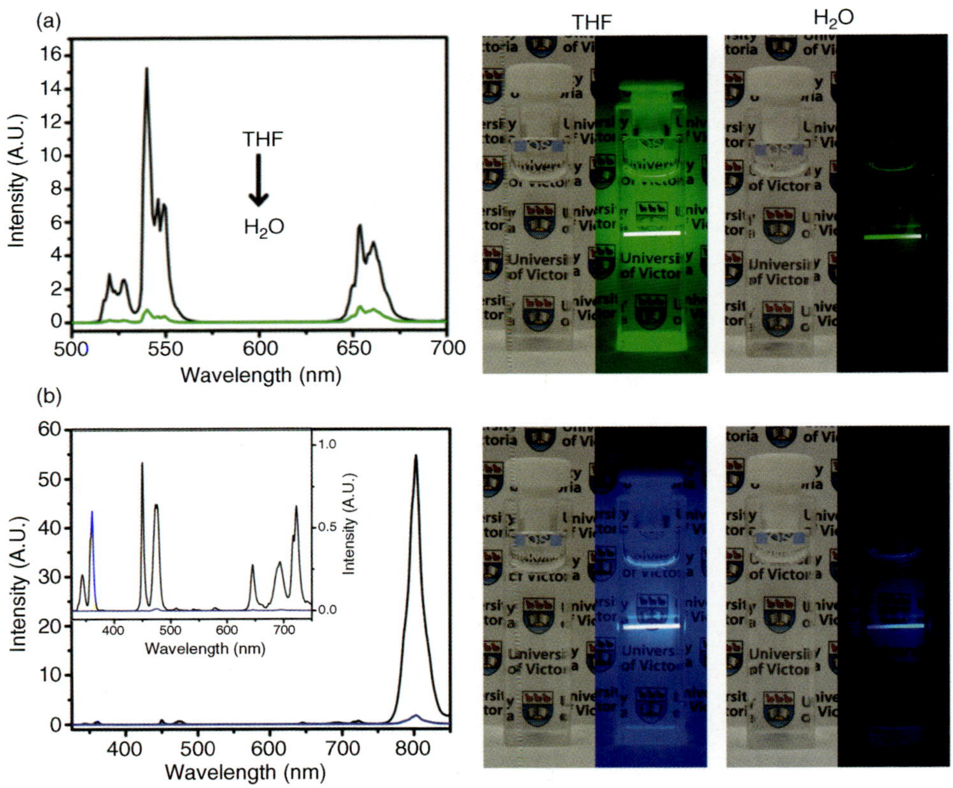


***Figure 7.1*** *Upconversion spectra of original oleate-stabilised NPs in THF. Reproduced with permission from [2]. Copyright © 2010, American Chemical Society*

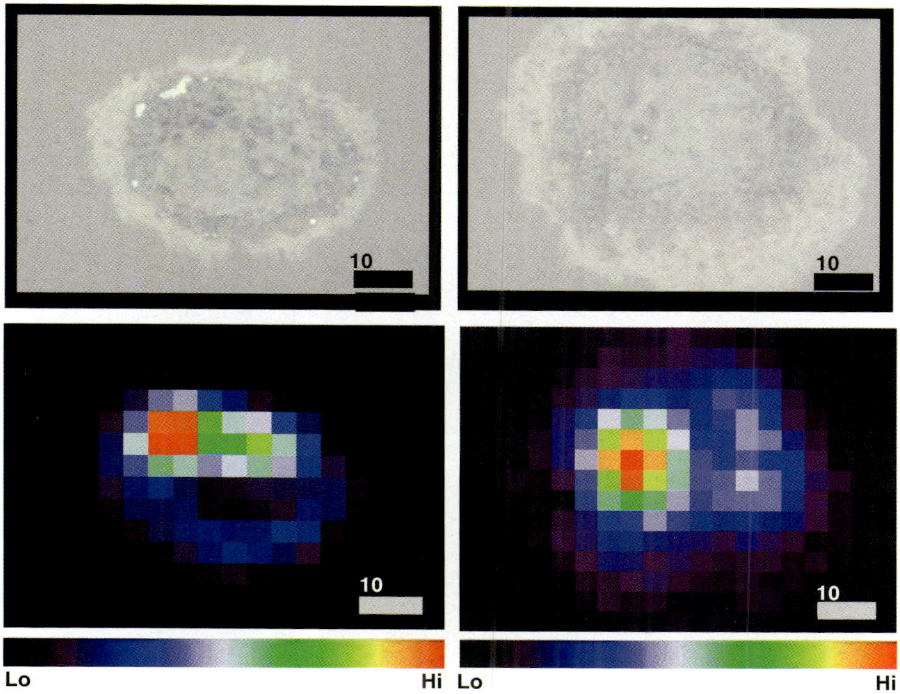


**Figure 7.7** *Upconversion imaging of ovarian cancer cells CaOV3 incubated with PEG(2000)-phosphate coated $NaYF_4$. Reproduced with permission from [2]. Copyright © 2010, American Chemical Society*

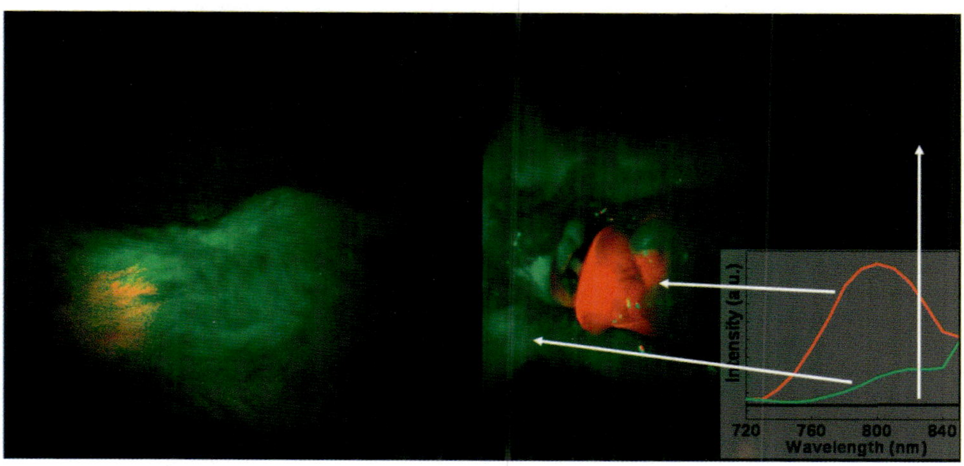


**Figure 7.9** *Whole body images of mouse injected iv with upconversion nanoparticles; intact mouse (left), same mouse after dissection (right). The red colour indicates emission from upconversion nanoparticles, green and black show background as indicated by the arrows. The inset presents the PL spectra corresponding to the spectrally unmixed components of the multispectral image obtained with the Maestro system. Reproduced with permission from [70]. Copyright © 2008, American Chemical Society*

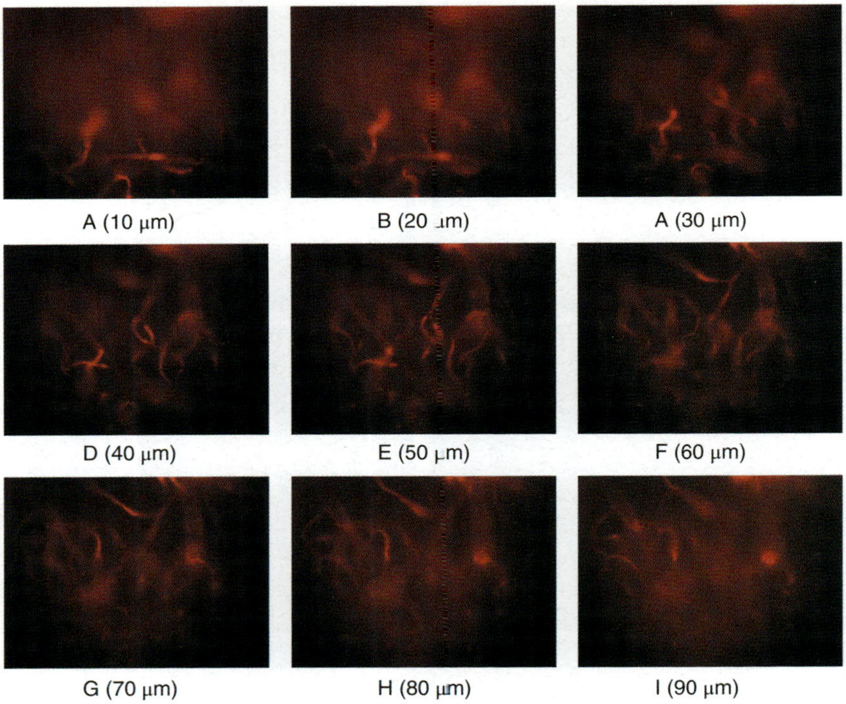


***Figure 7.11*** *Nine* in vivo *images of blood capillaries obtained at different depths inside the brain of a mouse after skull thinning using two-photon upconversion wide field microscopy (TPUWFM). The images were taken at 5 W cm$^{-2}$ with an exposure of 10 s at a gain of 34.7. The area of each image is 696 × 520 pixels. A 2 × 2 on-chip binning was performed. An Olympus 40 × 0.8 NA water lens was used for imaging the capillaries, with 800 nm emission upon 980 nm excitation from core/shell/shell. $NaYF_4$:Yb(20%):Tm(2%)/$NaYF_4$:Yb(20%): Tm(2%)/$NaYF_4$ nanocrystals were used for imaging. These images are false-coloured. Reproduced with permission from [89]. Copyright © 2011, American Chemical Society*

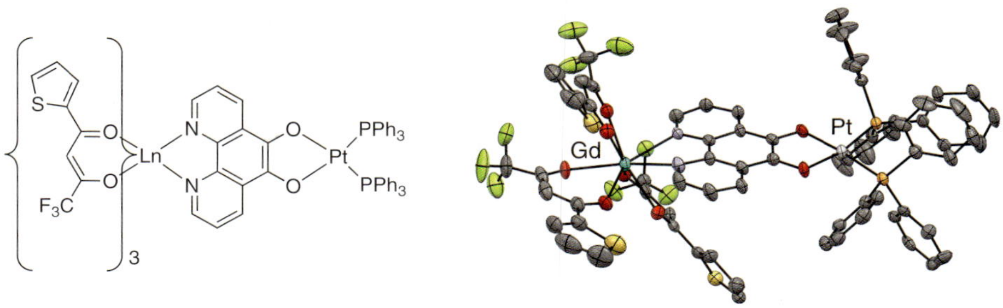


***Figure 9.3*** *Example of a solid state structure of a lanthanide β-diketonate d-f hybrid. Created using the Mercury 2.4 software from the crystallographic data in Reference [17]*

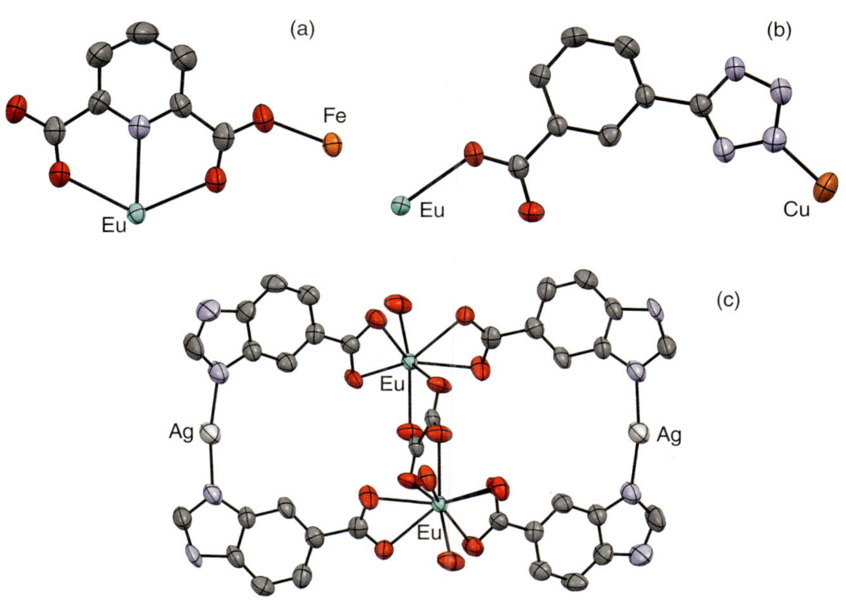


***Figure 9.4*** *Examples of solid state heterometallic architectures obtained using bridging organic ligands. Created using the Mercury 2.4 software from the crystallographic data in in the respective article (see text for references)*

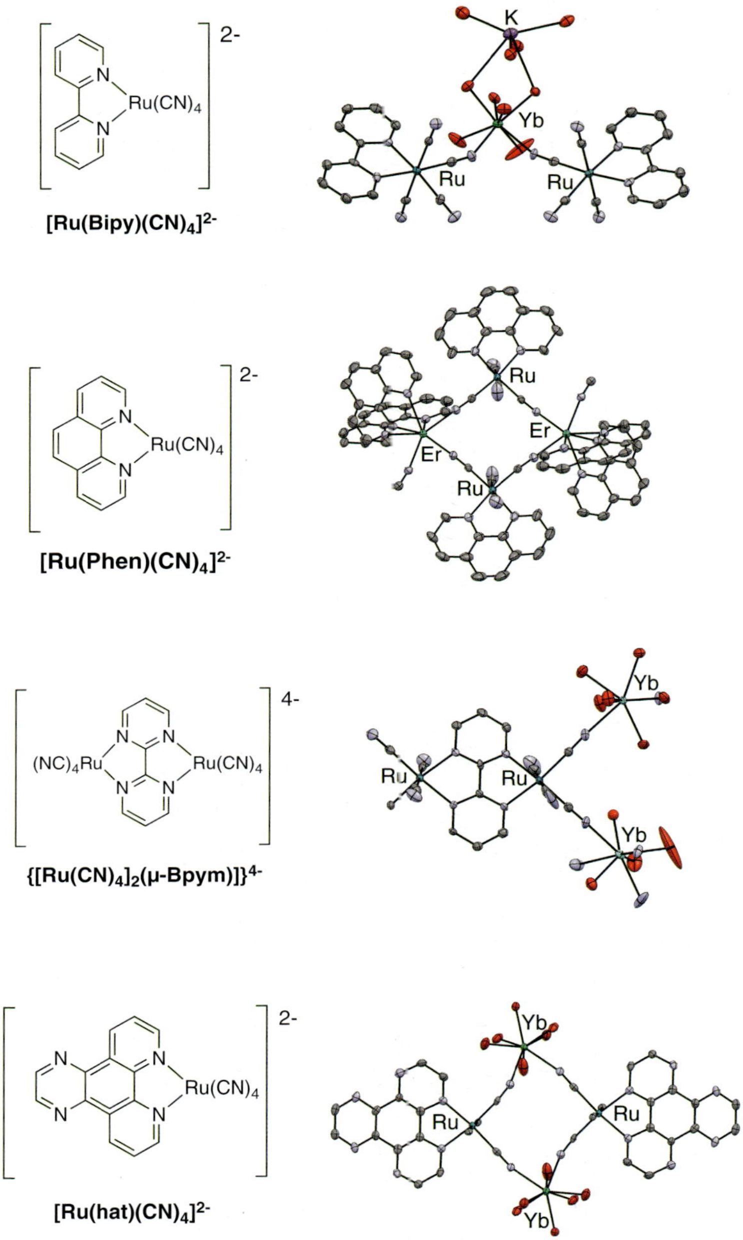


***Figure 9.5*** *Building blocks and solid structures of luminescent coordination polymers derived from complex anions. Created using the Mercury 2.4 software from the crystallographic data in the respective articles (see text for references)*

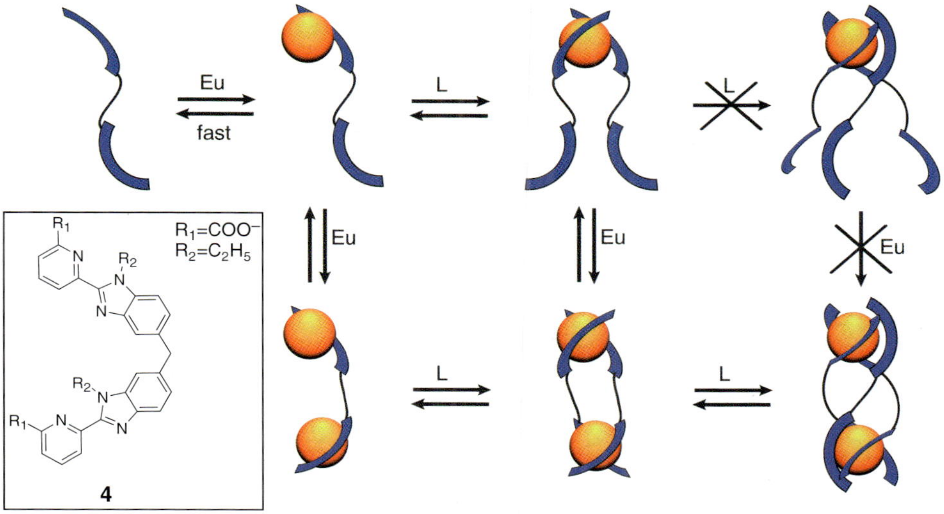


***Figure 9.6*** *Self-assembly mechanism for the formation of triple-stranded helicates as discussed by Elhabiri et al. [35]. Adapted with permission from [35]. Copyright © 2004, John Wiley*

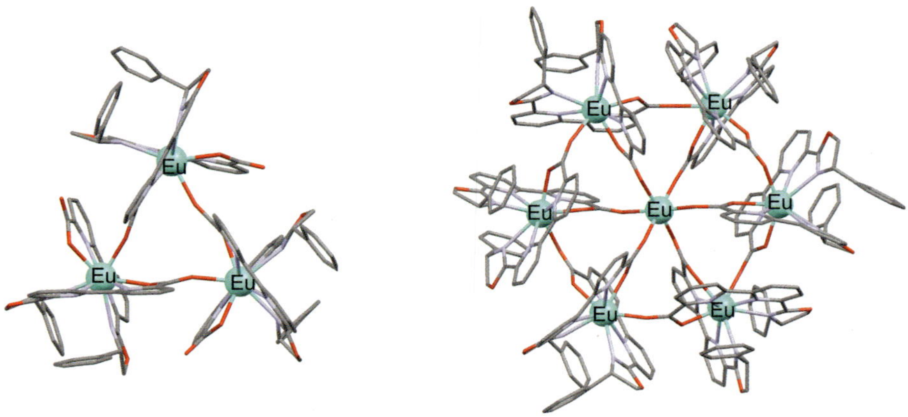


***Figure 9.9*** *Representation of the solid-state structure for the europium molecular triangle and wheel obtained by Mazzanti and co-workers [52,53]. Created using the Mercury 2.4 software from the crystallographic data in the respective articles (see text for references)*

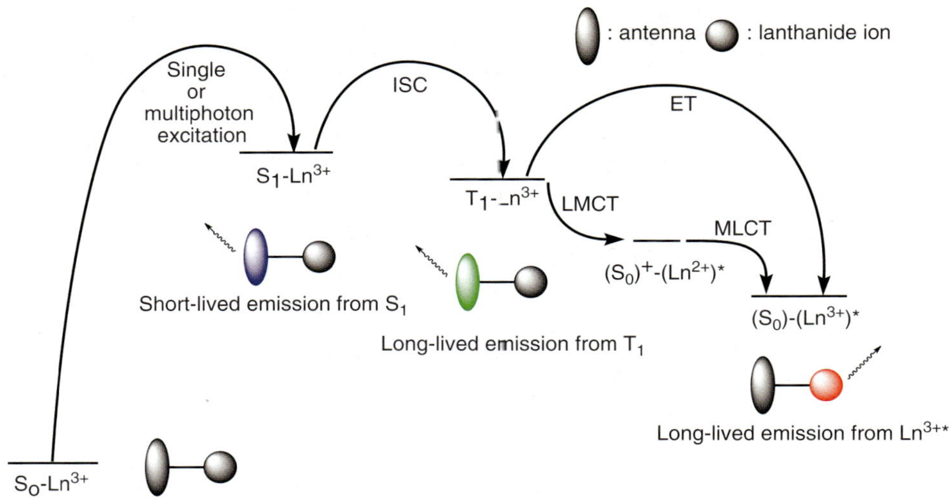


***Figure 9.19*** *Possible energy transfer processes that can occur in a lanthanide antenna assembly leading to luminescence from the lanthanide centre*

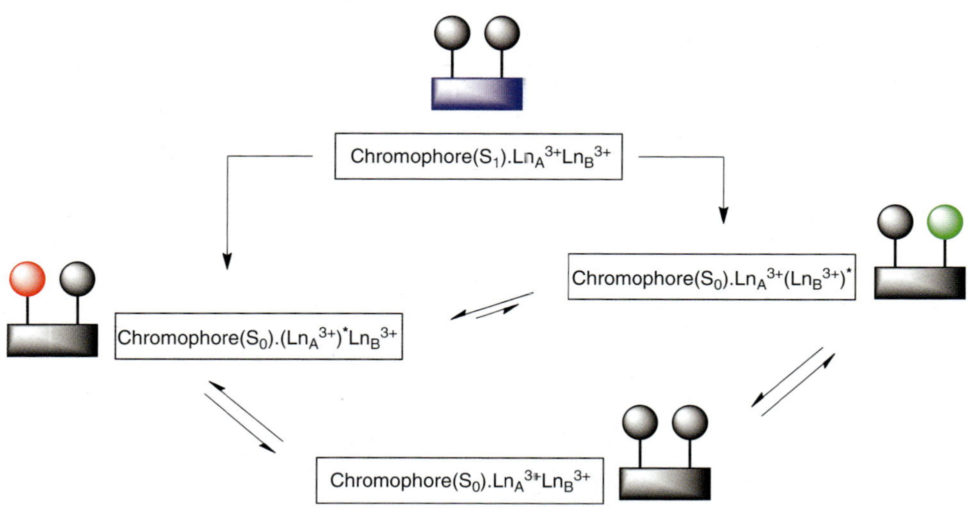


***Figure 9.21*** *Schematic representation of the possible energy transfer pathways in a bimetallic array*

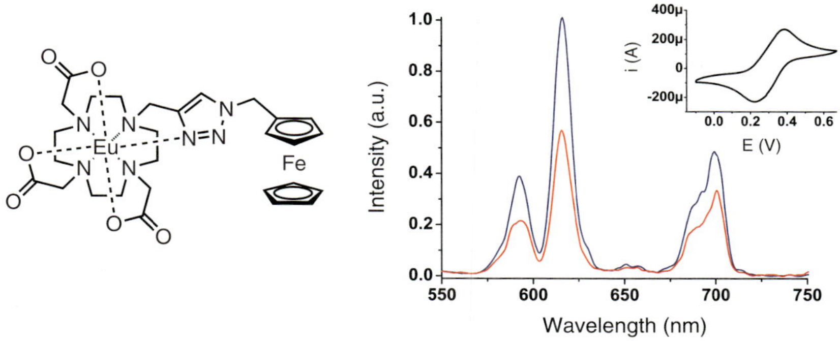


***Figure 9.22*** *A redox active d-f hybrid complex whose luminescence can be modulated by upon electrochemical oxidation (red line) and reduction (blue line) of the ferrocene unit*

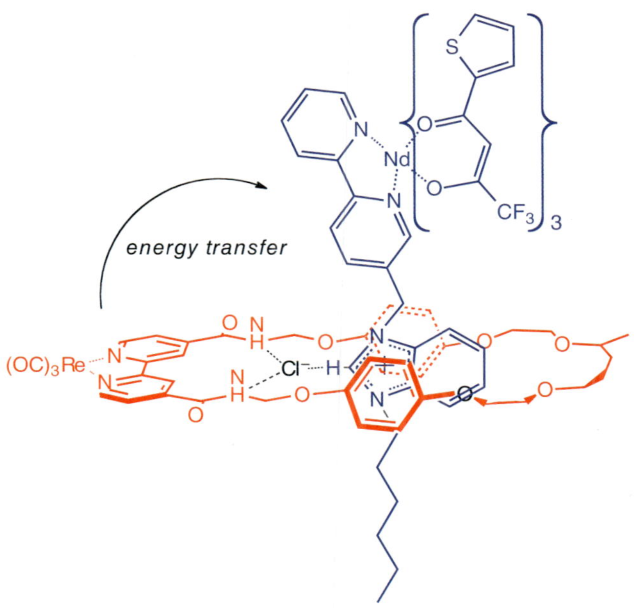


***Figure 9.24*** *Pseudorotaxane formation from a transition metal appended macrocycle and a lanthanide appended thread*

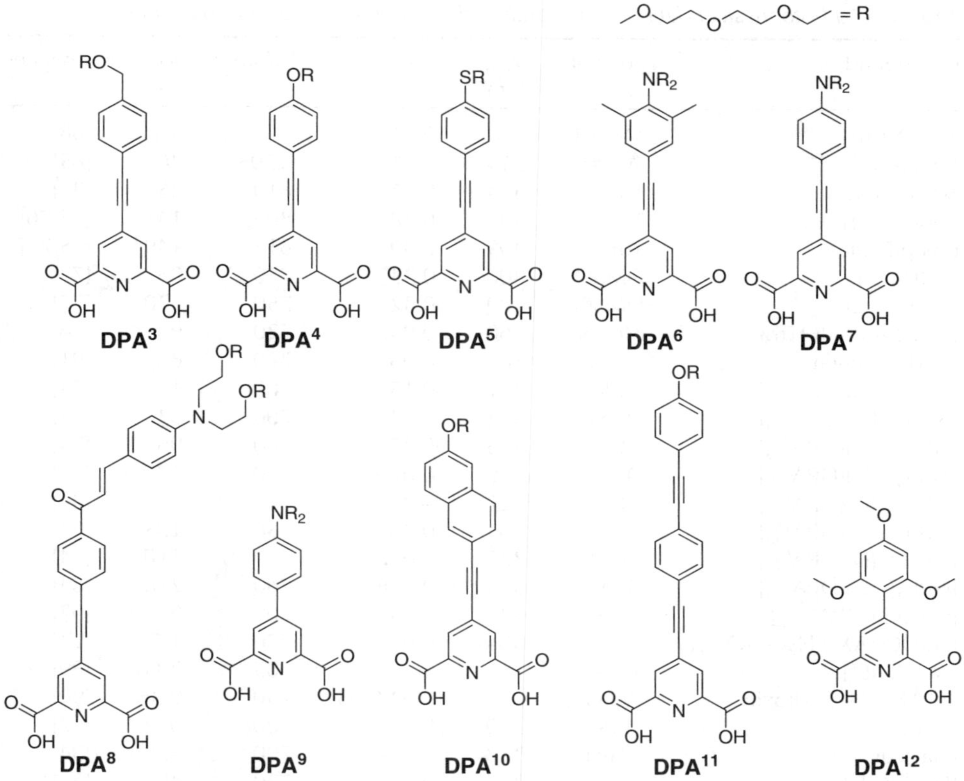


***Scheme 5.1*** *Structure of CT-functionalised dipicolinate ligands studied for structure-brightness relationship. Redrawn from References [82,84,85]*

the lowest energy transition was unambiguously established on the basis of TD-DFT calculations. Careful analysis of the photophysical properties indicates that the ligand $\mathbf{DPA^5}$ induces the most efficient sensitisation, with an excellent 42% luminescence quantum yield of the related europium complex, and that the sensitisation is possible up to 450 nm with ligand $\mathbf{DPA^7}$. In addition, the energy difference between the relaxed ILCT excited state and the $^5D_1$ excited state was producing the same results as those published by Takalo et al. [83] in the case of the triplet excited state sensitisation process, namely a decrease of the sensitisation efficiency when the ILCT excited state becomes closer to the europium $^5D_1$ accepting state.

### 5.3.5 Optimising Molecular Two-photon Cross Section: the Brightness Trade-off

Since the proof of concept for two-photon induced Ln(III) emission, huge efforts have been devoted to develop and optimise an antenna that could be used as an efficient two-photon sensitiser. As explained above, the use of delocalised charge transfer transitions allows an increase of the two-photon cross-section and the demonstration of such excited state

***Table 5.3*** *Photophysical data of some lanthanide complexes for 2PA applications*

| Compound | Solvent | $\lambda_{abs}$ (nm) | $\Phi$ | $\lambda^2$ (nm) | $\sigma_{2PA}$ (GM) | References |
|---|---|---|---|---|---|---|
| $TbL^1(NO_3)_3$ | MeOH | | 0.11 | | 3.1 | [88] |
| $[Tb(DPA^{12})_3]^{3-}$ | Water | 314 | 0.31 | <705 | 26 | [85] |
| $MkEu(fod)_3$ | Tol | 413 | 0.17 | 810 | 253 | [85] |
| $dbptEu(tta)_3$ | Tol | 406 | 0.58 | 808 | 141 | [75,76] |
| $dmbptEu(tta)_3$ | Tol | 409 | 0.59 | 812 | 149 | [75,76] |
| $dbitEu(tta)_3$ | DMSO | 400 | 0.33 | 750 | 98 | [77] |
| $dbHtEu(tta)_3$ | DMSO | 400 | 0.32 | 750 | 320 | [77] |
| $[Eu(DPAM)_3][OTf]_3$ | MeCN | 360 | 0.056 | 720 | 96 | [74] |
| $Eu(THA)_3Phen$ | DCM | 364 | 0.08 | 720 | 80 | [91] |
| $[NBu_4]_3[Eu(DPA^2)_3]$ | DCM | 321 | 0.15 | 700 | 14 | [84] |
| $[NBu_4]_3[Eu(DPA^3)_3]$ | DCM | 322 | 0.43 | 700 | 53 | [84] |
| $[NBu_4]_3[Eu(DPA^4)_3]$ | DCM | 318 | 0.27 | 740 | 96 | [84] |
| $[NBu_4]_3[Eu(DPA^5)_3]$ | DCM | 370 | 0.070 | 740 | 775 | [84] |
| $[NBu_4]_3[Eu(DPA^6)_3]$ | DCM | 427 | ---[a] | | | [84] |
| $[NBu_4]_3[Eu(DPA^7)_3]$ | DCM | 364 | 0.28 | 730 | 193 | [84] |
| $[NBu_4]_3[Eu(DPA^8)_3]$ | DCM | 335 | 0.090 | 700 | 110 | [84] |
| $[NBu_4]_3[Eu(DPA^9)_3]$ | DCM | 335 | 0.036 | 700 | 218 | [84] |
| $[Na]_3[Eu(DPA^{4G})_3]$ | Water | 332 | 0.15 | 700 | 92 | [92] |
| $[Eu(DO2AX^1)(H_2O)][OTf]_3$ | Water | 384 | 0.04 | 770 | 1.7 | [93] |
| $[Na]_2[EuL^2]$ | Water | | | 705 | 28.6 | [89] |
| $EuL^{3/}EuL^3$ + Saxitoxin | Water | --- | 0.03/0.08 | 750 | 9/36 | [90] |
| $[Eu_2(L^4)_3]$ | TRIS | 350 | 0.09 | 800 | 0.75 | [62] |
| $[Na][EuL^7]$ | Water | 338 | 0.284 | 700 | 4.2 | [94] |
| $[EuL^8][OTf]$ | DCM | 345 | 0.11 | 720 | 45 | [95] |
| $YbL^{Co}L^{6a}$ | DMSO | 430 | 0.02 | 860 | 319 | [96] |
| $YbL^{Co}L^{6b}$ | DMSO | 430 | 0.035 | 860 | 375 | [96] |
| $YbL^{Co}L^{6c}$ | Tol | 428 | 0.091 | 800 | 101 | [97] |
| $YbL^{Co}L^{6d}$ | Tol | 427 | 0.011 | 800 | --- | [97] |
| $YbL^{Co}L^{6e}$ | Tol | 428 | 0.016 | 800 | 145 | [97] |
| $YbL^{Co}L^{6f}$ | Tol | 403 | 0.020 | 800 | 247 | [97] |

Abbreviations: Tol = toluene, DCM = dichloromethane.
[a] No Eu emission is observed.

mediated sensitisation of Ln(III) allowed for a better design of systems. Molecular engineering rules for the optimisation of 2PA cross-section have been carefully established in the case of organic molecules or conjugated polymers [86–88]. The two-photon cross-section and photophysical properties of the Ln(III) complexes are reported in Table 5.3. The main effort has been devoted to Eu(III) complexes since the $^5D_1$ accepting excited state is low enough to enable sensitisation from rather strong push–pull structures and because of the ease in measuring emission in the visible range. On the other hand, limited studies have been done on Tb(III) complexes because the $^5D_4$ accepting excited state is high in energy, therefore limiting the access from CT excited states, which results in a rather small two-photon cross-section. More recently, some studies on Yb(III) complexes were published in which only sparse two-photon cross-sections are reported because of the difficulty of obtaining a reliable luminescence quantum yield and the absence of a two-photon cross-section reference in the NIR spectral range. Here, it has to be highlighted that most of the results have been obtained in organic solvents, precluding *in vivo*, *in vitro* or *in cellulo* measurements because of the low aqueous solubility or stability of those complexes.

Based on these established rules, new push–pull ligands have been designed and successfully used to sensitised Eu(III) luminescence by two-photon antenna effect. On the one hand, Werts [85] and Wang [75,76] independently reported dipolar tris-acetylacetonato europium complexes bound to Michler's ketone (**Mk**) or functionalised bis-pyrazolyl-triazine ligands (**dbpt, dmbpt, dbit, dbHt**), respectively [75–77]. On the other hand, our group described octupolar tris-functionalised pyridine dicarboxamide $[(\mathbf{DPAM})_3\mathrm{Eu}]^{3+}$ complexes (Fig. 5.11) [74]. All these complexes are luminescent and present a broad intense absorption band in the visible (>350 nm) responsible for the two-photon absorption located in the biologically transparent spectral window (700–1100 nm) with two-photon cross-sections in the 100–250 GM range. Also, complexes based on a CT-substituted acetylacetonate derivative (**THA**; see Fig. 5.18 below) have been studied, yielding rather modest values of 80 GM. To optimise the two-photon cross-section, our group undertook a structure–properties study where the influence of the strength of the donor moieties on the two-photon cross-section in substituted tris-dipicolinate structures was systematically studied. In this work we found, as expected, that the better donor yielded the better two-photon cross-section, with a value as high as 775 GM for $[\mathrm{NBu_4}]_3[\mathrm{Eu}(\mathbf{DPA^7})_3]$ (Scheme 5.1) [84].

To date, the water soluble and stable complexes have not yielded values as high as in organic solvents but their preparation has very often allowed microscopy studies, illustrating the interest to develop and optimise such structures for these applications. For instance, Charbonnière et al. reported a pyrazolyl-pyridine ligand ($\mathbf{L^2}$; Scheme 5.2) that possesses a two-photon cross-section of 28.6 GM at 705 nm [89]. In this context, Wong et al. published a pseudo on/off two-photon system ($\mathbf{L^3}$; Scheme 5.2) that enables the detection of Saxitoxin [90]. With such a ligand, the authors showed an increase of the two-photon

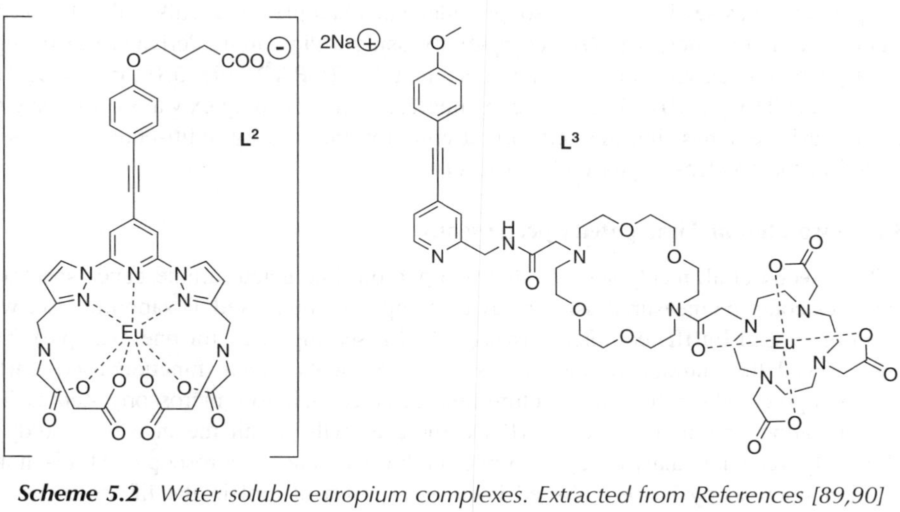


***Scheme 5.2*** *Water soluble europium complexes. Extracted from References [89,90]*

cross-section upon the interaction of Saxitoxin with the designed molecule from 9 to 36 GM that was followed also by an increase of the luminescence quantum yield resulting in a much better two-photon brightness. In addition, Bünzli et al. reported a dinuclear $[Eu_2(\mathbf{L^4})_3]$ complex that can be used for microscopy studies with a two-photon cross-section inferior to 1 GM [62]. The same type of low values were reported by Beeby et al. with DO2AX (Fig. 5.13a), but these complexes were used to establish the proof of concept of two-photon time-gated spectroscopy (see below).

As explained above, the preparation of Tb(III) complexes with a significant two-photon cross-section is sparse. To date, the best Tb(III) complex reported, ($[Tb(\mathbf{DPA^{12}})_3]^{3-}$, Scheme 5.1), exhibits a two-photon cross-section of 26 GM [85].

Whereas the two-photon sensitisation of Yb(III) was initially reported by Lakowicz et al. in 2001 using a DTPA chelated featuring an amino-fluorescein pendant arm [98], the molecular two-photon cross-section was not determined for instrumental reasons (Fig. 5.12). In this context, Wong and co-workers reported a porphyrinato-Yb(III) complex capped with a tripodal $[CpCo(OMe)_2P{=}O]^-$ anion ($\mathbf{L^{Co}}$) and featuring a pendant rhodamine B antenna (Fig. 5.12) [96]. The two-photon cross-section was determined at 375 GM using the open-aperture Z-scan method. Furthermore, using the same platform to chelate the Ln(III), Wong et al. also reported Yb(III) complexes sensitised by a porphyrin chromophore (free base or metalated [97]; Fig. 5.12) possessing a two-photon cross-section higher than 100 GM.

### 5.3.6 Two-photon Excited Luminescence in Solid Matrix

Multi-photon excitation has been performed on solid matrices such as sol-gel for instance. Preparation and one-photon properties of sol-gel containing Ln(III) have been published in an exhaustive review [99]. For those sol-gels, the high concentration of dyes allows easy measurement. Un-optimised systems, directly coated by triethoxysilane function, can therefore be used and two- or three-photon excitation can be performed as illustrated by Boyer et al. with Tb(III) and Eu(III) complexes on dipicolinate-diester chelator substituted by triazole [100]. In the latter example, no CT transition is present, resulting in low two- and three-photon cross-sections. We also provided an example of a bulk sol-gel material containing a more efficient Eu(III) complex by using a dipicolinic derivative featuring a alkoxy-phenylene-ethynylene two-photon antenna [92] ($\mathbf{DPA^{4G}}$; Fig. 5.17; $\sigma_{2PA} = 92$ GM at 700 nm in $CH_2Cl_2$) [101]. In this latter example, the Eu(III) complex was only doping the sol-gel matrix, emphasising the very bright emission induced by multi-photon excitation detected in the condensed phase (see below).

### 5.3.7 Two-photon Time-gated Spectroscopy

In 2007, Beeby et al. nicely showed that two-photon time gated luminescence spectra of complexes could be measured using a cavity-dumped mode locked Ti-sapphire laser with two un-optimised Eu(III) complexes ($\sigma_{2PA} = 4$ GM; see Fig. 5.13 for one example). This work allowed the authors to state that such experiments "make functionalised Ln(III) complexes potential candidates for future time-gated two-photon microscopy studies. The next generation of functionalised Ln(III) complexes will permit the in cellulo study of biologically relevant analytes by two-photon luminescence microscopy. Multi-photon excitation microscopy has established itself as a powerful tool in the life sciences; the

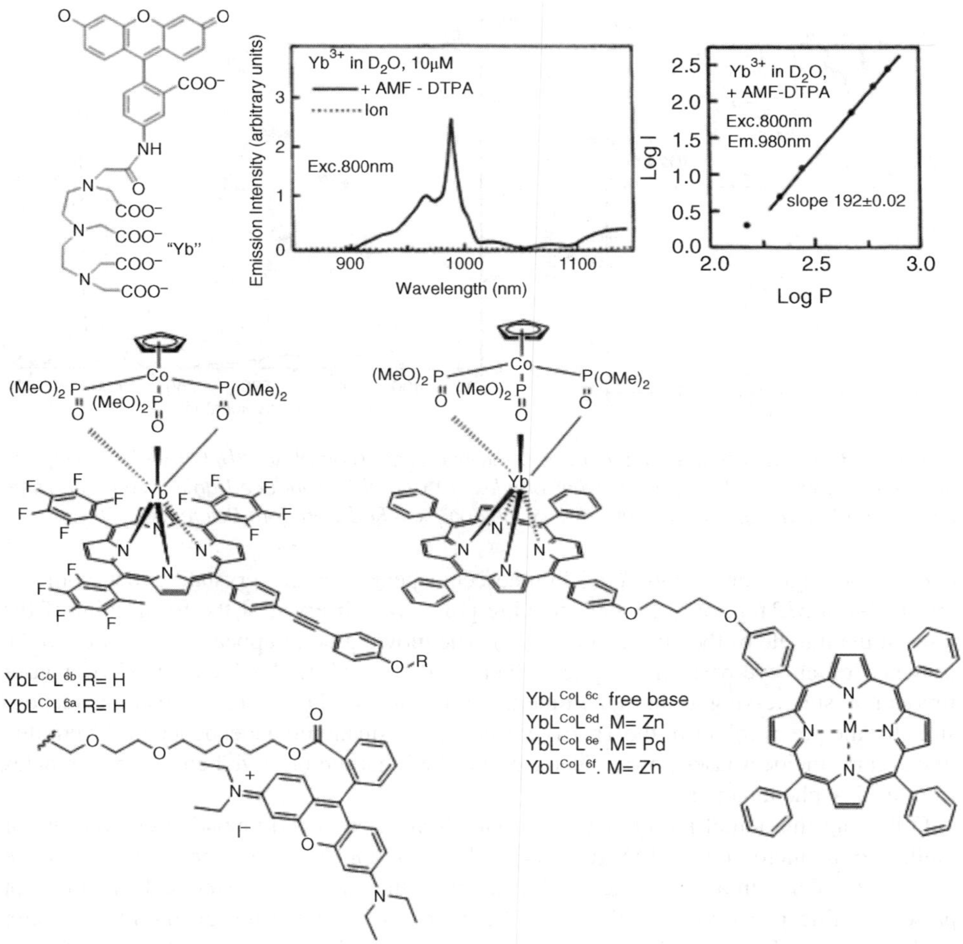


***Figure 5.12*** *Top: structure of aminofluorescein-DTPA (AMF-DTPA) ligand and 2PA induced Yb luminescence spectra and quadratic dependence of the luminescence intensity versus the incident laser power. Bottom: chemical structure of porphyrinato-ytterbium complexes. Redrawn from References [96,98]*

enhanced spatial resolution and reduced sample damage are particularly attractive attributes." We fully agree with this outlook [135].

## 5.4 Towards Biphotonic Microscopy Imaging

### 5.4.1 Proof of Concept

In 2007, our group reported the first microscopic observation of two-photon excited Tb(III) emission, in bulk crystals of tris-dipicolinate complex or in protein crystals (e.g., HEWL, hen eggs white lysozyme [102]) doped with the same complex supress crystals [60]. In spite of the very low 2PA efficiency of this complex, the microscopy experiment was successful

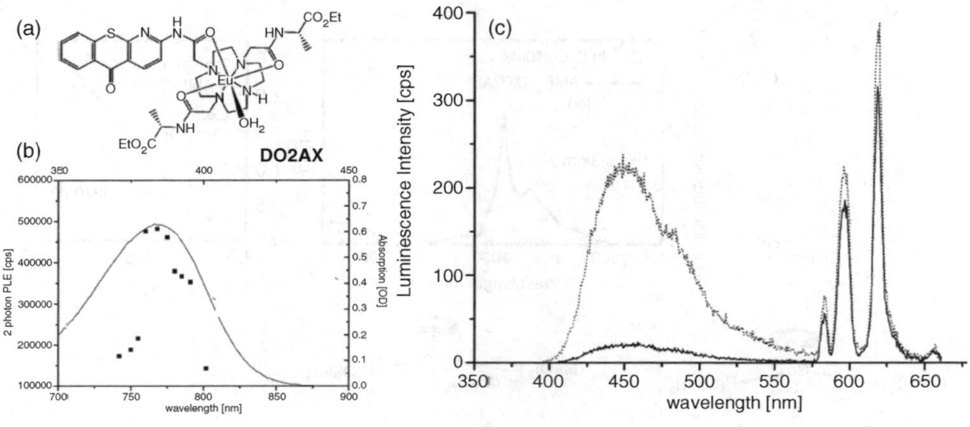


***Figure 5.13*** *(a) Molecular structure of one of the investigated complexes. (b) UV/visible absorption and two-photon absorption spectra of the complex. (c) Two-photon induced luminescence and time gated two-photon induced luminescence of the complex. Redrawn from Reference [93]*

due to the high concentration of luminescent species in the crystal; under confocal irradiation at 532 nm, only a small glowing point was observed at the focal point of the laser in the middle of the crystal (Fig. 5.14). The glowing point appears orange due to the presence of an eye-protecting optical filter used to cut off the incident 532 nm laser irradiation, suppressing at the same time the most intense $^5D_4 \rightarrow {}^7F_{6\text{-}5}$ transitions responsible for the green colour of the Tb(III) emission. The quadratic dependence of the emitted light versus incident laser power clearly established that the observed emission originates from a 2PA phenomenon.

Following this initial report, several groups have used already existing complexes for multi-photon microscopy [62,88,103,104]. All these complexes, described in Fig. 5.15, are stable and soluble in aqueous media. Furthermore, they are able to stain cells and do not present significant cell toxicity. But, since they were not especially designed for two-photon application (no CT excited state is present in the sensitising chromophore), their unoptimised two-photon cross-sections are low (<10 GM).

In particular, Wong and co-workers described *in vitro* cell imaging using three-photon excitation of $TbL^1(NO_3)$ [where $L^1$ designates a tripodal ligand *N*-[2-(bis{2-[(3-methoxybenzoyl)amino]-ethyl}amino)ethyl]-3-methoxybenzamide; Fig. 5.15] featuring $\sigma^{(2)} = 3.1$ GM and $\sigma^{(3)} = 1.9$ GM [88]. In spite of its oligomeric structure, this complex was described to be stable in water and three-photon microscopy imaging of various cells types was described. A few months later, the group of Parker reported the first practical utilisation of two-photon scanning microscopy using a terbium complex featuring an aza-xanthone antenna chromophore as a probe ($L^{5a\text{-}f}$) [103]. They showed that the cell toxicity, up-take kinetics and final localisation depend on the substitution of the complex by various organic, peptidic or even biologic moieties (R = ester, acid, amide, alkyl organic fragment or $LysArg_7$, $Arg_7$, guanidinium or human serum albumin, HSA, conjugated link). More recently, Bünzli and co-workers reported the two-photon scanning microscopy of HeLa cells incubated with a Eu(III) helicate complex $Eu_2L^4{}_3$ (Fig. 5.15), underlining the interest of such bimetallic derivatives for luminescent lanthanide bio-probe applications.

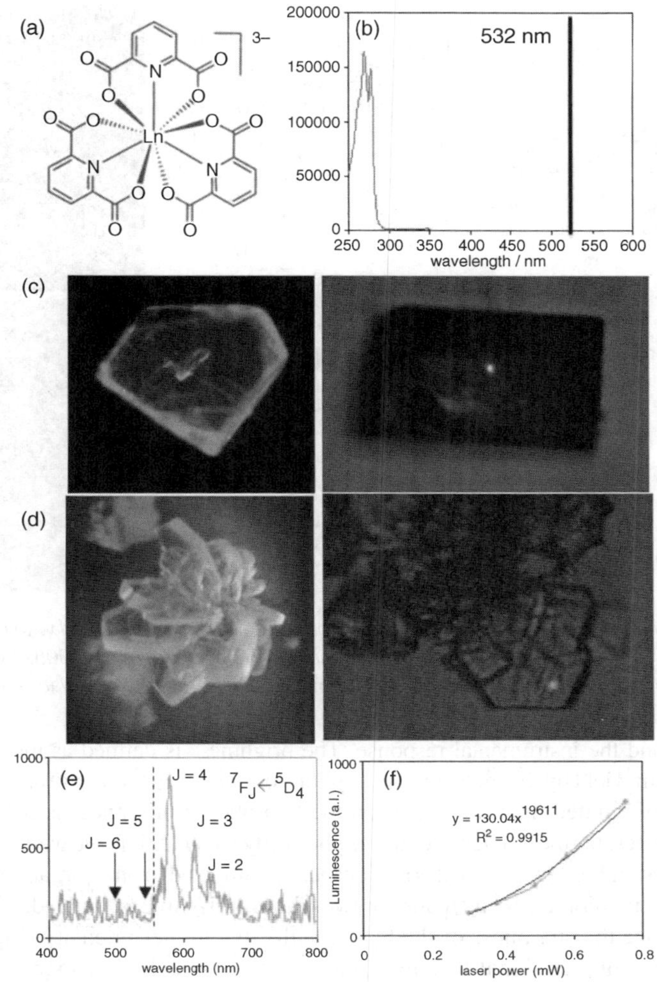


***Figure 5.14*** *Linear microscopy imaging (UV irradiation at 365 nm) of a lysosyme-derivative crystal containing terbium tris-dipicolinate (left), nonlinear two-photon microscopy under 532 nm laser irradiation (right). Reproduced from [60]. Copyright © 2007, John Wiley (See colour plate section)*

All those examples and some others [105] clearly illustrate the feasibility of multi-photon imaging microscopy using lanthanide complexes as probes. However, the very low nonlinear efficiency of these complexes is slightly compensated by the increase of the incident laser power intensity to obtain well resolved images, but this clearly remains an important drawback for any practical applications.

### 5.4.2 Towards the Design of an Optimised Bio-probe

The key parameter for luminescence microscopy is represented by the sensitivity, in other words the ability to detect the probe signal, in the inherent noise due to the auto fluorescence

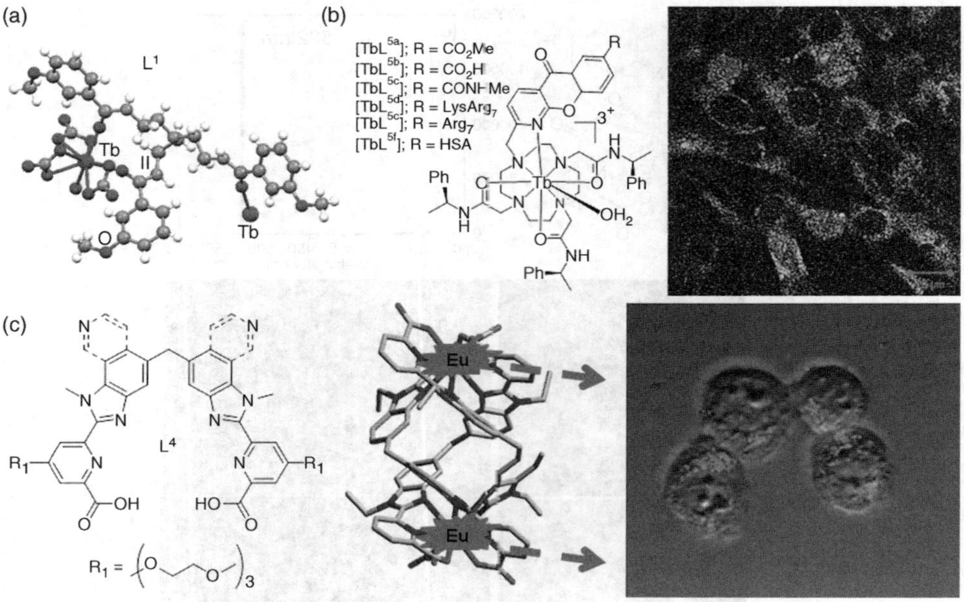

*Figure 5.15* Water soluble complex used for biphotonic cell imaging. Adapted with permission from [88]. Copyright © 2008, American Chemical Society, [103] Copyright © 2008, Royal Society of Chemistry, and Copyright © 2010, American Chemical Society (See colour plate section)

of the sample and the instrumental response. The brightness is defined as the product of the emission quantum yield by the molecular absorption coefficient, $B^{(1)} = \varepsilon \times \Phi$ and is proportional to the amount of emitted light; by extension the two photon brightness is defined by $B^{(2)} = \sigma_{2PA} \times \Phi$ and represents the efficiency of a two-photon dye. In addition, we want to highlight that lanthanide sharp luminescence provides advantages as compared to other organic chromophores featuring broad emission for a given one- or two-photon brightness. Indeed, its detection is much easier since the sharpness of the band results in a higher intensity (Fig. 5.16). This generally gives a higher signal-to-noise ratio, which is a real advantage for microscopy experiments. This observation should define an *effective brightness* that would take into account the full width at half maximum, $\omega_{1/2}$.

$$B^{(1\text{eff})} = (\varepsilon \times \Phi)/\omega_{1/2} \text{ or } B^{(2\text{eff})} = (\sigma_{2\text{PA}} \times \Phi)/\omega_{1/2}$$

In order to tackle the solubility and low two-photon brightness drawbacks, our group reported in 2008 the design of a europium tris-dipicolinate complex containing an extended π-conjugated functionalised ligand (**DPA$^{4G}$**). To counter-balance the strong lipophilic character of this antenna, the ligand was further decorated by hydrosolubilising moieties, here a polyethylene glycol fragment. The target complex presents remarkable spectroscopic properties in water ($\Phi = 0.157$, $\tau = 1.06$ ms, $\sigma_{2PA} = 92$ GM at 700 nm) resulting in high two-photon brightness in water ($B^{(2)} = 14.4$ GM) and allows reporting the feasibility of two-photon scanning microscopy of fixed cells with an excellent signal to noise ratio (Fig. 5.17) [92]. It is however worth noting that this compound presents two

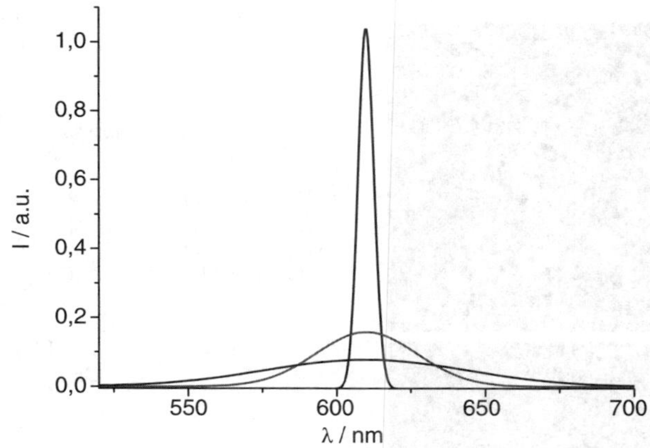


**Figure 5.16** *Variation of the luminescence profile for a given area with the full width at half maximum $\omega_{1/2}$ (black 5 nm, red 20 nm, blue 40 nm)*

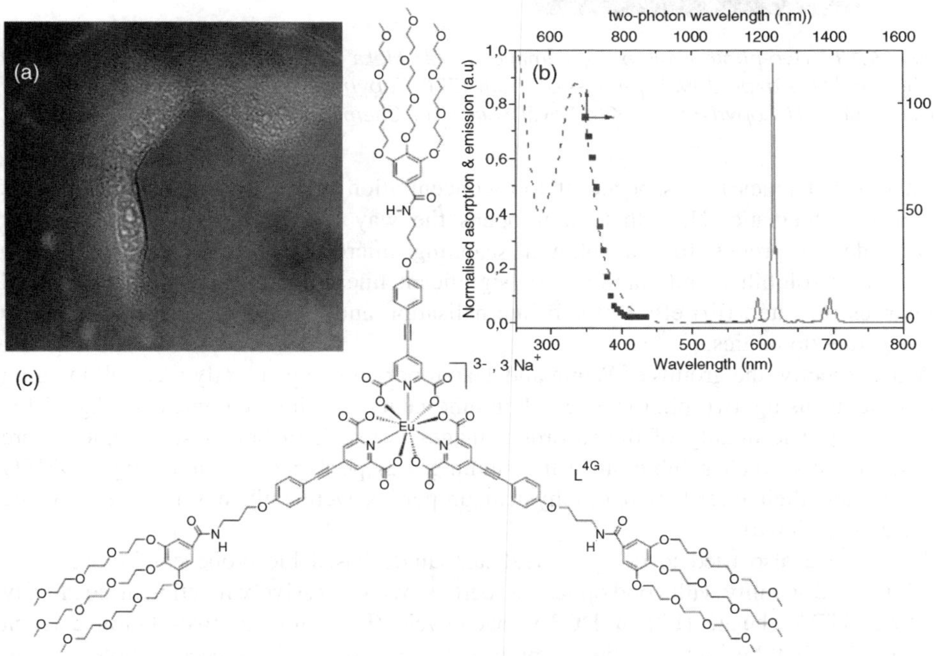


**Figure 5.17** *(a) Two-photon scanning image of T24 fixed cells using a 760 nm incident wavelength. (b) Spectroscopic properties in water (blue dashed line, absorption; blue squares two-photon absorption, red solid line, emission). (c) Structure of the probe. Adapted with permission from [92]. Copyright © 2008, American Chemical Society (See colour plate section)*

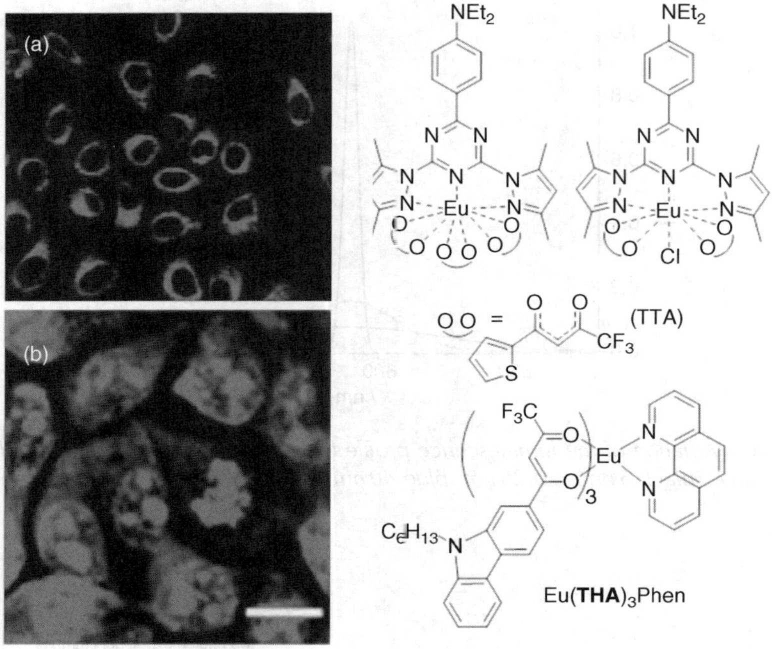


***Figure 5.18*** *Two-photon microscopy images of (a) HeLa and (b) MCF-7 cell, using Eu(III) probes (right). Adapted with permission from [77]. Copyright © 2011, American Chemical Society and [91] Copyright © 2008, Royal Society of Chemistry (See colour plate section)*

limitations: it tends to dissociate at low concentration and its internalisation in cells remains problematic. Nevertheless, it opens the way for the design of luminescent lanthanide bio-probes for two-photon scanning microscopy ideally combining: (i) good water solubility and stability, (ii) significant linear and nonlinear photophysical properties in water, (iii) efficient cell internalisation and low toxicity, (iv) targeting or sensing functionalities.

More recently, the groups of Wong and Tian reported independently nice cell imaging experiments using two-photon excited luminescent europium complexes (Fig. 5.18) [77,91,106]. The quality of the recorded images is excellent but these complexes are not sufficiently water-soluble since the staining was performed by injecting a DMSO solution and their excellent photophysical properties were only described in organic solvent (see above).

Further, we also tried to design a real lanthanide based bio-probe combining water solubility and stability with good optical properties. We successively functionalised already existing DTPA chelate (**L**$^{6}$) or DO3A macrocycle (**L**$^{7}$) with our two-photon antenna (Scheme 5.3) but the corresponding complexes did not present the expected photophysical properties due to either a quenched luminescence (by the coordination of one remaining water molecule) or other unexplained phenomenon yielding a low $\sigma_{2PA}$ value and hence extremely low two-photon brightness [$B^{(2)}$(Eu**L**$^{7}$) = 1 GM] in spite of a good 25% quantum yield [94]. Pursuing the same idea, we reported the design of bis(bipyridine) aza-macrocylic

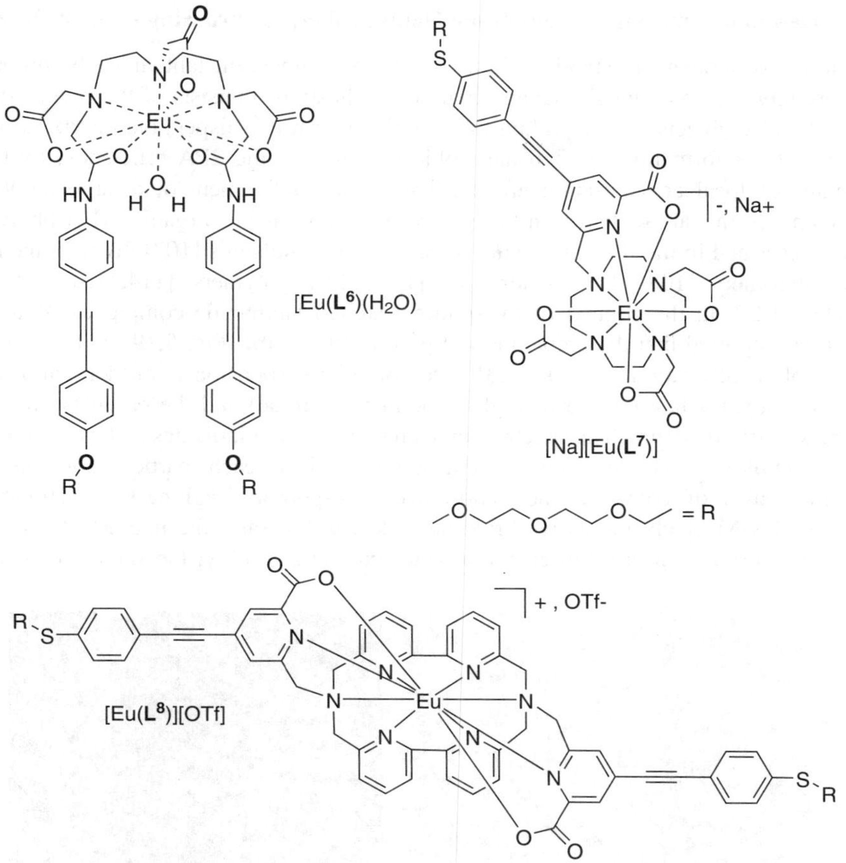


***Scheme 5.3*** *Structures of polydentate/macrocyclic complexes containing two-photon antenna [94,95]*

ligands featuring or not π-conjugated donor picolinic pendant arms (**L**[8]) [95]. It is worth noting that the nonfunctionalised ligand leads to the formation of a europium complex that is stable in water and that was strongly luminescent, but featuring, of course, a poor two-photon absorption cross-section. In contrast, the europium complex with a ligand based on the extended conjugated antenna presents a two-photon absorption cross-section of 45 GM at 700 nm but is not luminescent in water. This quenching is not due to the coordination of O-H vibrators to the metal and is more likely explained by an internal dynamic effect [95]. This study emphasises that functionalisation of the ligand by conjugated moieties may induce an unexpected and dramatic influence on the related complex photophysical properties. This phenomenon is not fully rationalised but has to be overcome when designing optimised lanthanide-based probes for two-photon scanning laser microscopy imaging, and significant improvements have to be done to prepare a real bioprobe suitable for practical bio-imaging application.

### 5.4.3 Design of Lanthanide containing Nano-probes, toward Single-object Imaging

An alternative strategy to stabilise the two-photon luminescent lanthanide bioprobe and therefore envisage biological relevant imaging applications consists of the preparation of nanometre size objects by embedding the complex in a water-dispersible matrix. It is well known that the formation of such nano-objects enhances the 2PA efficiency, by taking advantage of local concentration effects, that is the confinement of a large number of chromophores in nanoscale volumes. Several examples using organic NLO-phores are already described in the literature using micellar nano-emulsions [107], hyperbranched or cyclic oligomers [108–110], dendrimers [111–113], polymers [114,115] and silica NPs [116–122]. In the context of two-photon excited lanthanide complexes, Wang and co-workers reported first the preparation of pure Eu(tta)$_3$**dpbt** (Fig. 5.19) particles using a very simple precipitation process [123]. The colloidal suspension is further stabilised by addition of a surfactant (cetyltrimethyl ammonium bromide) and the particles present an average size of 30–40 nm. The spectroscopic characterisation indicates a clear stabilisation of the complex in aqueous media. Furthermore, since each particle contains circa 8500 molecules of complex, the nano-object two-photon brightness is estimated at ca. $3.2 \times 10^5$ GM. Such impressive brightness clearly illustrates the interest of such local confinement effects. In order to ensure a better bio-compatibility, the same complex was

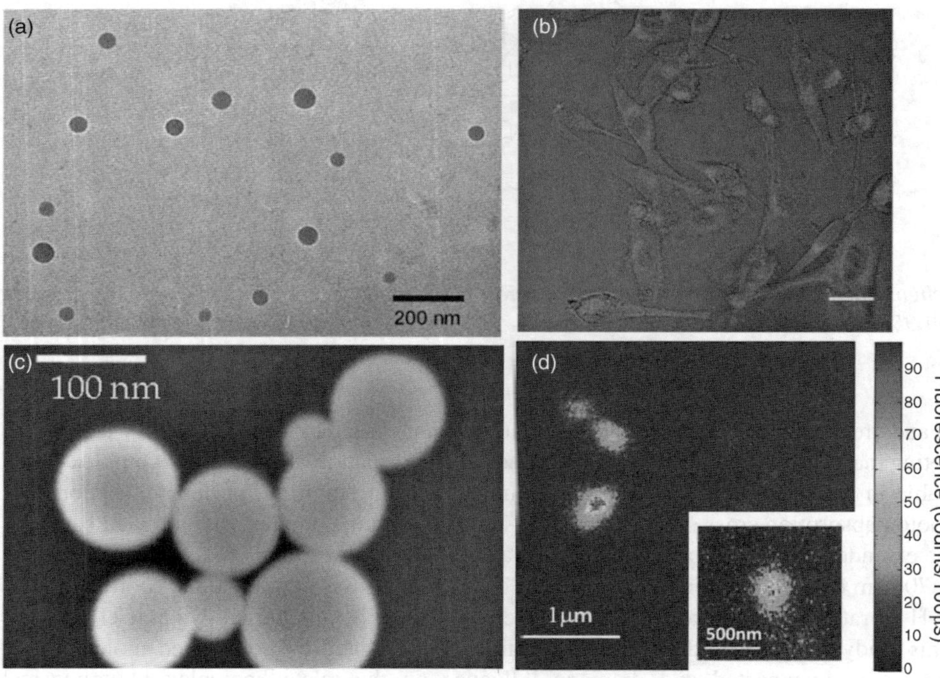


*Figure 5.19 TEM image of nano-spheres prepared by: (a) micro-emulsion or (c) sol-gel techniques. (b) Two-photon scanning microscopy of MDA-MB-231 cancer cells stained with anti-GFR conjugated nanoparticle. (d) Two-photon luminescence images of single silica nanoparticles. Adapted with permission from [124]. Copyright © 2010, John Wiley and [125] Copyright © 2008, Royal Society of Chemistry (See colour plate section)*

encapsulated in the hydrophobic core of an amphiphilic co-polymer (poly-methylmethacrylate-*co*-methacrylic acid) using the micellar nano-emulsions method [124]. The 50-nm diameter core-shell nano-particles were dispersible in water and were further conjugated with anti-epidermal growth factor monoclonal antibody (*anti*-EGFR) known to selectively recognise cancer cells. The nano-spheres (Fig. 5.19) contained ca. 3500 $Eu(tta)_3$**dpbt** molecules for a $1.2 \times 10^5$ GM two-photon brightness and presented high dispersion stability in water solution and excellent bio-compatibility. They selectively recognised cancer cells expressing a high level of EGFR, allowing a clear detection using two-photon scanning microscopy imaging (Fig. 5.19). Such a target-specific two-photon microscopy imaging experiment in cells opens the way for similar studies in thick tissues towards small animals.

In the same context, our group described the preparation of luminescent silicate NPs doped with a $[Eu(\mathbf{DPA^7})_3]^{3-}$ complex chosen for its very high 2PA efficiency ($\sigma_{2PA} = 775$ GM at 740 nm) [125]. In order to prepare a larger amount of nano-particles, an original spray-drying process was developed starting from sol-gel solutions containing specific silicon alkoxides, as silicate precursors, in order to control the hydrophilic character, porosity and water dispersion of the particles. The particles doped with 5 wt % of $[Eu(\mathbf{DPA^7})_3]^{3-}$ (Fig. 5.19) exhibit bright luminescence and are perfectly stable by dispersion in water even at very low concentration, clearly illustrating the strong stabilisation of the complex in the matrix. This bright luminescence (that can be sensitised by a two-photon excitation) combined with the sharpness of the Eu(III) emission at 615 nm afforded an exceptional signal-to-noise ratio in two-photon microscopy imaging experiments and enabled the detection of single nanoparticles (Fig. 5.19) even at low-incident laser intensity. A similar approach was also recently reported by Boyer et al. using triazole functionalised tris-dipicolinate-diester Eu(III) or Tb(III) grafted or doped in silica nano-spheres prepared by a micro-emulsion technique [126].

### 5.4.4 Towards NIR-to-NIR Imaging

As illustrated by this review, most of the research devoted to the two-photon sensitisation of lanthanide luminescence has focused on terbium and europium complexes and impressive results were obtained in bioimaging applications using two-photon microscopy. Whereas the two-photon excitation is located in the 700–900 nm spectral range, these ions present their strongest emission band below 650 nm and are therefore not ideal candidates for thick tissues or *in vivo* imaging. Indeed, it is well known that NIR wavelengths, located in the 700–1100 nm biological window, are less absorbed or scattered by biological tissues and that this NIR-to-NIR microscopic configuration (named according to the excitation and detection spectral ranges) is the most straightforward for in depth imaging [127–130]. Beyond organic NIR fluorophores or inorganic quantum dots, lanthanide NIR emitters [Yb (III), Nd(III) and Er(III)] present strong potentialities for this application [4,9,131]. Recently, up-converting phosphor particles $Y_2O_3$:$Er^{3+}$/$Yb^{3+}$ exhibited 1550 nm emission under 980 nm excitation; a convenient surface coating makes these particles biocompatible and suitable for *in vivo* imaging [132,133]. However, as mentioned in the Introduction, the up-conversion process is a multi-stepwise linear absorption phenomenon, fundamentally different from the nonlinear multi-photon photophysical process. Focusing on the nonlinear excitation process, the proof of concept of Yb(III) sensitisation by a two-photon excitation was initially reported by Lakowicz et al. and, about 10 years later, Wong and co-workers

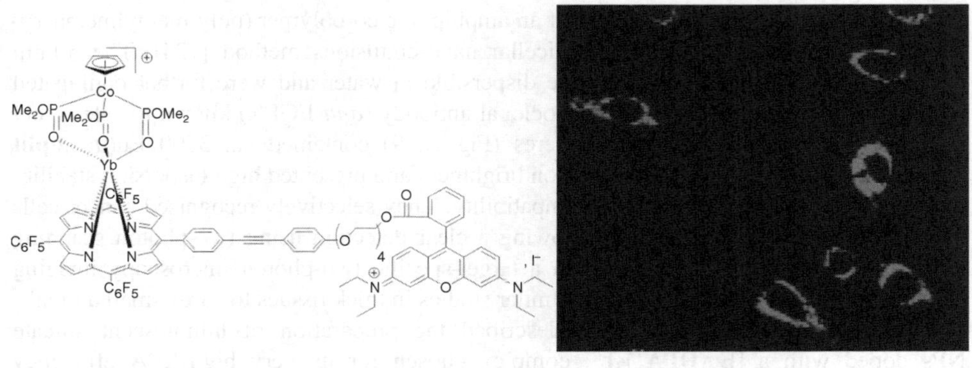

*Figure 5.20* *Structure of the complex and two-photon imaging of Hela cells ($\lambda_{ex}$ = 860 nm, $\lambda_{det}$ = 500–800 nm). Adapted with permission from [96]. Copyright © 2011, American Chemical Society*

reported the $\text{YbL}^{\text{Co}}\text{L}^{6b}$ complex featuring a pendant rhodamine B antenna (Figs. 5.13 and 5.20) and exhibiting very impressive photophysical properties in water, with a luminescence quantum yield of 2.5% and a lifetime of 1.1 μs in water (3.5% and 20.5 μs in DMSO), with a high two-photon cross-section of 375 GM, as highlighted above. According to the author the high stability in water is ensured by the tri-phosphoro cobalt complex, and it is worth noting that the energy transfer from the antenna seems to be not fully optimised, since an intense residual ligand-centred emission is still observed at around 600 nm. This complex successfully stained living HeLa cells and two-photon microscopy revealed a specific localisation within the mitochondria. In spite of the significant breakthrough made by this group, it is however important to note that the image was only recorded in the classical NIR-to-visible configuration (860 nm excitation; 500–800 nm detection) taking advantage of the residual ligand-centred emission and not the Yb(III) emission (Fig. 5.20).

In the same context, our group was recently interested in the design of macrocylic Yb(III) complexes based on the TACN platform further functionalised by CT-containing antenna (**TACN-DPA$^7$**; Fig. 5.21) and its tris-dipicolinate analogous (**DPA$^7$**; Fig. 5.17) [134]. These two complexes were able to efficiently sensitise the Yb(III) emission in water and, as a result

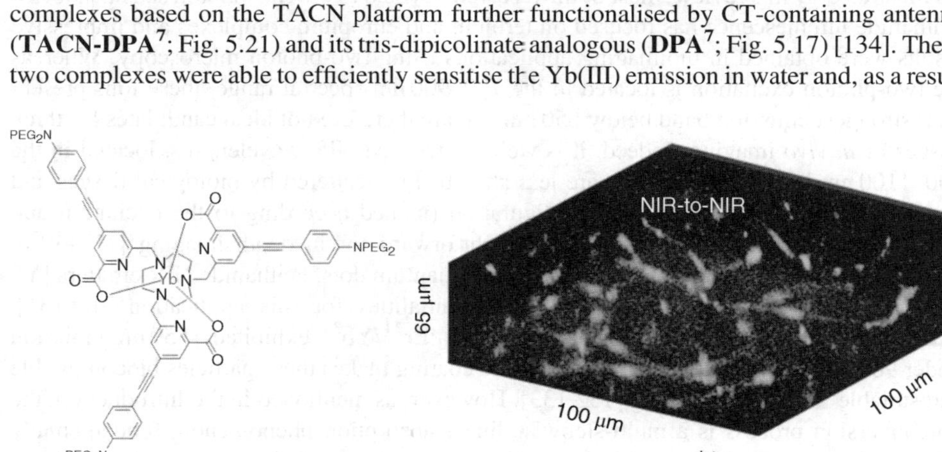


*Figure 5.21* *Structure of the complex and two-photon imaging of mouse brain slice ($\lambda_{ex}$ = 760 nm, $\lambda_{det}$ = 1000 nm). Adapted with permission from [134]. Copyright © 2012, John Wiley*

of its increased stability, the macrocyclic complex exhibited a NIR-luminescence lifetime of 3 μs, 10 times higher than that of [Yb(**DPA**$^7$)$_3$]$^{3-}$. The luminescence was induced by a two-photon antenna process but no two-photon cross-section could be measured due to: (i) the low quantum yield of the residual CT emission in the visible and (ii) the absence of any available standard in the NIR spectral range. However, the $\sigma_{2PA}$ value of compounds [Yb(**DPA**$^7$)$_3$]$^{3-}$ and Yb**TACN-DPA**$^7$ in Fig. 5.21 can be roughly estimated to be in the same range that of Eu(III) and Lu(III) analogs, determined to 775 and 500 GM, respectively [84]. We demonstrated that one-photon cell imaging can be efficiently performed in the Vis-to-Vis or Vis-to-NIR configuration taking advantage of the dual CT and Yb(III) emission. In addition, whereas two-photon cell imaging was successfully conducted in the NIR-to-Vis configuration (using the residual CT emission), it was impossible to record any image in the more interesting NIR-to-NIR configuration (using Yb(III) emission), since conventional biphotonic microscopes contain optical filtering schemes as an eye-protecting system which do not allow this spectral configuration. To tackle these limitations, we developed our own biphotonic microscopy set-up based on adequate optical filtering [134] that allows this NIR-to-NIR configuration. After validation experiments, the mouse brain vascular network was successfully imaged in depth with both excitation (760 nm) and detection (1000 nm) in the biological transparency window. This proof of concept experiment opens an interesting perspective for the development of thick tissue imaging towards biological relevant applications (for diagnostic or guided surgery).

## 5.5 Conclusions

In conclusion, this review illustrates the last decade (2001–2012) of research endeavour devoted to the sensitisation of lanthanide luminescence by a nonlinear two- (multi-) photon process. The motivation for this project is to combine the unique spectroscopic properties of lanthanide complexes (sharp transitions, long luminescence lifetime, environment sensitivity, circularly polarised luminescence, etc.) with the huge potentialities of the biphotonic excitation (NIR excitation, 3D resolution, etc.), in particular for bio-imaging applications. To that end, several scientific and instrumental breakthroughs have been performed almost simultaneously by several groups around the world:

1. The proof of concept of the two-photon antenna effect was established with Tb, Eu and Yb ions and two-photon time-gated spectroscopy has been evidenced.
2. The optimisation of the two-photon cross-section has triggered the discovery of a new sensitisation process mediated by a antenna charge transfer excited state.
3. Microscopy imaging using Tb, Eu and even Yb containing probes has been reported and, in the case of the NIR emitter, the development of a new biphotonic microscopy set-up was achieved.

However, to date, a real biphotonic luminescent lanthanide bio-probe suitable for biologically relevant microscopy imaging application remains to be discovered. Such probes must of course fulfill spectroscopic requirements like optimised two-photon brightness, but they have to present not only a good stability and solubility in biological media but also a functionalisation site for eventual bio-conjugation. One can anticipate that such an "ideal" lanthanide bio-probe will be prepared during the next decade [135].

## References


1. Jüstel T., H. Nikol, C. Ronda, *Angew. Chem., Int. Ed.* **1998**, *37*, 3084.
2. de Bettencourt-Dias A., *Dalton Trans.* **2007**, *2007*, 2229.
3. Kido J., Y. Okamoto, *Chem. Rev.* **2002**, *102*, 2357.
4. Faulkner S., S. J. A. Pope, B. P. Burton-Pye, *Appl. Spectro. Rev.* **2005**, *40*, 1.
5. Pandya S., J. Yu, D. Parker, *Dalton Trans.* **2006**, *2006*, 2757.
6. Gunnlaugsson T., J. P. Leonard, *Chem. Commun.* **2005**, *2005*, 3114.
7. Montgomery C. P., B. S. Murray, E. J. New, R. Pal, D. Parker, *Acc. Chem. Res.* **2009**, *42*, 925.
8. Moore E. G., A. P. S. Samuel, K. N. Raymond, *Acc. Chem. Res.* **2009**, *42*, 542.
9. Eliseeva S. V., J.-C. G. Bünzli, *Chem. Soc. Rev.* **2010**, *39*, 189.
10. Hemmilä I., S. Webb, *Drug Discovery Today* **1997**, *2*, 373.
11. Muller G., *Dalton Trans.* **2009** *2009*, 9692.
12. Andraud C., O. Maury, *Eur. J. Inorg. Chem.* **2009**, *2009*, 4357.
13. Ma Y., Y. Wang, *Coord. Chem. Rev.* **2010**, *254*, 972.
14. Sénéchal K., L. Toupet, I. Ledoux, J. Zyss, H. Le Bozec, O. Maury, *Chem. Commun.* **2004**, *2004*, 2180. K. Sénéchal-David, A. Hemeryck, N. Tancrez, L. Toupet, J. A. G. Williams, I. Ledoux, J. Zyss, A. Boucekkine, J.-P. Guégan, H. Le Bozec, O. Maury, *J. Am. Chem. Soc.* **2006**, *128*, 12243.
15. Tancrez N., C. Feuvrie, I. Ledoux, J. Zyss, L. Toupet, H. Le Bozec, O. Maury, *J. Am. Chem. Soc.* **2005**, *127*, 13474.
16. Law G.-L., K.-L. Wong, K.-K. Lau, S.-T. Lap, P. A. Tanner, F. Kuo, W.-T. Wong, *J. Mater. Chem.* **2010**, *20*, 4074–4079. Walton J. W., R. Carr, N. H. Evans, A. M. Funk, A. M. Kenwright, D. Parker, D. S. Yufit, M. Botta, S. De Pinto, K.-L. Wong *Inorg. Chem.* **2012**, *51*, 8042–8056.
17. Valore A., E. Cariati, S. Righetto, D. Roberto, F. Tessore, R. Ugo, I. L. Fragalà M. E. Fragalà G. Malandrino, F. De Angelis, L. Belpassi, I. Ledcux-Rak, K. H. Thi, J. Zyss, *J. Am. Chem. Soc.* **2010**, *132*, 4966.
18. Ayhan M. M., A. Singh, C. Hirel, A. G. Gürek, V. Ahsen, E. Jeanneau, I. Ledoux-Rak, J. Zyss, C. Andraud, *Y. Bretonnière, J. Am. Chem. Soc.* **2012**, *134*, 3655.
19. Goppert-Mayer M., *Ann. Phys.* **1931**, *9*, 273.
20. Kaiser W., C. G. B. Garrett, *Phys. Rev. Lett.* **1961**, *7*, 229.
21. Albota M., D. Beljonne, J.-L. Brédas, J. E. Ehrich, J. Y. Fu, A. A. Heikal, S. E. Hess, T. Kogej, M. D. Levin, S. R. Marder, D. MacCord-Maughon, J. W. Perry, H. Röckel, M. Rumi, G. Subbramaniam, W. W. Webb, X. L. Wu, C. Xu, *Science* **1998**, *281*, 1653.
22. Van Stryland E. W., M. Sheik-Bahae, *Opt. Eng.* **1998**, *60*, 655.
23. Sutherland R. L., in *Handbook of Nonlinear Optics*, Marcel Dekker Inc., New York, 2003, p. 583.
24. Xu C., W. W. Webb, *J. Opt. Soc. Am. B* **1996**, *13*, 481.
25. He G. S., L.-S. Tan, Q. Zheng, P. N. Prasad, *Chem. Rev.* **2008**, *108*, 1245.
26. Kwok A. S., A. Serpenguzel, W. F. Hsieh, R. K. Chang, J. B. Gillespie, *Opt. Lett.* **1992**, *62*, 3423.
27. Morel Y., A. Irimia, P. Najechalski, O. Stephan P. L. Baldeck, C. Andraud, *J. Chem. Phys.* **2001**, *114*, 5391.
28. Bouit P.-A., G. Wetzel, G. Berginc, B. Loiseaux, L. Toupet, P. Feneyrou, Y. Bretonnière, K. Kamada, O. Maury, C. Andraud, *Chem. Mater.* **2007**, *19*, 5325.
29. Kawata S., H.-B. Sun, T. Tanaka, K. Takada, *Nature* **2001**, *412*, 607.
30. Wang I. I., M. Bouriau, P. L. Baldeck, C. Martineau, C. Andraud, *Optics Lett.* **2002**, *27*, 1348.
31. Cumpston B. H., S. P. Ananthavel, S. Barlow, D. L. Dyer, J. E. Ehrlich, L. L. Erskine, A. A. Heikal, S. M. Kuebler, I.-Y. S. Lee, D. McCord-Maughon, J. Qin, H. Rockel, M. Rumi, X.-L. Wu, S. R. Marder, J. W. Perry, *Nature* **1999**, *398*, 51.

32. Denk W., J. H. Strickler, W. W. Webb, *Science* **1990**, *248*, 73.
33. Svoboda K., S. M. Block, *Annu. Rev. Biophys. Biomol. Sruct.* **1994**, *23*, 247.
34. Bashkatov A. N., E. A. Genina, V. I. Kochubey, V. V. Tuchin, *J. Phys. D: Appl. Phys.* **2005**, *38*, 2543.
35. Helmchen F., W. Denk, *Nat. Methods* **2005**, *2005*, 932.
36. Maurin M., L. Vurth, J. C. Vial, P. L. Baldeck, S. R. Marder, B. Van der Sanden, O. Stephan, *Nanotechnology* **2009**, *20*, 235102.
37. Xu C., W. Zipfel, J. B. Shear, R. M. Williams, W. W. Webb, *Proc. Natl. Acad. Sci. USA* **1996**, *93*, 10763.
38. Zipfel W. R., R. M. Williams, W. W. Webb, *Nat. Biotech* **2003**, *21*, 1369.
39. Shen Y. R., *The Principles of Nonlinear Optics*, John Wiley & Sons Inc., New York, 1984.
40. Prasad P. N., *Introduction to Biophotonics*, John Wiley & Sons Inc., New York, 2003.
41. Orr B. J., J. F. Ward, *Mol. Phys.* **1971**, *20*, 513.
42. Myung Kim H., B. Rae Cho, *Chem. Commun.* **2009**, 153.
43. Andraud C., R. Fortrie, C. Barsu, O. Stéphan, H. Chermette, P. L. Baldeck Advances in Polymer Science (eds: S. Marder, K.-S. Lee), 2008, 214, 149.
44. Chung S.-J., S. Zheng, T. Odani, L. Beverina, J. Fu, L. A. Padilha, A. Biesso, J. M. Hales, X. Zhan, K. Schmidt, A. Ye, E. Zojer, S. Barlow, D. J. Hagan, E. W. Van Stryland, Y. Yi, Z. Shuai, G. A. Pagani, J.-L. Brédas, J. W. Perry, S. R. Marder, *J. Am. Chem. Soc.* **2006**, *128*, 14444.
45. Blab G. A., P. H. M. Lommerse, L. Cognet, G. S. Harmsa, T. Schmidt, *Chem. Phys. Lett.* **2001**, *350*, 71.
46. Hayek A., A. Grichine, T. Huault, C. Ricard, F. Bolze, B. Van Der Sanden, J.-C. Vial, Y. Mély, A. Duperray, P. L. Baldeck, J.-F. Nicoud, *Microsc. Res. Tech.* **2007**, *70*, 880.
47. Krishna T. R., M. Parent, M. H. V. Werts, L. Moreaux, S. Gmouth, S. Charpak, A.-M. Caminade, J.-P. Majoral, M. Blanchard-Desce, *Angew. Chem. Int. Ed.* **2006**, *45*, 4645.
48. Tian Y., C. Y. Chen, Y. J. Cheng, A. C. Young, N. M. Tucker, A. K. Y. Jen, *Adv. Funct. Mater.* **2007**, *17*, 1691.
49. Barsu C., R. Cheaib, S. Chambert, Y. Queneau, O. Maury, D. Cottet, H. Wege, J. Douady, Y. Bretonnière, C. Andraud, *Org. Biomol. Chem.* **2010**, *8*, 142.
50. Allain C., F. Schmidt, R. Lartia, G. Bordeau, C. Fiorini-Debuisschert, F. Charra, P. Tauc, M. P. Teulade-Fichou, *ChemBioChem* **2007**, *8*, 424.
51. Eliseeva S. V., J.-C. G. Bünzli, Chapter 1 in *Lanthanide spectroscopy, Materials, and Bio-applications, Springer series on fluorescence, 7* (eds: P. Hännen, H. Härmä), Springer Verlag, Berlin, 2010.
52. Huang J., G. K. Liu, R. L. Cone, *Phys. Rev. B* **1989**, *39*, 6348.
53. Strucki J., W. Strek, *Chem. Phys.* **1990**, *143*, 347.
54. Berry A. J., R. G. Denning, I. D. Morrison, *Chem. Phys. Lett.* **1997**, *266*, 195.
55. Duan C.-K., P. A. Tanner, A. Meijerink, V. Makhov, *J. Phys. Chem. A* **2011**, *115*, 9188.
56. Lakowicz J. R., G. Piszczek, B. P. Maliwal, I. Gryczynski, *ChemPhysChem* **2001**, *2*, 247.
57. Shi M., M. Pan, F. Su, F. Wu, P. Han, D. Hao, H. Wang, *J. Lumin.* **2012**, *132*, 398.
58. Auzel F., *Chem. Rev.* **2003**, *104*, 139.
59. White G. F., K. L. Litvinenko, S. R. Meech, D. L. Andrews, A. J. Thomson, *Photochem. Photobiol. Sci.* **2004**, *3*, 47.
60. D'Aléo A., G. Pompidor, B. Elena, J. Vicat, P. L. Baldeck, L. Toupet, R. Kahn, C. Andraud, O. Maury, *ChemPhysChem* **2007**, *8*, 2125.
61. Piszczek G., B. P. Maliwal, I. Gryczynski, J. Dattelbaum, J. R. Lakowicz, *J. Fluo.* **2001**, *11*, 101.
62. Eliseeva S. V., G. Auböck, F. van Mourik, A. Cannizzo, B. Song, E. Deiters, A.-S. Chauvin, M. Chergui, J.-C. G. Bünzli, *J. Phys. Chem. B* **2010**, *114*, 2932.
63. Lai W. P.-W., W.-T. Wong, B. King-Fai Li, K.-W. Cheah, *New J. Chem.* **2002**, *26*, 576.

64. Luo L., W. P.-W. Lai, K.-L. Wong, W.-T. Wong, K.-F. Li, K.-W. Cheah, *Chem. Phys. Lett.* **2004**, *398*, 372.
65. Wong K.-L., W.-M. Kwok, W.-T. Wong, D. L. Phillips, K.-W. Cheah, *Angew. Chem., Int. Ed.* **2004**, *43*, 4659.
66. Wong K.-L., G.-L. Law, W.-M. Kwok, W.-T. Wong, D. L. Phillips, *Angew. Chem., Int. Ed.* **2005**, *44*, 3436.
67. Law G.-L., W.-M. Kwok, W.-T. Wong, K.-L. Wong, P. A. Tanner, *J. Phys. Chem. B* **2007**, *111*, 10858.
68. Ward M. D., *Coord. Chem. Rev.* **2007**, *251*, 1663.
69. Ward M. D., *Coord. Chem. Rev.* **2010**, *254*, 2634.
70. Chen F.-F., Z.-Q. Chen, Z.-Q. Bian, C.-H. Huang, *Coord. Chem. Rev.* **2010**, *254*, 991.
71. Chen Z., H. Xu, in *Rare Earth Coordination Chemistry: Fundamentals and Applications* (ed.: C. Huang), Wiley-VCH, Singapore, 2010, p. 473.
72. Kleinerman M., *J. Chem. Phys.* **1969**, *51*, 2370.
73. D'Aléo A., L. Ouahab, C. Andraud, F. Pointillart, M. O., *Coord. Chem Rev.* **2012**, *256*, 1604.
74. Picot A., F. Malvolti, B. Le Guennic, P. L. Baldeck, J. A. G. Williams, C. Andraud, O. Maury, *Inorg. Chem.* **2007**, *46*, 2659.
75. Hao R., M. Li, Y. Wang, J. Zhang, Y. Ma, L. Fu, X. Wen, Y. Wu, X. Ai, S. Zhang, Y. Wei, *Adv. Funct. Mater.* **2007**, *17*, 3663.
76. Fu L.-M., X.-F. Wen, X.-C. Ai, Y. Sun, Y.-S. Wu, J.-P. Zhang, Y. Wang, *Angew. Chem., Int. Ed.* **2005**, *44*, 747.
77. Lo W.-S., W.-M. Kwok, G.-L. Law, C.-T. Yeung, C. T.-L. Chan, H.-L. Yeung, H.-K. Kong, C.-H. Chen, M. B. Murphy, K.-L. Wong, W.-T. Wong, *Inorg. Chem.* **2011**, *50*, 5309.
78. Dadabhoy A., S. Faulkner, P. G. Sammes, *J. Chem. Soc., Perkin Trans. 2* **2000**, 2359.
79. Bretonnière Y., M. J. Cann, D. Parker, R. Slater, *Org. Bio. Chem.* **2004**, *2*, 1624.
80. Werts M. H. V., M. A. Duin, J. W. Hofstraat, J. W. Verhoeven, *Chem. Commun.* **1999**, 799.
81. Yang C., L.-M. Fu, Y. Wang, J.-P. Zhang, W.-T. Wong, X.-C. Ai, Y.-F. Qiao, B.-S. Zou, L.-L. Gui, *Angew. Chem., Int. Ed.* **2004**, *43*, 5010.
82. D'Aléo A., A. Picot, A. Beeby, J. A. Gareth Williams, B. Le Guennic, C. Andraud, O. Maury, *Inorg. Chem.* **2008**, *47*, 10258.
83. Latva M., H. Takalo, V.-M. Mukkala, C. Matachescu, J. C. Rodríguez-Ubis, J. Kankare, *J. Lumin.* **1997**, *75*, 149.
84. D'Aléo A., A. Picot, P. L. Baldeck, C. Andraud, O. Maury, *Inorg. Chem.* **2008**, *47*, 10269.
85. Werts M. H. V., N. Nerambourg, D. Pélégry, Y. Le Grand, M. Blanchard-Desce, *Photochem. Photobiol. Sci.* **2005**, *4*, 531.
86. Pawlicki M., H. A. Collins, R. G. Denning, H. L. Anderson, *Angew. Chem., Int. Ed.* **2009**, *48*, 3244
87. Kim H. M., B. R. Cho, *Chem. Commun.* **2009**, 153.
88. Law G.-L., K.-L. Wong, C. W.-Y. Man, W.-T. Wong, S.-W. Tsao, M. H.-W. Lam, P. K.-S. Lam, *J. Am. Chem. Soc.* **2008**, *130*, 3714.
89. Kadjane P., M. Starck, F. Camerel, D. Hill, N. Hildebrandt, R. Ziessel, L. J. Charbonnière, *Inorg. Chem.* **2009**, *48*, 4601.
90. He S., H. Li, Y.-W. Yip, C.-T. Yeung, Y. O. Fung, H.-K. Kong, H.-L. Yeung, G.-L. Law, K.-L. Wong, C.-S. Lee, M. H.-W. Lam, M. B. Murphy, P. K.-S. Lam, W.-T. Wong, *Org. Lett.* **2011**, *13*, 5036.
91. Hu Z.-J., X.-H. Tian, X.-H. Zhao, P. Wang, Q. Zhang, P.-P. Sun, J.-Y. Wu, J.-X. Yang, Y.-P. Tian, *Chem. Commun.* **2011**, *47*, 12467.

92. Picot A., A. D'Aléo, P. L. Baldeck, A. Grichine, A. Duperray, C. Andraud, O. Maury, *J. Am. Chem. Soc.* **2008**, *130*, 1532.
93. Palsson L.-O., R. Pal, B. S. Murray, D. Parker, A. Beeby, *Dalton Trans.* **2007**, *2007*, 5726.
94. D'Aléo A., M. Allali, A. Picot, P. L. Baldeck, L. Toupet, C. Andraud, O. Maury, *C. R. Chim.* **2010**, *13*, 681.
95. Bourdolle A., M. Allali, J.-C. Mulatier, B. Le Guennic, J. M. Zwier, P. L. Baldeck, J.-C. G. Bünzli, C. Andraud, L. Lamarque, O. Maury, *Inorg. Chem.* **2011**, *50*, 4987.
96. Zhang T., X. Zhu, C. C. W. Cheng, W.-M. Kwok, H.-L. Tam, J. Hao, D. W. J. Kwong, W.-K. Wong, K.-L. Wong, *J. Am. Chem. Soc.* **2011**, *133*, 20120.
97. Jiang F.-L., W.-K. Wong, X.-J. Zhu, G.-J. Zhou, W.-Y. Wong, P.-L. Wu, H.-L. Tam, K.-W. Cheah, C. Ye, Y. Liu, *Eur. J. Inorg. Chem.* **2007**, *2007*, 3365.
98. Piszczek G., I. Gryczynski, B. P. Maliwal, J. R. Lakowicz, *J. Fluo.* **2002**, *12*, 15.
99. Binnemans K., *Chem. Rev.* **2009**, *109*, 4283.
100. Chamas Z. E. A., X. Guo, J.-L. Canet, A. Gautier, D. Boyer, R. Mahiou, *Dalton Trans.* **2010**, *39*, 7091.
101. Maury O., A. D'Aléo, A. Picot, S. Brasselet, P. L. Baldeck, F. Chaput, S. Parola, C. Andraud, *J. Lumin.* **2013**, *133*, 175.
102. Pompidor G., A. D'Aléo, J. Vicat, L. Toupet, N. Giraud, R. Kahn, O. Maury, *Angew. Chem., Int. Ed.* **2008**, *47*, 3388.
103. Kielar F., A. Congreve, G.-L. Law, E. J. New, D. Parker, K.-L. Wong, P. Castreno, J. de Mendoza, *Chem. Commun.* **2008**, *2008*, 2435.
104. Kielar F., G.-L. Law, E. J. New, D. Parker, *Org. Bio. Chem.* **2008**, *6*, 2256.
105. Xiao X., J. P. Haushalter, K. T. Kotz, G. W. Faris, *Biomed. Opt. Express* **2011**, *2*, 2255.
106. Law G.-L., K.-L. Wong, C. W.-Y. Man, S.-W. Tsao, W.-T. Wong, *J. Biophotonics* **2009**, *2*, 718.
107. Gallavardin T., M. Maurin, S. Marotte, T. Simon, A.-M. Gabudean, Y. Bretonnière, M. Lindgren, F. Lerouge, P. L. Baldeck, O. Stéphan, Y. Leverrier, J. Marvel, S. Parola, O. Maury, C. Andraud, *Photochem. Photobiol. Science* **2011**, *10*, 1216.
108. Screen T. E. O., J. R. G. Thorne, R. G. Denning, D. G. Bucknall, H. L. Anderson, *J. Am. Chem. Soc.* **2002**, *124*, 9712.
109. Rath H., J. Sankar, R. Prabhu, V. T. K. Chandrashekar, A. Nag, D. Goswami, *J. Am. Chem. Soc.* **2005**, *127*, 11608.
110. Kim K. S., J. M. Lim, A. Osuka, D. Kim *J. Photochem. Photobiol. C* **2008**, *9*, 13.
111. Mongin O., T. R. Krishna, M. H. V. Werts, A.-M. Caminade, J.-P. Majoral, M. Blanchard-Desce, *Chem. Commun.* **2006**, *2006*, 915.
112. Mongin O., A. Pla-Quitana, F. Terenziani, D. Drouin, C. Le Droumaguet, A.-M. Caminade, J.-P. Majoral, M. Blanchard-Desce, *New J. Chem.* **2007**, *31*, 1354.
113. Terenziani F., V. Parthasarathy, A. Pla-Quitana, T. Maishal, A.-M. Caminade, J.-P. Majoral, M. Blanchard-Desce, *Angew. Chem. Int. Ed.* **2009**, *2009*, 8691.
114. Drobizhev M., Y. Stepanenko, A. Rebane, C. J. Wilson, T. E. O. Screen, H. L. Anderson, *J. Am. Chem. Soc.* **2006**, *128*, 12432.
115. Wu C., C. Szymanski, Z. Cain, J. McNeill, *J. Am. Chem. Soc.* **2007**, *129*, 12904.
116. Bertazza L., L. Celotti, G. Fabbrini, M. A. Loi, M. Maggini, F. Mancin, S. Marcuz, E. Menna, M. Muccini, U. Tonellato, *Tetrahedron* **2006**, *62*, 10434.
117. Kim S., H. E. Pudavar, P. N. Prasad, *Chem. Commun.* **2006**, *2006*, 2071.
118. Kim S., H. E. Pudavar, A. Bonoiu, P. N. Prasad, *Adv. Mater.* **2007**, *19*, 3791.
119. Kim S., T. Y. Ohulchanskyy, H. E. Pudavar, R. K. Pandey, P. N. Prasad, *J. Am. Chem. Soc.* **2007**, *129*, 2669.
120. Chelebaeva E., L. Raehm, J.-O. Durand, Y. Guari, J. Larionova, C. Guérin, A. Trifonov, M. Willinger, K. Thangavel, A. Lascialfari, O. Mongin, Y. Mir, M. Blanchard-Desce, *J. Mater. Chem.* **2010**, *20*, 1877.

121. Li L., Y. Tian, J. Yang, P. Sun, L. Kong, J. Wu, H. Zhou, S. Zhang, B. Jin, X. Tao, M. Jiang, *Chem. Commun.* **2010**, *46*, 1673.
122. Rendon N., A. Bourdolle, P. L. Baldeck, H. Le Bozec, C. Andraud, S. Brasselet, C. Coperet, O. Maury, *Chem. Mater.* **2011**, *23*, 3228.
123. Wen X., M. Li, Y. Wang, J. Zhang, L. Fu, R. Hao, Y. Ma, X. Ai, *Langmuir* **2008**, *24*, 6932.
124. Shao G., R. Han, Y. Ma, M. Tang, F. Xue, Y. Sha, Y. Wang, *Chem. Eur. J.* **2010**, *16*, 8647.
125. Philippot C., A. Bourdolle, O. Maury, F. Dubois, B. Boury, S. Brustlein, S. Brasselet, C. Andraud, A. Ibanez, *J. Mater. Chem.* **2011**, *21*, 18613.
126. Guo X., J.-L. Canet, D. Boyer, A. Gautier, R. Mahiou, *J. Mater. Chem.* **2012**, *22*, 6117.
127. Frangioni J. V., *Curr. Op. Chem. Biol.* **2003**, *7*, 626.
128. Kiyose K., H. Kojima, T. Nagano, *Chem. Asian J.* **2008**, *3*, 506.
129. Achilefu S., *Angew. Chem. Int. Ed.* **2010**, *49*, 9816.
130. Pansare V. J., S. Hejazi, W. J. Faenza, R. K. Prud'homme, *Chem. Mater.* **2012**, *24*, 812.
131. Bünzli J.-C. G., S. V. Eliseeva, *J. Rare Earths* **2010**, *28*, 824.
132. Soga K., K. Tokuzen, K. Tsuji, T. Yamano, H. Hyodo, H. Kishimoto, *Eur. J. Inorg. Chem.* **2010**, *2010*, 2673.
133. Kamimura M., N. Kanayama, K. Tokuzen, K. Soga, Y. Nagasaki *Nanoscale* **2011**, *3*, 3705.
134. D'Aléo A., A. Bourdolle, S. Bulstein, T. Fauquier, A. Grichine, A. Duperray, P. L. Baldeck, C. Andraud, S. Brasselet, O. Maury, *Angew. Chem. Int. Ed.* **2012**, *51*, 6622.
135. Grichine A., A. Haefele, S. Pascal, A. Duperray. R. Michel, C. Andraud, O. Maury, During the course of the publication of this book chapter, we reported the proof-of-concept of two-photon time-resolved microscopy experiment. *Chem. Science,* **2014**, DOI: 10.1039/C4SC00473F.

# 6

# Lanthanide Ion Complexes as Chemosensors

*Thorfinnur Gunnlaugsson*[1] *and Simon J.A. Pope*[2]

[1]*Trinity College, University of Dublin, Ireland*
[2]*School of Chemistry, Cardiff University, Wales, UK*

## 6.1 Photophysical Properties of $Ln^{III}$ Based Sensors

### 6.1.1 Emission Based Sensors

The luminescence from emissive $Ln^{III}$ originates from inner shell 4*f*-4*f* transitions, which are observably sharp in appearance and characteristic of the specific ion: $Ln^{III}$ ions can possess emission bands that usefully address the UV, visible or NIR regions. The associated Stokes' shifts can be large, but are a function of the particular pathway for populating the *f*-centred excited state. Since the 4*f*-4*f* transitions are symmetry (parity) forbidden they possess very low molar absorptivities for direct excitation, so the established strategy for overcoming this is to incorporate a sensitising chromophore (commonly known as an antenna group), which absorbs light and transfers energy to the 4*f* excited state (the origin of the apparent Stokes' shift) via a mechanism that often, but not exclusively, involves the triplet excited state of the sensitiser (Fig. 6.1). The antenna group can be covalently attached to the ligand architecture or introduced via other means (as a co-ligand, for example). This general feature of emissive $Ln^{III}$ complexes provides rich opportunities for the design of sensor type molecules based on such species. The intensity and/or integrated emission of the complex can be usefully exploited since the choice of lanthanide can place the luminescence profile beyond that associated, for example, with background and/or endogenous fluorescence of biological samples [1–4].

*Luminescence of Lanthanide Ions in Coordination Compounds and Nanomaterials*, First Edition.
Edited by Ana de Bettencourt-Dias.

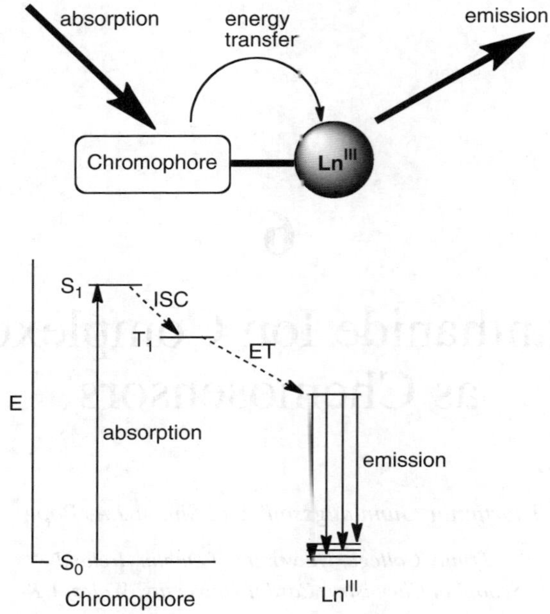


***Figure 6.1*** *Top: simplified model for sensitised lanthanide luminescence. Bottom: typical energy level diagram for an emissive chromophore-appended lanthanide complex sensitised via a ligand-centred triplet excited state*

As described earlier, the lifetimes and quantum yields of emissive $Ln^{III}$ complexes vary dramatically due to the extremely sensitive nature of the 4*f*-centred excited states to O-H, N-H and C-H vibrational manifolds, which can provide efficient, non-radiative deactivation pathways; the efficiency of energy transfer between the antenna and lanthanide ion also determines overall quantum yields. A classical approach to maximising the emissivity of $Ln^{III}$ complexes is to therefore inhibit the approach of water solvent to the inner coordination sphere (and where '*q*' denotes the number of coordinated solvent molecules); high denticity, metal ion encapsulating ligands with hydrophobic peripheries can achieve this very effectively, reducing *q* to zero [4].

### 6.1.2 Luminescence Lifetime

A probe with a long lifetime [typically regarded as luminescence lifetime ($\tau$) > 100 ns] is preferable, in fact arguably essential, in a biological context, since time-gating can effectively eliminate background signals of endogenous species (biological fluorescence is typically $\tau < 25$ ns) and thus enhances the sensitivity of a measurement and simplifies assignment of the signal [5–7]. More specifically, $\tau$ is an invaluable parameter for the consideration of emission-based sensors, since more detailed information can be obtained regarding the local concentration of the probe and thus the targeted analyte (via $K_{assoc}$); this is particularly valuable in a biological context, for example, when considering intracellular concentrations of metabolites and/or ions. Therefore changes in $Ln^{III}$ emission lifetime can be exploited to report on the local environment and give local information, unlike intensity-based measurements of standard fluorophores [8]. Binding events, changes in hydration/solvation and changes in ligand coordination mode to $Ln^{III}$ can each fundamentally affect the emission lifetime of the complex (inducing changes of hundreds of μs in some cases). Therefore one of the practical challenges in the design of $Ln^{III}$ lifetime-based probes is to ensure that complex stability is not compromised whilst exploiting lifetime as a measurable output. Such an approach is a valuable consideration in the design of responsive probes based on $Ln^{III}$ complexes.

### 6.1.3 Spectral Form, Hypersensitivity and Ratiometric Peaks

Generally, the room temperature emission spectra of $Ln^{III}$ species show incompletely resolved structure within the peaks. However, an advantageous attribute of luminescent $Ln^{III}$ complexes is the dependence of this emission spectral form on the specific coordination environment of the ion. This sensitivity arises from the selection rules associated with intraconfigurational (4*f*-4*f*) electronic transitions; the selection rules for forced electric dipole transitions are relaxed due to 5*d* and 4*f* orbital mixing. In reality the majority of the complexes included for discussion here are non-centrosymmetric, low symmetry species and the relative intensities of the 4*f*-4*f* transitions are generally determined by the induced electric dipole transition selection rules. It should also be noted that visibly emissive $Eu^{III}$ also possesses a magnetic dipole transition, $^5D_0 \rightarrow {}^7F_1$, whose intensity is relatively independent of the coordination environment [1,9].

For $Eu^{III}$, with sharp multiple-line emission bands $^5D_0 \rightarrow {}^7F_J$ ($J=0, 1, 2, 3, 4$), the majority ($J=0, 2, 3, 4$) are subtly sensitive to the nature and type of donor, and the coordination geometry at $Eu^{III}$. For example, the $^5D_0 \rightarrow {}^7F_2$, which is often referred to as the hypersensitive transition of $Eu^{III}$, is allowed as a forced electric dipole transition, whereas $^5D_0 \rightarrow {}^7F_J$ where $J=0$, 3 and 5 are not, and are therefore generally weaker. Ultimately, consideration of these details allow $Eu^{III}$ to be effectively utilised as a very sensitive luminescent probe, since binding events, changes in hydration and changes in ligand coordination mode to $Eu^{III}$ can each be reported through analysis of the spectral form (i.e. relative intensities) of the $^5D_0 \rightarrow {}^7F_J$ ($J=0, 1, 2, 3, 4$) transitions [1,9]. In practical terms, of the other $Ln^{III}$ ions that are useful from in this context, $Tb^{III}$ and $Nd^{III}$ have shown potential regarding the sensitivity of the spectral form to the local coordination environment (Fig. 6.2).

## 6.2 Sensor Design Principles

The general strategies for obtaining sensitised $Ln^{III}$ luminescence afford varied opportunities for the modulation of emission (i.e. a measurable change in output in response to an input) from such species. This has driven an imaginative diversity regarding sensor design based on luminescent lanthanide complexes [1,7,10–11]. However, some general design concepts hold true for the vast majority of notable examples and these can be encompassed in the following key areas.

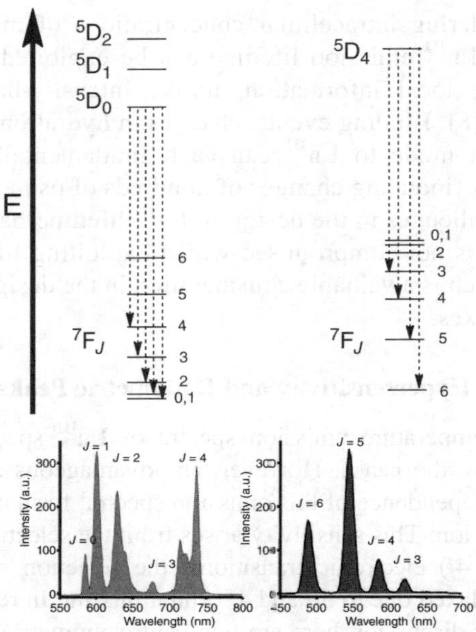


***Figure 6.2*** *Energy levels and main emissive transitions of Eu*$^{III}$ *and Tb*$^{III}$

### 6.2.1 The Design of Ln-receptor Sites and Antenna Components

The ligand requirements for binding Ln$^{III}$ ions are often dictated by the required application, and contextually the utility of Ln$^{III}$ complexes as sensors usually demands complex stability in solution (often aqueous). Biological compatibility [7,12] and robustness *in vitro* and *in vivo* are obvious requirements when considering the utility of complexes to report in real time in an imaging context (e.g. cellular studies using confocal fluorescence microscopy). As such the most popular binding sites for Ln$^{III}$ are polydentate ligands incorporating hard donors (e.g. polyaminocarboxylates), either in macrocylic form (e.g. DOTA type ligands based on cyclen) or acyclic form (e.g. DTPA type ligands). Variations on these ligand themes can include the incorporation of hetercaromatic moieties (e.g. 2,2′-bipyridine, 2,2′;6′,2″-terpyridine) into the ligand frameworks (Scheme 6.1), imparting added rigidity and acting dually as sensitisers for luminescence.

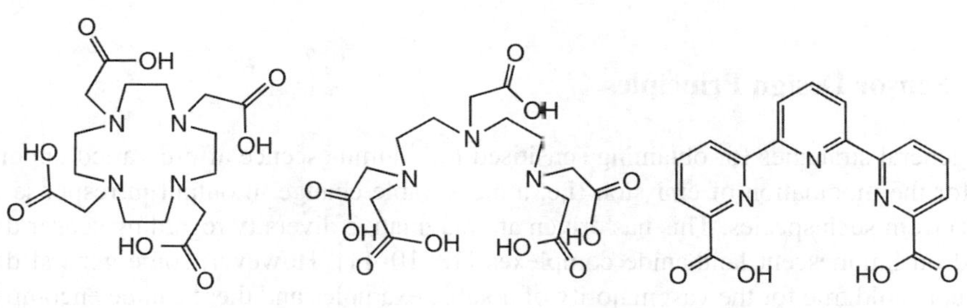

***Scheme 6.1*** *Common ligand types for Ln*$^{III}$*. Left to right: DOTA, DTPA, substituted terpy*

Assuming that the inclusion of an antenna unit is a pre-requisite for the design of a $Ln^{III}$-based luminescent sensor, there are a number of approaches to consider since the chromophore must be incorporated into the complex structure. The distance (e.g. a spacer unit or bridge) between the chromophore and the $Ln^{III}$ should be considered (with antenna coordination to the $Ln^{III}$ providing the minimum possible distance), since energy transfer between the photoactive components is an important parameter for efficient emission. The chromophore can also possess additional functionality that allows reportage of the local environment, such as: (i) peripheral binding sites (binding events may even induce structural changes), (ii) pH response, and (iii) tunable redox character and its implications for (photoinduced) electron transfer mechanisms. Many of these perturbations could be expected to alter the optical and photophysical characteristics of the antenna, which in turn (with reference to the simplified Jablonksi diagram; Fig. 6.1) can have profound implications for the photophysical parameters that describe the luminescence output from $Ln^{III}$.

### 6.2.2 Covalent versus Self-assembled Ln-receptor Design

Further to the comments above regarding incorporation of the antenna unit, two methods allow the formation of $Ln^{III}$-antenna species (Scheme 6.2). The preformation of covalently linked antenna–ligand arrays is commonly achieved via alkylation of a ligand core ($Ln^{III}$ is subsequently added to the structurally complete ligand).

An alternative approach relies upon self-assembly of the antenna with a coordinatively unsaturated $Ln^{III}$ complex yielding a new *ternary* species [13–15]. For example, the affinity of $Ln^{III}$ for various anions is dictated by electrostatics and the sterics at the metal centre, in turn governed by the polydentate ligand, especially the arm substituents of the ligand core (i.e. aromatic/heterocyclic sensitiser). Anion binding to the metal ion typically occurs through a reversible intermolecular process (Scheme 6.3), inducing reversible displacement of coordinated water (solvent) molecules from the $Ln^{III}$ centre. Therefore, for a heptadentate ligand, the resultant $Ln^{III}$ complex would typically expect to have 1–2 coordinated water molecules; coordination of mono- or bidentate anions will likely liberate one or both of these water molecules resulting in measurable changes in luminescent output (e.g. intensity and lifetime). An understanding of the anion binding affinities and of the resultant perturbation of the $Ln^{III}$ luminescence are critically

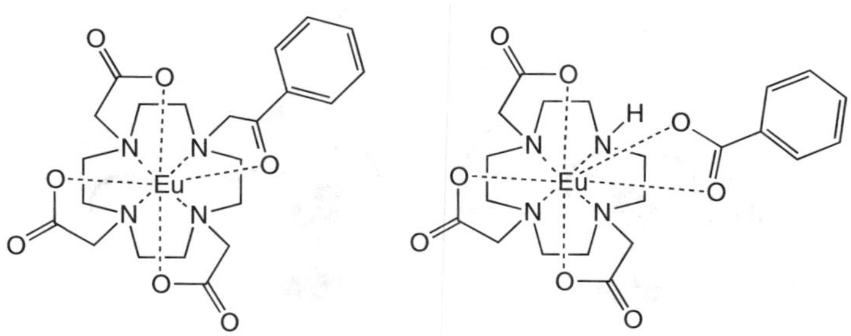


***Scheme 6.2*** *Comparison of covalently linked (left)* versus *self-assembled (right) aromatic chromophores*

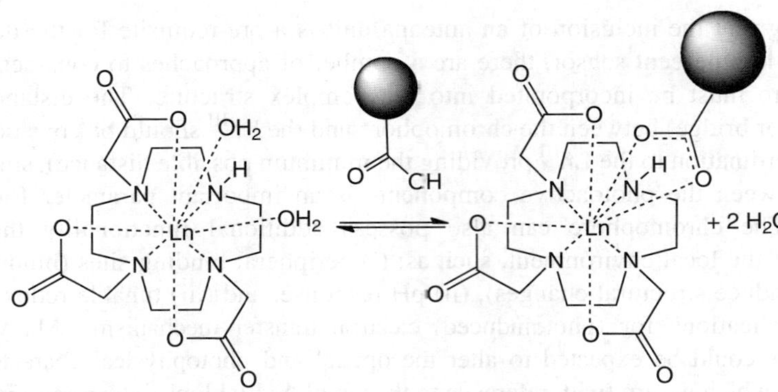


*Scheme 6.3* An example of reversible anion binding that induces a change in $Ln^{III}$ hydration

important in a biological context, since various anionic residues are available for binding *in vivo* (see later discussion).

Typically an antenna can bind to the $Ln^{III}$ via anionic sites such as carboxylates and phosphinates; it is important to note that such a process is reversible and the kinetic robustness of the assembly will be determined by the local environment and competing binding events with endogenous ions/species. However, this apparent limitation can be used advantageously if, for example, the target analyte is chromophoric (i.e. the binding event 'switches on' $Ln^{III}$ emission) or a non-chromophoric residue displaces an antenna unit (i.e. the binding event 'switches off' $Ln^{III}$ emission; Fig. 6.3).

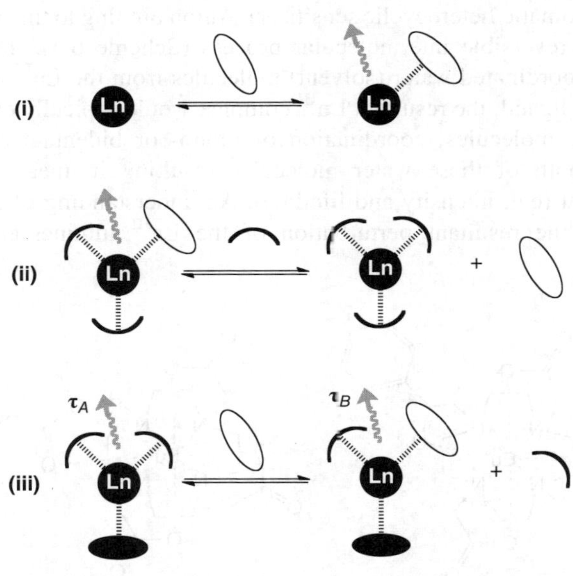


*Figure 6.3* Cartoon demonstrating: (i) binding of a chromophoric analyte; (ii) displacement of an antenna with a non-chromophoric analyte; (iii) binding of a chromophoric analyte to an emissive complex

Of course, other scenarios are possible whereby a coordinatively unsaturated $Ln^{III}$ complex already incorporates a covalently attached antenna. In such cases, binding events (displacement of coordinated water/solvent molecules) at the $Ln^{III}$ ion are reported more subtly via changes in lifetime, quantum yield and observable changes in the form of the structured emission profiles, due to the hypersensitivity of certain *f-f* transitions (e.g. $Eu^{III}$). Finally, it should be noted that reports of displacement assays based on lanthanide complexes are often conceptually directly related to the points above regarding reversible binding events at $Ln^{III}$: the principle of a $Ln^{III}$-based displacement assay owes much to the discussion of self-assembled arrays.

### 6.2.3 Sensors for Cations

#### *6.2.3.1 Groups 1 and 2; Alkali Metal Ions*

The motivation for detecting the mono- and divalent metal ions of Groups 1 and 2, respectively, resides in their biological, physiological or toxicological importance. The popular cationic target analytes are $Na^{I}$, $K^{I}$, $Ca^{II}$ and $Mg^{II}$. In many ways the ligand preferences/requirements for these ions actually share many commonalities with those designed with $Ln^{III}$ in mind. That is, ligands for relatively hard electropositive ions, with minimal ligand field preferences, high coordination numbers and coordinate bonding best described as predominantly electrostatic in nature.

The strategies for detecting such ions using emissive $Ln^{III}$ complexes usually requires incorporation of a secondary binding site, often based upon mixed aza crown ethers or aminocarboxylate ligands. Further, a common design approach involves conjugating the secondary binding site to the antenna unit (i.e. the antenna acts as a bridging moiety between both coordination spheres). In this way binding events can be transduced through the ligand architecture, potentially perturbing the $Ln^{III}$-centred emission. This design strategy is also important if one considers the mechanistic aspects of PeT (photoinduced electron transfer, Fig. 6.4) as a means of modulating luminescence output from $Ln^{III}$.

Some cyclen-based structures (Scheme 6.4) have been designed for the detection of Gp 1 metal ions [16,17], using aza crown units (ane$N_2O_3$ type [15] or ane$N_2O_4$ type [18]) as the binding sites. The $Tb^{III}$ complexes were most effective, showing modulation of emission in aqueous conditions (pH 7.4) upon addition of either $Na^+$ or $K^+$. The size of the aza crown also allowed some discrimination against other ions, such as $Li^+$. Since the free $Tb^{III}$ complex was inherently weakly emissive, with corresponding short lifetimes, the mode of signal response was defined as 'switching on'; this was attributed to the changes in the electronic nature of the aniline-type chromophore upon binding of metal ions.

A similar approach was reported by de Silva et al. [18] with mono-aza crown ethers attached to the peripheral sites of a functionalised 6,6-dicarboxy-terpyridine type ligand (Fig. 6.5); the $Ln^{III}$ is coordinated in a $N_3O_2$ binding site together with solvent molecules. The addition of Gp1 ions resulted in a dramatic enhancement of $Eu^{III}$ emission, attributed primarily to the suppression of PeT resulting from binding of the cation at the aza crown ether site. The analogous $Tb^{III}$ complex highlighted a cautionary note associated with low-lying, oxygen sensitive, triplet states, since these can provide an additional quenching pathway via a back-energy transfer mechanism from the $Tb^{III}$ to the ligand triplet state.

In a variation on this approach, Wong [19] reported a mono-aza crown ether covalently appended via alkyle linkers to triacetic acid derivatives of cyclen and cyclam, giving

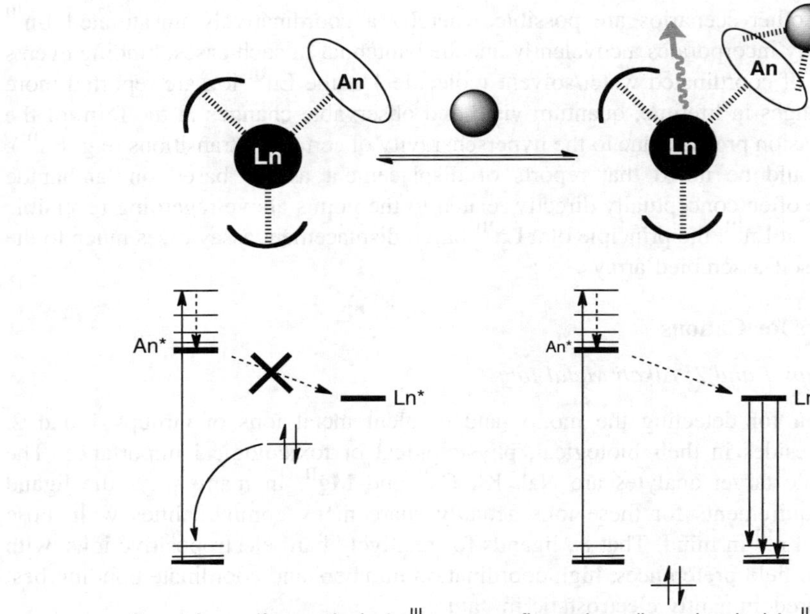


***Figure 6.4*** *Cartoon representing the influence of analyte binding on photoinduced electron transfer (PeT)*

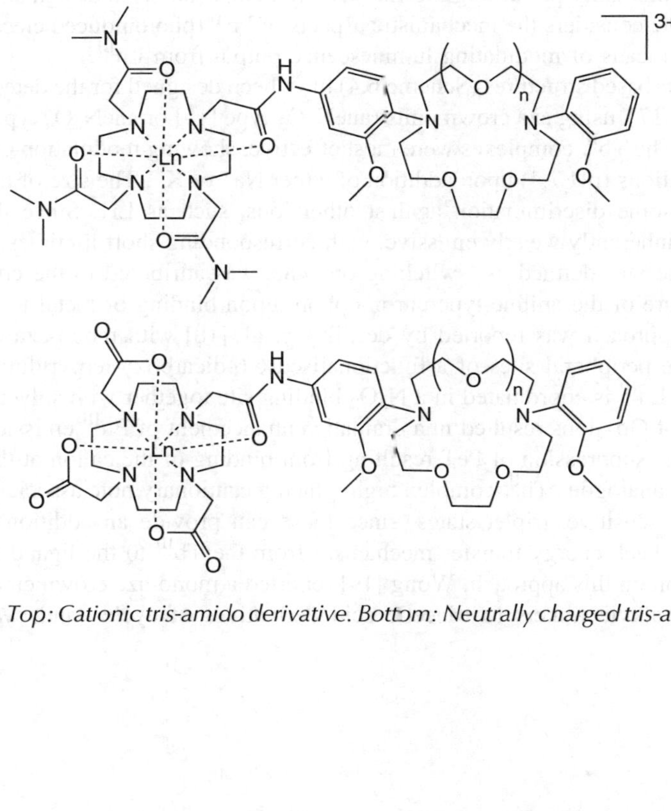


***Scheme 6.4*** *Top: Cationic tris-amido derivative. Bottom: Neutrally charged tris-acetate complex*

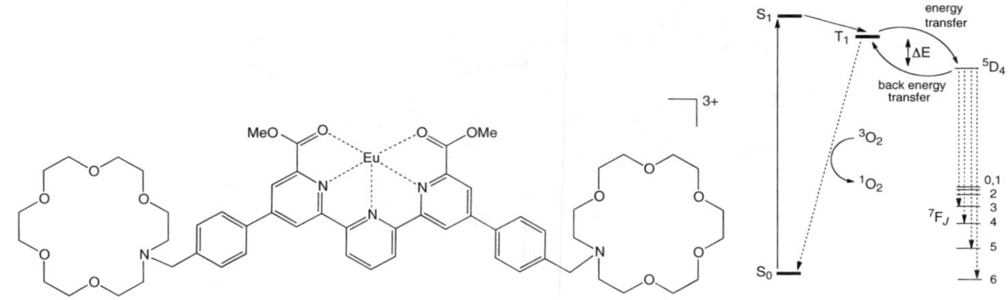


***Figure 6.5*** *Left: de Silva's complex. Right: simplified energy level diagram representing back energy transfer.*

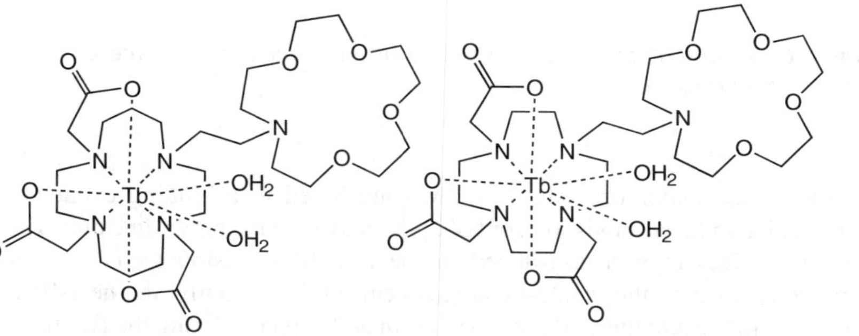


***Scheme 6.5*** *Tb-based cyclam (left) and cyclen (right) derived sensors for cationic alkali metal ions*

7-coordinate $Ln^{III}$ complexes (Scheme 6.5). Such species are not chromophorically enriched and therefore required the addition of an auxilliary antenna, self-assembled in solution (in a fashion akin to Scheme 6.3), using aryl carboxylates such as *p*-chlorobenzoate or salicylate (and confirmed using changes in the lifetime values). Whilst the co-addition of the antenna induced sensitised emission from the $Tb^{III}$ complex, a further enhancement of luminescence intensity was noted upon addition of $Li^+$, $Na^+$ and $K^+$, which coordinate to the mono-aza crown site. It should be noted that in these examples the addition of alkali metal ions does not lead to measurable displacement of the $Ln^{III}$ ion.

A different kind of ditopic ligand has been developed for the detection of $Ba^{2+}$ ions, wherein $Nd^{III}$ is held in a cryptate-like binding site and the analyte binding motif is a diaza-type crown ether (Scheme 6.6). A chromophoric 2,2′-bipyridine unit completes the coordination sphere of the $Ln^{III}$, but also acts as a bridging ligand with a crown-6 macrocycle tethered to the 6,6′-positions of the bipyridine antenna: the study shows that the bipyridine unit effectively sensitises the $Nd^{III}$, resulting in near-IR emission. The addition of alkali metal ions was first investigated using UV-vis. absorption studies, which showed that the $\pi \rightarrow \pi^*$ transition was most sensitive to $Ba^{2+}$, revealing a significant

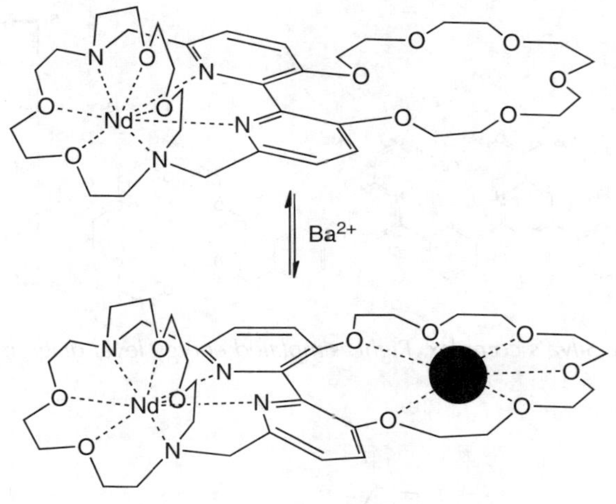


*Scheme 6.6 Representation of the influence of $Ba^{2+}$ binding upon the co-planarity of the bipyridine chromophore*

blue shift in $\lambda_{abs}$ maximum. This shift was attributed to a binding event in the crown receptor, which induced a twist (i.e. out of co-planarity) in the bipyridine unit. The change in absorption coefficient was transduced to the near-IR emission signal by choosing the appropriate $\lambda_{exc}$ where the greatest change occurred. Importantly, the near-IR lifetimes of $Nd^{III}$ showed that $q$ remains at 0 (i.e. no coordinated solvent) during the addition of analyte, suggesting that the coordination sphere does not vary dramatically [20].

### 6.2.3.2 Sensors for d-Block Metal Ions

The design of luminescent probes for d-block ions is again driven by an interest in the biological and toxicological aspects of such ions, although in many cases the sensitivity of the probes is still somewhat below that necessary for the detection of physiological concentrations of such ions. Nonetheless, the design of any luminescent probe targeting d-block ions requires an intimate appreciation of the coordination geometry and ligand requirements of the metal ion target in question. In this way, binding sites can be very specifically targeted and potentially allow excellent selectivity over other competitive cationic analytes.

For convenience this discussion will also include $Zn^{II}$, $Cd^{II}$ and $Hg^{II}$ as analytes of interest, although these are sometimes referred to as post-transition metals. In fact, due to its increasing importance in a biological and physiological context, $Zn^{II}$ has probably attracted the most attention as a target. $Zn^{II}$ can commonly adopt a variety of well-defined coordination geometries, with coordination numbers of 4–6, and has a strong preference for heteroaromatic N-donors and amino carboxylates; for example, tris-picolylamine (TPA) is a very popular motif for effectively targeting $Zn^{II}$ [21]. Again, the binding sites for such species are usually appended to the ligand-$Ln^{III}$ core.

Acyclic derivatives based upon the DTPA ligand core have been reported by Nagano and utilised for targeting and imaging intracellular zinc [22]. The design rationale for such a

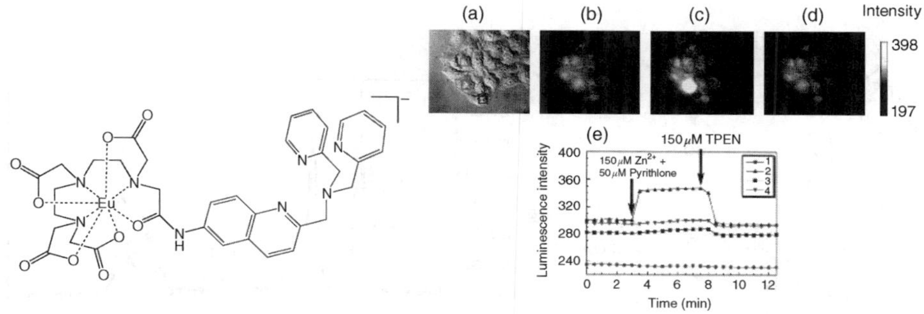


***Figure 6.6*** *$Eu^{III}$ complex structure (left) of the zinc-responsive probe showing the zinc binding site in red. Bright-field transmission and luminescence images of $Zn^{2+}$ in HeLa cells loaded with Eu complex in HBSS buffer. The luminescence at 580–654 nm, excited at 320–400 nm, was measured at 30 s intervals. (a) Bright-field transmission image (0 min). (b) Luminescence image of (a) (0 min). (c) Luminescence image (5 min) following an addition of 50 μM pyrithione (zinc ionophore) and 150 μM $ZnSO_4$ to the medium at 3 min. (d) Luminescence image (10 min) following an addition of 150 μM TPEN to the medium at 7.5 min. Luminescence images (b–d) correspond to the luminescence intensity data in (e), which shows the average intensity of the corresponding area or cell area (1, extracellular region; 2, intracellular region of the injected cell; 3, 4, intracellular regions of non-injected cells). Reproduced with permission from [22]. Copyright © 2004, American Chemical Society (See colour plate section)*

species requires a dual functioning binding site with good affinity for $Zn^{II}$, and an ability to sensitise the chelated $Eu^{III}$. The DTPA core was appended with a bridging quinoline-type chromophore, which was further functionalised with a dipicolylamine unit (Fig. 6.6). The quinoline unit provides a slightly longer wavelength of absorption in comparison to pyridine, facilitating excitation at 320 nm. The emissivity associated with the free anionic $Eu^{III}$ complex was quite poor (<1%), but binding $Zn^{II}$ induced a favourable 8.2-fold increase in quantum yield. The use of luminescence lifetime measurements, and water relaxivity measurements on the corresponding $Gd^{III}$ analogue, showed that the lanthanide coordination sphere was not perturbed by the $Zn^{II}$-binding event.

Parker also reported some of the earliest examples of zinc-responsive probes based upon the functionalised mono-amide DO3A core (Fig. 6.7) [23]. Under physiologically simulated conditions the addition of $Zn^{II}$ resulted in moderate enhancements (26–42%) in emission intensity from $Tb^{III}$ ($\lambda_{ex} = 262$ nm). Again, the suppression of PeT from the aniline moiety was thought to be a contributory factor and the binding was in a 1:1 stoichiometry. Although $Mg^{II}$ and $Ca^{II}$ also induced increases in $Ln^{III}$ emission, these required much higher concentrations, and therefore these probes can claim genuine selectivity for $Zn^{II}$ over other ions.

In a subtle variation on this ligand system, Gunnlaugsson reported [24] a $Tb^{III}$ complex incorporating a phenyl iminoacetate-based receptor that demonstrated luminescent responses to $Cu^{II}$ and $Hg^{II}$ (no effect was observed with $Cd^{II}$, $Zn^{II}$, $Ca^{II}$ or $Mg^{II}$). Again, the use of relatively high excitation energies ($\lambda_{ex} = 285$ nm) was required to sensitise emission from $Tb^{III}$. Addition of either $Cu^{II}$ or $Hg^{II}$ under aqueous conditions quenched the $Tb^{III}$ emission by 65% and 40%, respectively. In competitive experiments the higher affinity of the probe for $Cu^{II}$ allowed discrimination in a mixture with $Hg^{II}$.

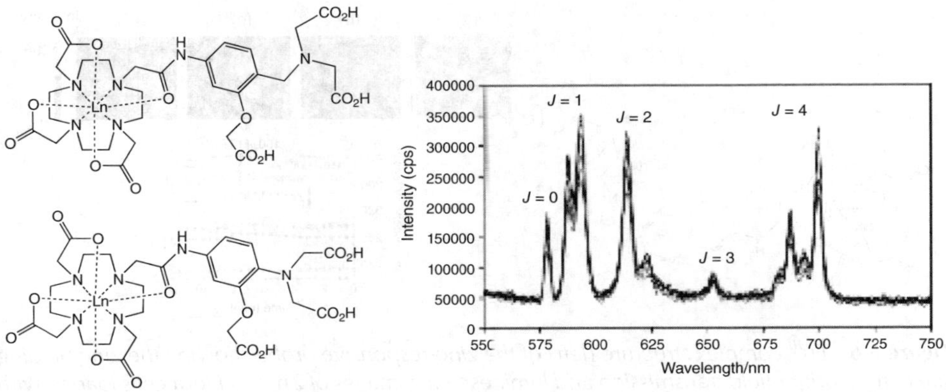


***Figure 6.7*** *Left: Parker's sensors utilising functionalised aniline and benzylamine chromophores. Right: The effect of adding $ZnCl_2$ to the $Eu^{III}$ complex of the benzylamine derivative*

The quenching effect of $Cu^{II}$ upon $Eu^{III}$ luminescence had been noted in a cyclen-based ligand [25] with a pendant 1,10-phenanthroline moiety, which has good affinity for $Cu^{II}$. The probe was responsive to $Cu^{II}$ in aqueous solution at pH 7.4, with an ionic background of various alkali metal ions ($\lambda_{ex} = 266$ nm). The stoichiometry of binding was 1:3, yielding a tetranuclear 'Eu-$Cu_3$' assembly; binding was found to be reversible. In these systems it appeared that the normal pathway for sensitising the $Eu^{III}$ was blocked by the binding event at the phenanthroline ligand; the excited singlet state of the antenna undergoes electron transfer to the $Cu^{II}$. The phenanthroline-derived probe also showed emissive responses to $Co^{II}$ and $Fe^{II}$, both of which demonstrated varying degrees of quenching, providing additional low-lying MLCT and ligand field states as effective non-radiative pathways [26]. In related work [27] a pendant terpyridine-type ligand was used dually as an antenna on a cyclen framework, and a binding site for $Fe^{II}$ (Scheme 6.7). Thus, upon sensing $Fe^{II}$, a trimetallic (i.e. 1:2, Eu–$Fe_2$) assembly was produced. Binding $Fe^{II}$ led to a quenching of the $Eu^{III}$ emission, again attributed to the formation of $M_{Fe}L_{tpy}$CT states that are able to quench the $^5D_0$ excited state of $Eu^{III}$. In all of these examples the first coordination sphere of the $Eu^{III}$ is essentially unaltered by the binding event and therefore the perturbation of the $Eu^{III}$ luminescence is largely associated with changes in intensity, rather than the structural form of the spectra (see Section 3.2.2 in Chapter 3).

A small handful of examples has been reported that does not utilise the popular cyclen framework. Highly emissive europium complexes of bidentate phosphine oxide ligands [28] containing a 2,2′-bipyridine framework have also been reported to show luminescent responses to both $Zn^{II}$ and $Cu^{II}$, albeit in acetone solution: both ions quench the emission from $Eu^{III}$. Terbium complexes [29] of a dual chelating BBATA-type ligand (Fig. 6.8) have shown a luminescence enhancement following addition of $Zn^{II}$: the utility of these systems has also been effectively demonstrated in an applied biological context, with $Zn^{II}$ imaged in living cells.

An attractive photophysical attribute of the emissive trivalent lanthanide ion is the large dependency of emission lifetime upon variation in the hydration/solvation of the ion in

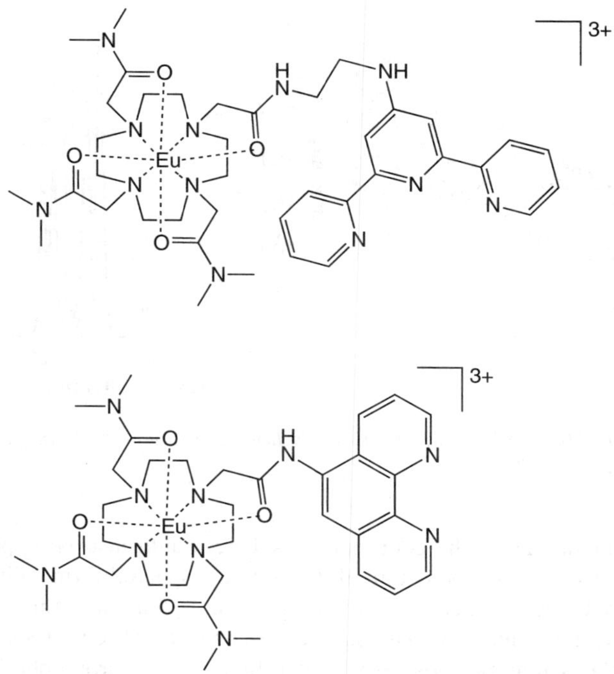

**Scheme 6.7** *Polypyridine receptors for d-block metal ions*

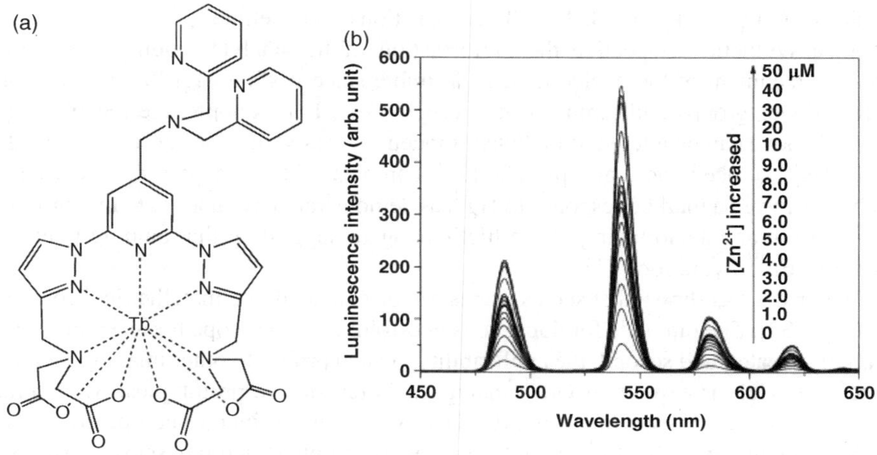


**Figure 6.8** *Left: BBATA-type ligand adorned with a $Zn^{2+}$ receptor site. Right: influence of $Zn^{2+}$ upon Tb(III) emission. Reproduced from [29], with permission from Elsevier*

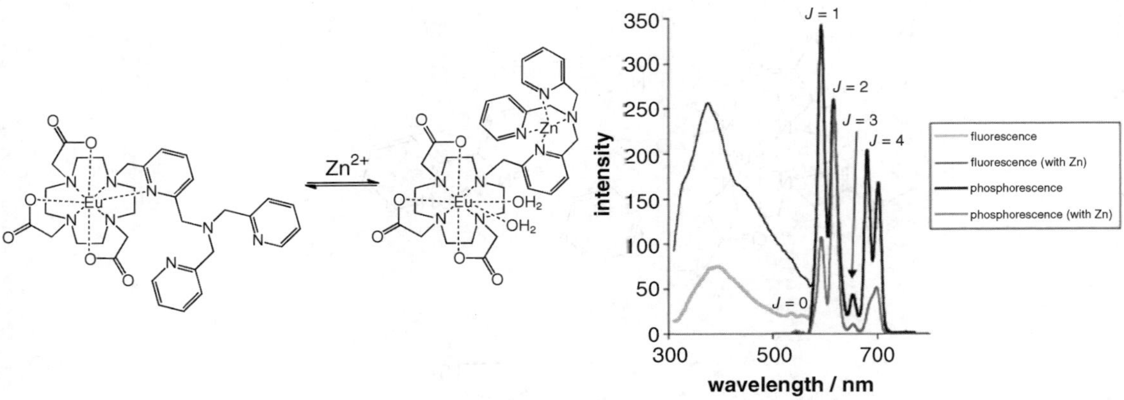


***Figure 6.9*** *The modulation of Eu$^{III}$ hydration as a function of analyte binding*

question. Such alterations to the coordination sphere would also be expected to induce measurable changes in the fine structure of the emission spectral form. With this in mind a probe was reported that specifically aimed to modulate hydration at Ln$^{III}$, whilst retaining complex stability and emissivity. The structure of the complex is based upon a mono-functionalised DO3A unit and possesses a dual-functioning chromophoric TPA unit for binding Zn$^{II}$ [30]. The study revealed that binding Zn$^{II}$ increased the intensity of ligand fluorescence and induced pronounced changes in the form of the emission spectra ($\lambda_{ex} = 280$ nm), with notable variation in the ratio of the $J = 1/J = 2$ transitions (Fig. 6.9). The lifetime of Eu$^{III}$ emission decreased from 1.0 to 0.4 ms in water, corresponding to an increased inner sphere hydration number of Eu$^{III}$ ($q$ increases from 0 to 2), presumably by opening up the coordination sphere of Eu$^{III}$, but without liberation of Eu$^{III}$. Therefore, the presence of Zn$^{II}$ was signalled by observable changes in the emission intensities of $^5D_0 \rightarrow {}^7F_J$ ($J = 0$, 1, 2, 3, 4) transitions and lifetime.

From a synthetic perspective the preformation of the analyte binding site, prior to covalent attachment of the cyclen unit, is desirable since it allows for the introduction of different donor groups and tuning of the receptor site. For example, the receptor design discussed above can be altered to include thiophene (for targeting 'softer' cationic analytes such as Hg$^{II}$) or acetic acid groups. The Eu$^{III}$ complex with a receptor site incorporating thiophene showed a marked response to Hg$^{II}$, as demonstrated by changes to the form of the Eu$^{III}$ spectrum and a shortening of the lifetime, again suggesting that binding induced an increase in Eu$^{III}$ hydration [31].

Related work has shown that such systems can be extended to bimetallic Eu$^{III}$ complexes (Fig. 6.10) [32]. The rationale for doing so is two-fold: first, the scope for receptor design is potentially greater, and second, the applicability of such probes to other imaging modalities is extended, such as exploiting Gd$^{III}$ analogues as responsive prototypical MRI contrast agents. In this context, two mono-pyridyl DO3A units were linked via a piperazine unit, with the resultant dimetallic Eu$^{III}$ complexes showing highly structured, sensitised emission ($\lambda_{ex} = 277$ nm). The addition of d-block metal ions in buffered water was investigated and revealed that Fe$^{III}$, Zn$^{II}$ and Cd$^{II}$ produced no effect, whilst Cu$^{II}$ quenches emission with no

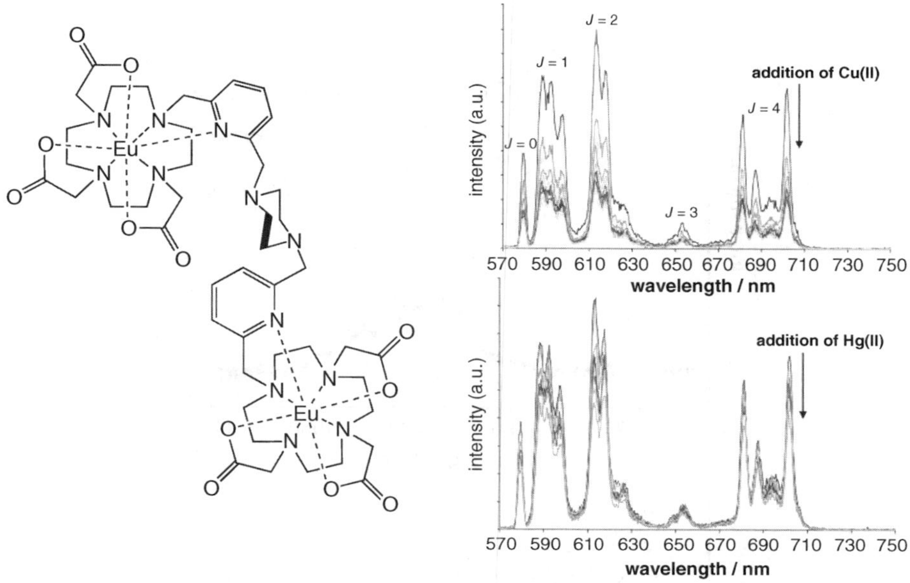


***Figure 6.10*** *A dimetallic $Eu^{III}$ complex that responds to $Cu^{II}$ and $Hg^{II}$. (See colour plate section)*

change in the ratios of the peak intensities. In contrast, $Hg^{II}$ produced a quenching effect together with alteration of the ratios of $^5D_0 \rightarrow {}^7F_J$ $J = 1/J = 2$, as well as notable changes in the fine structure of the $^5D_0 \rightarrow {}^7F_J$ transitions. The lifetimes of the $Eu^{III}$ emission were shortened, consistent with an increase in $Eu^{III}$ hydration upon $Hg^{II}$ addition. The companion experiments assessing the relaxivity behaviour of the $Gd^{III}$ analogues support the notion that only the addition of $Hg^{II}$ increases the hydration of the $Ln^{III}$. In fact it is noteworthy that for those researchers focused on the design of responsive agents based upon the modulated water relaxivities of $Gd^{III}$ complexes, additional confirmation regarding changes in $Gd^{III}$ hydration are often sought via companion luminescence measurements on the $Eu^{III}$ analogues. In this context the reader is directed to significant works representative of a broad discipline [33–35].

From a biological perspective the development of responsive $Ln^{III}$ probes for use in optical imaging applications inevitably requires the use of sensitising wavelengths that are bio-compatible, that is systems which do not require UV-excited chromophores. Although a number of $Eu^{III}$ and a handful of $Tb^{III}$ complexes utilising antennas such as tetraazatriphenylene, azaxanthone, azathiaxanthone, pyrazolyl-azaxanthone, and acridone have shown great utility in cellular imaging through confocal fluorescence microscopy studies [36,37], there is also a great interest in the use of near-IR emitting species for applied studies. However, to date very few examples exist of $Ln^{III}$-based responsive probes which exploit the potential of modulated near-IR emission.

Therefore, the incorporation of a long wavelength sensitising antenna into an analyte receptor site enables visible light sensitisation of near-IR emitting lanthanide ions such as $Nd^{III}$ and $Yb^{III}$, and the potential for single molecule bimodal luminescent/MR agents [38].

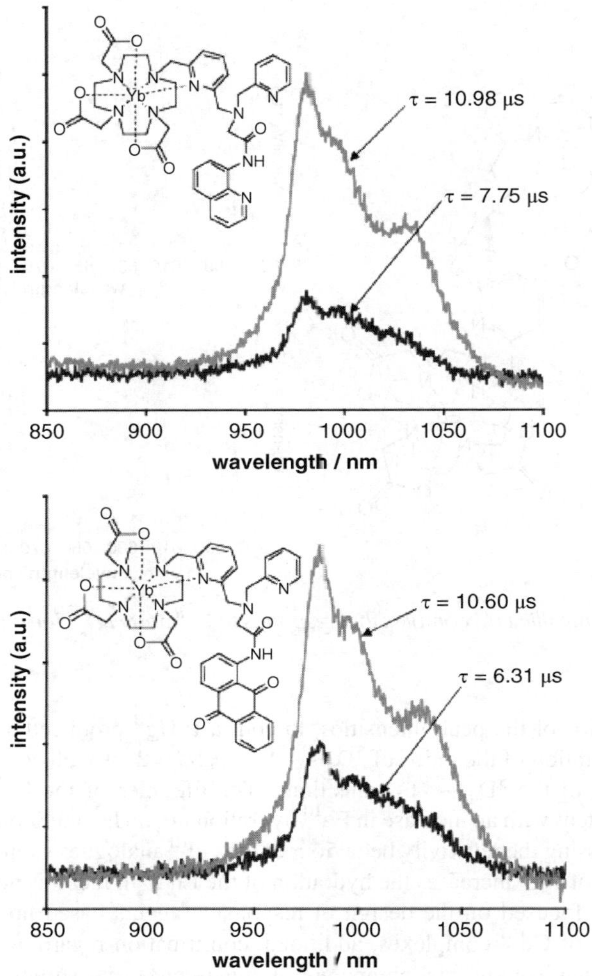


***Figure 6.11*** *Modulation of $Yb^{III}$ near-IR emission after addition of $Hg^{II}$ (red traces). (See colour plate section)*

For example, 1-amidoanthraquinone and 8-amidoquinoline derivatives (Fig. 6.11) each possess excitation wavelengths of 400–430 nm and promote emission from $Nd^{III}$ (ca. 880, 1055 and 1340 nm) and $Yb^{III}$ (ca. 980 nm). Assessment of the luminescence of $Yb^{III}$ complexes in the presence of $Cu^{II}$, $Zn^{II}$, $Cd^{II}$ and $Hg^{II}$ revealed that $Cu^{II}$ quenched emission, whereas $Hg^{II}$ induced significant enhancements in both the emission intensity and lifetime of $Yb^{III}$. Although $Zn^{II}$ and $Cd^{II}$ produced changes in the ligand fluorescence suggestive of a binding event, this was not transduced to the $Yb^{III}$. The $Hg^{II}$ induced enhancement of $Yb^{III}$ emission and lifetimes suggests that non-radiative deactivation pathways are perturbed upon $Hg^{II}$ binding (the integrated emission increases), which may be a consequence of enhanced rigidity within the ligand framework and/or an improved shielding of the $Yb^{III}$ ion from outer sphere solvent effects.

An example of a sensor system that does not incorporate a covalently linked chromophore has been reported by Gunnlaugsson and is essentially based upon a displacement assay. The system features a ternary $Eu^{III}$ complex with bathophenathroline disodium sulfonate (BPS) as a co-ligand, which also provides the antenna; the complex demonstrates classical $Eu^{III}$ emission following irradiation at 278 nm. Addition of $Fe^{II}$ to the ternary complex extricates the diimine ligand to form $[Fe(BPS)_3]^{2+}$ and thus liberates the antenna-free $Eu^{III}$ complex, resulting in a dramatic drop in the luminescence intensity [39].

### *6.2.3.3 Lanthanides as Luminescent Labels/Molecular Beacons*

The photophysical properties of emissive $Ln^{III}$ complexes render them extremely attractive for bio-related analyses [40,41]. This section briefly describes some examples where preformed $Ln^{III}$ complexes, as opposed to the chelation of free $Ln^{III}$ ions by short peptides, such as $Ca^{2+}$-binding loops [42], can be utilised as luminescent labels. Myriad options regarding complex functionality have dramatically broadened their appeal and utility in the areas of bioconjugates, bioanalyses and immunoassay. Typically, from a synthetic perspective, a known chelating moiety is adorned with a functional group (terminating amines, thiols, carboxylic acids [43] and thiocyanates [44] all provide useful coupling options) that then allows easy covalent attachment to a variety of substrates and/or vectors (Fig. 6.12). This section will briefly introduce the applicability of luminescent $Ln^{III}$ chelates as molecular beacons (predominantly in a sensing context), using a handful of examples to demonstrate the breadth of their appeal in bio-related applications.

Takalo and Rodriguez-Ubis [45] described the synthesis of $Tb^{III}$ chelates (based on functionalised terpyridine or analogous type ligands) as labels for proteins, as well as their luminescence properties as antibody conjugates. In particular, they highlight the necessary selection of a suitable chromophore possessing a triplet excited state of sufficient energy (tunable via the ligand architecture) to maximise the emissivity of the conjugates, providing suitable systems for bioaffinity assays. Protein labelling has also been achieved with $Tb^{III}$ and $Eu^{III}$ chelates for time-resolved bioaffinity assays [46]. Indeed, lanthanide chelates have also been applied to solid-phase peptide labelling [47].

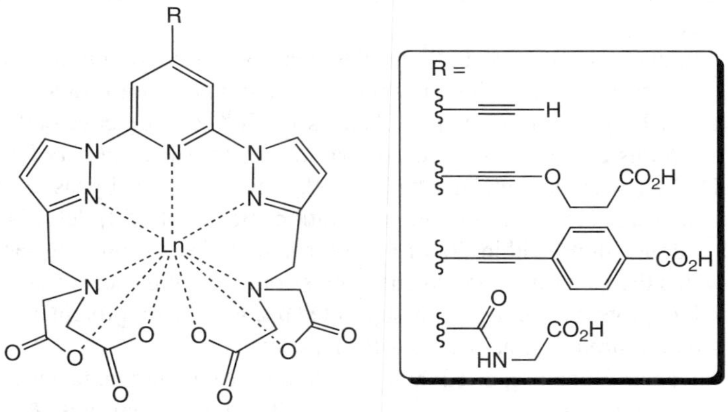


***Figure 6.12*** *Examples of terminating functionalities for covalent attachment of substrates and/or vectors*

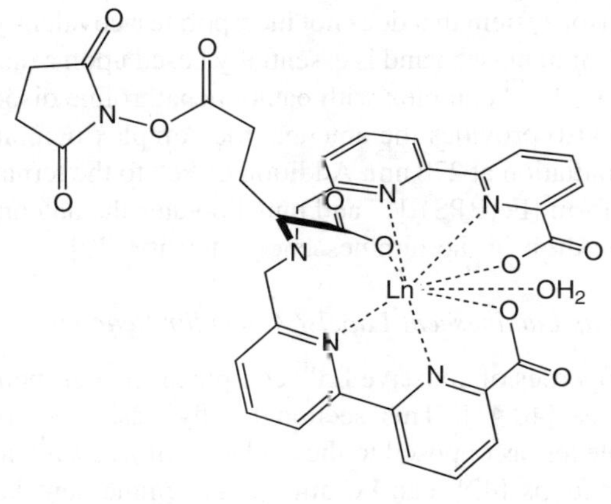

**Scheme 6.8** *Highly luminescent succinimide terminated $Ln^{III}$ chelate*

Highly luminescent, lanthanide-based tags (quantum yields up to ~30%) have been developed by Ziessel, Charbonnière and Roda, allowing complexes of bipy-based chelates (Scheme 6.8) to be covalently attached to proteins, such as bovine serum albumin [48]. The advantages of the (long-lived emission) $Ln^{III}$-based labelling were elegantly demonstrated via time-resolved luminescence microscopy studies and comparison to (short-lived emission) fluorescein-labelled analogues.

FRET can be used in various bioanalyses probing, for example, conformational changes in biological assemblies such as protein interactions, and DNA hybridisation: these approaches exploit the distance-dependence ($1/r^6$) of FRET. An additionally useful aspect of the energy transfer is invoked when donors with long excited-state lifetimes, such as $Ln^{III}$, are used. In a dynamic system, a donor–acceptor pair that is too far apart for efficient transfer may be brought into close proximity by a diffusion process while the donor is excited. Therefore energy transfer will be enhanced by diffusion if the distance diffused during the donor lifetime is comparable to, or greater than, the mean distance between donors and acceptors [8].

The long lifetimes of emissive $Ln^{III}$ species therefore provide opportunities for time-resolved RET experiments, allowing enhancement of signal to noise ratios and increased signal selectivity. Polyaminocarboxylate chelates of $Tb^{III}$ can be covalently attached to trimethoprim and this conjugate has been shown to bind to *Escherichia coli* dihydrofolate reductase (eDHFR) with nanomolar affinity [49]. This approach bypasses the need for specific antibodies and simplifies sample preparation. Bao et al. [50] described the use of $Eu^{III}$ and $Tb^{III}$ donor units within donor-acceptor assemblies for probing target oligonucleotides, towards the quantification of gene expression. Quinoline-sensitised $Ln^{III}$ chelates have been used for protein labelling on Ni-supported beads; protein–protein interactions [51] have been assessed through time-resolved RET assays.

In further advances Bünzli reported the use of luminescent lanthanide helicates as labels for 'lab-on-a-chip' detection of cancer biomarkers. The helicate species (Fig. 6.13) were conjugated to Avidin, Streptavidin and various antibodies. The advantages of the $Ln^{III}$-based system are the significant improvements of the signal to reference ratios in

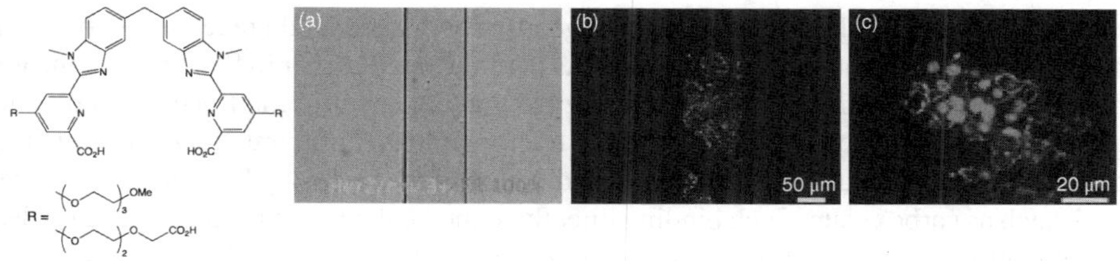


*Figure 6.13* An example of a Bünzli ligand for the bimetallic helical complexes (left), and 'on-chip' immunohistochemical detection of Her2/neu and ER in a breast cancer tissue sample. (a) Bright field image; (b) merged luminescent image, Her2/neu detected by a green-emitting $Tb^{III}$ complex and ER stained with a red-emitting $Eu^{III}$ complex; (c) magnified image. Reprinted with permission from [52]. Royal Society of Chemistry. (See colour plate section)

immunocytochemical analyses, an improvement in the time for analysis and a far more economic use of the biomarker [52].

An assessment of binding events at cell surfaces can also be probed (sensed) with luminescent lanthanide labels. $Tb^{III}$ and $Eu^{III}$ complexes of chromophore-appended DO3A-type ligands have been designed to target the lipophilic plasma membrane of cells by incorporating long, alkyl chains into the ligand framework. The amphiphilic complexes (Fig. 6.14) appear to bind to the cell surface, presumably via the alkyl chains, and can be imaged using confocal fluorescence microscopy. The use of both $Eu^{III}$ and $Tb^{III}$ complexes to dually label the cell provides further possibilities for interrogating the cell surface by studying the distance-dependent intermolecular energy transfer processes that can occur between the $Ln^{III}$ ions. In this case, the energetic pathway favours intermolecular $Tb^{III}$-sensitised-$Eu^{III}$ emission, through observation of the relative quenching of the donor component ($Tb^{III}$) versus the acceptor ($Eu^{III}$) [53].

### 6.2.4 Sensors for Anions

#### *6.2.4.1 Relevant Design Principles Revisited*

As discussed in the introduction to this chapter, the lanthanides are normally found in their 3+ oxidation sate and as such are strong Lewis acids. Consequently, these ions have high affinity for anions; hydroxide is one particular example, which leads to the precipitation of

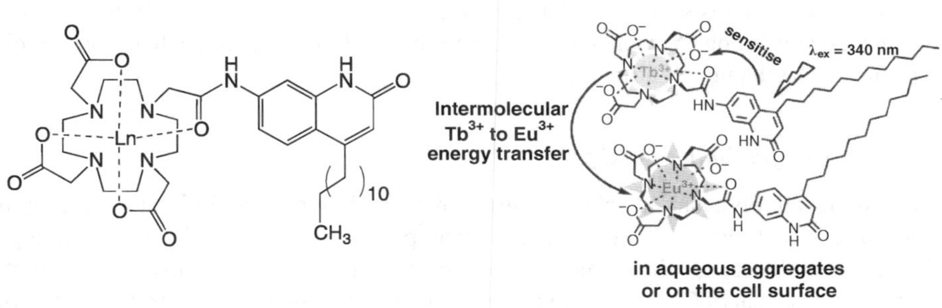


*Figure 6.14* Amphiphilic complexes as luminescent cell surface labels. Reprinted with permission from [53]. Copyright © 2011, American Chemical Society

the lanthanides as $Ln(OH)_3$. However, the high affinity that the lanthanides show for anions can also be capitalised on in a positive way. As demonstrated earlier in Scheme 6.2, the ions can be incorporated within a coordinating ligand that either fulfill the Ln(III) coordination requirements (usually of 8–9), or within a coordinatively unsaturated ligand which, as shown in Scheme 6.3, can then fulfill the coordination of these ions by ligation to an external ligand, such as carboxylate. Such binding directly to the lanthanide ion does not usually lead to precipitation, and hence this process can be used as a means of anion recognition and sensing of anions. This binding mode has been extensively used in luminescent sensing, as demonstrated in Fig. 6.3, and the field of anion binding to lanthanide ions in aqueous solution has recently been reviewed by Butler and Parker in an elegant manner, where, in particular, the binding modes shown in Fig. 6.3 are discussed using recent examples from the literature [54].

More in detail, the use of $Eu^{III}$ ions in such sensors has been highly attractive as binding, or coordination to the metal centre can have a significant effect on the lanthanide centred emission, arising from the deactivation of the $^5D_0$ excited state to the $^7F_J$ $J=2$ ground state. This transition is hypersensitive and as such reflects directly the changes in the local environment of the ion. These types of sensors have found use in the sensing of halides and carboxylates. As such, sensing is usually reversible and these types of sensors have also been developed for use in 'displacement assays', which is depicted in a cartoon manner in Fig. 6.3. Such assays involve competitive ligand binding, where for instance a coordinating antenna is replaced by a competitive anion. If such an anion lacks an antenna, or has an antenna that due to the nature of its excited state is unable to sensitise the lanthanide excited state, the recognition process gives rise to quenching of the lanthanide emission. However, these roles can be reversed, for example a displacement assay can simply involve the removal or exchange of a competitive binding unit and give rise to enhanced sensitisation of the lanthanide excited state and hence, enhancement in the lanthanide emission. Such recognition modes have found their greatest use in the sensing of carboxylate and phosphate anions. The attractiveness of the above sensing methods lies in their simple design principle: a self-assembly processes which often can be achieved using simple coordinating ligands. In contrast to this, anion recognition can also be achieved by modulation of the photophysical properties of the antenna upon binding at a receptor site, as depicted in Fig. 6.4. In contrast to the above binding modes, fewer examples of such sensors have been published to date, which reflects the fact that such sensing involves the incorporation of anion selective receptors, but unlike that shown above for the sensing of cations, such sensing is generally more challenging. In this section, both types of lanthanide–anion sensors will be discussed, where the greater emphasis will be on the former mode of sensing, where anions bind directly to the lanthanide centre.

#### *6.2.4.2 Sensors for Halides and pH*

Halides are negatively charged spherical structures which normally bind to anion receptors in a non-directional fashion. One of the first examples of the use of halides for binding to lanthanide complexes, by direct metal coordination within a coordinatively unsaturated environment, was that of Charbonnière et al. [55] who developed lanthanide complexes of a bis-bipyridine-phosphine oxide ligand (Scheme 6.9a), as luminescent anion sensors for halides. In this design, the bipyridine (bpy) units were expected to coordinate to the

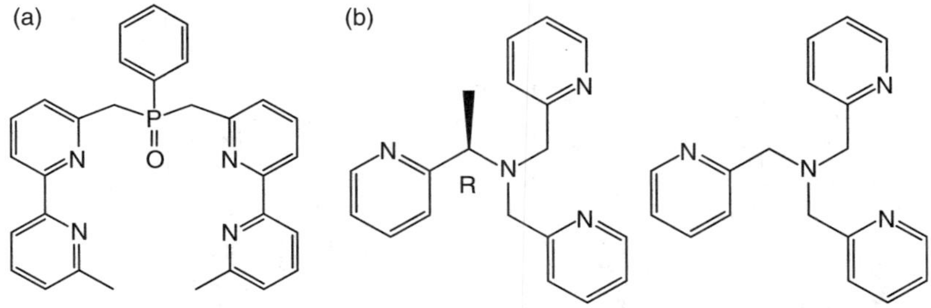


***Scheme 6.9*** *Ligands employed in lanthanide complexes that were used as halide and nitrate sensors*

lanthanide ion and function as sensitising antennas, while the strong coordinating ability of the P=O bond would increase the stability of the complex, despite the fact that it would remain coordinatively unsaturated, with two solvent molecules occupying the final coordination positions. It was found that in aqueous solution neither the $Eu^{III}$ nor the $Tb^{III}$ complex gave rise to any luminescence, indicating that neither of the bpy units were coordinated to the lanthanide ions under these conditions. However, in MeCN solution, after excitation of the bpy absorption bands (310 nm), the triflate salt complexes were found to display the characteristic lanthanide luminescence bands. Unfortunately, no luminescence changes were observed upon the addition of fluoride to a solution of these sensors. Moreover, the addition of acetate did not give rise to any luminescence enhancements. However, upon addition of tetrabutylammonium chloride, a 5.0- and 2.5-fold enhancement was observed for the emission arising from the $Eu^{III}$ and the $Tb^{III}$ complexes, respectively, demonstrating that these sensors could detect $Cl^-$ over $F^-$. However, when these sensors were titrated with nitrate, the emission from both the $Eu^{III}$ and the $Tb^{III}$ complexes was even further enhanced, exhibiting ca. 11- and sevenfold increases. For both anions, the mechanism was proposed to be due to displacement of coordinating solvent molecules in a similar manner to that described above in Scheme 6.3.

Tsukube and co-workers [56] have also reported in a similar manner luminescent anion sensors that were selective for chloride and nitrate. The $Eu^{III}$ complex of tris(2-pyridylmethyl)amine (Scheme 6.9b) was shown to be a luminescent sensor for nitrate, while the corresponding $Tb^{III}$ complex displayed selectivity for $Cl^-$, giving rise to ca. 4.9- and 5.4-fold enhancements in the metal centred emission, in acetonitrile solutions. Both the chiral and the achiral $Eu^{III}$ complexes were shown to display 1:1 stoichiometry with these anions, and were found to be stable enough for use in organic solution. Here, as above, the pyridine units fulfills the role of the sensitising antenna, absorbing at 260 nm. The addition of ions such as $I^-$, $Br^-$, $F^-$, $ClO_4^-$, $SCN^-$ and $HSO_4^-$ did not elicit a response.

The $Eu^{III}$ (Scheme 6.10a) bis-bipyridyl-based cryptate developed by Pina et. al. [57] is an example of a pH modulated sensor, which can also be described as an indirect sensor for hydroxyl anions. In aqueous solution the sensor exhibited characteristic $Eu^{III}$ emission with a maximum intensity around pH 7. However, below pH 7 the emission was almost 50% quenched, while in alkaline solution the amines become deprotonated, giving rise to significantly longer excited state lifetimes with higher emission intensities. At very high

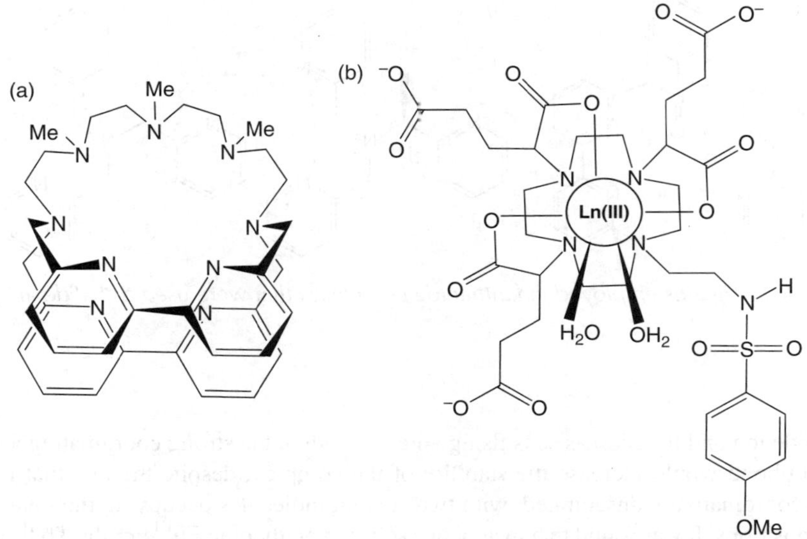


***Scheme 6.10*** *pH responsive sensors that involve indirect sensing of hydroxides*

pH (>11) loss of emission was observed, which was attributed to the formation of a hydroxylated Eu$^{III}$ species as discussed above. Parker et al. [58] developed several examples of pH sensors which also can be described as indirect anion sensors. These are based on the use of pH driven intramolecular self-assembly between a lanthanide complex and a sulfonate antenna. In alkaline solution, the sulfonamide is in its anionic form, which allows for coordination to the lanthanide centre to take place. This is accommodated by the displacement of the metal-bound water molecules, which gives rise to enhancement in the lanthanide emission, upon excitation at 332 nm. Recently, Parker et al. [59] extended this design principle by using related complexes for live cell imaging of lysosomal pH changes.

In water, most halides have a low affinity for lanthanide complexes of the type discussed in the previous example. However, the exception is fluoride. Recently, Charbonnière et al. [60] demonstrated two examples utilising eight-coordinate Eu$^{III}$ cyclen complexes (Scheme 6.11a) with a $q$ value of 1, as luminescent sensors for $F^-$ in aqueous solution. Upon $F^-$ binding a new species with $q = 0$ is formed, illustrating the displacement of an axial water molecule giving rise to an enhancement in the Eu$^{III}$ centred emission. In fact, the simple displacement of labile metal-bound water molecules by anions can have a significant effect on the luminescent and chiroptical properties of overall seven-, eight- or nine-coordinate cyclen complexes. This was first established by Parker and co-workers using seven-membered cyclen complexes bearing chiral phenyl or naphthalene antennas (Scheme 6.11b). In aqueous media, the heptadentate Tb$^{III}$ and Eu$^{III}$ complexes gave rise to weak lanthanide emission, due to the presence of two bound water molecules [61,62]. In aqueous solution it was observed that both water molecules could be displaced by ions such as phosphate, lactate, and carbonate, resulting in the formation of a 1:1 adduct. This ternary

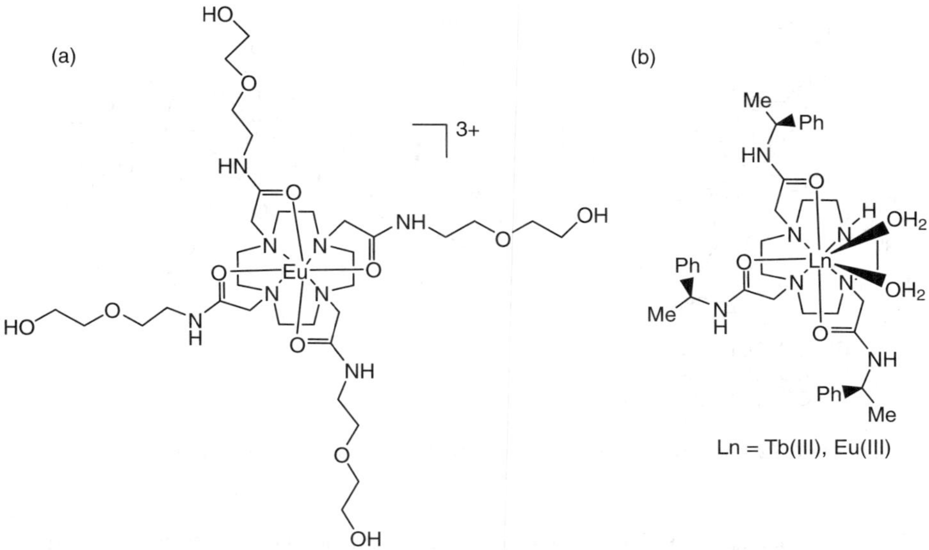


***Scheme 6.11*** *Sensors for halides and oxy-anions*

complex formation was signaled by large changes in the lifetimes of the $Eu^{III}$ and $Tb^{III}$ excited states. Like it was seen for the example in Scheme 6.11a, in the case of $F^-$, and $H_2PO_4^-$ and acetate, $q$ was determined to be ca. 1, implying the loss of a single bound water molecule upon coordination of these anions to the lanthanide ions. In contrast, the lifetimes remained unchanged for ions such as $Br^-$ and $Cl^-$.

The Parker group has explored the use of the direct binding of simple organic anions, as well as inorganic phosphates, to structures such as those shown in Scheme 6.11b, possessing heptadentate coordination environments. Examples of these are shown in Scheme 6.12. They have as well as used such structures for the analysis of cellular environments. The structures shown in Scheme 6.12a–c are based on a series of cationic, zwitterionic, and anionic $Eu^{III}$ complexes, and were developed with the view of forming luminescent sensors for the detection of bicarbonate [63,64]. Here, an acridone chromophore was used as the sensitising antenna, facilitating the population of the $Eu^{III}$ excited state upon excitation at 390–410 nm. This is a significant development as the excitation of $Eu^{III}$ has normally been limited by the use of relatively high energy antennas. Similarly, the $Tb^{III}$ complex shown in Scheme 6.12d is based on the use of systems possessing a related azaxanthone chromophore, as well as a naphthyl moiety [65]. The $Tb^{III}$ excited state was formed by excitation of the azaxanthone group at 355 nm. This system was developed as a potential anion binder, however, the relatively long-lived excited state of the naphthyl functionality allowed for successful singlet oxygen formation due to the quenching of its excited state by atmospheric oxygen. Each of these complexes (Scheme 6.12) was found to selectively bind bicarbonate at physiological pH, and the advantages of using such an antenna as a sensitiser was initially reported by Faulkner et al. [66].

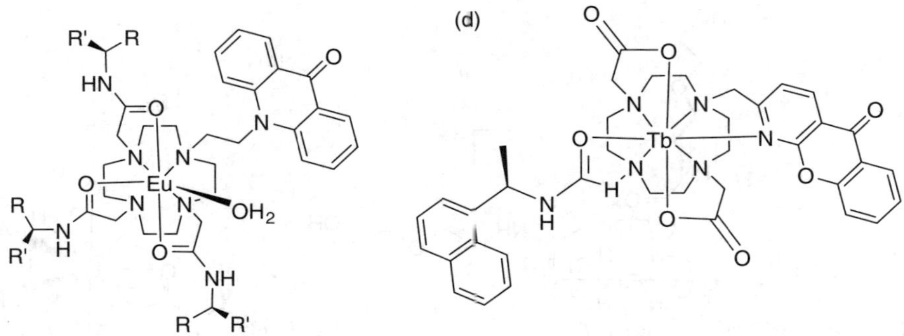

(a) R = CO2Et, R' = Me
(b) R = $CO_2^-$, R' = Me
(c) R = $CO_2^-$, R' = $(CH_2)_2CO_2^-$

***Scheme 6.12*** *Sensors and imaging agents developed by the Parker group [63,64]*

The use of such acridone $Eu^{III}$ complexes has been extended to the sensing of phosphate. The binding of selected phosphates, as well as phosphorylated amino acids such as Ser-OP and Tyr-OP, was investigated by observing the emission spectra of the complexes. Job plot analysis revealed the formation of 1:1 adducts, which displayed subtly different emission spectra with Tyr-OP showing the lowest $J=2/J=1$ intensity ratio. In each case, the obtained data was consistent with the formation of mono-aqua species with phosphate anion acting as a monodentate ligand [67]. The binding process was signaled by changes in the luminescence lifetimes and of the $Eu^{III}$ emission intensities, resulting in a modulation of $q$ from 1 to 0, in a similar manner to that seen above. These observations provide evidence for the displacement of the metal-bound inner sphere water molecule by the anion. Furthermore, by simply monitoring the ratio of the $Eu^{III}$ emission intensity at 618/588 nm and/or by using the 618/702 nm emission bands, the concentration of bicarbonate could be assessed in a background of competing anions such as lactate, citrate, and phosphate. This simple analysis enabled the generation of sensors possessing internal reference channels, minimising the error in assessing the anion concentration.

The idea of ratiometric anion sensing has also been expanded to the use of two or more metal ions of the same nature, or by using mixed-metal ion centres, such as those shown in Scheme 6.13a, b [68]. The former is based on the use of a dimetallic $Eu^{III}$-bismacrocyclic conjugate, and it was demonstrated that the sensor was able to bind organic anions such as dicarboxylates. Small dicarboxylic acids such as aspartic, malonic, succinic, and glutaric acid were demonstrated to bind to the sensor in solutions at pH 6.5. However, of these anions, the authors showed that only malonic acid gave rise to selective $Eu^{III}$ luminescence enhancements, as the emission intensity was reduced for all the other anions tested. The related $Eu^{III}/Tb^{III}$ complex shown in Scheme 6.13b, however, is based on the use of two $Eu^{III}$ centres and a single $Tb^{III}$ centre. As with previous examples and with many of those discussed above, the emission arising from these systems is caused by the displacement of labile metal-bound water molecules by these anions, giving rise to ternary complex formation. The example in Scheme 6.13b, in contrast, could consist of the use of three $Eu^{III}$ or $Tb^{III}$ ions, or by using a mixture of two $Tb^{III}$ ions, which were found to bind to the

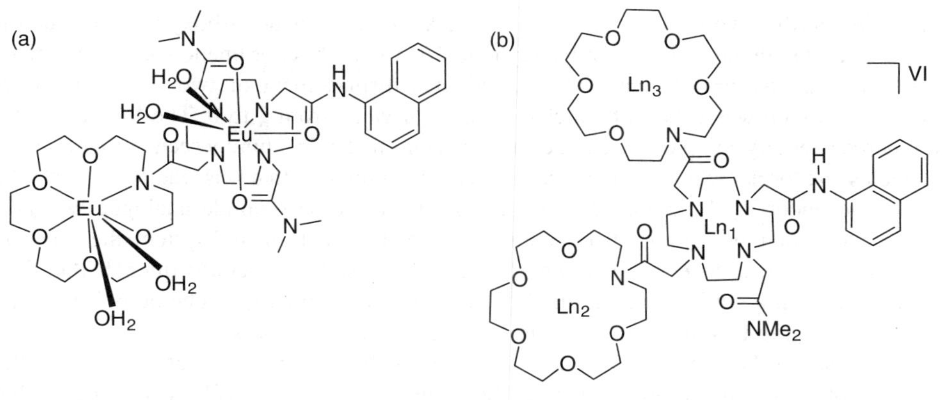


***Scheme 6.13*** *Sensors developed for di-carboxylates using more than one metal ion centre and two types of macrocyclic systems*

crown ethers, while the central cyclen moiety contained a Eu$^{III}$ ion. The authors demonstrated that ratiometric sensing of malonic acid was possible, by simply comparing the changes in the Tb$^{III}$ intensity to that of the Eu$^{III}$; that is, the Tb$^{III}$ emission provides a means for incorporating an internal reference channel for the changes in the Eu$^{III}$ emission. Interestingly, using the emission intensity changes at 545 nm as a reference for the changes in the Eu$^{III}$ emission at 593 nm or 617 nm enabled the authors to demonstrate a linear relationship for the Eu$^{III}$/Tb$^{III}$ emission ratio upon titrating this Eu$^{III}$/Tb$^{III}{}_2$ system with malonic acid.

The above examples demonstrate how simple anions such as halides, carboxylates and carbonates can bind directly to the lanthanide centres and concomitantly modulate the photophysical output of such complexes. In a similar manner, aryl-based anions can be employed, and the formation of luminescent ternary complexes can thus be achieved. Both Gunnlaugsson et al. [69] and Faulkner et al. [70] have demonstrated, for instance, that simple anionic antennas can be bound to Tb$^{III}$ and Yb$^{III}$ cyclen complexes (Scheme 6.14a, b, c). The former two Tb$^{III}$ examples (Scheme 6.14a, b) are based on the use of cationic complexes,

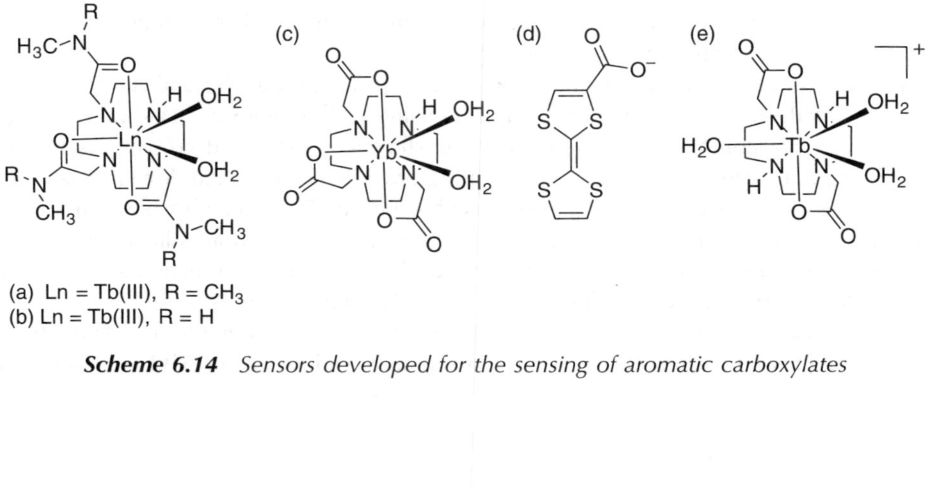


***Scheme 6.14*** *Sensors developed for the sensing of aromatic carboxylates*

where the coordinatively unsaturated $Tb^{III}$ complexes bind aromatic carboxylic anions such as *N,N*-dimethylaminobenzoic acid and salicylic acid in water. The coordination of these aromatic anions to the non-luminescent $Tb^{III}$ occurred via displacement of two Ln-bound water molecules, which was verified by a change in the $q$-value from 2 to 0. The formation of a luminescent ternary complex gave large enhancements in the $Tb^{III}$ luminescence. Of these two examples, the former was shown to give rise to intense emission that was readily visible to the naked eye under a UV lamp. The emission from the primary amide analogue was also investigated, and it was shown that in comparison to the *N,N*-di-methyl system, the emission was significantly reduced, indicating that the N-H oscillators could quench the $Tb^{III}$ excited state. In a similar manner the $Tb^{III}$ complex displayed in Scheme 6.5 was also shown to be able to bind to carboxylic-based aryl chromophores.

In contrast to these results, the $Eu^{III}$ complex of this system was not emissive in the presence of these anions; however, later the Gunnlaugsson group [71] demonstrated that β-diketonates could be used as the coordinating antenna, giving rise to luminescent complexes that emit in the red part of the electromagnetic spectrum. Furthermore, the authors were able to demonstrate the presence of the two metal-bound water molecules in this system using X-ray crystallography, which showed that these two molecules had a bite-angle similar to that proposed by Parker et al. in their earlier work discussed in Scheme 6.11b.

The example of Faulkner et al. [70], shown in Scheme 6.14c, is somewhat different from the one just discussed, as it is based on the use of an overall charge neutral cyclen complex. It was shown to be able to form ternary complexes with long wavelength absorbing antennas such as the tetrathiafulvalene carboxylate chormophore shown in Scheme 6.14d, and that excitation of the antennas gave rise to sensitised lanthanide emission in the near-IR region in methanol solution. As above, the authors demonstrated that the carboxylate coordinates through displacement of the two Ln-bound water molecules. More recently, this research group demonstrated the sensitisation of the $Nd^{III}$ analogue of this structure, using a pyrene carboxylate as the sensitising antenna [72]. Coordination of the pyrene carboxylate in water promoted intense characteristic $Nd^{III}$ emission bands at 1055 and 1350 nm. These results are of extreme importance for the development of lanthanide-based sensors and probes for biological applications. These near-IR emitting lanthanides (such as $Nd^{III}$, $Er^{III}$ and $Yb^{III}$) possess lower energy emissive states, and hence are able to make use of chromophores with longer absorption wavelengths (i.e. >400 nm). This renders them of great advantage over other lanthanide ions such as $Tb^{III}$ and $Eu^{III}$, when regarding *in vivo* applications. Nevertheless, since these ions emit at wavelengths greater than 800 nm, more specialised instruments and detection hardware are required.

As demonstrated above in Scheme 6.14a, b, anions such as salicylic acid can be detected in water with the formation of ternary complexes using $Tb^{III}$ complexes. In a similar manner, a mono-cationic complex (Scheme 6.14e) was shown by Gray and co-workers to bind salicylic acid as well as salicylurate in water. Meanwhile Georges and Arnaud reported a method of salicylic acid detection using a combination of $Tb^{III}$ and ethylenediaminetetra-acetic acid (EDTA). The presence of the EDTA prevents the formation of insoluble terbium hydroxides and also acts to exclude water molecules from the coordination sphere of the lanthanide ion, reducing quenching caused by O–H oscillators [73].

It was discussed earlier that the $Eu^{III}$ analogue of the cyclen structure in Scheme 6.14a was not luminescent; however, the use of β-diketonates yielded an emissive complex. This

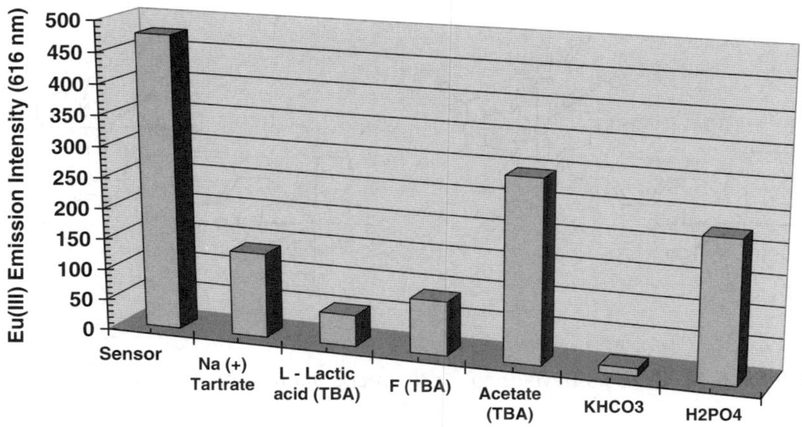


***Figure 6.15*** *The sensing of various anions using displacement assays*

selectivity has been further developed by Gunnlaugsson and co-workers in the design of novel luminescent sensors for anions that function as 'displacement assays'. Various anions, such as phosphates, were detected by this method of sensing [71] where the emission from the ternary complex was 'switched off', as demonstrated in Fig. 6.15. Furthermore, the formation of the ternary complex was shown to be pH dependent and that within the physiological pH window self-assembly was promoted, giving rise to a highly luminescent species.

The concept of 'displacement assays' for anions was further developed on the back of the aforementioned results, where the secondary amine within the cyclen structure was functionalised with a long alkyl-thiol chain for attachment to 5 nm gold nanoparticles (AuNPs). Since the $Eu^{III}$ complex (Scheme 6.15a) lacks any antenna, the AuNP-complex conjugate was found to be non-luminescent in the absence of the β-diketonate. However, in the presence of the antenna the AuNP conjugate became highly luminescent, with characteristic $Eu^{III}$ emission observed in buffered solution at pH 7.4 [74]. This system was then used to detect various anions, such as the phosphate flavin mononucleotide, which

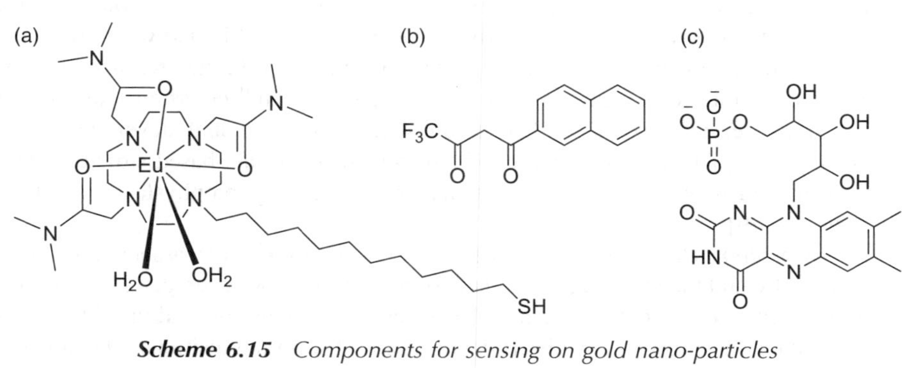

***Scheme 6.15*** *Components for sensing on gold nano-particles*

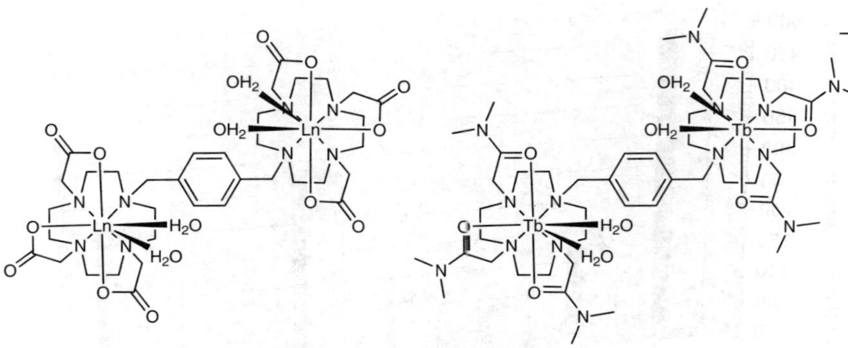

***Scheme 6.16*** *Sensors developed for the recognition of bis-carboxylates in water*

gave rise to almost complete displacement of the sensitising antenna with concommitant quenching of the Eu$^{III}$ centred emission. Other anions such as ADP, AMP, ATP and NADP displayed a 20–50% quenching effect. The Gunnlaugsson group has more recently demonstrated that this kind of system can bind to proteins and that the active site of the proteins can then be employed as a 'receptor' unit, giving rise to a new type of luminescent biosensors whereby the Eu$^{III}$ emission is modulated [75]. Moreover, the group has recently shown that NIR emitting lanthanides such as Yb$^{III}$ can also been used in AuNP-functionalised systems, by using visible absorbing antennas such as xylenol orange [76].

In the above examples, mono-nuclear complexes were generally employed and the anions sensed were monodentate or bidentate. Several examples where such anions have been detected by the use of bis-cyclen complexes, possessing either alkyl or aryl bridges, have also been developed. Examples of such designs are shown in Scheme 6.16. The former of these was developed by Pope and Faulkner, who used the charge neutral xylyl bridged bis-cyclen system, which was coordinatively unsaturated, giving each of the metal ion centres a *q*-value of 2 [77]. The related coordinatively unsaturated cationic Tb$^{III}$ dinuclear analogue was reported by the Gunnlaugsson group [78]. This latter complex showed the ability to detect the presence of mono- or bis-aromatic carboxylates, such as *N,N*-dimethylaminobenzoic acid and terephthalic acid, respectively, in buffered aqueous solution at physiological pH. The recognition of these ions was again found to occur through the binding of the carboxylates to the metal centre, via the displacement of the metal-bound water molecules. This gave rise to significant enhancements in the Tb$^{III}$ emission. In the presence of tartaric acid, the recognition process gave rise to quenching of the Tb$^{III}$ emission intensity. The Faulkner group recently extended their effort within this area and developed several examples of bis-cyclen structures that can bind to bis-carboxylate-based structures (such as demonstrated in Scheme 6.17) with significant changes in the photophysical properties of the complexes [79].

The above examples have focussed on the sensing of anions by direct association with the lanthanide centre and the advantages in this design approach have been discussed. Usually such systems give rise to enhancements of the lanthanide emission and/or allow for the monitoring of the sensing event by observing the ratios of several transitions. The drawback

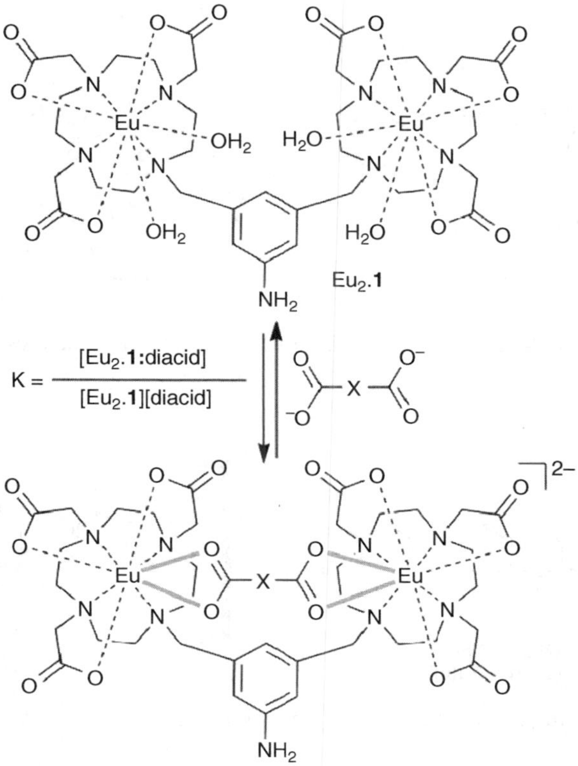


***Scheme 6.17*** *Recent example of recognition of bis-carboxylates from the Faulkner group*

to such a design is, however, that the selectivity is usually not optimised, and that only narrow classes of anions can be detected. An alternative approach to this design is to incorporate an anion recognition moiety into the antenna. In such systems, the anion recognition at the antenna site is reflected in its ability to populate the lanthanide excited state, leading to luminescence enhancement or quenching of the lanthanide emission.

Parker et al. developed some of the first examples of these types of sensors, shown in Scheme 6.18 [80]. These cationic structures detect halide ions, where the anion interacts with the phenanthridinium antenna through electrostatic interactions. For both of the $Eu^{III}$ complexes the anion 'binding' leads to a decrease in the intensity of the phenanthridinium fluorescence (centred at 405 nm) and an associated decrease in the $Eu^{III}$ luminescence intensity by a factor of four. The $Cl^-$ quenching effect was found to be independent of added lactate, phosphate, citrate, and hydrogencarbonate. The efficiency of quenching followed the order $I^- > Br^- > Cl^-$ and was thought to involve charge transfer from the halide ion to the excited singlet of the chromophore. An extension of this design is shown in Scheme 6.19, where well known anion recognition sites such as urea were incorporated into lanthanide complexes as antennas. The former of these, Scheme 6.19a, b, is based on the use of cationic cyclen complexes, of which the $Tb^{III}$ complex was shown to be able to

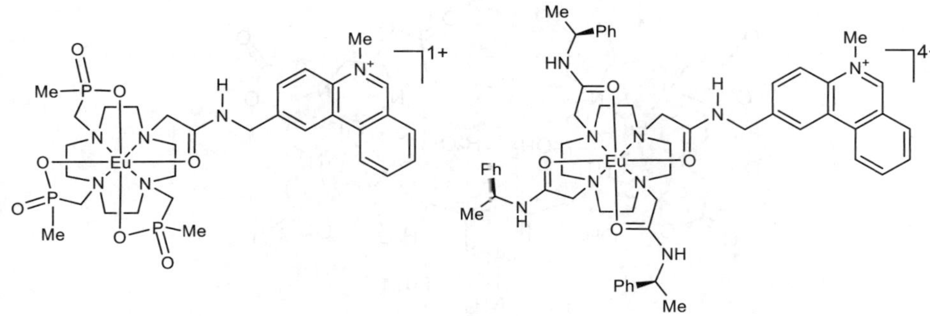

***Scheme 6.18*** *Sensors developed for the recognition of halides in water*

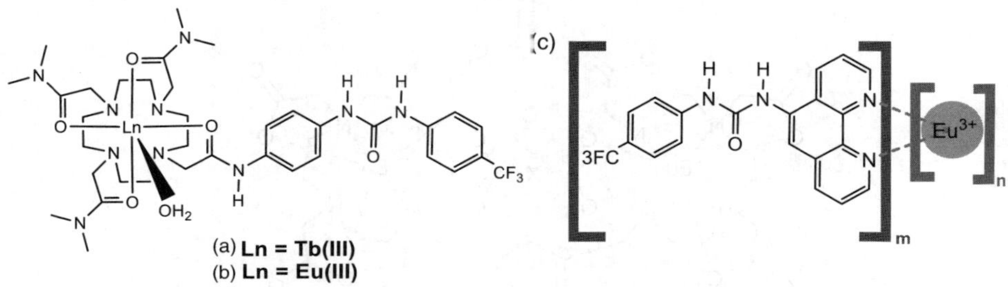


***Scheme 6.19*** *Structures developed for anion sensing using a receptor based antenna*

detect anions such as phosphate and acetate [81,82]. However, analysis of the changes in the $Tb^{III}$ emission spectra using non-linear regression analysis demonstrated that, in the case of phosphate, the sensor was able to interact with the anion at the urea receptor site via hydrogen bonding in organic solution, and also through the displacement of metal-bound water, which gave rise to multiple anion:sensor stoichiometries. The same principle was used in Scheme 6.19c, where the urea recognition site was incorporated into a phenanthroline (phen) derived ligand, which was then used to form coordination complexes with various lanthanides [83]. The addition of various anions modulated the sensitisation process from the phen ligand to the $Eu^{III}$ excited state. However, analysis of the ground state changes in the antenna suggested that, while the anion was bound through hydrogen bonding to the urea site, this binding most likely induces the dissociation of the $Eu^{III}$ from the complex. Hence, this latter example can be viewed as being an extension to the displacement assay sensing mode discussed above.

## 6.3 Interactions with DNA and Biological Systems

While the above examples have focused on the sensing of structurally simple and biologically relevant anions, several examples of lanthanide complexes have emerged in the literature in the past decade where the sensing of large biologically relevant anionic

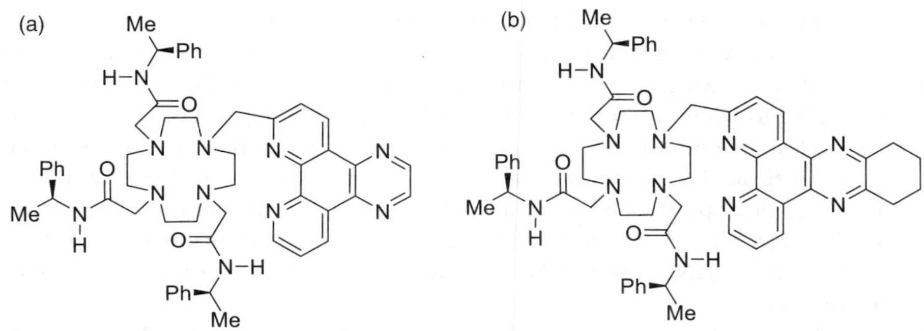


***Scheme 6.20*** *Structures developed by Parker et al. for interacting with DNA [84–86]*

structures, such as DNA, have been demonstrated [84–86]. Parker et al. developed many excellent examples of cyclen-based complexes, where the antenna was also an intercalator [84–86]. The group showed that upon binding to DNA the emission from the lanthanide centre was modulated dramatically. Scheme 6.20 shows two of the earlier examples of cyclen-based ligands developed within the Parker group.

Gunnlaugsson et al. developed a near-IR emitting sensor for DNA where both $Nd^{III}$ and $Yb^{III}$ cyclen complexes were conjugated via a short spacer to a $Ru^{II}$ polypyridyl centre (structures shown in Scheme 6.7) [87]. In these systems the polypyridyl complex is a well-known DNA binder, the study of which has been explored by changes in the photophysical properties of its $^3$MLCT excited state. As these units also function as sensitising antennas, the binding to DNA was expected to also be transduced to the lanthanide-centred emission, as the energy transfer process should be modulated upon DNA binding. As can be seen in Figure 6.16, this was indeed the case, where both the $^3$MLCT emission (occurring in the visible region) was slightly enhanced, while the $Nd^{III}$ emission was affected to a much greater extent, being 'switched off' in the process. Analysis of the changes in these

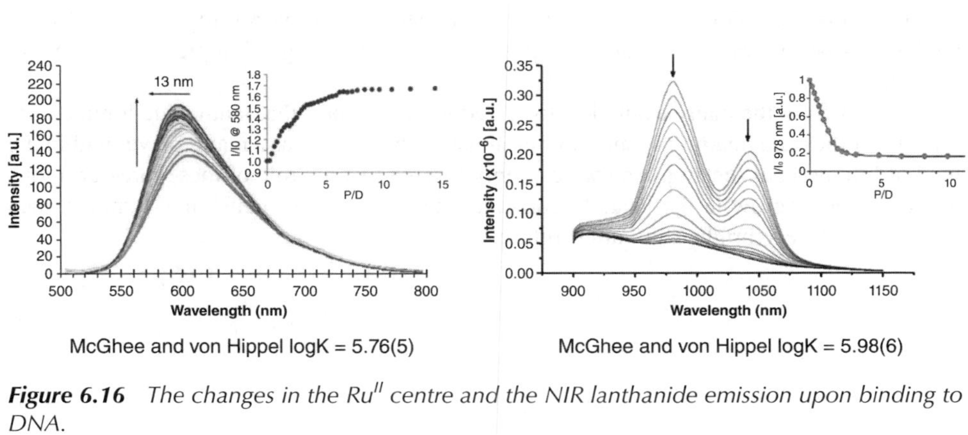


***Figure 6.16*** *The changes in the $Ru^{II}$ centre and the NIR lanthanide emission upon binding to DNA.*

respective emission spectra was done using conventional methods and from these, the binding affinity was determined. The results demonstrated that both centres gave the same binding constant showing that in this case the lanthanide part of the complex was not directly binding to DNA, but was perturbed by the DNA binding interactions at the $Ru^{II}$ centre. This design demonstrates that two channel systems can be employed to observe binding to nucleic acids, and that this offers a new avenue for the development for novel imaging agents that operate within the near-IR region. DNA targeting agents have also been developed by Pikramenou et al. [88] where the changes in lanthanide emission were modulated upon the DNA intercalation of two antennas, by a hairpin-shaped hetero-metallic complex.

Recently Pope et al. reported examples of $Gd^{III}$ and near-IR emissive $Nd^{III}$ and $Yb^{III}$-based cyclen systems for use in DNA binding analyses (Figure 6.17) [89]. These structures are based on the use of amino-anthraquinones that are well known to intercalate with DNA, as visible light absorbing antennas. In these structures the anthraquinone unit also functions as a bridge which preorganises the two cyclen centres, making the two structures highly desirable for binding to nucleic acids as was demonstrated using various spectroscopic techniques including luminescence and proton relaxivity, as well as by using molecular modeling (Figure 6.17b).

The sensing of biologically relevant anionic structures has become an important area of lanthanide luminescence sensing. This work has been extended to the development of novel imaging agents for application in cells and tissues. This area was recently elegantly reviewed by Parker et al. [90]. The interest in the sensing of charge neutral molecules, or in monitoring chemistry in real time using such probes has also increased. While this area of sensing will not be covered in this review, we will conclude with the example shown in Scheme 6.21, developed by the Gunnlaugsson group for the monitoring of enzymatic reactions in real time. Here the $Tb^{III}$ cyclen complex has a malimide antenna that is also a good Michael acceptor. It was shown that, upon titrating the complex with thiols, the thiol reacted with the α,β-unsaturated system, making a covalent bond, and in the process greatly improving the sensitisation process to the $Tb^{III}$ excited state [91]. The authors then demonstrated that glutathione could be detected in competitive media by using the enzymatic reduction of the oxidised form by glutathione reductase, with the reduction being 'sensed' by the enhancement in the lanthanide-centred emission. This example nicely demonstrates the versatility of the use of lanthanide-based complexes and conjugates for monitoring biological reactions in real time.

It is clear from the many examples discussed in this chapter that lanthanide luminescent sensors, probes and imaging agents have established themselves as a major player within the field of luminescent sensing. And we can only foresee that the role of these rare-earth ions will continue to grow in the years to come, and that the field of lanthanide luminescence sensing has a very bright and varied future.

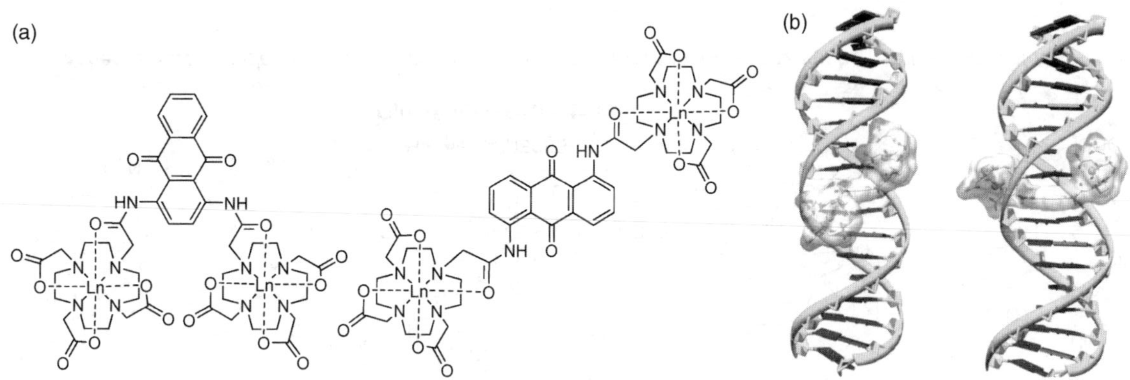


**Figure 6.17** *(a) Structures developed by Pope et al. [89] as DNA targeting agents. (b) Molecular modelling of the binding of the agents to DNA. Reprinted with permission from [89]. Royal Society of Chemistry*

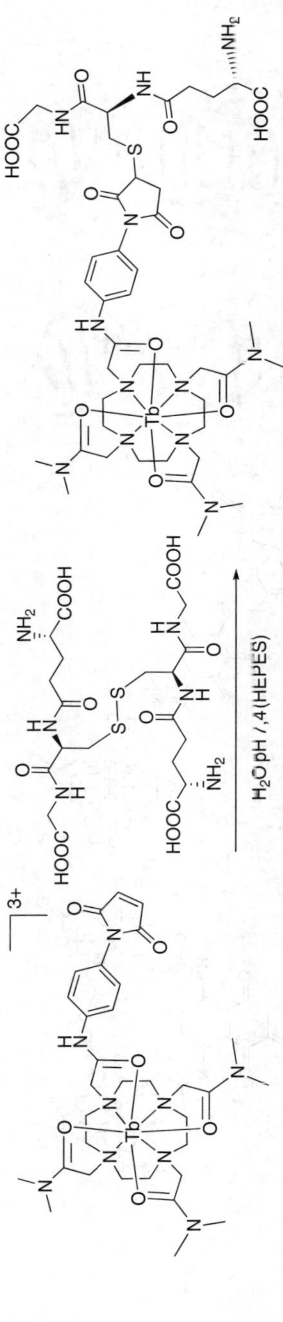


***Scheme 6.21*** *Cyclen based $Tb^{III}$ complex developed for monitoring enzymatic reactions in real time*

## Abbreviations

| | |
|---|---|
| bpy | 2,2′-bipyridine |
| phen | 1,10-phenanthroline |
| terpy | 2,2′;6′,2″-terpyridine |
| DPPZ | dipyrido-[3,2-a:2′,3′-c]-phenazine |
| FLIM | fluorescence lifetime imaging microscopy |
| LMCT | ligand to metal charge transfer |
| MLCT | metal to ligand charge transfer |
| Near-IR | near infra red |
| U.V. | ultra violet |
| Cyclen | 1,4,7,10-tetraazacyclododecane |
| DOTA | 1,4,7,10-tetraazacyclododecane-1,4,7,10-tetraacetic acid |
| DO3A | 1,4,7,10-tetraazacyclododecane-1,4,7-triacetic acid |
| DTPA | diethylenetriaminepentaacetic acid |
| BBATA | N,N,N1,N1-{2,6-bis(3″2-aminomethyl-1″2-pyrazolyl)-4-[N,N-bis(2-picolyl) amino-methylenepyridine]} tetrakis(acetic acid) |
| FRET | Förster resonance energy transfer |

## References


1. Bünzli J.-C.G., C. Piguet, *Chem. Soc. Rev.*, 2005, 34:1048.
2. Klink S.I., L. Grave, D. N. Reinhoudt, F.C.J.M. van Veggel, *J. Phys. Chem. A*, 2000, 104:5457.
3. Reifenberger J.G., G.E. Snyder, G. Baym, P.R. Selvin, *J. Phys. Chem. B*, 2003, 107:12862.
4. Beeby A., I.M. Clarkson, R.S. Dickens, S. Faulkner, D. Parker, L. Royle, A.S. de Sousa, J.A. G. Williams, M. Woods, *J. Chem. Soc., Perkin Trans. 2*, 1999, 1999:493.
5. Faulkner S., S.J.A. Pope, B.P. Burton-Pye, *Appl. Spec. Rev.*, 2005, 40:1.
6. Beeby A., S.W. Botchway, I.M. Clarkson, S. Faulkner, A.W. Parker, D. Parker, J.A.G. Williams, *J. Photochem. Photobiol. B*, 2000, 57:83.
7. Pandya S., J. Yu, D. Parker, *Dalton Trans.*, 2006, 2006:2757–2766.
8. Lakowicz J.R., *Principles of Fluorescence Spectroscopy*, 2006, Springer, New York.
9. Tanner P.A., C.-L. Duan, *Coord. Chem. Rev.*, 2010, 115:9188.
10. dos Santos C.M.G., A.J. Harte, S.J. Quinn, T. Gunnlaugsson, *Coord. Chem. Rev.*, 2008, 252:2512.
11. New E.J., D. Parker, D.G. Smith, J.W. Walton, *Curr. Op. Chem. Biol.*, 2010, 14:238.
12. Moore E.V., A.P.S. Samuel, K.N. Raymond, *Acc. Chem. Res.*, 2009, 42:542.
13. Faulkner S., B.P. Burton-Pye, T. Khan, L.R. Martin, S.D. Wray, P.J. Skabara, *Chem. Commun.*, 2002, 2002:1668.
14. Pope S.J.A., B.P. Burton-Pye, R. Berridge, T. Khan, P.J. Skabara, S. Faulkner, *Dalton Trans.*, 2006, 2006:2907.
15. Gunnlaugsson T., A.J. Harte, J.P. Leonard, M. Nieuwenhuyzen, *Chem. Commun.*, 2002, 2002:2134.
16. Gunnlaugsson T., J.P. Leonard, *Chem. Commun.*, 2003, 2003:2424.
17. Gunnlaugsson T., J.P. Leonard, *Dalton Trans.*, 2005, 2005:3204.
18. de Silva A.P., H.Q.N. Gunaratne, T.E. Rice, S. Stewart, *Chem. Commun.*, 1997, 1997:1891.
19. Li C., W-T. Wong, *Chem. Commun.*, 2002, 2002:2034.
20. Coldwell J.B., C.E. Felton, L.P. Harding, R. Moon, S.J.A. Pope, C.R. Rice, *Chem. Commun.*, 2006, 2006:5048.

21. Dai Z., J.W. Canary, *New J. Chem.*, 2007, 31:1708.
22. Hanaoka K., K. Kikuchi, H. Kojima, Y. Urano, T. Nagano, *J. Am. Chem. Soc.*, 2004, 126:12470.
23. Reany O., T. Gunnlaugsson, D. Parker, *J. Chem. Soc., Perkin Trans*, 2000, 2:1819.
24. McMahon B.K., T. Gunnlaugsson, *Tet. Lett.*, 2010, 51:5406.
25. Gunnlaugsson T., J.P. Leonard, K. Senechal, A.J. Harte, *Chem. Commun.*, 2004, 2004:782.
26. Nonat A.M., A.J. Harte, K. Senechall-David, J.P. Leonard, T. Gunnlaugsson, *Dalton Trans.*, 2009, 2009:4703.
27. Senechal-David K., J.P. Leonard, S.E. Plush, T. Gunnlaugsson, *Org. Lett.*, 2006, 8:2727.
28. Hasegawa Y., R. Hieda, T. Nakagawa, T. Kawai, *Helv. Chim. Acta*, 2009, 92:2238.
29. Ye Z., G. Wang, J. Chen, X. Fu, W. Zhang, J. Yuan, *Biosens. Bioelectr.*, 2010, 26:1043–1048.
30. Pope S.J.A., R.H. Laye, *Dalton Trans.*, 2006, 2006:3108.
31. Andrews M., B.M. Kariuki, S.J.A. Pope, *Polyhedron*, 2011, 30:2055.
32. Andrews M., A.J. Amoroso, L.P. Harding, S.J.A. Pope, *Dalton Trans.*, 2010, 39:3407.
33. Major J.L., R.M. Boiteau, T.J. Meade, *Inorg. Chem.*, 2007, 31:1354.
34. Woods M., G.E. Kiefer, S. Bott, A. Castillo-Muzquiz, C. Eshelbrenner, L. Michaudet, K. McMillan, S.D.K. Mudigunda, D. Ogrin, G. Tircso, S. Zhang, P. Zhao, A.D. Sherry, *J. Am. Chem. Soc.*, 2004, 126:9248.
35. Que E.L., C.J. Chang, *J. Am. Chem. Soc.*, 2006, 47:8022–8024.
36. Montgomery C.P., B.S. Murray, E.J. New, R. Pal D. Parker, *Acc. Chem. Res.*, 2009, 42:925–937.
37. Atkinson P., K.S. Findlay, F. Kielar, R. Pal, D. Parker, R.A. Poole, H. Puschmann, S.L. Richardson, P.A. Stenson, A.L. Thompson, J. Yu, *Org. Biomol. Chem.*, 2006, 4:1707.
38. Andrews M., J.E. Jones, L.P. Harding, S.J.A. Pope, *Chem. Commun.*, 2011, 47:206.
39. Kotova O., S. Comby, T. Gunnlaugsson, *Chem. Commun.*, 2011, 47:6810.
40. Bünzli J-C. G., *Chem. Rev.*, 2010, 110:2729.
41. Shinoda S., H. Tsukube, *Analyst*, 2011, 136:431.
42. Barthelmes K., A. M. Reynolds, E. Peisach, H. R. A. Jonker, N. J. DeNunzio, K. N. Allen, B. Imperiali, H. Schwalbe, *J. Am. Chem. Soc.*, 2011, 133:808–819.
43. Starck M., P. Kadjane, E. Bois, B. Darbouret, A. Incamps, R. Ziessel, L.J. Charbonniere, *Chem. Eur. J.*, 2011, 17:9164.
44. Maindron N.M., S. Poupart, M. Hamon, J-B. Langlois, N. Ple, L. Jean, A. Romieu, P-V. Renard, *Org. Biomol. Chem.*, 2011, 9:2357.
45. Takalo H., V-M. Mukkala, L. Merio, J.C. Rodriguez-Ubis, R. Sedano, O. Juanes, E. Brunet, *Helv. Chim. Acta*, 1997, 80:372.
46. Reddy D. R., L.E. Pedro Rosa, L.W. Miller, *Bioconj. Chem.*, 2011, 22:1402.
47. Shabanpoor F., F. Separovic, J.D. Wade, *J. Pept. Sci.*, 2011, 17:169.
48. Weibel N., L.J. Charbonnière, M. Guardigli, A. Roda, R. Ziessel, *J. Am. Chem. Soc.*, 2004, 126:4888.
49. Rajapakse H.E., D.R. Reddy, S. Mohandessi, N.G. Butlin, L.W. Miller, *Angew. Chem. Int. Ed.*, 2009, 48:4990.
50. Tsourkas A., M.A. Behlke, Y. Xu, G. Bao, *Anal. Chem.*, 2003, 75:3697.
51. Kim S.H., P. Ge, J.A. Katzenellenbogen, *Chem. Commun.*, 2009, 2009:183.
52. Fernández-Moreira V., B. Song, V. Sivagnanam, A-S. Chauvin, C. D.B. Vandevyver, M. Gijs, I. Hemmilä, H-A. Lehr, J-C.G. Bünzli, *Analyst*, 2010, 135:42–52.
53. Lee M., M.S. Tremblay, S. Jockusch, N.J. Turro, *Org. Lett.*, 2011, 13:2802.
54. Butler S.J., D. Parker, *Chem. Soc. Rev.*, 2012, 41:7673. DOI: 10.1039/c2cs35144g.
55. Carbonnière, L. J.; Ziessel, R.; Montalti, M.; Prodi, L.; Zaccheroni, N.; Boehme, C.; Wipff, G.; *J. Am. Chem. Soc.*, 2002, 124:7779.
56. Yamada, T.; Shinoda, S.; Tsukube, H.; *Chem. Commun.*, 2002, 2002:1218.
57. Bazzicalupi C., A. Bencini, A. Bianchi, C. Giogi, V. Fusi, A. Masotti, B. Valtancoli, A. Roque and F. Pina, *Chem. Commun.*, 2000, 2000:561–562.

58. Lowe M.P., D. Parker, *Inorg. Chim. Acta*, 2001, 317:163–173.
59. Smith D.G., B.K. McMahon, R. Pal, D. Parker, *Chem. Commun.*, 2012, 48:8520–8522.
60. Tripier R., C. Platas-Iglesias, A. Boos, J.-F. Morfin and L. Charbonniere, *Eur. J. Inorg. Chem.* 2010, 2010:2735; Lima L. M. P., A. Lecointre, J.-F. Morfin, A. de Blas, D. Visvikis, L. J. Charbonniere, C. Platas-Iglesias and R. Tripier, *Inorg. Chem.* 2011, 50:12508.
61. Dickins, R. S.; Gunnlaugsson, T.; Parker, D.; Peacock, R. D. *Chem. Commun.* 1998, 1998:1643–1644.
62. Bruce, J. I.; Dickins, R. S.; Govenlock, L. J.; Gunnlaugsson, T.; Lopinski, S.; Lowe, M. P.; Parker, D.; Peacock, R. D.; Perry, J. J. B.; Aime, S.; Botta, M. *J. Am. Chem. Soc.* 2000, 122:9674–9684.
63. Bretonnière, Y.; Cann, M. J.; Parker, D.; Slater, R. *Chem. Commun.* 2002, 2002:1930–1931.
64. Bretonniere, Y.; Cann, M. J.; Parker, D.; Slater, R. *Org. Biomol. Chem.* 2004, 2:1624–1632.
65. Law, G.-L.; Pal, R.; Palsson, L. O.; Parker, D.; Wong, K.-L. *Chem. Commun.* 2009, 2009:7321.
66. Dadabhoy, A.; Faulkner, S.; Sammes, P. G. *J. Chem. Soc., Perkin Trans. 2* 2002, 2002:348–357.
67. Atkinson, P.; Bretonniere, Y.; Parker, D. *Chem. Commun.* 2004, 2004:438–439.
68. Plush, S. E.; Gunnlaugsson, T. *Org. Lett.* 2007, 9:1919–1922;. Plush, S. E.; Gunnlaugsson, T. *Dalton Trans*, 2008, 2008:3801–3804.
69. Gunnlaugsson, T.; Harte, A. J.; Leonard, J. P.; Nieuwenhuyzen, M. Supramolecular Chemistry 2003, 15:505–519;. Gunnlaugsson, T.; Harte, A. J.; Leonard, J. P.; Nieuwenhuyzen, M. *Chem. Commun.* 2002, 2002:2134–2135.
70. Faulkner, S.; Burton-Pye, B. P.; Khan, T.; Martin, L. R.; Wray, S. D.; Skabara, P. J. *Chem. Commun.* 2002, 2002:1668–1669.
71. Leonard J.P., C.M.G. dos Santos, S.E. Plush, T. McCabe, T. Gunnlaugsson, *Chem. Commun.* 2007, 2007:129–131.
72. Faulkner, S.; Carrie, M.-C.; Pope, S. J. A.; Squire, J.; Beeby, A.; Sammes, P. G. *Dalton Trans.* 2004, 2004:1405–1409.
73. Esplin T. L., M. L. Cable, H. B. Gray and A. Ponce, *Inorg. Chem.*, 2010, 49:4643; Arnaud, N.; Georges, J.; *Analyst*, 1999, 124:1075.
74. Massue, J.; Quinn, S. J.; Gunnlaugsson, T. *J. Am. Chem. Soc.* 2008, 130:6900.
75. Comby, S.; Gunnlaugsson, T. *ACSNano*, 2011, 9:7184–7197.
76. Truman, L. K.; Comby, S.; Gunnlaugsson, T. *Angew. Chem. Int. Ed.* 2012, 51:9624–9627.
77. Pope, S. J. A.; Kenwright, A. M.; Boote, V. A.; Faulkner, S. *Dalton Trans.* 2003, 2003:3780–3784.
78. Harte, A. J.; Jensen, P.; Plush, S. E.; Kruger, P. E.; Gunnlaugsson, T. *Inorg. Chem.* 2006, 45:9465–9474.
79. James A. Tilney, Thomas Just Sørensen, Benjamin P. Burton-Pyec and Stephen Faulkner, *Dalton Trans.* 2011, 40:12063–12066.
80. Parker, D.; Senanayake, P. K.; Williams, J. A. G. *J. Chem. Soc., Perkin Trans. 2* 1998, 1998:2129–2140.
81. dos Santos C.M.G., T. Gunnlaugsson, *Dalton Trans.*, 2009, 2009:4712–4721.
82. dos Santos C.M.G., P.B. Fernández, S.E. Plush, J.P. Leonard, T. Gunnlaugsson, *Chem. Commun.*, 2007, 2007:3389–3391.
83. dos Santos C.M.G., T. Gunnlaugsson, *Supramol. Chem.*, 2009, 21:173–180.
84. Bobba G., J. C. Frias and D. Parker, *Chem. Commun.*, 2002, 2002:890–891.
85. Bobba G., Bretonniere Y., Frias J.C., Parker D., *Org. Biomol. Chem.* 2003, 1:1870–1872.
86. Bobba, G.; Kean, SD.; Parker, D.; Beeby, A. Baker, G. *J. Chem. Soc. Perkin 2*, 2001, 2001:1738–1741.
87. Nonat A. M., S. J. Quinn and T. Gunnlaugsson, *Inorg. Chem.*, 2009, 48:4646–4648.
88. Glover, PB; Ashton, PR; Childs, LJ; Rodger, A; Kercher, M; Williams, RM; De Cola, L; Pikramenou, Z. *J. Am. Chem. Soc.* 2003, 125:9918–9919.
89. Jones J. E., A. J. Amoroso, I. M. Dorin, G. Parigi, B. D. Ward, N. J. Buurma and S. J. A. Pope, *Chem. Commun.*, 2011, 47:3374–3376.
90. Montgomery C. P., B. S. Murray, E. J. New, R. Pal and D. Parker, *Acc. Chem. Res.*, 2009, 42:925.
91. McMahon B.K., T. Gunnlaugsson, *J. Am. Chem. Soc.* 2012, 134:10725–10728.

# 7

# Upconversion of $Ln^{3+}$-based Nanoparticles for Optical Bio-imaging

*Frank C.J.M. van Veggel*

*Department of Chemistry, University of Victoria, Canada*

## 7.1 Introduction

The intriguing optical and magnetic properties of the trivalent lanthanide ions, the elements 58–70 that follow lanthanum in the Periodic Table (of the Elements), keep fascinating mankind and keep delivering new kinds of materials, even if the actual amounts of them in there can be quite small. These new materials seem very disparate in properties, ranging from new lasers and optical amplifiers, new catalysts, new supermagnets, to new probes for optical and magnetic resonance bio-imaging. The latter will be the focus of this chapter with a particular emphasis on new optical probes based on nanoparticles using a process called upconversion (more on this below). However, the magnetic resonance imaging will also be discussed, but rather briefly, because it is a "natural" extension and it is often combined with the optical probe (in so-called dual- and multi-imaging nanoparticles).

The lanthanides are also known as the rare-earths, but then scandium, yttrium, and lanthanum are included. The name rare-earth is clearly a misnomer and a leftover of the time when these elements were considered rare. They were and are certainly rather difficult to distinguish from each other because their chemistry is quite similar (but not their optical and magnetic properties), because the valence $4f$ electrons are shielded by filled $5s$ and $5p$ orbitals. They are not rare and large deposits have been found in China (with at the moment more than 90% of their "production"), Canada, and the United States. Their "production" is mostly a physical separation method, with common contaminants being zirconium and

*Luminescence of Lanthanide Ions in Coordination Compounds and Nanomaterials*, First Edition.
Edited by Ana de Bettencourt-Dias.

hafnium. The separation of the rare earth ions is a technological challenge because the chemical properties are so similar.

The upconversion process is a process in which two or more low-energy photons are converted into one of higher energy (more details on various mechanisms are described in Section 7.2). This process of converting two or more low-energy photons into one higher-energy photon is not restricted to lanthanides. One such other non-linear process is multi-photon absorption, via a quantum mechanically defined virtual energy level, followed by emission of one higher-energy photon (see Chapter 5). This process requires very high photon fluxes typically from femtosecond lasers, that is power densities of $10^5$ W $cm^{-2}$, because the two or more photons have to arrive on time for the species to undergo the electronic transition. A third method to produce higher-energy photons from lower-energy ones is via second- or third-harmonic generation in a non-linear medium (e.g. lithium niobate [1]). The best known upconversion based on lanthanides is arguably based on the $Er^{3+}/Yb^{3+}$ duet. The $Yb^{3+}$ ions are excited by 980 nm light, from for example an inexpensive CW 980 nm diode laser, and act as sensitisers of the $Er^{3+}$ ions. With two non-radiative energy transfer steps, the $Er^{3+}$ will be in an excited state from which green emission around 545 nm or red emission around 670 nm can occur (see Fig. 7.1).

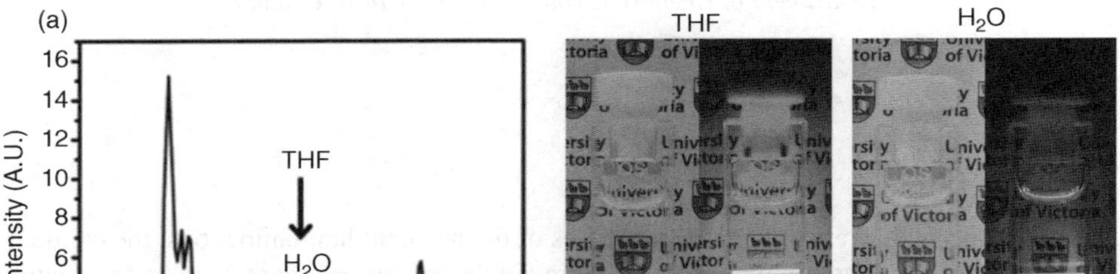


*Figure 7.1 Upconversion spectra of original oleate-stabilised NPs in THF. Reproduced with permission from [2]. Copyright © 2010, American Chemical Society. (See colour plate section)*

This non-linear process is possible at power densities of as little as several W $cm^{-2}$ and saturation occurs already at power densities of hundreds of W $cm^{-2}$. Saturation occurs because the system runs into a rate-limiting step, although often which one it actually is, is not investigated. The lower power densities are potentially very attractive because their use could avoid overheating of the tissue.

The second best case of $Ln^{3+}$-based upconversion is based on the $Tm^{3+}/Yb^{3+}$ duet, which can produce photoluminescence at 800 nm, through a two-photon process, and blue emission between 450 and 500 nm, through a three-photon process, and even ultraviolet emissions through multi-photon processes.

The blue and ultraviolet emissions are of little use for optical bio-imaging, because they would: (i) be absorbed, (ii) lead to autofluorescence (by excited fluorescent molecules in the medium), (iii) be heavily scattered thus blurring the image. The "real" prospect of optical bio-imaging stems from the $Tm^{3+}/Yb^{3+}$ duet, because excitation at 980 nm and emission at 800 nm are both in the biological window, where tissue is more transparent than to UV-Vis light. We probably all know this from experience. If we place our hand in front of a white light source the red light gets through where the skin is thin, because the blue and green get absorbed and scattered. I have also noticed this effect by aiming 980 nm light on the pulled skin of my elbow and seeing a near-infrared-sensitive card light up at the other side. The laser beam was maybe 1–2 mm in diameter, but the area that lit up on the card was more than 1 cm in diameter, so there was a significant amount of scatter of the excitation light of 980 nm. I estimated the thickness of my skin in this experiment to be about 5 mm, so with the significant observed scatter this points to a clear limitation in deep-tissue imaging if optical sectioning, that is lateral and depth resolution, is an essential factor in the bio-imaging of for example the blood capillaries in the brain of small animals (more on this later). If you would like to repeat this experiment be careful not to crank up the power of your diode laser too much because it will hurt you. It eventually hurts, also from personal experience, because enough 980 nm is absorbed to lead to an appreciable temperature increase, which leads to the usual "alarm" signals. Clearly, there is a limit to the power density of excitation that can safely be used in animal and human imaging. There are, however, limitations to the transparency window, mainly because water has several broad absorptions in the near-infrared, from overtones and combinations bands, with one notable at ~1450 nm. Thus, the "biological window" is often defined as being between 700 to 1100 or 1300 nm. The upper limit is rather ill-defined, but some people work with excitation at 1250 nm where there is a reduced absorption. Another potential advantage of working in the biological window is that there is no autofluorescence, because no fluorescent bio-organic moieties absorb in this wavelength range. An accompanying potential advantage is also that phototoxicity is absent. Yet another advantage is that the photocycle of the trivalent lanthanide ions is absolutely robust, that is, there is no photobleaching unlike organic molecules and polymers, inorganic complexes, and quantum dots and the like. This is because the optical properties of the trivalent lanthanide ions are in essence ionic properties, that is the energy levels of the ion determine the electronic transitions, despite the fact that the crystal field splitting has a very small effect and the ions can interact with each other. An example of the latter is the non-radiative energy transfer from an excited $Yb^{3+}$ ion to a (nearby) ground state $Er^{3+}$ ion.

## 7.2 Photophysical Properties of $Ln^{3+}$ Ions

The rich electronic structure of the trivalent lanthanide ions stems from three effects within the partially filled 4*f* orbitals. The major effect, in the order of $10^4$ cm$^{-1}$ is from spin–spin repulsion, followed by spin–orbit coupling, in the order of $10^3$ cm$^{-1}$, and then followed by the effect of the crystal field splitting, in the order of $10^2$ cm$^{-1}$. Hence, the electronic transitions, absorption and emission, are almost invariant to the crystal field, but fine splitting and changes in relative intensities do occur. The labelling of the energy levels of the $Ln^{3+}$ ions follows the Russell–Saunders coupling scheme and has the common form of $^{2S+1}\Gamma_J$, with the superscript 2S+1 being the spin multiplicity and S the total spin, the term Γ the total angular momentum, and the subscript J the total spin-orbit coupling quantum number (see Chapter 1). The rich electronic energy level structure of the trivalent lanthanide ions gives rise to transitions from the (deep) UV all the way down to the far-infrared. An important energy range is the near-infrared, which includes the telecommunication window between 1.33 and 1.55 μm, with the low energy being the most important (called the C-band) [3]. In 2002, we published the first $Ln^{3+}$-doped $LaF_3$ nanoparticles with efficient emissions in the near-infrared [4,5], followed by proof-of-principle of a polymer-based optical amplifier that operates in the telecommunication window (i.e. at 1.33 μm) [6].

An important thing to note is that the intra-4*f* electronic transitions are parity forbidden and should thus not be observed. However, the formal selection rules are "relaxed" by spin–orbit coupling, which is significant because the lanthanides have high atomic numbers. Another way of saying that the selection rules are relaxed is that they "steal" intensity from allowed transitions (e.g. the ground state $4f^n$ to 5*d* transitions). This forbidden nature of the intra-4*f* electronic transitions has two major consequences. One is the low molar extinction coefficient (or absorption cross section) with the $^2F_{7/2}$ to $^2F_{5/2}$ transition of $Yb^{3+}$ being one of the "strongest" at ~10 M$^{-1}$ cm$^{-1}$ (or ~$10^{-20}$ cm$^2$). The other major consequence is that the excited state lifetimes are long, from tens of microseconds to several milliseconds. The world record is probably the $^4I_{13/2}$ excited state of $Er^{3+}$, which has a radiative life time between 10 and 20 ms! These two effects are of course coupled in a sense, because if "up" is forbidden, "down" is also forbidden. This combined effect leads directly to a potential serious drawback for optical bio-imaging, where brightness is key to success. The brightness of an optical probe is defined as the number of photons that can be emitted in a given time slot with an excitation power density that does not lead to adverse effects, such as (local) heating. This will be discussed in Section 7.6.

## 7.3 Basic Principles of Upconversion

There are a few basic mechanisms involving $Ln^{3+}$ ions that can lead to upconversion. They will only briefly be discussed here with reference to an excellent review by Auzel [7]. The first is based on the ground state and excited state absorption with one $Ln^{3+}$ ion that leads, maybe with some loss in energy through a non-radiative process, to an excited state that can radiatively go to a lower excited state or the ground state by emission of the higher-energy photon (see Fig. 7.2, mechanism 1). $Er^{3+}$ can be excited with 980 nm light to produce green and red emission. A second mechanism (2, Fig. 7.2) is

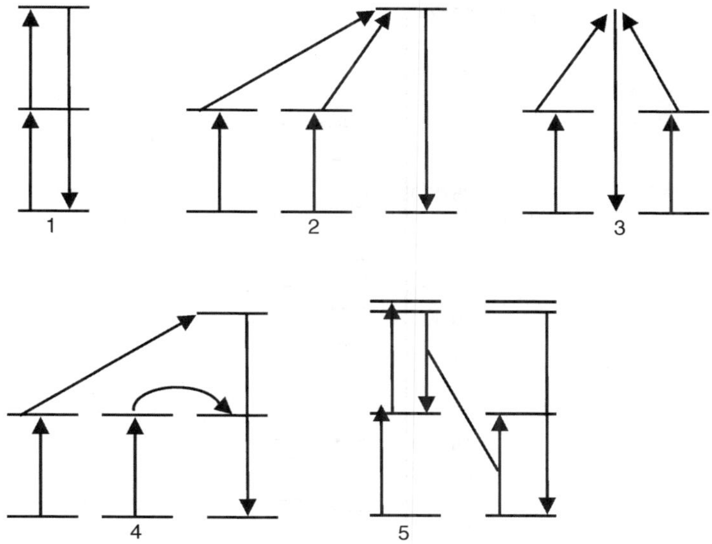


***Figure 7.2*** *Upconversion schemes; Reproduced with permission from [7]. Copyright © 2004, American Chemical Society*

the co-operative non-radiative energy transfer of two excited $Ln^{3+}$ ions to another third, nearby $Ln^{3+}$ ion, which can then decay radiatively. $Eu^{3+}$ and $Tb^{3+}$ can be excited through this upconversion mechanism using $Yb^{3+}$ ions as sensitiser. A third mechanism (3, Fig. 7.2) is the co-operative emission of two excited $Ln^{3+}$ ions. An example is the emission around 490 nm of $Yb^{3+}$-only doped materials when excited with enough 980 nm light. The fourth mechanism (4, Fig. 7.2), and the one most relevant to this review, is the consecutive non-radiative energy transfer from an excited $Ln^{3+}$ ion, the sensitiser or donor, to the other co-doped $Ln^{3+}$ ion, the acceptor, from which the higher-energy photon results.

The $Yb^{3+}/Er^{3+}$ case is as follows (see Fig. 7.3):

1. An excited $Yb^{3+}$ is produced through the absorption of a 980 nm photon giving the $^2F_{5/2}$ state.
2. This excited $Yb^{3+}$ interacts with a nearby ground state $Er^{3+}$ ($^4I_{15/2}$ state) which gets excited through a resonant, non-radiative energy transfer (mostly through a dipolar mechanism, but higher dipoles can also be involved), resulting in the $^4I_{11/2}$ state.
3. Another or the same $Yb^{3+}$ is excited (during the lifetime of the excited $Er^{3+}$).
4. This excited $Yb^{3+}$ ion interacts with the excited $Er^{3+}$ and, again, a non-radiative energy transfer process occurs leaving the $Yb^{3+}$ in the ground state and the $Er^{3+}$ ion in a higher excited state (more specifically the $^4F_{7/2}$ state, which relaxes non-radiatively to the $^2H_{11/2}$ and the $^4S_{3/2}$ states, the latter two being in thermal equilibrium at room temperature).

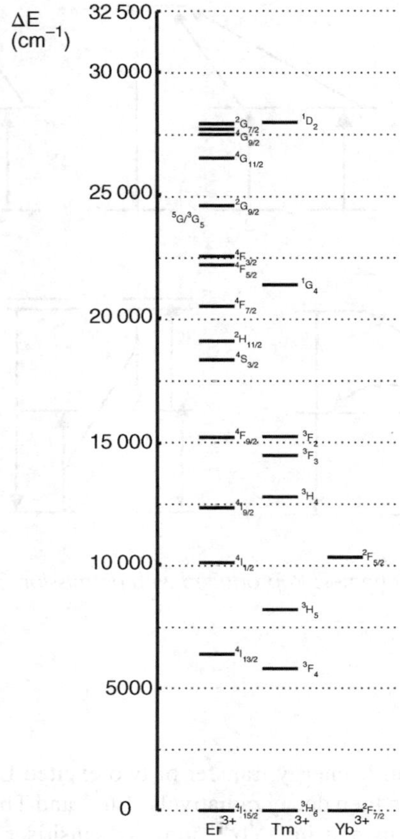


*Figure 7.3* *Part of the electronic energy levels of $Er^{3+}$, $Tm^{3+}$, and $Yb^{3+}$. Data on all $Ln^{3+}$ ions can be found in Reference [8]*

5. The $^2H_{11/2}$ and the $^4S_{3/2}$ states decay radiatively to the ground state by the emission of photon in the green (~525 and ~545 nm, respectively). These two states can relax non-radiatively to the $^4F_{9/2}$ state, from which red emission at ~ 670 nm can occur.

There is also another pathway to arrive at the $^4F_{9/2}$ state that goes via the $^4I_{13/2}$ excited state of $Er^{3+}$, generated through relaxation from the $^4I_{11/2}$ state, through a non-resonant non-radiative energy transfer step by interaction with a nearby excited $Yb^{3+}$ ion. Which process is dominant is often a subtle result of several effects, such as crystal field splitting, and so on, and would require detailed (time-resolved and temperature-dependent) photoluminescence studies to unravel. Further discussion is beyond the scope of this review.

The process for $Tm^{3+}$-based 800 nm emission through upconversion starts with a non-resonant energy transfer that leads to the $^3H_5$ excited state of $Tm^{3+}$, which then non-radiatively relaxes to the lowest excited state (i.e. $^3F_4$ state). The next non-radiative energy transfer step gives the $^3F_2$ state that relaxes, through the $^3F_3$, to the $^3H_4$ state from which 800 nm emission can occur.

The fifth (5 in Fig. 7.2), and somewhat more special, mechanism is the photo-avalanche mechanism that involves a non-resonant ground state absorption, a resonant excited state absorption, and a cross-relaxation step that produces the excited state from which excited state absorption occurs. The classical photo-avalanche mechanism is characterised by a clear threshold value for the power density because of the non-resonant ground state absorption. A variation on this classical scheme has been found by us [9].

Because of the very rich electronic structure of the $Ln^{3+}$ and because these ions are not used in vacuum many competing processes occur. These include:

1. Quenching processes, for example by surface bound ligands that stabilise the nanoparticles and/or (residual) absorbed water. The OH vibration of water is the most efficient quencher of excited $Ln^{3+}$ ions [10].
2. Unwanted cross-relaxation processes that destroy the key excited states.
3. Quenching by the host matrix, also referred to as phonon quenching.
4. Quenching by (surface) defects; unwanted photoluminescence from the excited $Ln^{3+}$ ions, for example the energy transfer efficiency of the $Yb^{3+}/Er^{3+}$(or $Tm^{3+}$) systems is not 100%, in part because there is $Yb^{3+}$ emission at ~1000 nm.

Number 1 in this list is typically reduced by increasing the distance between the quenchers and $Ln^{3+}$ ions by a core–shell architecture of the nanoparticles. Number 2 in this list, if it occurs, is hard to avoid because increasing the distance between the $Ln^{3+}$ ions by reducing the dopant concentration would likely also decrease the overall brightness. Nevertheless, often a maximum in the brightness can be found by systematic variation of the dopant concentration(s). In order to avoid that number 3 in this list is a dominant process, one has to choose low-phonon matrices, which is exactly the reason why most of the work is done by doping fluoride salts, for example $NaYF_4$, because they typically have phonon energies in the range of 300–350 $cm^{-1}$. The lattice phonon is due to the energy needed to excite a lattice vibration. The lower the phonon energy the more quanta are needed to quench an excited $Ln^{3+}$ ion (through the conservation of energy law) and the more quanta are needed the less likely this process becomes and thus becomes less efficient. Number 4 in this list is a little bit of a black box. (Surface) defects are often postulated as quenchers but nobody actually shows what the nature of this defect is. This could possibly be studied by using the world's most advanced electron microscopes. Defects in the interior of the nanoparticles are probably less of an issue, because the syntheses at 300 °C produce highly crystalline materials, but one never knows.

Based on the above, it will not come as a surprise that the quantum yield for upconversion is low. The quantum yield (of upconversion) is defined as follows:

$$\text{QY} = (\text{number of photons emitted})/(\text{number of photons absorbed})$$

It runs from 0 to 100% for a one-photon process, from 0 to 50% for a two-photon process, from 0 to 33% for a three-photon process, from 0 to 200% for a "half-photon" process such as in a quantum cutter, and so on. The essence of the method just involves two measurements. One is the photoluminescence intensity and the remainder of the excitation light of the sample and a reference measurement to determine the launched excitation light in the absence of any absorption. In practice, it is always wise to aim for just a little percentage absorption to avoid effects of re-absorption. Please note that some authors use quantum efficiency as their measure, but often this is done on an energy scale not a photon counting scale and consequently gives different numbers.

The best bulk material, $NaY_4$ doped with $Er^{3+}$ and $Yb^{3+}$, has an absolute upconversion quantum yield of 4% [11]. This is, do not get me wrong here, pretty high for a non-linear process, but far from the >80–90% quantum yields that organic dyes [12] and quantum dots have [13–17]. We were the first to report absolute upconversion quantum yields for the green emission of $Er^{3+}/Yb^{3+}$ doped $NaYF_4$ nanoparticles, stabilised with oleates and thus measured as dispersion in an apolar solvent [18]. The 100 nm nanoparticles have an absolute upconversion quantum yield for the green emission of $0.30 \pm 0.10\%$ at power densities of 150 W $cm^{-2}$, which is close to the start of the saturation! So, to call this class of nanoparticles efficient and bright seems totally inappropriate and I would like to urge everybody to avoid such misleading terms (more on the brightness in Section 7.6). A few remarks must be made here. The first is that our measurements were done using an $BaSO_4$-coated integrating sphere that was placed in the sample chamber of an Edinburgh Instrument's FLS 920 fluorimeter, with 980 nm excitation from a CW diode laser. The second is that, because the excitation (that part the light that is not absorbed) was measured on a near-infrared PMT (photomultiplier tube) and the green emission on a red-sensitive PMT, the system had to be calibrated for the relative sensitivity (which can be done with a calibrated tungsten lamp). The third is that reported (absolute) quantum yields of a non-linear process must be accompanied with the used power density. The upconversion intensity (I) scales with the power density (PD) to the power $n$, where $n$ is the number of photons involved, so the quantum yield is a function of the power density. This gives a convenient way to determine the number of photons involved by plotting the logI versus logPD, which gives a slope of $n$, because if $I \sim PD^n$ then $\log I \sim \log PD^n = n\log PD$. It needs to be stressed here that it is the power density (W $cm^{-2}$) that counts for a non-linear process not the launched excitation power (W). The fourth point to make here is that one should stay away from the saturation regime for quantum yield measurements because it would lead to erroneous values. Eventually the slope in the logI versus logPD plot will reach zero, which in the limit of very high PDs will give a quantum yield of 0%. The fifth point to make here is that one should always be aware of scatter effects. We noted that for nanoparticles larger than 30 nm, the solvent was not a good reference and we therefore used undoped samples with the same size and size dispersion. The reason for this effect is that the reflection of the $BaSO_4$ coating is not 100%, although pretty high and about the same over the wavelength scale, so one must mimic the average reflection path of the photon between sample and reference. I expect this to be the case for Teflon type coatings as well, which is another common coating for integrating spheres. This is also the reason why one would prefer only a little percentage absorption. The last point is that there is no accepted standard for upconversion to do relative measurements [19,20]. The 100 nm $Er^{3+}/Yb^{3+}$-doped $NaYF_4$ sample discussed above could serve as such reference [18], but the reported absolute quantum yield has to be confirmed by independent measurements. In addition, the absolute numbers for the red emission of $Er^{3+}/Yb^{3+}$-doped samples and the 800 nm emission from $Tm^{3+}/Yb^{3+}$-doped samples have yet to be determined by many other groups, although I came across one article that reported this number (see below) [21].

## 7.4 Synthesis of Core and Core–Shell Nanoparticles

### 7.4.1 Syntheses in Organic Solvent

There are two basic synthesis routes for making the fluoride salts of the lanthanides, that is $LnF_3$ and $MLnF_4$ (M = Li, Na, K), with most of the upconversion work concentrated on doped $NaYF_4$ core and core–shell nanoparticles, with $NaGdF_4$ being a popular second matrix because of the combination of optical and magnetic resonance imaging modalities. Almost all syntheses use oleic acid and/or oleyl amine as the stabilising ligands (the first as oleates!) in a high boiling solvent like octadecene and basically rely on the (relative) insolubility of the fluoride salts. Oleate and oleyl amine stabilise the formed nuclei and the important parameters to control are the ratio of the reagents to oleic acid and/or oleyl amine, the reaction temperature, often around 300 °C, and duration (of up to a couple of hours). The first route uses trifluoroacetate salts as the source of fluoride through their thermal decomposition. The second route typically uses the formation of "precursor" complexes/compounds that decompose at higher temperature to form nuclei that subsequently grow into nanoparticles. These precursor compounds could be the $Ln(OA)_n$ complexes ($n$ is possibly 3, because of the large excess of oleic acid used), but this aspect of the exact mechanism is not well documented.

The core–shell architecture gives a large variety of possible nanoparticles, because: (i) the shell could be the same as the core material but undoped (e.g. $NaYF_4$ on a doped $NaYF_4$ core, (ii) the shell could be the same as the core materials but doped with different $Ln^{3+}$ ions (e.g. a $Ce^{3+}/Tb^{3+}$ doped $NaYF_4$ shell on a $Er^{3+}/Yb^{3+}$ doped $NaYF_4$ core), (iii) the (doped) shell could be a different material than the core matrix material (e.g. a $NaGdF_4$ shell on a $Er^{3+}/Yb^{3+}$ doped $NaYF_4$ core). The usual strategy is to add the respective salts of the anticipated shell material to the dispersed core nanoparticles under otherwise similar conditions as for the core synthesis. Recently, we published a new method for the controlled growth of a shell, based on the deliberate defocussing of a nanoparticles dispersion by addition of smaller nanoparticles that form the source of the shell [22].

Isolation of the nanoparticles is typically done by adding a non-solvent, for example absolute ethanol, to the cooled reaction mixture, thus precipitating the nanoparticles, which can then be collected by centrifuge and subsequently washed to remove excess oleic acid and/or oleyl amine and other contaminants.

### 7.4.2 Syntheses in Aqueous Media

Similar strategies described in the previous paragraph are also used for the syntheses in aqueous media, but obviously with a water-soluble stabilising ligand (e.g. citrate [23]). If one works with ordinary laboratory glassware for synthesis, then the reaction temperature is naturally restricted to 100 °C, but one can use autoclaves to work at higher temperature. It is generally believed that the synthesis in organic solvents at ~300 °C leads to more crystalline nanoparticles, which is advantageous for their optical properties. This may also imply that the lower temperature syntheses lead to more defects that could potentially be quenchers of the $Ln^{3+}$ excited states involved in the (upconversion) process.

### 7.4.3 Surface Modification

Most of the anticipated uses of upconversion $Ln^{3+}$-based nanoparticles require dispersibility in biological media, with blood arguably being the most challenging medium, because of the presence of salts, protein, and so on that may lead to instability of the colloidal dispersion. It is thus no wonder that considerable effort has been devoted to surface modification of the oleate-stabilised nanoparticles. There are four basic strategies:

1. Silica coating by using a modified Stöber process. The Stöber process [24] is the controlled hydrolysis of tetraalkoxysilanes in aqueous medium.
2. Ligand exchange processes that swap the oleates (and amines) for water dispersible ones, the latter including (small) proteins and polymers as well.
3. Chemical conversion of the double bonds (of the oleates and/or oleyl amines) by oxidation to carboxylic acids or epoxides.
4. Intercalation of amphiphilic molecules or polymers into the alkenyl chains of the stabilising ligands. This approach is based on hydrophobic forces that "force" the lipophylic (= hydrophobic) part(s) of the amphiphilic molecule or polymer into the hydrophobic monolayer on the nanoparticles. The hydrophilic parts then make these nanoparticles water dispersible. This method has been reported using PEG-oleate [25] and derivatised PMAO [poly(maleic anhydride-*alt*-octadecene)] [26–29].

## 7.5 Characterisation

### 7.5.1 Basic Techniques

There are a handful of basic spectroscopic techniques to characterise these (core–shell) nanoparticles. The two almost always reported in article are (high-resolution) transmission electron microscopy [(HR-)TEM] [30] and X-ray (powder) diffraction (XRD). A basic TEM analysis relies on the interaction of the electron beam with the material to give a contrast (difference) in the bright field. There are multiple processes going on when the electrons pass the sample, with absorption, the production of characteristic X-rays, and elastically and inelastically scattered electrons among them [30]. From the contrast (and magnification) the average size and size dispersion can be determined. At higher magnification the nanoparticles may show lattice fringes that should have a periodicity that matches the crystal structure for one particular orientation of the nanoparticles relative to the electron beam. Because this relative orientation is not controlled, nanoparticles usually show different lattice fringes or none at all. Crystalline nanoparticles are pretty much the norm these days, with lattice fringes from one side of the nanoparticle to the other consistent with a single (nano)crystal. Such contrast is usually obtained but I would like to note that, in my own experience, the $LnF_3$ nanoparticles give much reduced contrast relative to $NaYF_4$ nanoparticles of the same size. This suggests to me that the $Ln^{3+}$ ions interact less with the electron beam than the $Na^+$ ions. If the nanomaterial is crystalline, an XRD pattern can be recorded which provides information of the present phase(s), and their relative amounts in case of more than one, and any impurity (crystalline) phase present. The presence of an amorphous phase as shell material (e.g. silica) or as impurity phase may appear as broad diffraction peaks but they may not show up at all, so some care has to be taken here.

The characterisation of core–shell structures by (HR-)TEM is rather straightforward if the shell materials are very different from the core materials. For instance, there is ample contrast difference if the shell material is silica or gold on a doped $Ln^{3+}$-based core. However, a challenge exists if the shell and core are both based on the lanthanide fluorides, because there is very little contrast in standard (HR-)TEM analyses. The core–shell nanoparticles may all seem to be larger than the core, but in my view this does not lead to the conclusion that "proper" core–shell nanoparticles have successfully been made (see below for an example from my own research). By "proper" core–shell nanoparticles I mean that the shell is completely surrounding the core and is (nearly) equally thick everywhere. I have also noticed that the core–shell is often more anisotropic in shape than the core nanoparticles, suggesting that the shell is not equally thick everywhere. Much to my surprise, this aspect is often ignored by my colleagues and they happily conclude that they made "proper" core–shell nanoparticles because the core–shell size is, on average, larger than the core size.

Another often used characterisation technique is energy-dispersive X-ray spectroscopy (EDS or EDX) in a scanning electron microscope (SEM), which provides the ensemble elemental composition. Inductively coupled plasma–mass spectrometry (ICP-MS) provides the same information but with higher accuracy.

### 7.5.2 Advanced Techniques

In addition to the above discussed basic spectroscopic techniques, I will only highlight some of the most-used advanced characterisation techniques and would like to refer to the literature cited for more information. One such advanced technique is high-angle annular dark field (HAADF) imaging in a TEM (Fig. 7.4). This method is based on the fact that some electrons will be scattered elastically with a certain scattering angle. These scattered electrons are detected by an annular detector that is situated below the sample. The observed contrast is, to a first approximation, proportional to the square of the atomic number Z of the element responsible for this scattered electron. Hence this technique is also known as Z-contrast imaging. However, other factors may also be part of the contrast such as strain and amount of materials (e.g. thin shell vs. core), so great care has to taken to make inferences about shell thickness, sharpness of the interface, and so on. Qualitatively, conclusions about core–shell structures based on a difference in contrast in the HAADF image seem justified.

A second technique in a HR-TEM that can also be operated in scanning mode is electron energy-loss spectroscopy (EELS) that is characterised by the loss of some of the electron's energy when interacting with the sample; hence this is an inelastic process. The energy loss can be measured in a kinetic analyser and is element specific. The two basic ways to execute this are as EELS line scans of individual nanoparticles (Fig. 7.4) or as two-dimensional EELS mapping (from which line scan profiles can be constructed) [31,32]. The latter naturally contains much more information, but does require that the nanoparticles be stable in the beam for a long time. Beam damage, especially in high-end TEMs, is a significant problem because of electron energies that go up to 300 keV.

An alternative to this technique is EDS line scans of individual nanoparticles. Element specific X-rays are produced as a result of the fact that some electrons of the beam will kick out a (core) electron, and this electron hole is filled by an electron of a higher shell with the concurrent release of the "excess" energy as a photon (as one of the possible pathways).

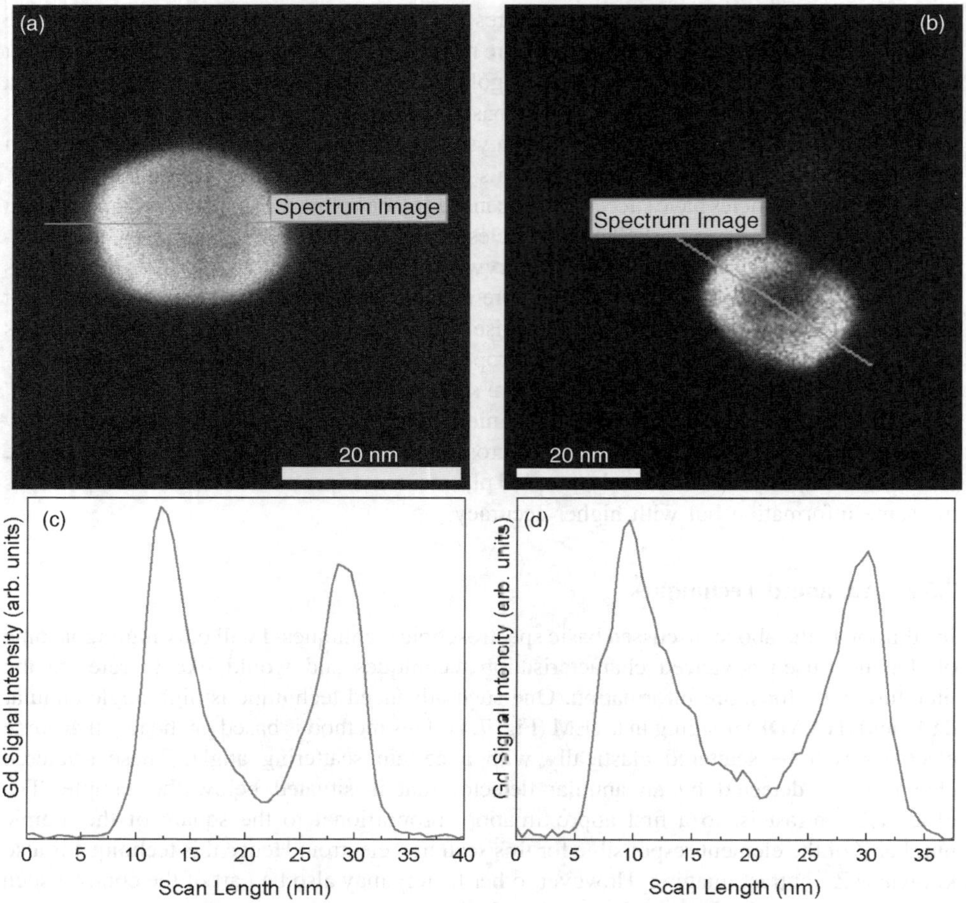


*Figure 7.4 (a, b) HAADF images of single $NaYF_4/NaGdF_4$ core/shell nanoparticles and (c, d) the corresponding EELS line scans showing Gd residing predominantly in the shell. Reproduced with permission from [31]. Copyright © 2011, American Chemical Society.*

An example is given in Fig. 7.5. In principle, EDS line scans provide the same information as EELS line scans but it is our experience that EDS is less sensitive, has a lower spatial resolution, and requires longer acquisition times thus increasing the demands of stability of the nanoparticles in the electron beam. However, there are constant improvements in detector sensitivity, spatial resolution, and so on, so this qualitative observation may already be obsolete. Beam damage can be suppressed significantly by working in a cryo-TEM [30]. Often just one line scan is being shown to prove the architecture of the nanoparticles, but in my view this is not sufficient to make the case.

A last set of advanced techniques is based on synchrotron radiation. In a synchrotron, a beam of electrons is accelerated to relativistic speeds and "bent" by so-called bending magnets or undulators. In order to conserve momentum, radiation through photons has to occur (this is a

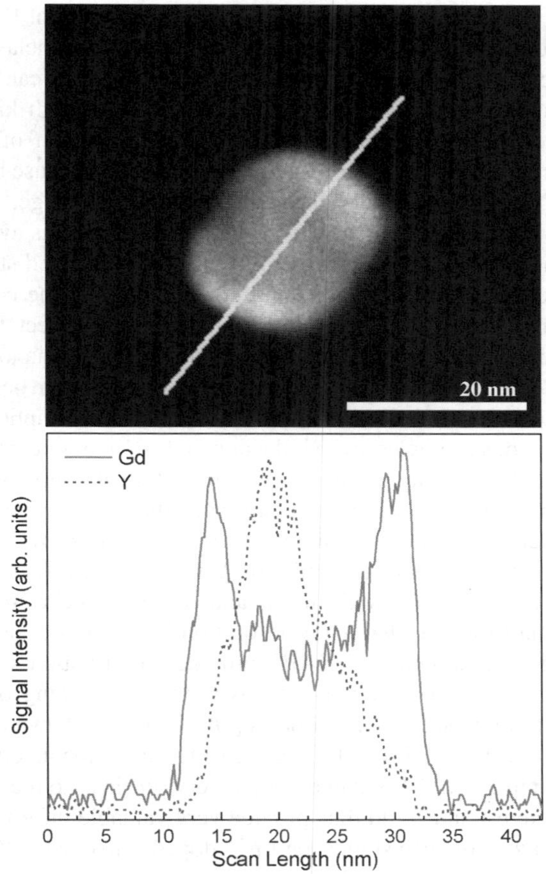


***Figure 7.5*** *HAADF image (top) and EDS line scan (bottom) across a single $NaYF_4/NaGdF_4$ core/shell NC showing Y in the core of the particle and Gd located in the shell. Reproduced with permission from [31]. Copyright © 2011, American Chemical Society.*

classical explanation that is by its very definition incomplete). This radiation occurs from infrared to hard gamma radiation and typically an "end station" or beam line is dedicated to a particular technique in a limited energy range. One technique we have exploited to address some fundamental questions is based on energy-dependent X-ray photo-electron spectroscopy (XPS). The basic phenomenon in XPS is the ejection of a (core) electron by absorption of an X-ray photon, the kinetic energy of which is determined in an analyser. This kinetic energy, $E_{\text{kinetic}}$, is related to the energy of the X-ray photon, $E_{\text{X-ray}}$, as follows:

$$E_{\text{kinetic}} = E_{\text{X-ray}} - E_{\text{binding}} + \Phi$$

$E_{\text{binding}}$ is the binding energy of the (core) electron and $\Phi$ is the work function of the particular instrument (which calibrates one instrument to another). The latter is usually a very small number relative to the other three quantities and does not need to be determined per se because one can use a well-defined reference for calibration. Because this method relies on the escape of (kinetic) electrons without losing energy via inelastic processes, it is a surface-sensitive technique in the sense that the (kinetic) electrons can only escape over a depth of several nanometres without losing their energy (although kinetic electrons are formed much deeper into the sample because of the penetration of the X-rays). This nanometre length scale gives the power to study nanoparticles because by tuning the X-ray energy, the "escape" length can be modulated at the nanometre range, so a depth profiling appropriate to study (spherical) core and core–shell nanoparticles is available. This nanometre depth profiling can be used to provide evidence for core–shell structures if the core $Ln^{3+}$ is different from the shell $Ln^{3+}$ ions and doping in either the core or shell is high enough to produce a sufficient signal to noise ratio [33,34]. One aspect that we have studied in detail is how the dopant $Ln^{3+}$ ions are distributed throughout the nanoparticles [35,36]. It seems reasonable to assume that the dopant ion distribution is random because the $Ln^{3+}$ ions are so similar in size and chemical reactivity. However, this assumption has been proven wrong by us for numerous examples, even if the dopant $Ln^{3+}$ ion is next to the host $Ln^{3+}$ ion in the Periodic Table. By the way, a very interesting historical overview has recently been published by Scerri [37]. It is somewhat surprising that this aspect is almost entirely overlooked in the field, whereas it seems obvious to want to know this if one wants to optimise an upconversion processes that relies on $Yb^{3+}$ as the sensitiser of other $Ln^{3+}$ ions!

Two other techniques that one could use available on a synchrotron are EXAFS and XANES. EXAFS stands for extended X-ray absorption fine structure and XANES for X-ray absorption near-edge fine structure. EXAFS provides information on the coordination environment of an specific atom or ion and has been used, among others, to study the coordination number of bimetallic nanoparticles [38,39]. I am not aware of such studies on $Ln^{3+}$-doped $LnF_3$-, $MLnY_4$- (M = Li-K), and $Ln_2O_3$-based upconversion nanoparticles. The spectrum measured by XANES is in principle very sensitive to the actual speciation of the element/ion of interest and could thus provide very valuable information. Also for this technique, I am not aware of such studies on $Ln^{3+}$-doped $LnF_3$-, $MLnY_4$- (M = Li-K), and $Ln_2O_3$-based upconversion nanoparticles.

There is one part of the puzzle that has not yet received much attention, which is the full characterisation of the stabilising monolayer. Questions that come to mind are how well packed this monolayer is when dispersed (in organic solvents or pH-buffered saline solutions, etc.) or how dynamic the binding of the stabilising moieties to the surface of the nanoparticles. We have addressed the first issue by performing Small Angle Neutron Scattering (SANS), showing that the oleates are highly solvated at least when they are on a PbSe quantum dot of ~5 nm in diameter [40]. Such studies have yet to be reported for $Ln^{3+}$-based nanoparticles.

One part of the full characterisation puzzle that has not received much attention at all is the interface of core and shell. Is this interface atomically sharp or is there a gradient of core $Ln^{3+}$ to shell $Ln^{3+}$ across this interface and, if so, how many monolayers is its thickness? This question is important I think, because in principle cation exchange could happen during the synthesis of the shell (thus scrambling to some extent the respective core and shell $Ln^{3+}$ ions or dopant ions). Cation exchange occurs in aqueous solution when nanoparticles with one $Ln^{3+}$ ion are exposed to another $Ln^{3+}$ ion (with no intent to grow a shell) [32,41]. This cation exchange process has been exploited by us to make core–shell nanoparticles of doped $NaLnF_4$ cores with a thin $NaGdF_4$ shell, as potential dual mode imaging contrast agents [32]. It seems to me that the interface is relatively sharp, that is not more than a few monolayers, based on some recent excellent EELS studies in a cryo-TEM [42]. Knowing the exact structure of this interface will tell us something important about how these core–shell structures are formed and maybe the nanoparticle formation by itself as well.

## 7.6 Bio-imaging

### 7.6.1 Basics

There are two basic ways to do animal imaging. The first is laser scanning in which the sample (stage) is rastered over a tightly focussed laser of which the focus point in the sample is controlled by the optics (see Fig. 7.6).

A set of *x-y* scans is collected at different focus depths and then combined by software to give a reconstruction of the three-dimensional image. The best spatial (= lateral and depth) resolution is obtained by using multi-photon (absorption) processes. The non-linear nature of the excitation gives rise to the fact that enough power density (= photon flux, so the units are Watts per area) is available only in the focus of the beam, thus providing the spatial resolution (several μm has been achieved) [43]. This spatial resolution is also called optical sectioning. An alternative to this method is the wide field mode, in which the whole area of interest is illuminated but now a position sensitive detector, for example a CCD camera, does the spatial resolving. This method is faster but the spatial resolution is lower.

The following is a discussion of the articles I found by doing a Web of Knowledge search in June of 2012 using the following key-words and Booleans: (nanoparticle* or nanocrystal* or nanophosphor*) and (upconversion or upconverting or up-conversion or up-converting) and (bio-imaging or bioimaging or imaging or bio-labeling or biolabelling or labelling) and [(cell studies or cell study or cellular) or (small animal*)].

### 7.6.2 Cell Studies

The first reports on cell and animal studies were published in 2008. If animal studies are also discussed in the cited work, they will be discussed in the next section. Das and Tan [44] used amine-functionalised $Y_2O_3$ co-doped with, among others, $Er^{3+}$ and $Yb^{3+}$ and showed that the (acute) cytotoxicity to human hepatocellular carcinoma (Hep-G2) cells is very low.

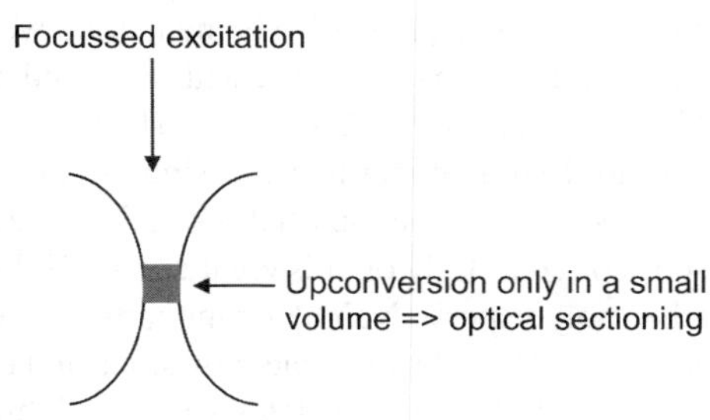


*Figure 7.6* *Only in the focus of the excitation beam (indicated as a dark square) is the power density (or photon flux) high enough to drive the non-linear processes*

Unfortunately, no optical cell or animal studies were reported. Li et al. [45] used a strategy in which they performed a chemical conversion of the double bonds in the oleate-stabilised $LaF_3$ co-doped with upconversion $Er^{3+}/Yb^{3+}$ to epoxides, using *m*-chloroperbenzoic acid, followed by ring opening with methoxy-terminated polyethylene glycol (mPEG-OH). The upconversion, upon 980 nm excitation (no power density stated), was dominated by the green emission from $Er^{3+}$. Human nasopharyngeal epidermal carcinoma cells (KB cells) showed little loss (up to ~20%) of cell viability and incubated cells were visualised by confocal microscopy and the authors concluded that the nanoparticles had been internalised. Jalil and Zhang reported an extensive study on silica-coated upconversion nanoparticles in a rat model [46]. Cell toxicity (viability) of rat skeletal myoblasts and bone marrow-derived mesenchymal stem cells were exposed to up to 100 μg ml$^{-1}$ of nanoparticles, with little adverse effects at the lower doses (i.e. 1 μg ml$^{-1}$), but considerable cell death (up to 35–40%) for the highest dose. They also studied the biodistribution, with fast accumulation in the heart, lungs, spleen, kidney, and liver, but near-complete clearance after 7 days. Shan et al. [47] reported on the cytotoxicity and cell uptake of $Er^{3+}/Yb^{3+}$ doped $NaYF_4$ nanoparticles with positive and negative functional groups on the surface. The former were synthesised by coating first with silica and then reacting the silica with 3-aminopropyltrimethoxysilane (APTMS), the latter by adding an amphiphilic polyacrylic acid. Sizes were in the range of 50–100 nm as revealed by TEM. Cell viability studies showed a small negative effect on the HOS (human osteosarcoma) cell line. Nanoparticles with carboxylic acids on their surface were taken up by cells, but no similar studies were done for the amino group terminated nanoparticles. Zako et al. [48] modified $Er^{3+}$ doped $Y_2O_3$ nanoparticles with APTMS and then reacted the (surface) amino groups with NHS- (*N*-hydroxysuccinimide) activated carboxy-terminated PEG. The size distribution by SEM was between 20 and 140 nm, with the maximum at 40 nm. The nanoparticles showed a noisy upconversion spectrum with 980 nm excitation, giving the characteristic green and red emission, with the red dominant (no power density was given). These particles showed no detrimental effects on the cell viability of PC12 cells (a clonal line of rat pheochromocytoma). Haase [49] showed the uptake of $NaYF_4{:}Er^{3+},Yb^{3+}$ nanoparticles by the roots of *Phalaenopsis* and *Arabidopsis* plants. The characteristic upconversion was observed by confocal laser scanning microscopy in the shoot and leaves of the plants. Jiang et al. [50] showed cell targeting by modifying silica-coated $Er^{3+}/Yb^{3+}$-doped $NaYF_4$ of ~20 nm overall size with folic acid and anti-Her2 (human epidermal growth factor receptor 2). These label HT-29 and SK-BR-2, respectively. They also showed the attachment of small interfering RNA (siRNA) to the anti-Her2 coated nanoparticles for targeted delivery. They observed a gene silencing effect of 45%. Cell viability was hardly affected. The nematode worm *C. elegans* takes up sub-10 nm unpcoverting nanoparticles ($Y_2O_3$ doped with $Er^{3+}$ and $Yb^{3+}$ and modified with 3-aminotrithoxysilane, APTES) as shown by Lim et al. [51] The worms showed almost no adverse effects by exposure to these nanoparticles. A similar study was reported by Chen et al. [52] with $Er^{3+}/Yb^{3+}$-doped $NaYF_4$ nanoparticles of 35 nm size from a synthesis in an ionic liquid. Yan [53] also reported a study on this creature and HeLa cells, showing uptake and low cytotoxicity. Rather polydisperse $NaYbF_4$ nanoparticles, doped with $Er^{3+}$, $Tm^{3+}$, or $Ho^{3+}$ (with approximate size of 15 nm) were coated by silica and used in some cell studies by Wang et al. [54]. They coated these nanoparticles with rabbit anti-CEA8 to label HeLa cells. Specificity was supported by a control experiment using amino-functionalised nanoparticles that showed no HeLa cell labelling. U87MG cancer cells were labelled by

Zako et al. [55] using upconverting $Y_2O_3$-based nanoparticles coated with cyclic RGD for specificity. The cyclic RGD was added via a Michael addition to pendant maleimide groups on PEG units that were covalently bound to the APTES -modified $Y_2O_3$ nanoparticles. Human breast cancer cells (MCF-7) were used as the negative control. We showed that water-dispersible upconversion nanoparticles (co-doped with $Tm^{3+}/Yb^{3+}$ and $Er^{3+}/Yb^{3+}$ and ~20 nm in size) can be made via simply replacing the oleate groups with PEG-phosphate. The upconversion intensity of the water-dispersed nanoparticles is much reduced compared to the oleate-stabilised nanoparticles dispersed in THF. A core–shell structure restored the intensity to some extent. CaOV3 cells (an ovarian cancer cell line) were successfully labelled, through the mapping of the 800 nm emission upon 980 nm excitation of the $Tm^{3+}/Yb^{3+}$ couple, and the uptake, likely through ordinary endocytosis, was shown by TEM analysis of sliced fixed cells. The cells did not show adverse effects during their incubation with nanoparticles (see Figs 7.7 and 7.8) [2].

Relative to polyethylene glycol and silica coating, polyvinyl pyrrolidone (PVP) is seldom used despite its excellent biocompatibility and ease of synthesis of end-group functionalised versions [56]. Wolfbeis [57] reported the use of so-called click chemistry to modify their silica-coated upconversion nanoparticles bearing azide groups, to introduce biotin and maleimide. Ryu et al. [58] doped $NaGdF_4$ of 10–40 nm diameter with $Tm^{3+}/Yb^{3+}$ and $Er^{3+}/Yb^{3+}$ for dual imaging use. They made their nanoparticles water dispersible by coating an

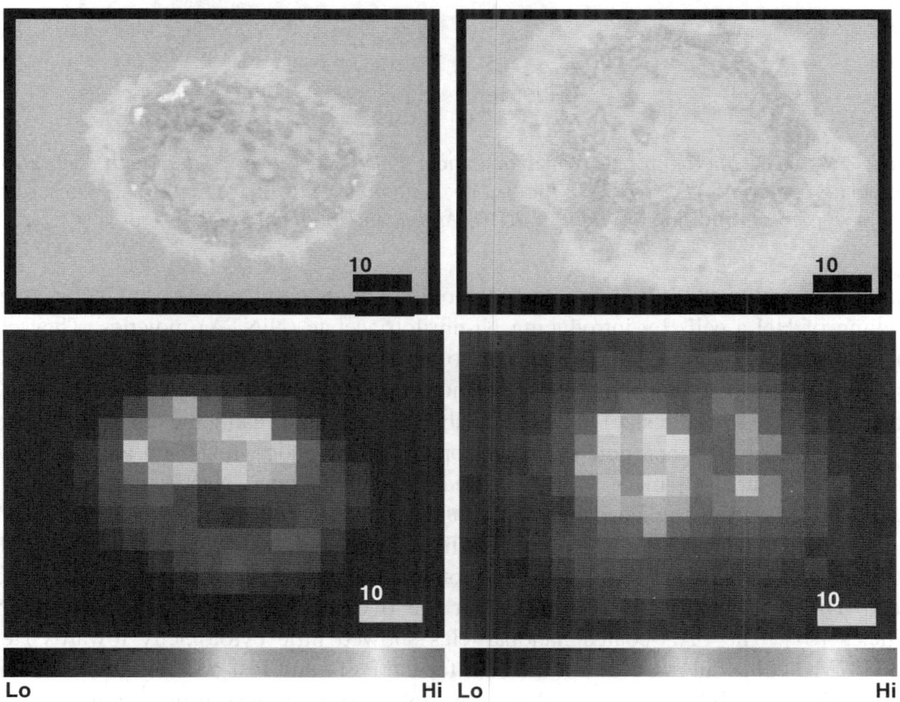


*Figure 7.7 Upconversion imaging of ovarian cancer cells CaOV3 incubated with PEG(2000)-phosphate coated $NaYF_4$. Reproduced with permission from [2]. Copyright © 2010, American Chemical Society. (See colour plate section)*

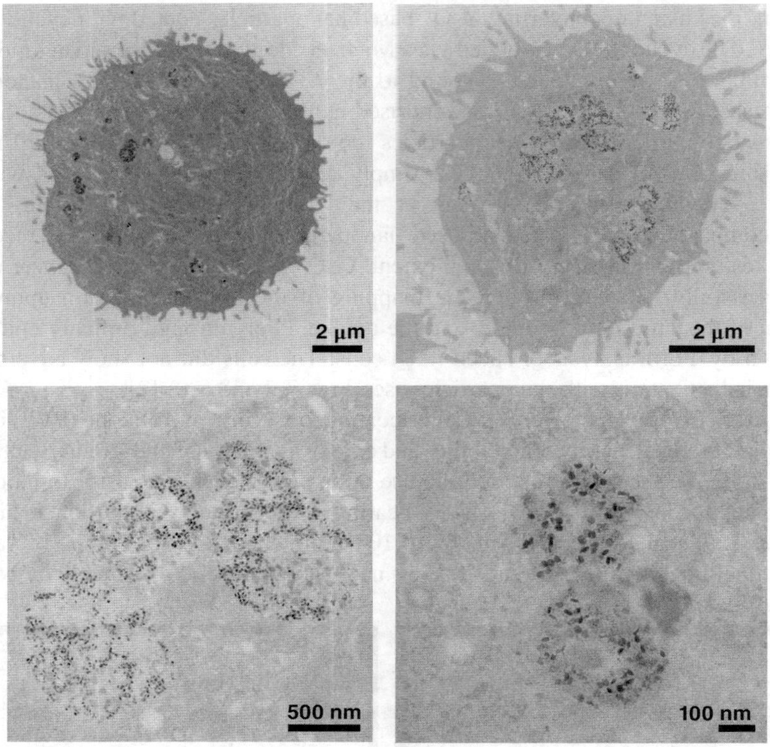


***Figure 7.8*** *TEM micrographs of ovarian cancer cells CaOV3 incubated with PEG(2000)-phosphate coated $NaYF_4$: $Tm^{3+}$ 0.5 mol%, $Yb^{3+}$ 30 mol% NPs for 24 h at 37°C. Reproduced with permission from [2]. Copyright © 2010, American Chemical Society*

amphiphilic polymer, poly(acrylic acid) modified with octyl amine, and showed cell labelling of HeLa cells by introducing Ni-nitrilotriacetate (NiNTA) moieties. They concluded that these nanoparticles were toxic to the HeLa cells. All the MRI results will be discussed below. Capobianco [59] used "non-functionalised" $Er^{3+}/Yb^{3+}$-doped $NaYF_4$ nanoparticles (~18 nm in size) to image incubated HeLa cells, using the red emission from $Er^{3+}$. The HeLa cells incorporated the nanoparticles by endocytosis and appeared to be healthy. Jin et al. [60] made PVP-stabilised $Er^{3+}/Yb^{3+}$-doped $NaYF_4$ nanoparticles of roughly 50 nm size and replaced the PVP with PEI (polyethyleneimine) or PAA (polyacrylic acid) to control the surface charge from positive to negative, respectively. The PVP-, PEI- and PAA-coated nanoparticles had a zeta-potential of 10.2, 51.1, and −22.5 mV, respectively. The zeta-potential is a measure for the overall charge of the dispersed nanoparticles [61]. The PAA-coated nanoparticles showed little cytotoxicity towards HeLa and U87MG cells, but this was not the case for the PVP- and PEI-coated ones, which had cell viabilities of 80% or less after 24 h of exposure to up to 500 µg $ml^{-1}$. Only the PEI-coated nanoparticles were taken up by HeLa cells, by clarthin-mediated endocytosis, with very little if any by the other two types of nanoparticles. This was confirmed by TEM analysis. Nagarajan and Zhang studied silica-coated $Er^{3+}/Yb^{3+}$- and $Tm^{3+}/Yb^{3+}$-doped

$NaYF_4$ nanoparticles in a phantom that mimics tissue. The green upconversion from $Er^{3+}$ was measurable from a depth of up to 1 cm, but the 800 nm upconversion from $Tm^{3+}$ was measurable only up to 6 mm. The latter reduced depth was due to a lower detector sensitivity, because theoretically it should be better. A long-term cell tracking study was published by Nam et al. [62] using 30 nm $NaYF_4{:}Er^{3+},Yb^{3+}$ nanoparticles that were made water dispersible by coating a second layer of PEG-phospholipid. Lin [63] showed an interesting pH-induced thermal release of a drug in $Er^{3+}/Yb^{3+}$-doped $NaYF_4$ nanoparticles with a hydrogel as coating. The hydrogel was poly[(*N*-isopropylacrylamide)-co-(methacrylic acid)] (P(NIPAM-co-MAA)) (prepared using 5 mol% of MAA). The drug released from the hydrogel was doxorubicin hydrochloride. Cell uptake by HeLa cells was also shown by Wang et al. [64] for $NaYF_4{:}Er^{3+},Yb^{3+}$ nanoparticles. $YF_3$-based upconversion particles larger than 100 nm were reported by Zhong et al. [65] and they conclude that the HeLa cells took up these particles, but I am not convinced by their image. PVP was used by Yin et al. [66] to make $Ln^{3+}$ doped $GdVO_4$ nanoparticles that were coated with silica. Standard luminescence and upconversion studies were shown on HeLa cells showing uptake and some cytoxicity. Zhang and Stucky and their co-workers reported "nano-rattles" for *in vitro* and *in vivo* upconversion studies [67]. The nano-rattle could be loaded with doxorubicin for cancer treatment. Cell death was appreciably higher for the loaded nano-rattles compared to free doxorubicin, which could be due to increased uptake by the cancer cells. Not strictly belonging to this review, but the work by Speghini with upconversion $CaF_2$-based nanoparticles deserves mentioning [68]. They show that penetration depth of up to 2 mm can be achieved with 800 nm emission, which is four times deeper than the visible light from $Er^{3+}$.

### 7.6.3 Animal Studies

Chatterjee et al. [69] reported on the use of $NaYF_4$ nanoparticles doped with $Tm^{3+}/Yb^{3+}$ and $Er^{3+}/Yb^{3+}$ of ~50 nm in diameter with PEI, polyethyleneimine, as stabilising coating. The $Tm^{3+}/Yb^{3+}$ co-doping gave the usual blue emission upon excitation with 980 mm light. But the 800 nm emission, which can be quite strong relative to the blue emission, was not recorded. The $Er^{3+}/Yb^{3+}$ doped nanoparticles, upon 980 nm excitation, gave green (540 nm) and red (653 nm) emissions. They showed that no degradation of the upconversion intensity occurred, but the power density was not stated. Mouse biodistribution studies, after tail injection, showed an initial accumulation in the lungs, heart, spleen, kidney, and liver, with very little remaining after 7 days. They showed specific uptake of folic acid coated nanoparticles by human adenocarcinoma cells (HT-29 cells) but hardly any by ovarian carcinoma cells (OVCAR3), both overexpressed with folate receptors. The authors estimated that the emission from upconverting $Er^{3+}/Yb^{3+}$ nanoparticles up to 10 mm deep can be obtained from rats with local injection; the depth was estimated from the needle penetration and hair was shaved off. Prasad et al. [70] used rather ill-defined $Tm^{3+}/Yb^{3+}$ doped $NaYF_4$ nanoparticles in the size range of 20 nm, in which the initial oleates had been replaced by incubation and stabilised by 3-mercaptopropionic acid. Excitation with 980 nm led to 800 nm light as the most intense emission, along with blue and red emissions. They used the 800 nm light for the animal studies. Exposure of human pancreatic cancer cells (Panc 1) to these nanoparticles showed about 10% loss of cell viability at 2.0 mg $ml^{-1}$ levels. The cell studies by microscopy showed a good correspondence between the upconversion signal and the ordinary light image, suggesting uptake by the cells. They

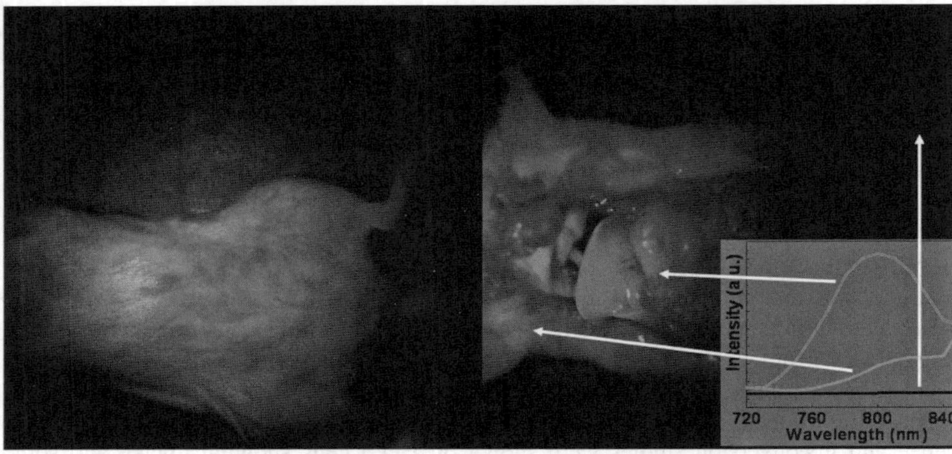


***Figure 7.9*** *Whole body images of mouse injected iv with upconverting nanoparticles; intact mouse (left), same mouse after dissection (right). The red colour indicates emission from upconverting nanoparticles, green and black show background as indicated by the arrows. The inset presents the PL spectra corresponding to the spectrally unmixed components of the multispectral image obtained with the Maestro system. Reproduced with permission from [70]. Copyright © 2008, American Chemical Society. (See colour plate section)*

were able to visualise the nanoparticles in a Balb-c mouse with a CCD camera through skin with hair, upon exposure to 975 nm light from a fibre coupled diode laser that was defocused (Fig. 7.9).

Post mortem analysis showed high uptakes in the liver and spleen. Unfortunately, power densities, not even the used power, are stated, so it is unclear if and to what extent heating had occurred. Salthouse et al. [71] transferred "red emitting" yttrium oxide to water by coating with polyacrylic acid, which was modified by EDC coupling with PEG-amine, giving a hydrodynamic size of 175 nm as determined by dynamic light scattering (DLS). Excitation with 980 nm light gave an emission between 650 and 700 nm (most likely from $Er^{3+}$, but the green was not recorded; it also not clear if the samples were co-doped with $Yb^{3+}$). Injection of this material into a nude mouse and dissection of the animal after one day showed uptake in the liver, with little to nothing in the spleen and kidney. Zhang [72] showed the myoblast cell uptake of silica-coated $Er^{3+}/Yb^{3+}$-doped $NaYF_4$ nanoparticles (50 nm in overall size) and showed their flow through blood vessels in a mouse. $Y_2O_3$ nanoparticles co-doped with $Er^{3+}$ and $Yb^{3+}$ and coated with polyacrylic acid that was modified with 2000 MW mPEG-$NH_2$ were contrasted against the Cy5.5 dye for depth penetration in both phantoms and mice [73]. The Cy5.5 gave emission at 710 nm upon 680 nm excitation and the usual red emission at 670 nm upon 980 nm excitation. These studies clearly showed that the Cy5.5 emission is much less attenuated by an optically tissue-equivalent phantom than the nanoparticles' emission over a range of 1 to almost 10 mm. This shows that using the green and red upconversion from the $Er^{3+}/Yb^{3+}$ couple is hardly beneficial if deep-tissue imaging is required. Nonetheless, the superior photostability of $Ln^{3+}$-based luminescent labels remains an advantage for long-term studies. By insertion of a glass capillary into the esophagus of a mouse, which was squeezed to about

12 mm thickness for the measurements, it could be shown that the upconversion nanoparticles did not display autofluorescence and that the shape of the capillary could be reproduced. The analogous Cy5.5 experiments clearly showed autofluorescence, thus blurring the image. In my search I found only one report, by Liu et al. [74], that describes the use of fluorescence diffuse optical tomography using the emission of upconverting nanoparticles. The authors used $NaYF_4$:$Tm^{3+}$,$Yb^{3+}$ in a gelatine phantom with 480 mW $cm^{-2}$ excitation at 980 nm, and could reconstruct the embedded glass tube. Significant improvement was observed through the non-linear character of the upconversion versus a one-photon probe (Rhodamine 6G) at 85 mW $cm^{-2}$ excitation at 532 nm. Tian et al. [75] reported the imaging of the subcutaneous injection of $NaYF_4$ nanoparticles doped with $Tm^{3+}$/$Yb^{3+}$, $Er^{3+}$/$Yb^{3+}$, $Ho^{3+}$/$Yb^{3+}$/$Ce^{3+}$ of ~100 nm in diameter with 974 nm excitation (20 W $cm^{-2}$). Unfortunately, it is unclear from what depth the signals originated, but it is likely just below the skin. Li [76] studied the biodistribution and clearance of poly(acrylic acid)-stabilised doped $NaYF_4$ nanoparticles of 11.5 nm in diameter from mice. They found heavy localisation in liver and spleen and some in the lungs, and relatively slow clearance, that is, it took at least 115 days to remove nearly all injected materials from the mice. A neuro-toxicity study with rod-shaped, doped $NaYF_4$ nanoparticles of 25–55 $nm^2$ on Balb-c nude mice was reported by Yu et al. [77] They conjugated chlorotoxin (CTX) to the surface amino groups, from PEI, on the nanoparticles. The used rat glioma C6 cells showed cell viability of more than 80% and the initial uptake after 24 h in lung, liver, spleen, kidney, and heart fell down quickly to levels below 20 μg $g^{-1}$ after 7 days. They imaged a xenographed tumour injected subcutaneously in a mouse predominantly through the red upconversion of $Er^{3+}$/$Yb^{3+}$-doped $NaYF_4$ nanoparticles, with very little of the green coming through the tissue. The intrinsic green to red ratio was about one. The power density at the animal was 500 mW $cm^{-2}$. Li [78] used triply doped, rather ill-defined $NaGdF_4$ nanoparticles for both optical and MR imaging. The doping was done with $Er^{3+}$, $Tm^{3+}$ and $Yb^{3+}$ showing the expected emission in the green and red from $Er^{3+}$ and around 800 nm from $Tm^{3+}$ upon 980 nm excitation. They converted the oleate-stabilised nanoparticles by oxidation of the double bonds to carboxylic acids. Cell viability of KB cells was still more than 80% after 12 h of exposure at 500 μg/ml. They were able to get a clear upconversion signal by collecting the red emission from $Er^{3+}$, but hardly through collecting the green emission from $Er^{3+}$ after intravenous injection. Post-mortem analysis showed high concentrations in the spleen and liver. A similar study by Li using $Er^{3+}$/$Yb^{3+}$- and $Tm^{3+}$/$Yb^{3+}$-doped $LaF_3$ was recently published [79]. Targeting was done with surface-bound folic acid of KB cells, with MCF-7 as negative controls. Cheng et al. [80] published a nice extension of FRET (fluorescence resonance energy transfer), which they dubbed LRET (luminescence resonance energy transfer). They physisorbed dyes into the stabilising organic layer of $Er^{3+}$/$Yb^{3+}$- and $Tm^{3+}$/$Yb^{3+}$-doped $NaYF_4$ nanoparticles. The organic layer was composed of pegylated PMAO [poly(maleic anhydride-*alt*-octadecene)] that was intercalated into the oleate layer on the nanoparticles (~34 nm in diameter). The $Er^{3+}$/$Yb^{3+}$-doped nanoparticles showed green and some red upconversion upon 980 nm excitation. The upconversion and LRET to rhodamine B and 6G, and TQ1 was used to image athymic nude mice that were injected subcutaneaously. Imaging was done with 0.2 W $cm^{-2}$ 980 nm excitation to show proof of principle for multiplexing, by injection at different locations in the back of the mice. Another approach for multiplexing was reported by Kim [81] who injected subcutaneously HeLa cells labelled with different quantum dots and $NaYF_4$ nanoparticles co-doped with

$Tm^{3+}$ and $Yb^{3+}$ into the backs of hairless mice. No power density nor launched power were given. Li [82] published a trimodal imaging approach combining upconversion, fluorescence, and positron emission tomography. The upconversion was from $Er^{3+}/Yb^{3+}$- or $Tm^{3+}/Yb^{3+}$-doped $NaYF_4$ nanoparticles, the fluorescence close to 600 nm from an Os(II) complex, and the positron emission tomography (PET) from $^{18}F^-$; the latter was introduced by simply exposing the "cold" nanoparticles to $^{18}F^-$ and isolation by centrifugation. Some standard HeLa cell experiments were done, showing some loss in cell viability at 800 μg $ml^{-1}$ of the nanoparticles and uptake by them. Furthermore, they showed the *in vivo* imaging of the lymphatic system by upconversion and the uptake in the liver and spleen by PET after intravenous injection. A similar study was also reported on tri-modal imaging, that is upconversion, PET through $^{18}F^-$ and MRI (the latter will be discussed below) [83]. The same group also reported on $Gd^{3+}/Yb^{3+}/Tm^{3+}$-doped $NaLuF_4$ nanoparticles and suggested that an upconversion signal from 2 cm deep in a mouse can be obtained [21]. Unfortunately, the used power density was not given. Following our method [18], they measured an absolute quantum yield of 0.47% for the 800 nm upconversion of $Tm^{3+}$. The used power density for this experiment was 17 W $cm^{-2}$, but it is unclear to me if this is close to the saturation because the log(I) versus log(PD) plot was not given. Yet another, quite similar study was reported on the same host doped with $Yb^{3+}$ and $Er^{3+}$ or $Tm^{3+}$ [84]. Rather non-monodisperse $Er^{3+}/Yb^{3+}$-doped $NaYF_4$ nanoparticles of about 60 nm size were used by Wang et al. [85] and were made water dispersible by using 3-mercaptopropionic acid, 6-aminocaproic acid, and poly(ethylene glycol) methyl ether. Studies with HeLa and human lung carcinoma A549 cells showed uptake and little cytotoxicity. One mouse was shown with subcutaneously injected nanoparticles that lit up upon 980 nm excitation. As dual-mode imaging nanoparticles, Li [86] also made upconverting nanoparticles with an iron oxide shell, the latter for $T_2$ contrast. The usual cell and small animal studies were done with respect to the upconversion (from the $Tm^{3+}/Yb^{3+}$ couple in $NaYF_4$). As a way to reduce the heating of the tissue, by absorption of the 980 nm light (mostly) by water, Zhan et al. [87] used 915 nm as the excitation source for the upconversion process for $Tm^{3+}/Yb^{3+}$-doped $NaYF_4$ nanoparticles. They showed convincingly that the heating and thus cell death is significantly reduced when using 915 nm compared to 980 nm excitation. The water absorption peak is at 974 nm, but the nanoparticles show considerable absorption just over 900 nm. It would be interesting to know what the absolute quantum yields of upconversion are for the two excitation wavelengths. HeLa cells were also imaged and an *in vivo* proof of principle that 915 nm light can be used as the excitation source was shown with subcutaneously injected nanoparticles into a nude mouse. Targeted *in vivo* imaging with upconverting $NaYF_4:Yb^{3+}$, $Er^{3+}$ nanoparticles was reported by Hischemöller et al. [88] by coupling anti-MUC-1 to the nanoparticles, which is specific to single chain Fv fragments. They replaced the initial hydroxyethyl-diphosphonic acid by polyacrylic acid and bio-conjugation was done with EDC as coupling agent. These authors showed a proof of principle with nude mice.

### 7.6.4 Discussion

A lot of the optical work published would be hard to reproduce because the used power densities are not given, and sometimes not even the launched powers. The power density matters and not the launched power because the upconversion processes are non-linear. This would be especially important for *in vivo* experiments, because heating of the tissue limits the maximum power density (at the surface of the animal). This point was nicely discussed by Zhan et al. [87]. Relatively little work has been done with the 800 nm emission from $Tm^{3+}$, which could be because of hardware issues. Most systems have detectors geared towards maximum sensitivity in the green, because many fluorescent dyes and fluorescent proteins work in this range. Many of my colleagues, and us initially but not anymore, claim that their upconverting nanoparticles are bright and/or efficient, but this seems a very overstated claim. This point of view can be illustrated by defining a figure of merit for brightness, that is the number of photons per second per nanoparticle:

$$\mathrm{FOM}_{\mathrm{brightness}} = [\text{number of photons absorbed} \times \text{QY of UC}]/\text{life time of excited } \mathrm{Ln}^{3+}$$

with QY of UC standing for quantum yield of upconversion. The absorption of $Yb^{3+}$ around 980 nm is only 10 $M^{-1}\,cm^{-1}$, which is low compared to dyes with molar absorptivities of up to 100 000 $M^{-1}\,cm^{-1}$. The quantum yield of upconversion for 100 nm $Er^{3+}/Yb^{3+}$-doped $NaYF_4$ nanoparticles is only 0.30% (for the green emission). Finally, the excited state lifetimes of the $Ln^{3+}$ are several hundreds of microseconds to milliseconds in these low-phonon matrices. If one integrates all this in this simple equation, then it seems hard to argue that these nanoparticles are bright. If one is restricted to $Yb^{3+}$ as the absorber and sensitiser, then the absorption cannot be altered much. The quantum yield of upconversion is probably limited to just a few percent, because the bulk doped $NaYF_4$ has 4% UC efficiency and this host is considered the most efficient for upconversion [11]. This leaves the lifetime to be changed, preferably by orders of magnitude. In principle this could be done by combining the upconverting nanoparticles with plasmonic (metallic) nanoparticles, but this has yet to be shown. Greatly increasing the excitation power is not going to be without concern, because of over-heating and the fact that optical sectioning will be lost. The optical sectioning will be lost, because now also in the volumes above and below the focal point will the power density be high enough to drive the upconversion process (Fig. 7.10).

We have shown the limitations clearly in a recent, very critical article on the "promise" of these nanoparticles for deep-tissue optical imaging [89]. The laser scanning mode became inordinately long, because one had to wait for at least 500 μs before one could go to the next pixel, because otherwise the emission from the previous pixel would also be picked up (an effect the experts call "bleeding") and so spoil the optical sectioning. In this work, we also

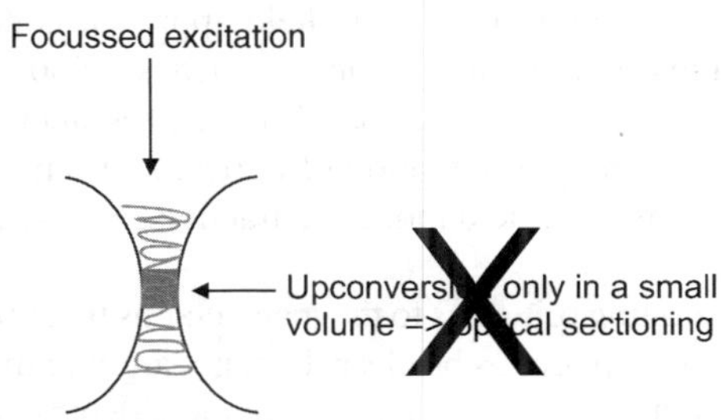


***Figure 7.10*** *Because the power density (or photon flux) is so high on each side of the focus (the dark square) the power density is high enough to drive non-linear processes; the result is loss of optical sectioning. The curly line indicates where upconversion occurs*

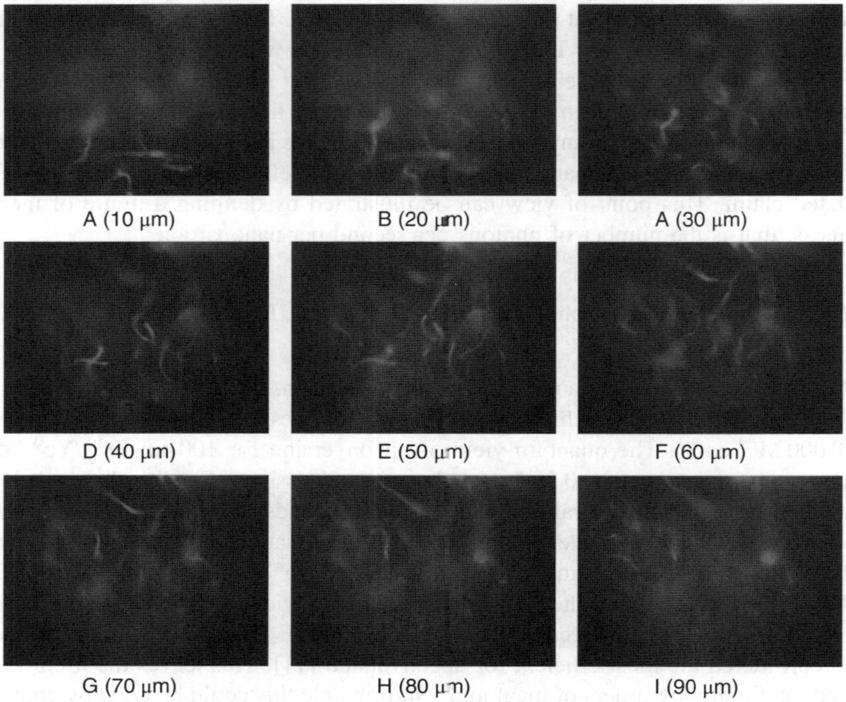


***Figure 7.11*** *Nine in vivo images of blood capillaries obtained at different depths inside the brain of a mouse after skull thinning using two-photon upconversion wide field microscopy (TPUWFM). The images were taken at 5 W cm$^{-2}$ with an exposure of 10 s at a gain of 34.7. The area of each image is 696×520 pixels. A 2×2 on-chip binning was performed. An Olympus 40×0.8 NA water lens was used for imaging the capillaries, with 800 nm emission upon 980 nm excitation from core/shell/shell. $NaYF_4$:Yb(20%):Tm(2%)/$NaYF_4$:Yb(20%):Tm(2%)/$NaYF_4$ nanocrystals were used for imaging. These images are false-coloured. Reproduced with permission from [89]. Copyright © 2011, American Chemical Society. (See colour plate section)*

showed that the optical sectioning in the wide field approach is not better than what one could achieve in the best two-photon microscopes, which is ~600 μm deep and still see, for example, blood capillaries in brain tissue, one of the most scattering tissues in mammals. We estimate a depth of 400 μm with retention of optical sectioning with the upconverting nanoparticles (Fig. 7.11). The wide field mode is attractive because it is much faster than the laser scanning mode.

This loss of optical sectioning brings us to the many claims that imaging can be done up to several mm and even 2 cm deep. Let us be clear, I am not arguing that light that is generated at these depths cannot be detected. So yes, tumours at these depths can be visualised, but the actual position, size, and shape cannot be determined because excitation and emission are heavily scattered over these distances, even in the biological window. Scatter is an intrinsic phenomenon of tissue, hence the physics are set. However, not knowing the exact location, size, and shape is not very helpful to clinicians and surgeons. In the imaging stage one would for instance not be able to resolve multiple (smaller) tumours if they are close. Furthermore, it is not very helpful to the surgeon who has to know the exact location, size, and shape because they want to cut "a little extra but not make a hole in the patient". A bit more on this in the next section.

In general, there is fairly positive data on cytotoxicity and clearance from the body, but the often rapid accumulation in lungs, heart, and so on is not a good sign. However, long-term studies have yet to be conducted. Some of the many aspects I discussed here, and many others, have also been addressed by others [90–114].

## 7.7 Upconversion and Magnetic Resonance Imaging

Because of the fact that the trivalent lanthanides come with optical and magnetic properties, it is actually quite obvious to combine more than one imaging mode into one nanoparticle. So far, the imaging modes that have been combined are optical (especially upconversion), magnetic resonance imaging (MRI), positron emission tomography (PET), and computed tomography (CT, through X-ray attenuation/contrast). It seems obvious to combine more than one modality as a way to offset limitations of one with the other, but the actual proof that this is the case has not yet been seen by me. So far, I have seen proof of principles of multimodal imaging, but the actual advantages in a clinical setting have yet to be shown. For instance, it is not immediately obvious why one would have to combine optical imaging and MRI. The first has severe depth penetration limits (several millimetres at best), whereas the latter has in principle no depth limitation; it all depends on how big an instrument engineers can build. However, I do see one possible advantage in this combination, which is MRI pre-surgery and optical imaging in-surgery. The pre-surgery imaging should provide the surgeon with high resolution data of location and shape of the tumour. The in-surgery imaging could allow the surgeon to cut "just a little bit more" than the tumour and show if cancerous cells are loose because of the cutting. Catching loose cancer cells is highly desirable, because they may go "on the move" and metastasise. Even here, one should consider why this combination, because the MRI may be done days before the actual surgery, thus requiring the (hopefully) retained accumulation in the tumour. Why not optimise separately the two modalities?

MRI is considered a powerful, non-invasive imaging technique in which the image contrast is based on differences in the $T_1$ or $T_2$ relaxation times of the protons of water. The $T_1$ is the characteristic time scale for the spin-lattice relaxation after the nuclear spins of the protons have been moved out of their equilibrium distribution in the magnetic field with a radio-frequency pulse. Similarly, the $T_2$ is the spin–spin relaxation time. In order to increase the contrast, a contrast agent can be administered. The two classes are obviously the ones that affect $T_1$ or $T_2$, respectively. The prototypical positive $T_1$ contrast agents for $T_1$ are $Gd^{3+}$ complexes for use at 1.5 and 3.0 T instruments, which are the current clinically used magnetic fields [115,116]. The prototypical negative $T_2$ contrast agent could arguably be superparamagnetic iron oxide, which is a vast field in itself [117–120]. I shall be very brief on the use of $Ln^{3+}$-based nanoparticles for MRI, because this is not the focus of this review. There is clearly a need for more potent MRI contrast agents, especially at magnetic fields higher than 4.0 T, where almost all small animal studies are being done. There are no optimised contrast agents for these fields. However, drawing some general conclusions is virtually impossible because of the lack of comparable data. For instance, important information is lacking on the size and shape of the nanoparticles and the surface coating matter. The biggest obstacle in comparing data is the fact that they are often reported at different field strengths. In addition, I would like to argue that relaxivity data should be presented on a *per (magnetic)* $Ln^{3+}$, a *per mass*, and a *per nanoparticle* basis, as a way to make comparisons possible. It is not always clear if the *per ion* or *per nanoparticle* relaxivity values are reported!

In an early study we showed the potential of $Gd^{3+}$-based nanoparticles for magnetic resonance (MRI) and computed tomography (CT) [121]. In this article we speculated that $Gd^{3+}$ ions not at the surface could be in part responsible for the high relaxivities. The $r_1$ ($r_1 = 1/T_1$) and $r_2$ ($r_2 = 1/T_2$) values of the citrate $GdF_3$ nanoparticles at 3.0 T gave mass relaxivities of $8.8 \pm 0.2$ and $9.4 \pm 0.2$ (mg ml$^{-1}$)$^{-1}$ s$^{-1}$. However, a later systematic study by us with sub-10 nm $NaGdF_4$ [122] proved this wrong. For the latter the respective $r_1$ numbers are (at 1.5 T and 8.0 nm nanoparticles): 3.0 mM$^{-1}$ s$^{-1}$, 12 (mg ml$^{-1}$)$^{-1}$ s$^{-1}$, and 11 000 mM$^{-1}$ s$^{-1}$. ICP-MS is a very accurate and sensitive method to determine the $Ln^{3+}$ content of the dispersion. This latter study also showed a small effect of the tumbling time of the nanoparticles on the *per nanoparticle* relaxivity. Knowing that the surface $Gd^{3+}$ was almost entirely responsible for the $r_1$, it seems worthwhile to have only $Gd^{3+}$ ions there in a core–shell architecture, where the core could have another function (e.g. upconversion). To this end, we used cation exchange [41] to make doped $NaYF_4$-$NaGdF_4$ core–shell nanoparticles of overall ~20 nm diameter made water-dispersible by changing the oleates for PVP [32]. The respective $r_1$ relaxivities are (at 9.4 T and ~20 nm particles of which 0.6 nm is the shell thickness) 2.33 mM$^{-1}$ s$^{-1}$ and 1.56 $10^4$ mM$^{-1}$ s$^{-1}$ (the *per mass* basis was not reported). The $r_2$ relaxivity per $Gd^{3+}$ ion measured at 9.4 T is in the range of 49–160 mM$^{-1}$ s$^{-1}$, which is significantly higher than $r_1$ as expected. Suh and Hyeon and their co-workers [123] also reported on doped $NaGdF_4$ nanoparticles of 20 and 41 nm size. The $r_1$ values at 1.5 T are 1.40 and 1.05 mM$^{-1}$ s$^{-1}$, which must be the *per ion* values, as is clear from their discussion. Ryu et al. [58] also reported on doped $NaGdF_4$ nanoparticles for MRI purposes. The *per ion* values at 4.7 T for 10 and 40 nm nanoparticles are 0.99 and 0.47 mM$^{-1}$ s$^{-1}$. For the silica-coated $GdVO_4$ nanoparticles the *per ion* $r_1$ value at 4.7 T is 0.438 mM$^{-1}$ s$^{-1}$ [66]. The $r_1$ value at 3.0 T for PEI-coated $Gd^{3+}$/$Ce^{3+}$/$Tb^{3+}$-doped $NaYF_4$ nanoparticles of ill-defined size and shape (of 10–20 by up to nearly 100 nm$^2$) was 2.87 mM$^{-1}$ s$^{-1}$, and I suspect this to be the *per ion* value [124]. Three articles by Li also reported on the use of MRI. The doped $NaGdF_4$ nanoparticles show $r_1$ relaxivities at 3.0 T of 5.60 mM$^{-1}$ s$^{-1}$ [78]. For the triply doped $NaYF_4$ nanoparticles they reported a *per ion* $r_1$ value at 3.0 T of 0.405 mM$^{-1}$ s$^{-1}$ [83]. They also showed small animal MRI data. The iron oxide coated nanoparticles gave an $r_2$ value on a *per iron* basis at 3.0 T of 189 mM$^{-1}$ s$^{-1}$ [86]. They further showed a $T_2$-weighed contrast in a rat. Recently, we published a first report on water-dispersible, mono-disperse $NaDyF_4$ nanoparticles (5.4, 9.8, and 20.3 nm in size) as $T_2$ contrast agent [125]. The largest nanoparticles show an $r_2$ at 9.4 T of 101 mM$^{-1}$ s$^{-1}$, 392 (mg ml$^{-1}$)$^{-1}$ s$^{-1}$, and 6479 mM$^{-1}$ s$^{-1}$, respectively. At 3T the $r_2$ is 15.8 mM$^{-1}$ s$^{-1}$, 61 (mg ml$^{-1}$)$^{-1}$ s$^{-1}$, and 102 mM$^{-1}$ s$^{-1}$, respectively. The *per ion* $r_1$ at 9.4 and 3.0 T was nearly invariant across the size range at 0.2–0.3 mM$^{-1}$ s$^{-1}$. For the 20.3 nm nanoparticles the $r_2/r_1$ is 306 at 9.4 T and 61 at 3.0 T. These are very encouraging numbers.

Work on $Ln_2O_3$ nanoparticles by Peters et al. has also appeared [126,127] with an excellent recent review on $Dy^{3+}$- and $Ho^{3+}$-based MRI contrast agents [128].

## 7.8 Conclusions and Outlook

Have I been the devil's advocate on the usefulness of upconverting nanoparticles for bio-imaging? Probably I have, because I do think the field is a bit hyped. If one is not interested in the resolution of fine structure, such as blood capillaries or the exact location and shape of a tumour, then these nanoparticles could do a fine job, maybe even up to 1 or 2 cm deep. Otherwise, I think they are not yet good enough, as shown by the figure of merit for brightness ($FOM_{brightness}$) defined above. Absorption, quantum yield of upconversion, and excited state life times all work against a high value of the $FOM_{brightness}$. Clearly, we have to find ways to boost the $FOM_{brightness}$ by two to three orders of magnitude. Moreover, I argue that we should show how multimodal imaging, in whatever combination, is going to be an improvement in animal studies and hopefully the clinic. To this end, I have seen very encouraging work but much remains to be done.

## References


1. Kokanyan E. and E. Dieguez, New perspectives of lithium niobate crystals, *J. Optoelectron. Adv. Mater.*, **2**, 205–214 (2000).
2. Boyer J. C., M. P. Manseau, J. I. Murray, and F. C. J. M. van Veggel, Surface Modification of Upconverting $NaYF_4$ Nanoparticles with PEG-Phosphate Ligands for NIR (800 nm) Biolabeling within the Biological Window, *Langmuir*, **26**, 1157–1164 (2010).
3. Polman A. and F. C. J. M. van Veggel, Broadband sensitisers for erbium-doped planar optical amplifiers: review, *J. Opt. Soc. Am. B-Opt. Phys.*, **21**, 871–892 (2004).
4. Stouwdam J. W. and F. C. J. M. van Veggel, Near-infrared emission of redispersible $Er^{3+}$, $Nd^{3+}$, and $Ho^{3+}$ doped $LaF_3$ nanoparticles, *Nano Lett.*, **2**, 733–737 (2002).
5. Hebbink G. A., J. W. Stouwdam, D. N. Reinhoudt and F. C. J. M. van Veggel, Lanthanide(III)-doped nanoparticles that emit in the near-infrared, *Adv. Mater.*, **14**, 1147–1150 (2002).
6. Dekker R., D. J. W. Klunder, A. Borreman, M. B. J. Diemeer, K. Wörhoff, A. Driessen, J. W. Stouwdam and F. C. J. M. van Veggel, Stimulated emission and optical gain in $LaF_3$: Nd nanoparticle-doped polymer-based waveguides, *Appl. Phys. Lett.*, **85**, 6104–6106 (2004).
7. Auzel F., Upconversion and anti-stokes processes with f and d ions in solids, *Chem. Rev.*, **104**, 139–173 (2004).
8. Stein G. and E. Wurzberg, Energy-gap Law in Solvent Isotope-effect on Radiationless Transitions of Rare-earth Ions, *J. Chem. Phys.*, **62**, 208–213 (1975).
9. Sivakumar S., F. C. J. M. van Veggel and P. S. May, Near-infrared (NIR) to red and green up-conversion emission from silica sol-gel thin films made with $La_{0.45}Yb_{0.50}Er_{0.05}F_3$ nanoparticles, hetero-looping-enhanced energy transfer (Hetero-LEET): A new up-conversion process, *J. Am. Chem. Soc.*, **129**, 620–625 (2007).
10. Oude Wolbers M. P., F. C. J. M. van Veggel, B. H. M. Snellink-Rüel, J. W. Hofstraat, F. A. J. Geurts and D. N. Reinhoudt, Novel preorganised hemispherands to encapsulate rare earth ions: Shielding and ligand deuteration for prolonged lifetimes of excited $Eu^{3+}$ ions, *J. Am. Chem. Soc.*, **119**, 138–144 (1997).

11. Page R. H., K. I. Schaffers, P. A. Waide, J. B. Tassano, S. A. Payne, W. F. Krupke and W. K. Bischel, Upconversion-pumped luminescence efficiency of rare-earth-doped hosts sensitised with trivalent ytterbium, *J. Opt. Soc. Am. B-Opt. Phys.*, **15**, 996–1008 (1998).
12. Murov S. L., I. Carmichael and G. L. Hug, *Handbook of Photochemistry*, 2nd edn, Marcel Dekker, New York, 1993.
13. Zhuang Z. B., Q. Peng and Y. D. Li, Controlled synthesis of semiconductor nanostructures in the liquid phase, *Chem. Soc. Rev.*, **40**, 5492–5513 (2011).
14. Smith A. M. and S. M. Nie, Semiconductor Nanocrystals: Structure, Properties, and Band Gap Engineering, *Acc. Chem. Res.*, **43**, 190–200 (2010).
15. El-Sayed M. A., Small is different: Shape-, size-, and composition-dependent properties of some colloidal semiconductor nanocrystals, *Acc. Chem. Res.*, **37**, 326–333 (2004).
16. Nirmal M. and L. Brus, Luminescence photophysics in semiconductor nanocrystals, *Acc. Chem. Res.*, **32**, 407–414 (1999).
17. Wise F. W., Lead salt quantum dots: The limit of strong quantum confinement, *Acc. Chem. Res.*, **33**, 773–780 (2000).
18. Boyer J. C. and F. C. J. M. van Veggel, Absolute quantum yield measurements of colloidal $NaYF_4$: $Er^{3+}$, $Yb^{3+}$ upconverting nanoparticles, *Nanoscale*, **2**, 1417–1419 (2010).
19. Eaton D. F., Reference Materials for Fluorescence Measurement, *Pure Appl. Chem.*, **60**, 1107–1114 (1988).
20. Sillen A. and Y. Engelborghs, The correct use of "average" fluorescence parameters, *Photochem. Photobiol.*, **67**, 475–486 (1998).
21. Liu Q., Y. Sun, T. S. Yang, W. Feng, C. G. Li and F. Y. Li, Sub-10 nm Hexagonal Lanthanide-Doped $NaLuF_4$ Upconversion Nanocrystals for Sensitive Bioimaging in Vivo, *J. Am. Chem. Soc.*, **133**, 17122–17125 (2011).
22. Johnson N. J. J., A. Korinek and F. C. J. M. van Veggel, Self-focusing by Ostwald Ripening: A Strategy for Layer-by-Layer Epitaxial Growth on Upconverting Nanocrystals *J. Am. Chem. Soc.*, **134** (2012).
23. Sudarsan V., S. Sivakumar, F. C. J. M. van Veggel and M. Raudsepp, General and convenient method for making highly luminescent sol-gel derived silica and alumina films by using $LaF_3$ nanoparticles doped with lanthanide ions ($Er^{3+}$, $Nd^{3+}$, and $Ho^{3+}$), *Chem. Mater.*, **17**, 4736–4742 (2005).
24. Stöber W., A. Fink and E. Bohn, Controlled Growth of Monodisperse Silica Spheres in Micron Size Range, *J. Coll. Interf. Sci.*, **26**, 62–69 (1968).
25. Pichaandi J., J. C. Boyer, K. R. Delaney and F. C. J. M. van Veggel, Two-Photon Upconversion Laser (Scanning and Wide-Field) Microscopy Using Ln(3+)-Doped NaYF4 Upconverting Nanocrystals: A Critical Evaluation of their Performance and Potential in Bioimaging, *J. Phys. Chem. C*, **115**, 19054–19064 (2011).
26. Jiang G. C., J. Pichaandi, N. J. J. Johnson, R. D. Burke and F. C. J. M. van Veggel, An Effective Polymer Cross-Linking Strategy To Obtain Stable Dispersions of Upconverting $NaYF_4$ Nanoparticles in Buffers and Biological Growth Media for Biolabeling Applications, *Langmuir*, **28**, 3239–3247 (2012).
27. Zhao H. G., D. F. Wang, T. Zhang, M. Chaker and D. L. Ma, Two-step synthesis of high-quality water-soluble near-infrared emitting quantum dots via amphiphilic polymers, *Chem. Commun.*, **46**, 5301–5303 (2010).
28. Zhao H. G., D. F. Wang, M. Chaker and D. L. Ma, Effect of Different Types of Surface Ligands on the Structure and Optical Property of Water-Soluble PbS Quantum Dots Encapsulated by Amphiphilic Polymers, *J. Phys. Chem. C*, **115**, 1620–1626 (2011).
29. Zhao H. G. M. Chaker and D. L. Ma, Self-selective recovery of photoluminescence in amphiphilic polymer encapsulated PbS quantum dots, *Phys. Chem. Chem. Phys.*, **12**, 14754–14761 (2010).

30. Williams D. B. and C. B. Carter, *Transmission Electron Microscopy*, 2nd ed., Springer, New York, 2009.
31. Abel K. A., J. C. Boyer, C. M. Andrei and F. C. J. M. van Veggel, Analysis of the Shell Thickness Distribution on $NaYF_4$/$NaGdF_4$ Core/Shell Nanocrystals by EELS and EDS, *J. Phys. Chem. Lett.*, **2**, 185–189 (2011).
32. Dong C. H., A. Korinek, B. Blasiak, B. Tomanek and F. C. J. M. van Veggel, Cation Exchange: A Facile Method To Make $NaYF_4$:Yb,Tm-$NaGdF_4$ Core-Shell Nanoparticles with a Thin, Tunable, and Uniform Shell, *Chem. Mater.*, **24**, 1297–1305 (2012).
33. Abel K. A., J. C. Boyer and F. C. J. M. van Veggel, Hard Proof of the $NaYF_4$/$NaGdF_4$ Nanocrystal Core/Shell Structure, *J. Am. Chem. Soc.*, **131**, 14644–14645 (2009).
34. Abel K. A., J. C. Boyer and F. C. J. M. van Veggel, Hard Proof of the $NaYF_4$/$NaGdF_4$ Nanocrystal Core/Shell Structure (vol 131, pg 14644, 2009), *J. Am. Chem. Soc.*, **132**, 5533–5533 (2010).
35. Dong C., J. Pichaandi, T. Regier and F. C. J. M. van Veggel, Nonstatistical Dopant Distribution of $Ln^{3+}$-Doped $NaGdF_4$ Nanoparticles, *J. Phys. Chem. C*, **115**, 15950–15958 (2011).
36. Dong C. H., J. Pichaandi, T. Regier and F. C. J. M. van Veggel, The unexpected structures of "core-shell" and "alloy" $LnF_3$ nanoparticles as examined by variable energy X-ray photoelectron spectroscopy, *Nanoscale*, **3**, 3376–3384 (2011).
37. Scerri E. R., *The Periodic Table; Its Story and Its Significance*, 1st ed., Oxford University Press, New York, 2007.
38. Balcha T., J. R. Strobl, C. Fowler, P. Dash and R. W. J. Scott, Selective Aerobic Oxidation of Crotyl Alcohol Using AuPd Core-Shell Nanoparticles, *ACS Catal.*, **1**, 425–436 (2011).
39. Dash P., T. Bond, C. Fowler, W. Hou, N. Coombs and R. W. J. Scott, Rational Design of Supported PdAu Nanoparticle Catalysts from Structured Nanoparticle Precursors, *J. Phys. Chem. C*, **113**, 12719–12730 (2009).
40. Abel K. A., P. A. FitzGerald, T. Y. Wang, T. Z. Regier, M. Raudsepp, S. P. Ringer, G. G. Warr and F. C. J. M. van Veggel, Probing the Structure of Colloidal Core/Shell Quantum Dots Formed by Cation Exchange, *J. Phys. Chem. C*, **116**, 3968–3978 (2012).
41. Dong C. H. and F. C. J. M. van Veggel, Cation Exchange in Lanthanide Fluoride Nanoparticles, *ACS Nano*, **3**, 123–130 (2009).
42. Zhang F., R. Che, X. Li, C. Yao, J. Yang, D. Shen, P. Hu, W. Li and D. Zhao, Direct Imaging the Upconversion Nanocrystal Core/Shell Structure at the Subnanometer Level: Shell Thickness Dependence in Upconverting Optical Properties, *Nano Lett.*, **12**, 2852–2858 (2012).
43. Denk W., J. H. Strickler and W. W. Webb, 2-Photon Laser Scanning Fluorescence Microscopy, *Science*, **248**, 73–76 (1990).
44. Das G. K. and T. T. Y. Tan, Rare-earth-doped and codoped $Y_2O_3$ nanomaterials as potential bioimaging probes, *J. Phys. Chem. C*, **112**, 11211–11217 (2008).
45. Hu H., M. X. Yu, F. Y. Li, Z. G. Chen, X. Gao, L. Q. Xiong and C. H. Huang, Facile Epoxidation Strategy for Producing Amphiphilic Up-Converting Rare-Earth Nanophosphors as Biological Labels, *Chem. Mater.*, **20**, 7003–7009 (2008).
46. Jalil R. A. and Y. Zhang, Biocompatibility of silica coated $NaYF_4$ upconversion fluorescent nanocrystals, *Biomaterials*, **29**, 4122–4128 (2008).
47. Shan J. N., J. B. Chen, J. Meng, J. Collins, W. Soboyejo, J. S. Friedberg and Y. G. Ju, Biofunctionalization, cytotoxicity, and cell uptake of lanthanide doped hydrophobically ligated $NaYF_4$ upconversion nanophosphors, *J. Appl. Phys.*, **104** (2008).
48. Zako T., H. Nagata, N. Terada, M. Sakono, K. Soga and M. Maeda, Improvement of dispersion stability and characterization of upconversion nanophosphors covalently modified with PEG as a fluorescence bioimaging probe, *J. Mater. Sci.*, **43**, 5325–5330 (2008).

49. Hischemöller A., J. Nordmann, P. Ptacek, K. Mummenhoff and M. Haase, In-Vivo Imaging of the Uptake of Upconversion Nanoparticles by Plant Roots, *J. Biomed. Nanotechnol.*, **5**, 278–284 (2009).
50. Jiang S., Y. Zhang, K. M. Lim, E. K. W. Sim and L. Ye, NIR-to-visible upconversion nanoparticles for fluorescent labeling and targeted delivery of siRNA, *Nanotechnology*, **20**, 155101 (2009).
51. Lim S. F., R. Riehn, C. K. Tung, W. S. Ryu, R. Zhuo, J. Dalland and R. H. Austin, Upconverting nanophosphors for bioimaging, *Nanotechnology*, **20**, 405701 (2009).
52. Chen J., C. R. Guo, M. Wang, L. Huang, L. P. Wang, C. C. Mi, J. Li, X. X. Fang, C. B. Mao and S. K. Xu, Controllable synthesis of $NaYF_4$:Yb,Er upconversion nanophosphors and their application to in vivo imaging of Caenorhabditis elegans, *J. Mater. Chem.*, **21**, 2632–2638 (2011).
53. Zhou J. C., Z. L. Yang, W. Dong, R. J. Tang, L. D. Sun and C. H. Yan, Bioimaging and toxicity assessments of near-infrared upconversion luminescent $NaYF_4$:Yb,Tm nanocrystals, *Biomaterials*, **32**, 9059–9067 (2011).
54. Wang M., C. C. Mi, Y. X. Zhang, J. L. Liu, F. Li, C. B. Mao and S. K. Xu, NIR-Responsive Silica-Coated $NaYbF_4$:Er/Tm/Ho Upconversion Fluorescent Nanoparticles with Tunable Emission Colors and Their Applications in Immunolabeling and Fluorescent Imaging of Cancer Cells, *J. Phys. Chem. C*, **113**, 19021–19027 (2009).
55. Zako T., H. Nagata, N. Terada, A. Utsumi, M. Sakono, M. Yohda, H. Ueda, K. Soga and M. Maeda, Cyclic RGD peptide-labeled upconversion nanophosphors for tumor cell-targeted imaging, *Biochem. Biophys. Res. Commun.*, **381**, 54–58 (2009).
56. Johnson N. J. J., N. M. Sangeetha, J. C. Boyer and F. C. J. M. van Veggel, Facile ligand-exchange with polyvinylpyrrolidone and subsequent silica coating of hydrophobic upconverting beta-$NaYF_4$:$Yb^{3+}$/$Er^{3+}$ nanoparticles, *Nanoscale*, **2**, 771–777 (2010).
57. Mader H. S., M. Link, D. E. Achatz, K. Uhlmann, X. H. Li and O. S. Wolfbeis, Surface-Modified Upconverting Microparticles and Nanoparticles for Use in Click Chemistries, *Chem.-Eur. J.*, **16**, 5416–5424 (2010).
58. Ryu J., H. Y. Park, K. Kim, H. Kim, J. H. Yoo, M. Kang, K. Im, R. Grailhe and R. Song, Facile Synthesis of Ultrasmall and Hexagonal $NaGdF_4$: $Yb^{3+}$, $Er^{3+}$ Nanoparticles with Magnetic and Upconversion Imaging Properties, *J. Phys. Chem. C*, **114**, 21077–21082 (2010).
59. Vetrone F., R. Naccache, A. J. de la Fuente, F. Sanz-Rodriguez, A. Blazquez-Castro, E. M. Rodriguez, D. Jaque, J. G. Sole and J. A. Capobianco, Intracellular imaging of HeLa cells by non-functionalized $NaYF_4$: $Er^{3+}$, $Yb^{3+}$ upconverting nanoparticles, *Nanoscale*, **2**, 495–498 (2010).
60. Jin J. F., Y. J. Gu, C. W. Y. Man, J. P. Cheng, Z. H. Xu, Y. Zhang, H. S. Wang, V. H. Y. Lee, S. H. Cheng and W. T. Wong, Polymer-Coated $NaYF_4$:$Yb^{3+}$, $Er^{3+}$ Upconversion Nanoparticles for Charge-Dependent Cellular Imaging, *ACS Nano*, **5**, 7838–7847 (2011).
61. Doane T. L., C.-H. Chuang, R. J. Hill and C. Burda, Nanoparticle $\zeta$-Potentials, *Acc. Chem. Res.*, **45**, 317–326 (2011).
62. Nam S. H., Y. M. Bae, Y. I. Park, J. H. Kim, H. M. Kim, J. S. Choi, K. T. Lee, T. Hyeon and Y. D. Suh, Long-Term Real-Time Tracking of Lanthanide Ion Doped Upconverting Nanoparticles in Living Cells, *Angew. Chem.-Int. Edit.*, **50**, 6093–6097 (2011).
63. Dai Y. L., P. A. Ma, Z. Y. Cheng, X. J. Kang, X. Zhang, Z. Y. Hou, C. X. Li, D. M. Yang, X. F. Zhai and J. Lin, Up-Conversion Cell Imaging and pH-Induced Thermally Controlled Drug Release from $NaYF_4$:$Yb^{3+}$/$Er^{3+}$ Hydrogel Core-Shell Hybrid Microspheres, *ACS Nano*, **6**, 3327–3338 (2012).
64. Wang Z.-L., J. H. Hao, H. L. W. Chan, W.-T. Wong and K.-L. Wong, A Strategy for Simultaneously Realizing the Cubic-to-Hexagonal Phase Transition and Controlling the Small Size of $NaYF_4$:$Yb^{3+}$,$Er^{3+}$ Nanocrystals for In Vitro Cell Imaging, *Small*, **8**, 1863–1868 (2012).

65. Zhong S. L., Y. Lu, M. R. Gao, S. J. Liu, J. Peng, L. C. Zhang and S. H. Yu, Monodisperse Mesocrystals of $YF_3$ and $Ce^{3+}/Ln^{3+}$(Ln=Tb, Eu) Co-Activated $YF_3$: Shape Control Synthesis, Luminescent Properties, and Biocompatibility, *Chem.-Eur. J.*, **18**, 5222–5231 (2012).
66. Yin W. Y., L. J. Zhou, Z. J. Gu, G. Tian, S. Jin, L. Yan, X. X. Liu, G. M. Xing, W. L. Ren, F. Liu, Z. W. Pan and Y. L. Zhao, Lanthanide-doped $GdVO_4$ upconversion nanophosphors with tunable emissions and their applications for biomedical imaging, *J. Mater. Chem.*, **22**, 6974–6981 (2012).
67. Zhang F., G. B. Braun, A. Pallaoro, Y. C. Zhang, Y. F. Shi, D. X. Cui, M. Moskovits, D. Y. Zhao and G. D. Stucky, Mesoporous Multifunctional Upconversion Luminescent and Magnetic "Nanorattle" Materials for Targeted Chemotherapy, *Nano Lett.*, **12**, 61–67 (2012).
68. Dong N. N., M. Pedroni, F. Piccinelli, G. Conti, A. Sbarbati, J. E. Ramirez-Hernandez, L. M. Maestro, M. C. Iglesias-de la Cruz, F. Sanz-Rodriguez, A. Juarranz, F. Chen, F. Vetrone, J. A. Capobianco, J. G. Sole, M. Bettinelli, D. Jaque and A. Speghini, NIR-to-NIR Two-Photon Excited $CaF_2$:$Tm^{3+}$,$Yb^{3+}$ Nanoparticles: Multifunctional Nanoprobes for Highly Penetrating Fluorescence Bio-Imaging, *ACS Nano*, **5**, 8665–8671 (2011).
69. Chatteriee D. K., A. J. Rufalhah and Y. Zhang, Upconversion fluorescence imaging of cells and small animals using lanthanide doped nanocrystals, *Biomaterials*, **29**, 937–943 (2008).
70. Nyk M., R. Kumar, T. Y. Ohulchanskyy, E. J. Bergey and P. N. Prasad, High Contrast in Vitro and in Vivo Photoluminescence Bioimaging Using Near Infrared to Near Infrared Up-Conversion in $Tm^{3+}$ and $Yb^{3+}$ Doped Fluoride Nanophosphors, *Nano Lett.*, **8**, 3834–3838 (2008).
71. Salthouse C., S. Hildebrand, R. Weissleder and U. Mahmood, Design and demonstration of a small-animal up-conversion imager, *Opt. Express*, **16**, 21731–21737 (2008).
72. Idris N. M., Z. Q. Li, L. Ye, E. K. W. Sim, R. Mahendran, P. C. L. Ho and Y. Zhang, Tracking transplanted cells in live animal using upconversion fluorescent nanoparticles, *Biomaterials*, **30**, 5104–5113 (2009).
73. Vinegoni C., D. Razansky, S. A. Hilderbrand, F. W. Shao, V. Ntziachristos and R. Weissleder, Transillumination fluorescence imaging in mice using biocompatible upconverting nanoparticles, *Opt. Lett.*, **34**, 2566–2568 (2009).
74. Liu H. C., C. T. Xu and S. Andersson-Engels, Multibeam fluorescence diffuse optical tomography using upconverting nanoparticles, *Opt. Lett.*, **35**, 718–720 (2010).
75. Tian Z., G. Y. Chen, X. Li, H. J. Liang, Y. S. Li, Z. G. Zhang and Y. Tian, Autofluorescence-free in vivo multicolor imaging using upconversion fluoride nanocrystals, *Lasers Med. Sci.*, **25**, 479–484 (2010).
76. Xiong L. Q., T. S. Yang, Y. Yang, C. J. Xu and F. Y. Li, Long-term in vivo biodistribution imaging and toxicity of polyacrylic acid-coated upconversion nanophosphors, *Biomaterials*, **31**, 7078–7085 (2010).
77. Yu X. F., Z. B. Sun, M. Li, Y. Xiang, Q. Q. Wang, F. F. Tang, Y. L. Wu, Z. J. Cao and W. X. Li, Neurotoxin-conjugated upconversion nanoprobes for direct visualization of tumors under near-infrared irradiation, *Biomaterials*, **31**, 8724–8731 (2010).
78. Zhou J., Y. Sun, X. X. Du, L. Q. Xiong, H. Hu and F. Y. Li, Dual-modality in vivo imaging using rare-earth nanocrystals with near-infrared to near-infrared (NIR-to-NIR) upconversion luminescence and magnetic resonance properties, *Biomaterials*, **31**, 3287–3295 (2010).
79. Cao T. Y., Y. Yang, Y. A. Gao, J. Zhou, Z. Q. Li and F. Y. Li, High-quality water-soluble and surface-functionalized upconversion nanocrystals as luminescent probes for bioimaging, *Biomaterials*, **32**, 2959–2968 (2011).
80. Cheng L. A., K. Yang, M. W. Shao, S. T. Lee and Z. A. Liu, Multicolor In Vivo Imaging of Upconversion Nanopaticles with Emissions Tuned by Luminescence Resonance Energy Transfer, *J. Phys. Chem. C*, **115**, 2686–2692 (2011).

81. Jeong S., N. Won, J. Lee, J. Bang, J. Yoo, S. G. Kim, J. A. Chang, J. Kim and S. Kim, Multiplexed near-infrared in vivo imaging complementarily using quantum dots and upconverting $NaYF_4:Yb^{3+},Tm^{3+}$ nanoparticles, *Chem. Commun.*, **47**, 8022–8024 (2011).
82. Liu Q., M. Chen, Y. Sun, G. Y. Chen, T. S. Yang, Y. Gao, X. Z. Zhang and F. Y. Li, Multifunctional rare-earth self-assembled nanosystem for tri-modal upconversion luminescence/fluorescence/positron emission tomography imaging, *Biomaterials*, **32**, 8243–8253 (2011).
83. Zhou J., M. X. Yu, Y. Sun, X. Z. Zhang, X. J. Zhu, Z. H. Wu, D. M. Wu and F. Y. Li, Fluorine-18-labeled $Gd^{3+}/Yb^{3+}/Er^{3+}$ co-doped $NaYF_4$ nanophosphors for multimodality PET/MR/UCL imaging, *Biomaterials*, **32**, 1148–1156 (2011)
84. Yang T. S., Y. Sun, Q. Liu, W. Feng, P. Y. Yang and F. Y. Li, Cubic sub-20 nm $NaLuF_4$-based upconversion nanophosphors for high-contrast bioimaging in different animal species, *Biomaterials*, **33**, 3733–3742 (2012).
85. Wang Z. L., J. H. Hao, H. L. W. Chan, G. L. Law, W. T. Wong, K. L. Wong, M. B. Murphy, T. Su, Z. H. Zhang and S. Q. Zeng, Simultaneous synthesis and functionalization of water-soluble up-conversion nanoparticles for in-vitro cell and nude mouse imaging, *Nanoscale*, **3**, 2175–2181 (2011).
86. Xia A., Y. Gao, J. Zhou, C. Y. Li, T. S. Yang, D. M. Wu, L. M. Wu and F. Y. Li, Core-shell $NaYF_4:Yb^{3+},Tm^{3+}@Fe_xO_y$ nanocrystals for dual-modality $T_2$-enhanced magnetic resonance and NIR-to-NIR upconversion luminescent imaging of small-animal lymphatic node, *Biomaterials*, **32**, 7200–7208 (2011).
87. Zhan Q. Q., J. Qian, H. J. Liang, G. Somesfalean, D. Wang, S. L. He, Z. G. Zhang and S. Andersson-Engels, Using 915 nm Laser Excited $Tm^{3+}/Er^{3+}/Ho^{3+}$-Doped $NaYbF_4$ Upconversion Nanoparticles for in Vitro and Deeper in Vivo Bioimaging without Overheating Irradiation, *ACS Nano*, **5**, 3744–3757 (2011).
88. Hischemöller A., C. Walter, V. Weiler, H. Hummel, T. Thepen, M. Huhn, S. Barth, W. Hoheisel, K. Köhler, D. Dimova-Landen, C. Bremer, M. Haase and J. Waldeck, Labeling of Anti-MUC-1 Binding Single Chain Fv Fragments to Surface Modified Upconversion Nanoparticles for an Initial in Vivo Molecular Imaging Proof of Principle Approach, *Int. J. Mol. Sci.*, **13**, 4153–4167 (2012).
89. Pichaandi J., J. C. Boyer, K. R. Delaney and F. C. J. M. van Veggel, Two-Photon Upconversion Laser (Scanning and Wide-Field) Microscopy Using $Ln^{3+}$-Doped $NaYF_4$ Upconverting Nanocrystals: A Critical Evaluation of their Performance and Potential in Bioimaging, *J. Phys. Chem. C*, **115**, 19054–19064 (2011).
90. Jiang S., M. K. Gnanasammandhan and Y. Zhang, Optical imaging-guided cancer therapy with fluorescent nanoparticles, *J. R. Soc. Interface*, **7**, 3–18 (2010).
91. Perfezou M., A. Turner and A. Merkoci, Cancer detection using nanoparticle-based sensors, *Chem. Soc. Rev.*, **41**, 2606–2622 (2012).
92. Lee D.-E., H. Koo, I.-C. Sun, J. H. Ryu, K. Kim and I. C. Kwon, Multifunctional nanoparticles for multimodal imaging and theragnosis, *Chem. Soc. Rev.*, **41**, 2656–2672 (2012).
93. Swierczewska M., G. Liu, S. Lee and X. Chen, High-sensitivity nanosensors for biomarker detection, *Chem. Soc. Rev.*, **41**, 2641–2655 (2012).
94. Ghosh Chaudhuri R. and S. Paria, Core/Shell Nanoparticles: Classes, Properties, Synthesis Mechanisms, Characterization, and Applications, *Chem. Rev.*, **112**, 2373–2433 (2011).
95. Mout R., D. F. Moyano, S. Rana and V. M. Rotello, Surface functionalization of nanoparticles for nanomedicine, *Chem. Soc. Rev.*, **41**, 2539–2544 (2012).
96. Taylor A., K. M. Wilson, P. Murray, D. G. Fernig and R. Levy, Long-term tracking of cells using inorganic nanoparticles as contrast agents: are we there yet?, *Chem. Soc. Rev.*, **41**, 2707–2717 (2012).
97. Doane T. L. and C. Burda, The unique role of nanoparticles in nanomedicine: imaging, drug delivery and therapy, *Chem. Soc. Rev.*, **41**, 2885–2911 (2012).

98. Zhou J., Z. Liu and F. Y. Li, Upconversion nanophosphors for small-animal imaging, *Chem. Soc. Rev.*, **41**, 1323–1349 (2012).
99. Fedorov P. P., A. A. Luginina, S. V. Kuznetsov and V. V. Osiko, Nanofluorides, *J. Fluor. Chem.*, **132**, 1012–1039 (2011).
100. Wang M., G. Abbineni, A. Clevenger, C. B. Mao and S. K. Xu, Upconversion nanoparticles: synthesis, surface modification and biological applications, *Nanomed.-Nanotechnol. Biol. Med.* **7**, 710–729 (2011).
101. Bouzigues C., T. Gacoin and A. Alexandrou, Biological Applications of Rare-Earth Based Nanoparticles, *ACS Nano*, **5**, 8488–8505 (2011).
102. Wang G. F., Q. Peng and Y. D. Li, Lanthanide-Doped Nanocrystals: Synthesis, Optical-Magnetic Properties, and Applications, *Acc. Chem. Res.*, **44**, 322–332 (2011).
103. Zhang J., Y. Li, X. Hao, Q. Zhang, K. Yang, L. Li, L. Ma, S. Wang and X. Li, Recent Progress in Therapeutic and Diagnostic Applications of Lanthanides, *Mini-Rev. Med. Chem.*, **11**, 678–694 (2011).
104. Haase M. and H. Schafer, Upconverting Nanoparticles, *Angew. Chem.-Int. Edit.*, **50**, 5808–5829 (2011).
105. Chatterjee D. K., M. K. Gnanasammandhan and Y. Zhang, Small Upconverting Fluorescent Nanoparticles for Biomedical Applications, *Small*, **6**, 2781–2795 (2010).
106. Mader H. S., P. Kele, S. M. Saleh and O. S. Wolfbeis, Upconverting luminescent nanoparticles for use in bioconjugation and bioimaging, *Curr. Opin. Chem. Biol.*, **14**, 582–596 (2010).
107. Zhang C., L. D. Sun, Y. W. Zhang and C. H. Yan, Rare earth upconversion nanophosphors: synthesis, functionalization and application as biolabels and energy transfer donors, *J. Rare Earths*, **28**, 807–819 (2010).
108. Jacobsohn L. G., C. J. Kucera, T. L. James, K. B. Sprinkle, J. R. DiMaio, B. Kokuoz, B. Yazgan-Kukouz, T. A. DeVol and J. Ballato, Preparation and Characterization of Rare Earth Doped Fluoride Nanoparticles, *Materials*, **3**, 2053–2068 (2010).
109. Wang F., D. Banerjee, Y. S. Liu, X. Y. Chen and X. G. Liu, Upconversion nanoparticles in biological labeling, imaging, and therapy, *Analyst*, **135**, 1839–1854 (2010).
110. Wang F. and X. G. Liu, Recent advances in the chemistry of lanthanide-doped upconversion nanocrystals, *Chem. Soc. Rev.* **38**, 976–989 (2009).
111. Dosev D., M. Nichkova and I. M. Kennedy, Inorganic lanthanide nanophosphors in biotechnology, *J. Nanosci. Nanotechnol.*, **8**, 1052–1067 (2008).
112. Vetrone F. and J. A. Capobianco, Lanthanide-doped fluoride nanoparticles: luminescence, upconversion, and biological applications, *Int. J. Nanotechnol.*, **5**, 1306–1339 (2008).
113. Yan Z. G. and C. H. Yan, Controlled synthesis of rare earth nanostructures, *J. Mater. Chem.*, **18**, 5046–5059 (2008).
114. Pansare V. J., S. Hejazi, W. J. Faenza and R. K. Prud'homme, Review of Long-Wavelength Optical and NIR Imaging Materials: Contrast Agents, Fluorophores, and Multifunctional Nano Carriers, *Chem. Mater.*, **24**, 812–827 (2012).
115. Hanaoka K., Development of Responsive Lanthanide-Based Magnetic Resonance Imaging and Luminescent Probes for Biological Applications, *Chem. Pharm. Bull.*, **58**, 1283–1294 (2010).
116. Terreno E., W. Dastru, D. D. Castelli, E. Gianolio, S. G. Crich, D. Longo and S. Aime, Advances in Metal-Based Probes for MR Molecular Imaging Applications, *Curr. Med. Chem.*, **17**, 3684–3700 (2010).
117. Rosen J. E., L. Chan, D. B. Shieh and F. X. Gu, Iron oxide nanoparticles for targeted cancer imaging and diagnostics, *Nanomed.-Nanotechnol. Biol. Med.* **8**, 275–290 (2012).
118. Rumenapp C., B. Gleich and A. Haase, Magnetic Nanoparticles in Magnetic Resonance Imaging and Diagnostics, *Pharm. Res.*, **29**, 1165–1179 (2012).
119. Slabu I., G. Guntherodt, T. Schmitz-Rode, M. Hodenius, N. Kramer, H. Donker, G. A. Krombach, J. Otto, U. Klinge and M. Baumann, Investigation of Superparamagnetic Iron

Oxide Nanoparticles for MR-Visualization of Surgical Implants, *Curr. Pharm. Biotechnol.*, **13**, 545–551 (2012).

120. Yigit M. V., A. Moore and Z. Medarova, Magnetic Nanoparticles for Cancer Diagnosis and Therapy, *Pharm. Res.*, **29**, 1180–1188 (2012).
121. Cheung E. N. M., R. D. A. Alvares, W. Oakden. R. Chaudhary, M. L. Hill, J. Pichaandi, G. C. H. Mo, C. Yip, P. M. Macdonald, G. J. Stanisz, F. C. J. M. van Veggel and R. S. Prosser, Polymer-Stabilized Lanthanide Fluoride Nanoparticle Aggregates as Contrast Agents for Magnetic Resonance Imaging and Computed Tomography, *Chem. Mater.*, **22**, 4728–4739 (2010).
122. Johnson N. J. J., W. Oakden, G. J. Stanisz, R. S. Prosser and F. C. J. M. van Veggel, Size-Tunable, Ultrasmall $NaGdF_4$ Nanoparticles: Insights into their $T_1$ MRI Contrast Enhancement (vol 23, pg 3714, 2011), *Chem. Mater.*, **23**, 4377–4877 (2011).
123. Park Y. I., J. H. Kim, K. T. Lee, K. S. Jeon, H. Bin Na, J. H. Yu, H. M. Kim, N. Lee, S. H. Choi, S. I. Baik, H. Kim, S. P. Park, B. J. Park, Y. W. Kim, S. H. Lee, S. Y. Yoon, I. C. Song, W. K. Moon, Y. D. Suh and T. Hyeon, Nonblinking and Nonbleaching Upconverting Nanoparticles as an Optical Imaging Nanoprobe and $T_1$ Magnetic Resonance Imaging Contrast Agent, *Adv. Mater.*, **21**, 4467–4471 (2009).
124. Ren W. T., L. B. Liang, F. Qi, Z. B. Sun, Z. Y. Yang, X. Q. Huang, Q. Z. Wu, H. Y. Zhu, X. F. Yu, H. Quan and Q. Y. Gong, Bimodal Fluorescence and Magnetic Resonance Imaging Using Water-Soluble Hexagonal $NaYF_4$:Ce,Tb,Gd Nanocrystals, *J. Nanomater.*, 531217 (2011).
125. Das G. K., N. J. J. Johnson, J. Cramen, B. Blasiak, P. Latta, B. Tomanek and F. C. J. M. van Veggel, $NaDyF_4$ Nanoparticles as $T_2$ Contrast Agents for Ultrahigh Field Magnetic Resonance Imaging, *J. Phys. Chem. Lett.*, **3**, 524–529 (2012).
126. Norek M., E. Kampert, U. Zeitler and J. A. Peters, Tuning of the size of $Dy_2O_3$ nanoparticles for optimal performance as an MRI contrast agent, *J. Am. Chem. Soc.*, **130**, 5335–5340 (2008).
127. Norek M., G. A. Pereira, C. F. G. C. Geraldes, A. Denkova, W. Z. Zhou and J. A. Peters, NMR transversal relaxivity of suspensions of lanthanide oxide nanoparticles, *J. Phys. Chem. C*, **111**, 10240–10246 (2007).
128. Norek M. and J. A. Peters, MRI contrast agents based on dysprosium or holmium, *Prog. Nucl. Magn. Reson. Spectrosc.*, **59**, 64–82 (2011).

# 8

# Direct Excitation Ln(III) Luminescence Spectroscopy to Probe the Coordination Sphere of Ln(III) Catalysts, Optical Sensors and MRI Agents

*Janet R. Morrow and Sarina J. Dorazio*

*University at Buffalo, State University of New York, USA*

## 8.1 Introduction

### 8.1.1 Luminescence Spectroscopy for Defining the Ln(III) Coordination Sphere

Luminescence spectroscopy is an extremely powerful tool for probing the intricacies of the Ln(III) coordination sphere of complexes in solution. Information on properties that are crucial to Ln(III) solution chemistry, including the hydration number (number of bound water ligands), the number of different species present in solution, the mode of interaction of the complex with anions, and the formation of dinuclear complexes, dimers and aggregates can be obtained [1–3]. It is little wonder that Ln(III) ions are considered to be among the most versatile spectroscopic probes of metal ion coordination sphere.

Our focus in this chapter is on the application of direct excitation luminescence spectroscopy to understand the hydration number and detailed information about inner- and outer-sphere interactions with anions. Much can be learned about Ln(III) catalysts, optical sensors, and MRI contrast agents by studying these properties in aqueous solution. For example, to better understand the efficiency of Ln(III) catalysts, it is important to define the coordination sphere and characterise the strength of binding interactions to substrate and

*Luminescence of Lanthanide Ions in Coordination Compounds and Nanomaterials*, First Edition.
Edited by Ana de Bettencourt-Dias.

transition state analogs through luminescence spectroscopy [4,5]. Luminescence studies of Ln(III) complexes of biologically relevant anions are beneficial for the design of optical sensors [6–8]. Finally, the development of Ln(III) complexes as smart MRI contrast agents which respond to changes in pH or anion binding can be assisted through monitoring changes in the coordination sphere [8,9]. In this chapter, the application of Ln(III) luminescence to Ln(III) complexes that respond to changes in the environment is reviewed. This chapter complements two other chapters in this book that have a focus on Ln(III) coordination complexes for bio-imaging (Chapter 4) and for development as chemosensors (Chapter 6).

The most common method to produce Ln(III) luminescence is through excitation of a ligand chromophore that behaves as an antenna [1,10,11]. Upon excitation of the ligand, energy transfer leads to excitation of the Ln(III) centre that produces luminescence. This method benefits from the large extinction coefficient of the ligand (typically $\varepsilon = 3000$–$50\,000\ \mathrm{mM^{-1}\,cm^{-1}}$). However, the complexity of the emission spectra, as well as the numerous pathways in which luminescence can be quenched, make it challenging to use luminescence to assess the coordination chemistry of the Ln(III) centre. An alternative approach to produce Ln(III) luminescence is by direct photoexcitation of Ln(III) electronic transitions in the 4*f*-manifold [12,13]. Direct excitation methods require powerful (mJ $\mathrm{pulse^{-1}}$) laser-based instrumentation because the electronic transitions between states with a $4f^n$ configuration are electric dipole (parity) and spin forbidden ($\varepsilon < 1\ \mathrm{mM^{-1}\,cm^{-1}}$) [1,14]. Furthermore, the luminescence of Ln(III) ions is quenched in aqueous solution by non-radiative processes largely through OH oscillators [15,16]. This makes it challenging to study direct excitation Ln(III) luminescence under biologically relevant conditions. Nonetheless, laser-induced Ln(III) luminescence was studied extensively in several groups including Horrocks and Choppin in the 1980s and 1990s by using tunable dye laser technology [2,13]. More recently in the Morrow laboratory, a spectroscopic system was built based on an injection seeded frequency tripled ($\lambda = 355$ nm) Nd:YAG pump laser coupled with an optical parametric oscillator (master oscillator power oscillator or MOPO) which offers continuously tunable output in the UV (230–400 nm), visible (420–690 nm), and IR (735–1800 nm) regions of the spectrum with approximately 50 mJ $\mathrm{pulse^{-1}}$ at 580 nm, the wavelength region at which Eu(III) is directly excited [17]. This system facilitates the study of Eu(III) at 1000-fold lower detection limits than previously reported. Furthermore, the tunable MOPO system facilitates excitation experiments of several members of the Ln(III) series over a single day. Excitation wavelength can be readily varied over the useful visible range (440–600 nm). This feature is especially useful for excitation of Eu(III) *f*-*f* transitions and for energy transfer experiments between selected Ln(III) ions. The low concentration of Eu(III) that can be studied in experiments that use the laser-MOPO system, even with no strongly binding ligands to shield the metal ion, facilitates studies of Ln(III) binding to biopolymers such as DNA and RNA [18].

## 8.2 Direct Excitation Lanthanide Luminescence

### 8.2.1 Luminescence Properties of the Lanthanide Ions

Luminescence properties of Ln(III) are reviewed only briefly here. The reader is encouraged to read Chapter 1 for a more detailed treatment.

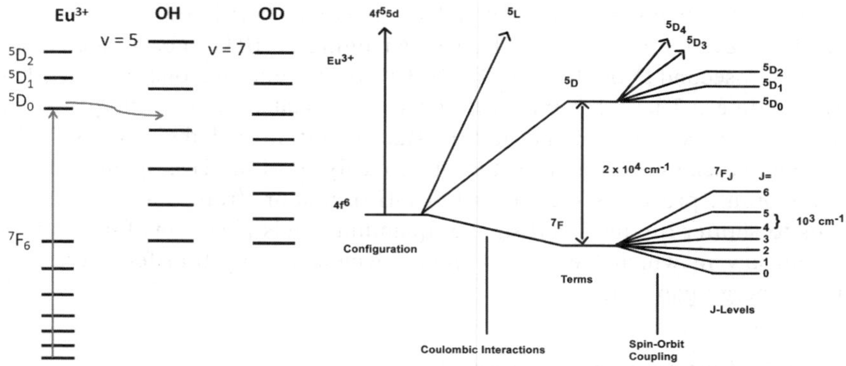


***Figure 8.1*** *Energy level diagram of Eu(III)*

The atomic line-like spectra of Ln(III) result from isolation of the inner 4*f* orbitals from the surrounding environment by the filled outercore 5*s* and 5*p* electrons [14]. Furthermore, electrons in the 4*f* orbitals do not participate appreciably in covalent bonding. The lack of interaction of the *f*-orbitals with the environment or crystal field leads to relatively pure *f*-orbitals so that selection rules for *f*-*f* transitions are largely adhered to. As a consequence, *f*-*f* transitions in absorbance and emission are weak ($\varepsilon < 1\ mM^{-1}\ cm^{-1}$) [1,14].

Energy diagrams for Ln(III) are useful for interpretation of basic elements of luminescence spectra. Figure 8.1 shows the energy level diagram for Eu(III) as the Ln(III) ion featured most frequently in this chapter because of its favourable luminescence properties [10,19].

The ground state $^7F$ manifold and the first three excited states of Eu(III) are shown. The energy differences between these terms are largely attributed to differences in Coulombic energy from interelectronic repulsion. Terms are further split by spin–orbit coupling. The resultant J levels are shown for the ground state manifold of Eu(III) which has seven J levels [14,20]. Ligand donors may split each of the J levels into further sublevels. From this description, the reader might expect that the absorbance and luminescence emission bands of the Ln(III) are highly complicated. This is indeed the case for most of the Ln(III) with the exception of Eu(III). For Eu(III), both the ground state ($^7F_0$) and the first excited state ($^5D_0$) are non-degenerate and thus not split by interaction with ligand donor groups. Thus, the $^7F_0 \rightarrow {}^5D_0$ transition gives a single unsplit peak for each different Eu(III) species. This feature makes Eu(III) a valuable probe of solution speciation for Eu(III) complexes as well as a surrogate for similar-sized Ln(III). The atomic radii of Ln(III) decrease by about 14% across the series, so that early and late Ln(III) are larger and smaller, respectively, than Eu(III) [14]. Similarity of the chemical properties and ionic radii of Ln(III) in the middle of the series makes it feasible to use Eu(III) as a substitute.

A second useful spectroscopic property of Eu(III) is its relatively long (millisecond) luminescence lifetime [1,19]. This relatively long lifetime is shared by Tb(III) and Gd(III). Such long lifetimes facilitate experiments that involve time gating of luminescence to obtain a better signal by removing short-lived fluorescence from biologically relevant fluorophores [11]. The remaining Ln(III) can be divided into those that have microsecond [Yb(III), Sm(III), Dy(III)] or sub-microsecond [(Er(III), Tm(III), Ho(III)] lifetimes. These lifetime differences are related to the energy difference between the highest level of the ground state manifold and the first excited state. The smaller the energy difference, the more easily the Ln(III) is quenched by non-radiative vibronic processes such as OH or NH oscillators from water or ligands, and this leads to shorter lifetimes and smaller quantum yields [21]. This follows from the energy gap law in which the rate of intramolecular energy transfer decreases with increasing energy gap [19].

### 8.2.2 Ln(III) Excitation Spectroscopy

Ln(III) excitation spectroscopy on the laser-MOPO system gives spectroscopic information that is similar to that obtained by absorbance spectroscopy. However, a low (μM) concentration of Ln(III) may be studied using the laser-MOPO system while achieving higher resolution and better signal to noise [17]. Figure 8.2 shows the excitation spectra of representative trivalent lanthanide ions [Eu(III), Tb(III), Dy(III), Sm(III)] that excite and emit in the visible region of the spectrum.

These spectra are typically collected by using a narrow band pass filter to monitor one of the emission peaks of the Ln(III). For example, a band pass filter of $628 \pm 27$ nm is used for Eu(III) to collect the $^5D_0 \rightarrow {}^7F_2$ emission. This strategy allows a higher energy flux and larger signal-to-noise than that obtained by passing the emitted light through a monochromator.

For Ln(III) other than Eu(III), the ground state or excited states are non-degenerate. This leads to excitation peaks that are split into multiple components that are quite complex (Fig. 8.2) [17]. It is difficult to deconvolute these peaks if there are multiple species in solution. In contrast, many of the excitation peaks of Eu(III) are relatively simple, especially for those transitions arising from the non-degenerate $^7F_0$ ground state. Of special simplicity is the $^7F_0 \rightarrow {}^5D_0$ transition which falls at 578–581 nm (Fig. 8.3).

A single peak is observed for each distinct Eu(III) species in solution. This is displayed more clearly in Fig. 8.3 where the $^7F_0 \rightarrow {}^5D_0$ excitation peaks of several different Eu(III) complexes are shown both in solution and solid state. In solution, the full width at half-maximum (FWHM) for each Eu(III) complex is typically around 0.30 nm, but is narrower in the solid state. The position of the $^7F_0 \rightarrow {}^5D_0$ Eu(III) excitation peak has been related to the type of donor atom and its charge in an empirical correlation [22,23]. For example, relative to the excitation peak of the Eu(III) aqua complex, anionic ligands such as carboxylates give rise to red-shifted peaks while neutral donors such as amides shift the excitation peak to a lesser extent. The position of the peak also depends on the Eu(III) coordination number in which a reduction in coordination number results in a blue shift of the excitation peak. Changes in the energy of the excitation peak have been rationalised based on the nephelauxetic effect of various donor groups [22,24]. Ligands that lead to red-shifting of the excitation peak decrease the electronic repulsion parameters within the Eu(III) ion. Nephelauxetic parameters may be related to differences in covalency of donor group bonds to Eu(III).

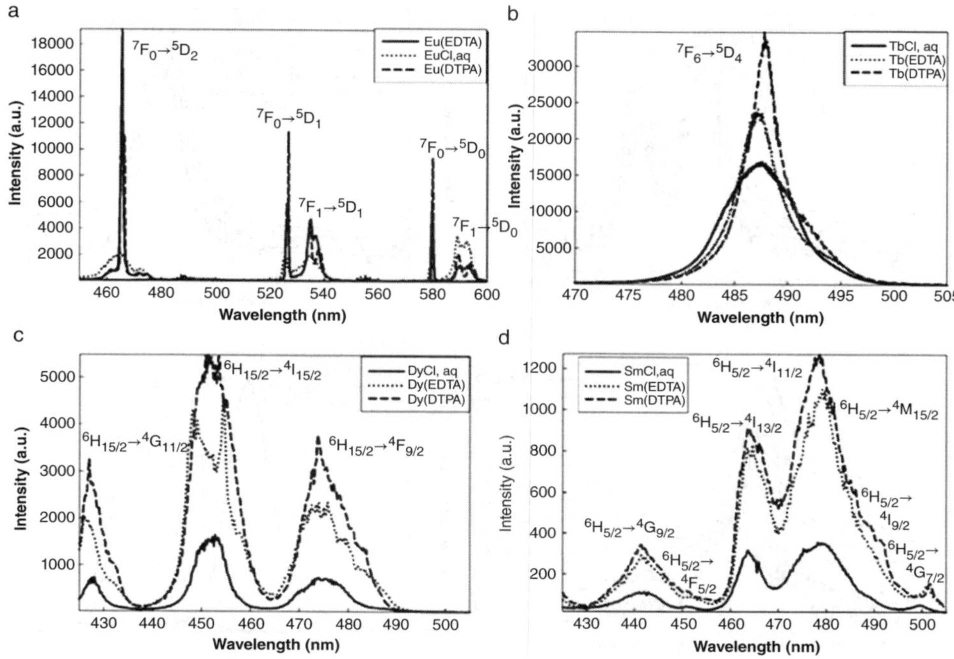


*Figure 8.2* *Excitation spectra of 1.00 mM Ln(III) complexes in 20 mM MES buffer, pH 6.5, I = 0.1 M (NaCl): (a) Eu(EDTA), Eu(DTPA), and $EuCl_3$(aq), (b) Tb(EDTA), Tb(DTPA), and $TbCl_3$(aq), (c) Dy(EDTA), Dy(DTPA), and $DyCl_3$(aq), (d) Sm(EDTA), Sm(DTPA), and $SmCl_3$(aq). Reproduced with permission from [17]. © 2009, Society for Applied Spectroscopy*

### 8.2.3 Ln(III) Emission Spectroscopy

Ln(III) emission spectroscopy is reviewed in several other chapters and will only be briefly touched on here. As before, the emphasis is on Eu(III) spectroscopy. Similar to other Ln(III) ions, the higher level Eu(III) states are quenched in solution so that most of the luminescence arises from the lowest energy excited state, the $^5D_0$ state [25]. The electric dipole induced transitions, including the $^5D_0 \rightarrow {}^7F_2$ ($\Delta J = 2$) and the $^5D_0 \rightarrow {}^7F_4$ ($\Delta J = 4$), exhibit hypersensitive character in which the intensities change more dramatically in response to differences in ligand field than do other emission bands [1,13]. Another useful transition is the $^5D_0 \rightarrow {}^7F_0$ ($\Delta J = 0$) emission peak which reflects the number of Eu(III) environments, similar to the $^7F_0 \rightarrow {}^5D_0$ excitation peak. The $^5D_0 \rightarrow {}^7F_1$ ($\Delta J = 1$) band (a magnetic dipole transition) splits into as many as three components in the presence of strong ligand fields. The splitting pattern may reflect the absence or presence of axial symmetry. The presence of multiple emission peaks, some of which are sensitive to the coordination sphere and some of which are unperturbed, facilitates the development of Eu(III) ratiometric sensors [26,27]. Figure 8.4a shows overlaid emission spectra of three Eu(III) complexes. Similarly, direct

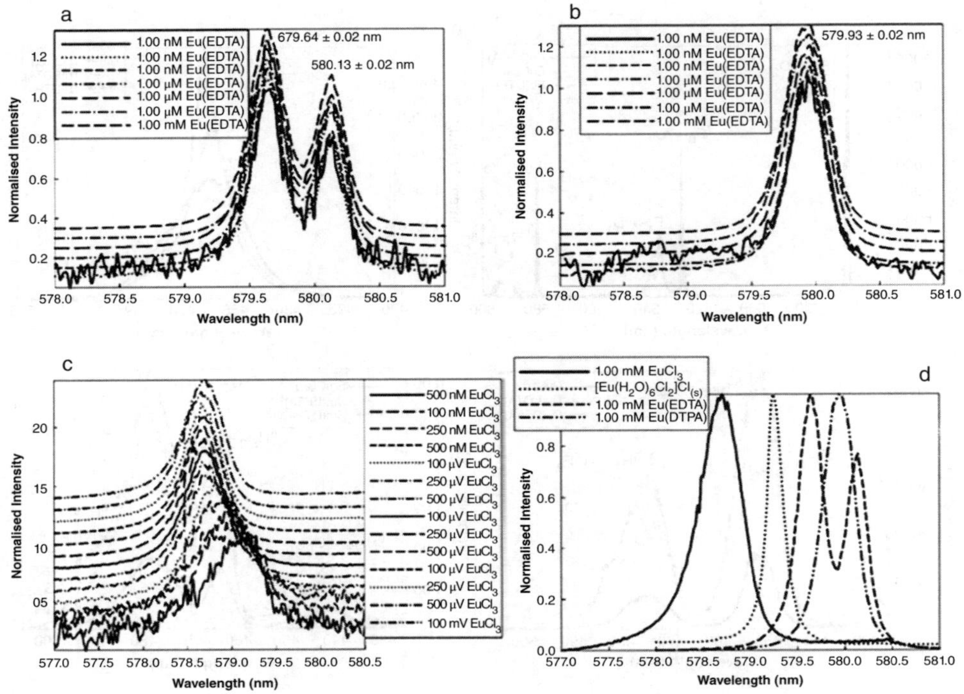


***Figure 8.3*** *Excitation spectra ($\lambda_{em} = 628 \pm 27$ nm) as a function of concentration for: (a) Eu (EDTA), (b) Eu(DTPA), (c) $EuCl_3$(aq), (d) comparison of 1.0 mM solutions of Eu(EDTA), Eu (DTPA), $EuCl_3$(aq), and solid $EuCl_3 \cdot 6(H_2O)$ excitation spectra. All complexes were prepared in 20 mM MES buffer, pH 6.50, I = 0.1 M (NaCl). Spectra are offset for clarity. Reproduced with permission from [17]. © 2009, Society for Applied Spectroscopy*

excitation of Tb(III), Dy(III) and Sm(III) leads to multiple emission peaks in the visible region (Fig. 8.4b,c,d).

### 8.2.4 Time-Resolved Ln(III) Luminescence Spectroscopy

Determining the number of bound waters in Ln(III) complexes by correlation of luminescence lifetimes is one of the most popularised and useful of the luminescence techniques [15,16]. This technique monitors the large non-radiative quenching contribution by OH groups in water or alcohols. Additional groups that quench luminescence include the NH of amides and amines. Substitution of protons by deuterons to give OD or ND groups gives much less effective quenching because energy must transfer to higher lying vibrational overtones. Luminescence decay experiments conducted alternatively in water and in $D_2O$ give information about the OH groups bound to the Ln(III) centre. Equation (8.1) was empirically derived from compilation

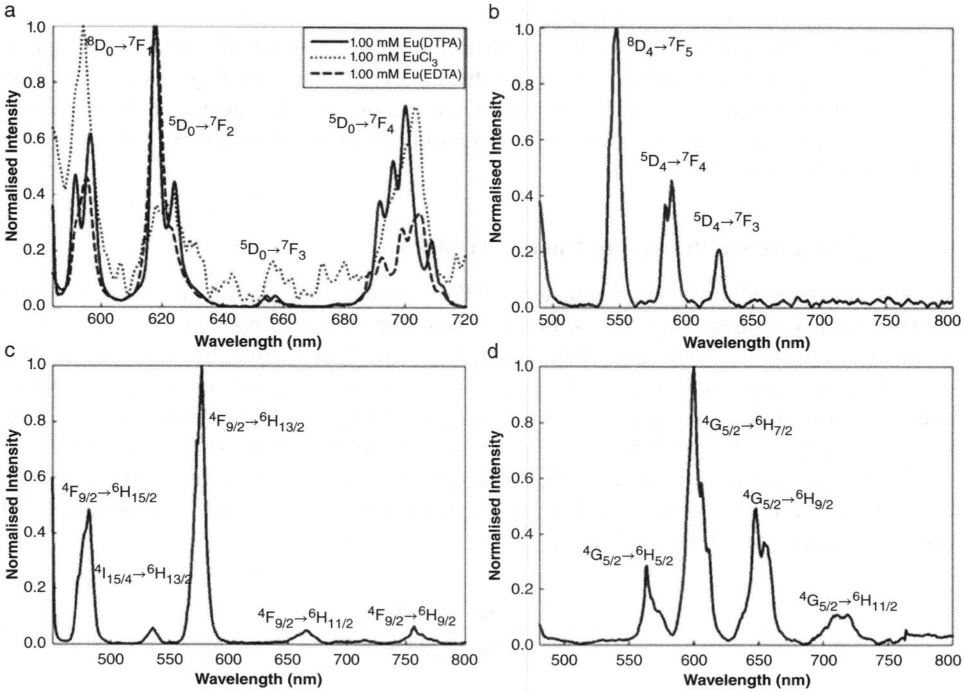


***Figure 8.4*** *Emission spectra of 1.0 mM Ln(III) in 20 mM MES buffer, pH 6.5, I = 0.1 M (NaCl): (a) Eu(EDTA), Eu(DTPA), and EuCl$_3$(aq), (b) Tb(DTPA), (c) Dy(DTPA), (d) Sm(DTPA). Reproduced with permission from [17]. © 2009, Society for Applied Spectroscopy*

of luminescence data of a large number of Eu(III) complexes. Eu(III) macrocyclic complexes have slightly different parameters compared to linear chelates that predominantly contain anionic oxygen donors. The following contributions are generally used for macrocyclic Eu(III) complexes: $\alpha = 0.25\ \text{ms}^{-1}$ for outersphere contributions, $\beta = 0.45\ \text{ms}^{-1}$ for alcohol OH, $\delta = 0.075\ \text{ms}^{-1}$ for amide NH, A = 1.2 ms where $n$ is the number of quenchers of a particular type [16].

$$q = A[k_{H_2O} - k_{D_2O} - k_{XH}]$$
$$k_{XH} = \alpha + \delta n_{O=CNH} + \beta n_{OH} \tag{8.1}$$

Despite the usefulness of this approach, it is important to appreciate that there are many uncertainties from other contributions to quenching and from quenching parameters that are not well-determined. For example, the magnitude of outersphere quenching has not been determined for many types of Ln(III) coordination complexes other than the type described above. Outersphere quenching is known to vary based on the charge and type of complex [16]. The efficiency of quenching is proportional to the distance of the quencher from the

Ln(III) and the vibrational overtone that matches the energy gap of the Ln(III), yet there is no correction for different geometries or coordination numbers that would modulate quenching by ligand groups through different bond lengths. Finally, other quenching processes such as back energy or electron transfer are not taken into account in the equation [28,29]. Nonetheless, time resolved luminescence is a powerful tool for probing OH or NH oscillators in the Ln(III) coordination sphere.

### 8.2.5 Luminescence Resonance Energy Transfer

Luminescence resonance energy transfer (LRET) proceeds by radiationless dipole–dipole coupled energy transfer from an excited luminophore through space to another luminophore [30]. There are several criteria that must be met for LRET. First, the emission band of the excited luminophore donor must overlap the absorbance band of the acceptor luminophore. The distance between the donor and acceptor molecules can be measured by using Equation 8.2 [31]. In Equation 8.2, the efficiency of energy transfer is $E$, $\tau_{DA}$ is the time-resolved luminescence lifetime of the donor–acceptor pair, $\tau_D$ is the lifetime of the donor, $R_0$ is the distance for 50% energy transfer to occur (or Förster distance), and $r$ is the calculated distance between the donor and acceptor.

$$E = 1 - \frac{\tau_{DA}}{\tau_D} = \frac{R_0^6}{R_0^6 + r^6} \tag{8.2}$$

LRET between Ln(III) and organic fluorophores [32,33] or between two Ln(III) ions [34,35] is a useful way to determine distances between luminescent species and has many applications. LRET between Ln(III) ions in aqueous solution has been used to study dimerisation of Ln(III) ions through bridging ligands or to determine distances between Ln(III) ions in dinuclear systems [36–39]. LRET has been used to approximate the distance between metal ion binding sites in biomolecules, especially calcium binding proteins [34]. These studies entail using two different lanthanide cations to estimate distances between metal ion binding sites, typically the Eu(III)-Nd(III) donor–acceptor pair. Nd(III) as acceptor has one of the largest extinction coefficients ($\varepsilon \sim 10\,\text{mM}^{-1}\,\text{cm}^{-1}$) of all the Ln(III), facilitating LRET over longer distances. The most useful donor is Eu(III), because the $^7F_0 \rightarrow {}^5D_0$ excitation peak is not split by ligand fields and this allows for precise control over which Eu(III) species is excited. One drawback in LRET between two Ln(III) ions is that the small overlap integral leads to a small $R_0$, the distance at which energy transfer is 50%. A typical $R_0$ for a Eu(III)/Nd(III) couple in a dinuclear complex is 6 Å, but only 3 Å for Eu(III)/Tb(III) [39]. This suggests that only LRET between closely spaced Ln(III) can be detected. Dinuclear Ln(III) complexes with bridging anions are sufficiently close to give rise to LRET. However, only closely spaced Ln(III) in biopolymer studies that involve multiple metal ion binding sites demonstrate LRET [18]. In contrast, LRET between sensitised luminescent Ln(III) donor and organic fluorophore acceptor typically can be observed over much longer distances (25–100 Å) [32,40,41]. Such LRET systems have applications in the study of protein-protein interactions, ion channels in cells [32], and mapping Ln(III) ion binding sites in complex RNA molecules [42,43].

## 8.3 Defining the Ln(III) Ion Coordination Sphere through Direct Eu(III) Excitation Luminescence Spectroscopy

### 8.3.1 Eu(III) Complex Speciation in Solution: Number of Excitation Peaks

The $^7F_0 \rightarrow {}^5D_0$ excitation peak of Eu(III) is a powerful tool for determining the number of different types of complex species in solution. Both the $^7F_0$ and $^5D_0$ levels are non-degenerate and not split by ligand fields. In principle then, each Eu(III) species would give rise to a single excitation peak. Although the position of this peak does not shift by a large extent (577–581 nm), the narrow band width (0.25 nm FWHM) of the excitation peaks and the high spectral resolution of the laser-MOPO facilitates the resolution of multiple excitation peaks (Fig. 8.3). The number of different species in solution is directly related to the number of different excitation peaks, with the caveat that excitation peaks might overlap if the environment of the Eu(III) species are not sufficiently different. The species that give rise to these excitation spectra can be further characterised by monitoring their luminescence lifetime decays upon excitation at each peak. This powerful technique can be applied to analyse different forms of Eu(III) complexes. In the following sections we discuss examples of: (i) Eu(III) coordination isomers, (ii) Eu(III) speciation differences based on ligand ionisation and other innersphere coordination changes, and (iii) Eu(III) complex dimerisation or aggregation.

### 8.3.2 Excitation Spectra of Geometric Isomers

Isomeric forms of Eu(III) complexes may give rise to multiple excitation peaks if the Eu(III) coordination sphere is sufficiently different. For example, Eu(TCMC) [refer to Fig. 8.5 for this and other Ln(macrocycle) structures] has two different diastereomeric forms in solution that result from distinct orientations of pendent groups and macrocyclic ring (Fig. 8.6) [44,45]. Note that the Ln(III) complexes are specified without water ligands [e.g. Eu(TCMC)] unless a particular coordination complex species is discussed [e.g. Eu(TCMC)($OH_2$)].

One diastereomer has a twisted square antiprismatic (TSAP) geometry and the other has a square antiprismatic (SAP) geometry [46]. The distinct arrangement of donor groups gives rise to two Eu(III) excitation peaks which are assigned as arising from two diastereomers in solution. This is based on comparison to the excitation spectrum of a crystalline form of [Eu(TCMC)($OH_2$)]($CF_3SO_3$)$_3$ which showed both diastereomeric forms in equal proportion. Luminescence experiments on a single crystal of [Eu(TCMC)($OH_2$)]($CF_3SO_3$)$_3$ showed two $^7F_0 \rightarrow {}^5D_0$ excitation peaks at approximately the same wavelength as that observed for the complex in solution. Further evidence for the diastereomeric relationship came from the nearly identical luminescence lifetimes upon excitation of either peak. This suggests that either (i) both diastereomers have the same number of bound water molecules and outersphere quenching, or (ii) interconversion of the two isomers is fast on the luminescence timescale. Similarly, Eu(TrAC) shows two excitation peaks, one major and one minor, that are assigned to two different diastereomers present in solution (Fig. 8.7) [47].

Another more familiar example of two isomeric forms in solution is Eu(EDTA) (Fig. 8.3a). The two different excitation peaks purportedly arise from Eu(EDTA) complexes with two different coordination numbers, one with two water ligands and one with three ligands. This assignment is based in part on the equilibrium behaviour of the two species as a function of pressure [48,49].

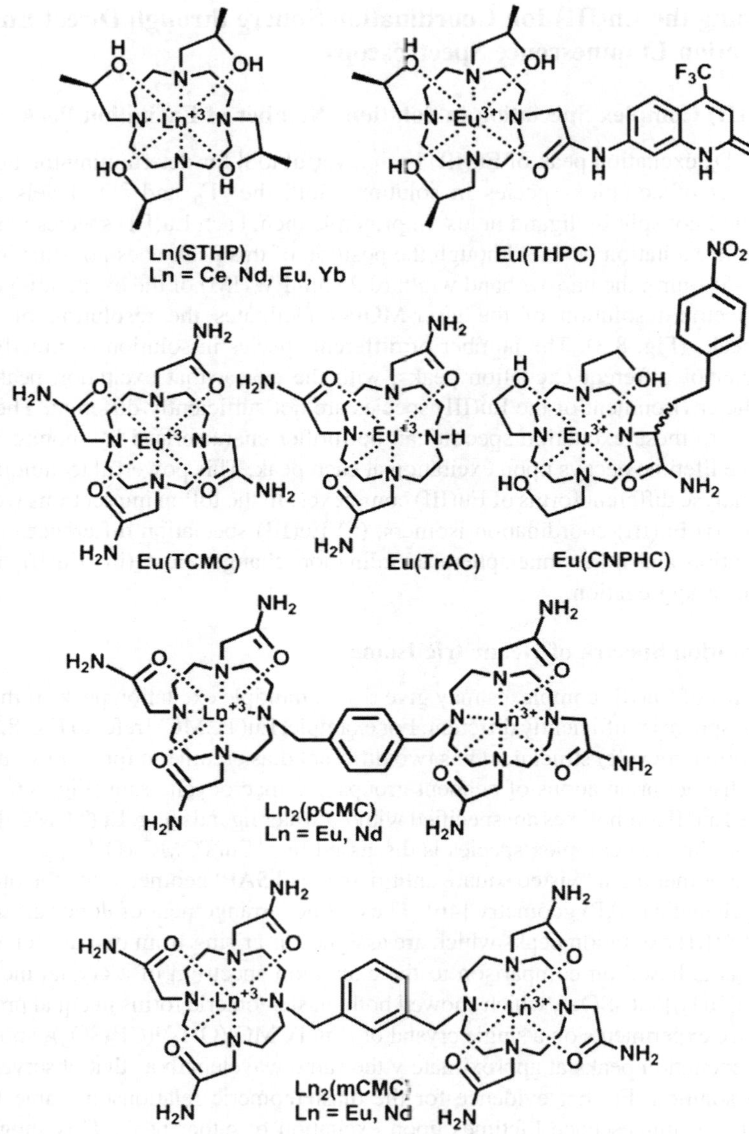


*Figure 8.5* *Macrocyclic lanthanide complexes discussed in this chapter. Ln(III) complexes are specified without water ligands unless a particular coordination complex species is discussed*

### 8.3.3 Innersphere Coordination of Anions

It is extremely challenging to define solution speciation based on Eu(III) luminescence data in solutions that only contain weak ligands such as buffer, or counter ions such as nitrate, chloride, or triflate. This is in large part due to the complexity of the solution chemistry of

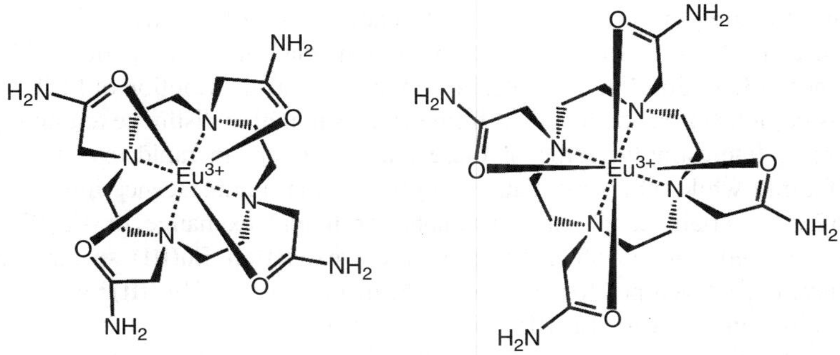


***Figure 8.6*** *Diastereomeric forms of macrocyclic complexes*

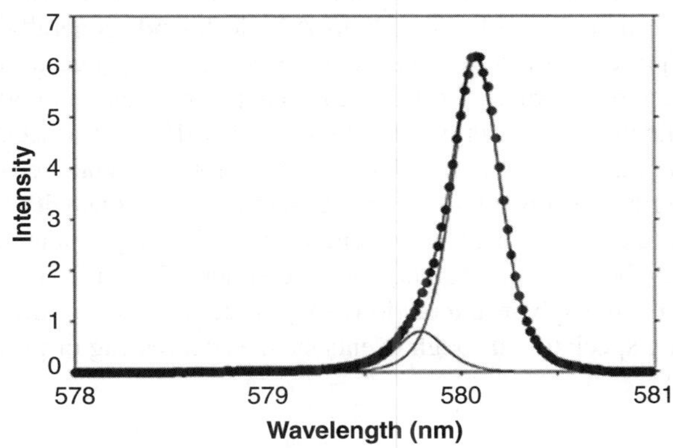


***Figure 8.7*** $^7F_0 \rightarrow {}^5D_0$ *excitation spectrum* ($\lambda_{em} = 628 \pm 27$ *nm) for Eu(TrAC) at pH 7.0, buffer = 20 mM, I = 0.10 M* $NaNO_3$. *Peak fit data is consistent with two different Eu(III) species. Reproduced with permission from [47]. Royal Society of Chemistry*

Ln(III) aqua complexes under these conditions. Attempts have been made to use luminescence lifetime data to determine whether simple anions such as perchlorate bind innersphere to Eu(III) [50–52]. Differences in lifetimes are difficult to interpret in light of competing phenomena such as anion quenching and speciation changes. Our work shows that certain anions such as nitrate and fluoride form innersphere complexes because the excitation peak shifts markedly upon binding to Eu(III) at both pH 5.0 and 6.5 [53]. However, solution speciation complications such as hydroxo complex formation or equilibria involving multinuclear complexes made interpretation very difficult for weakly binding anions, as discussed below. This highlights the difficulty of interpreting Ln(III) speciation in solutions that lack strong ligands due to extremely complicated solution chemistry.

More straightforward are studies that use Eu(III) excitation spectroscopy to monitor binding of multidentate chelates to Eu(III) with replacement of aqua ligands. Horrocks and

Choppin were among those who pioneered the application of Eu(III) excitation spectroscopy to the characterisation of complex solution speciation upon titration with strongly bound ligands [13,54–56]. For example, information on the distribution of Ln(III) in mono, bis, and tris complexes can be obtained from systems that give distinct excitation peaks for each complex. Interestingly, some of these complexes are in rapid exchange and give average lifetimes while others are static and give distinct luminescence lifetimes for each species [57]. In certain cases, rate constants for ligand exchange or Eu(III) complex formation constants can be obtained from direct excitation Eu(III) spectroscopy [58]. Methods have been developed for monitoring the displacement of Eu(III) by other Ln(III) to yield formation constants for Ln(III) complexes [55].

Studies that follow binding of biologically relevant anions to Eu(III) centres that are nearly encapsulated by macrocyclic ligands give luminescence data that leads to detailed models of anion binding [8,59]. In these complexes, most of the coordination sphere is controlled by the macrocyclic ligand and a single coordination site is perturbed by replacement of a water ligand by the anion. The limited number of coordination sites that are accessible makes interpretation of luminescence data more straightforward. Innersphere binding of a ligand, nearly always an anionic ligand, generally results in a new Eu(III) excitation peak with a lengthened luminescence decay due to a decrease in OH oscillator quenching from water displacement. There are many important examples of innersphere coordination changes in the field of Ln(III) luminescent sensors, anion responsive MRI contrast agents, or Ln(III) catalysts. More unusual are outersphere ligand interactions as described below. Outersphere interactions of anions with Eu(III) complexes do not affect the excitation peak wavelength, nor do they appreciably change the luminescence decay because the number of innersphere OH or NH quenchers remains the same. However, outersphere interactions may affect the appearance of the excitation spectrum by altering speciation through events such as depressing or promoting ionisation (see Section 8.5.2).

### 8.3.4 Ligand Ionisation

Eu(III) excitation spectroscopy is extremely useful for monitoring ionisation of Eu(III) complexes as a function of pH. An example of this is shown in Fig. 8.8 where a blue shifted excitation peak grows in with increasing pH for Eu(CNPHC)($OH_2$) [60]. The p$K_a$ of this group is 7.5 and correlates to that measured by pH potentiometric titrations. It is not certain whether this new species arises from deprotonation of an alcohol group or of the water ligand.

Unexpectedly, water counting experiments showed a more highly quenched Eu(III) luminescence. This is inconsistent with simple deprotonation and suggests a more fully hydrated species or perhaps increased hydrogen bonding to outersphere waters. A second unexpected observation is that the excitation peak for the deprotonated complex is blue-shifted relative to that of the Eu(CNPHC)($OH_2$) aqua complex. Again, this suggests that the coordination sphere has undergone unusual changes because binding of anionic ligands to Eu(III) typically gives rise to peaks that are red-shifted relative to the aqua complex [22]. Similar luminescence properties were observed for the related complex, Eu(STHP) ($OH_2$) [60]. Further information on this species is based on its reactivity. The deprotonated form of the complex reacts through an alkoxide nucleophile to displace good leaving groups

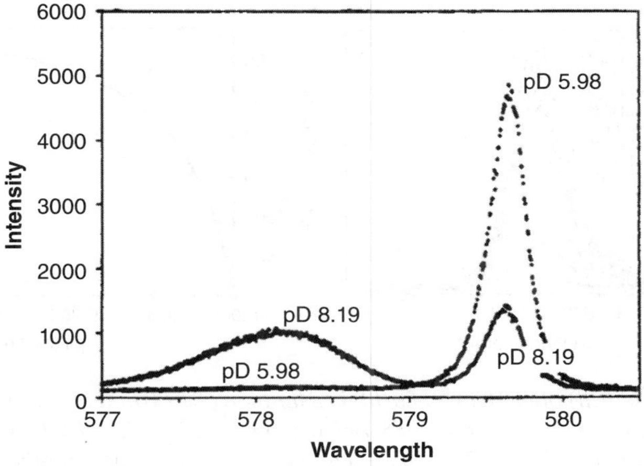


***Figure 8.8*** *$^7F_0 \rightarrow {}^5D_0$ excitation spectra ($\lambda_{em}$ at 628 ± 27 nm) for 0.01 mM solution of Eu (CNPHC) at pD 5.98 and pD 8.19, 20 mM buffer, 0.10 M NaCl. Reproduced with permission from [60]. © 1998, American Chemical Society*

on either phosphate diesters or on phosphoric anhydrides. This supports at least partial ionisation of the alcohol pendent group [61].

Ionisation of Eu(III) in water to give hydroxide complexes generally gives complicated solution chemistry with too many excitation peaks to be resolved. For example, the $^7F_0 \rightarrow {}^5D_0$ excitation spectrum of Eu(III) salts in aqueous solution as a function of pH attests to the complicated solution chemistry of Eu(III) under these conditions (Fig. 8.9).

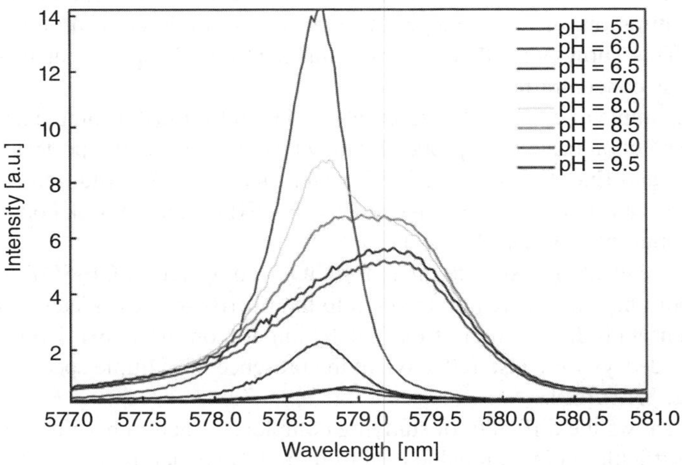


***Figure 8.9*** *$^7F_0 \rightarrow {}^5D_0$ excitation spectra of 25 μM Eu(III) ($\lambda_{em}$ at 628 ± 27 nm), I = 0.1 M (NaCl), 20 mM buffer showing pH dependence. Reproduced with permission from [53]. © 2009, John Wiley & Sons*

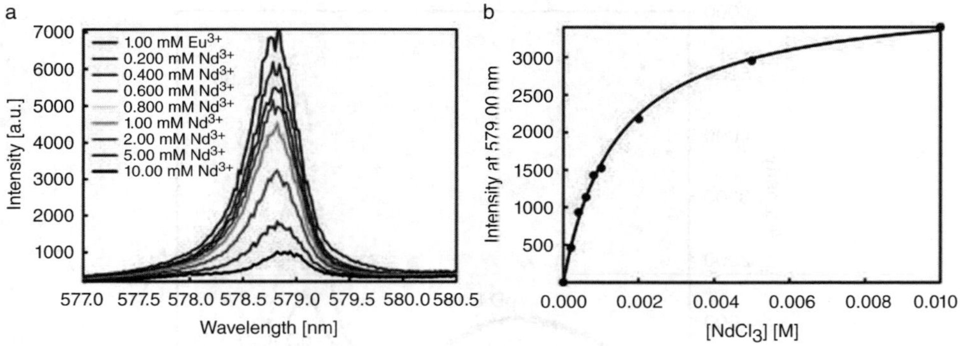


***Figure 8.10*** *(a) Nd(III) quenching of Eu(III) in a titration of 1.00 mM Eu(III) with Nd(III).* $^7F_0 \rightarrow {}^5D_0$ *Excitation spectra in 20.0 mM MES pH 6.5, I = 0.1 M NaCl. (b) Binding curve of the absolute change of intensity at 579.0 nm as a function of Nd(III) concentration, I = 0.1 M NaCl,* $K_d$ *1.43 ± 0.07 mM. Reproduced with permission from [53]. © 2009, John Wiley & Sons*

The excitation peak increases in intensity from pH 5 to 6, reaching a maximum at pH 7. This increase in intensity likely corresponds to the formation of hydroxo complexes that would have fewer quenching OH interactions and exhibit reduced symmetry of the Eu(III) in comparison to $Eu(OH_2)_9$. A further increase in pH leads to the appearance of new red-shifted excitation peaks that presumably correspond to the formation of new hydroxo complexes, perhaps polymeric in form. Surprisingly, even at pH 6.5, there is a substantial percentage of multinuclear Eu(III) complexes in solution [53]. This is shown by a titration of 1.0 mM $EuCl_3$ with $NdCl_3$ as an efficient quencher of Eu(III) luminescence. The decrease in the excitation peak is attributed to the formation of mixed multinuclear Eu-Nd complexes in solution (static rather than dynamic quenching) as shown by saturation of the change in intensity at high concentrations of Nd(III) (Fig. 8.10). This shows that there are multinuclear Ln(III) species at near neutral pH and millimolar concentrations.

A further example of complicated solution speciation for Eu(III) complexes as a function of increasing solution basicity as probed by direct excitation Eu(III) spectroscopy is shown for $Eu(ClO_4)_3$ and $Eu(CF_3SO_3)_3$ in DMSO/water mixtures [62]. Addition of $Me_4NOH$ to these solutions leads to a very broad asymmetric Eu(III) excitation peak consistent with the formation of multiple species (Fig. 8.11).

Curiously, the excitation spectra differ for $ClO_4^-$ compared to $CF_3SO_3^-$ salts, possibly due to the coordination of perchlorate anion to the Eu(III) species. Water counting studies were consistent with dehydration of the Eu(III) ion at more basic pH. However, luminescence lifetime decays were also reflective of the presence of multiple species with different lifetimes (Fig. 8.12) [62].

This study shows the difficulty of studying complex speciation by lifetime decays alone because it is difficult to obtain reliable data from multi-exponential fits. LRET studies with the Eu(III)/Tb(III) couple in DMSO were consistent with multinuclear species as shown by energy transfer. The presence of dinuclear hydroxo complexes was also supported by pH potentiometric titration data [62].

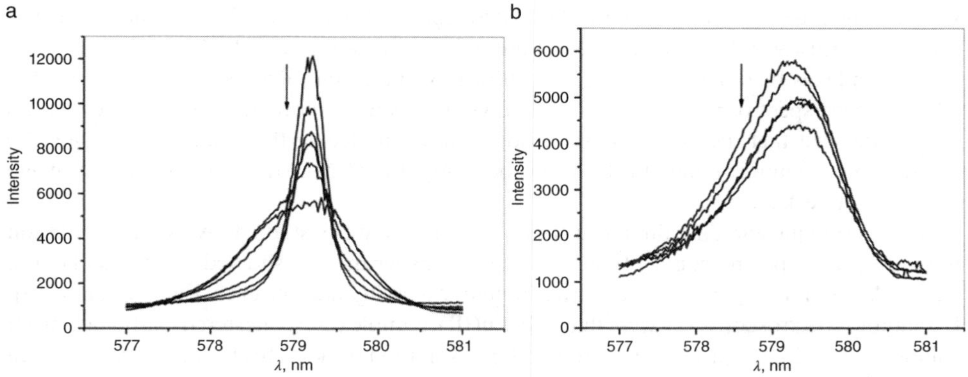


*Figure 8.11* $^7F_0 \rightarrow {^5D_0}$ excitation spectra ($\lambda_{em}$ at 628 ± 27 nm) of 1.0 mM $Eu(CF_3SO_3)_3$ (a) or $Eu(ClO_4)_3$ (b) in 80% DMSO at variable amounts of added $Me_4NOH$ from 0 to 2.5 mM. Reproduced with permission from [62]. Royal Society of Chemistry

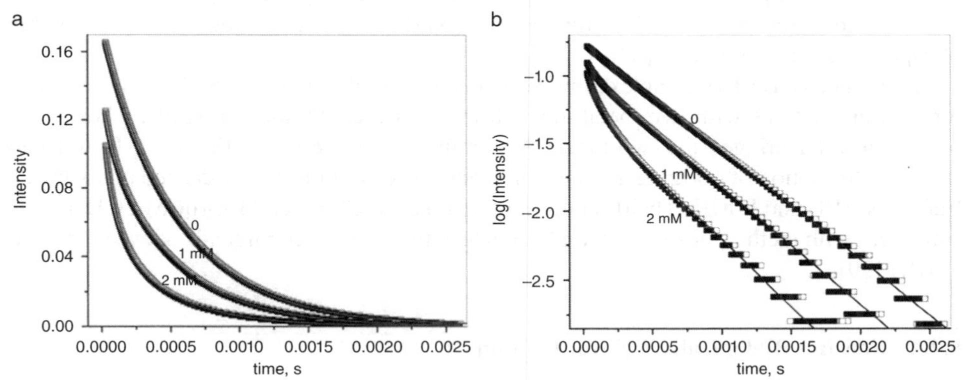


*Figure 8.12* Time-resolved intensity decays at 579.22 nm for 1.0 mM $Eu(CF_3SO_3)_3$ in the presence of 0, 1.0 and 2.0 mM $Me_4NOH$ in 80% DMSO. Lines show fitting to a single exponent curve for decay in the absence of added base and to a two exponent curve for decays in the presence of added base: (a) linear scale, (b) semilogarithmic scale. Reproduced with permission from [62]. Royal Society of Chemistry

## 8.4 Luminescence Studies of Anion Binding in Catalysis and Sensing

### 8.4.1 Phosphate Ester Binding and Cleavage

Interactions with phosphate esters can be monitored by utilising Eu(III) excitation spectroscopy to elucidate catalytic properties of the complexes [4,5,47]. For example, the efficiency of Ln(III) complexes as catalysts for the cleavage of phosphate diesters and RNA correlates, in part, to the strength of binding of the Ln(III) complex to phosphate esters [5,47]. Catalysis of phosphodiester cleavage involves the direct binding and interaction of the Ln(III) complex with the phosphate diester through water ligand replacement [63]. Ln(III) complexes such as Eu(TrAC), $Eu_2$(pCMC) or $Eu_2$(mCMC) are effective catalysts for RNA cleavage [5]. All three complexes bind phosphate esters as shown by kinetic inhibition studies and by Eu(III) luminescence spectroscopy. Notably, Eu(TCMC) does not coordinate to phosphate diesters despite the fact that the complex has a bound water molecule that could theoretically be replaced by an anionic ligand [46]. Correspondingly, Eu(TCMC) does not catalyse cleavage of phosphate esters.

Both mononuclear and dinuclear Ln(III) catalysts were studied by direct excitation luminescence spectroscopy. Water counting experiments showed that Eu(TrAC), $Eu_2$(pCMC) and $Eu_2$(mCMC) had two bound water ligands on each Eu(III) centre [5]. Excitation spectroscopy showed that the Eu(III) complexes sequestered carbonate from solutions over time to produce new red-shifted excitation peaks. This is not surprising given that the Eu(III) catalysts were designed to be strong anion binders in order to promote catalytic cleavage of phosphate diesters. The complexity of the solution chemistry highlights the difficulty of characterising the solution chemistry of Eu(III) complexes by using less informative methods such as pH potentiometric titrations. All complexes bound diethylphosphate and methylphosphate with replacement of a water molecule. Stronger binding of methylphosphate compared to diethylphosphate by the Eu(III) catalysts corresponds to the stronger interactions that occur upon development of negative charge on the phosphorane transition state [63].

The dinuclear Eu(III) complexes were prepared with the goal of accelerating cleavage through interactions with two metal ions. However, we could not distinguish whether the phosphate ester bridged the two metal ions from our luminescence titrations. In order to study the formation of bridging anionic complexes, LRET studies were carried out between Eu(III)/Nd(III) and Eu(III)/Tb(III) couples in both $Ln_2$(mCMC) and $Ln_2$(pCMC). Data was consistent with methylphosphate and phosphate acting as bridging ligands between the two Ln(III) [39].

### 8.4.2 Sensing Biologically Relevant Anions

Ln(III) complexes have been developed as sensors for biologically relevant anions such as phosphate, carbonate, and citrate. Parker's group has shown that Ln(III) macrocyclic complexes that contain an attached antenna can be used to sense biologically relevant anions in cell culture [7]. Most of these complexes are related to Eu(TrAC) which contains two available coordination sites for binding anions. In particular, carbonate is bound strongly in a bidentate interaction that displaces two bound water ligands [6]. Binding of carbonate to Eu(III) complexes gave rise to an increase in the hypersensitive $^5D_0 \rightarrow {}^7F_2$ emission peak while the $^5D_0 \rightarrow {}^7F_1$ emission peak changed very little. Monitoring both emission peaks is useful for ratiometric sensing of carbonate [26].

We recently reported on Eu(STHP) and Eu(THPC) as dual optical/MRI sensors for anions including phosphate, methylphosphate, lactate, acetate, carbonate and citrate [8]. These complexes are unusual in that they have a single bound water molecule and are nearly encapsulated by the macrocyclic ligand. Binding selectivity to anions is quite different than it is for macrocyclic complexes related to Eu(TrAC) that have two available coordination sites. An interesting feature of these complexes is that the hydroxyl groups of the macrocycle form outersphere interactions with ligands through hydrogen bonding [64].

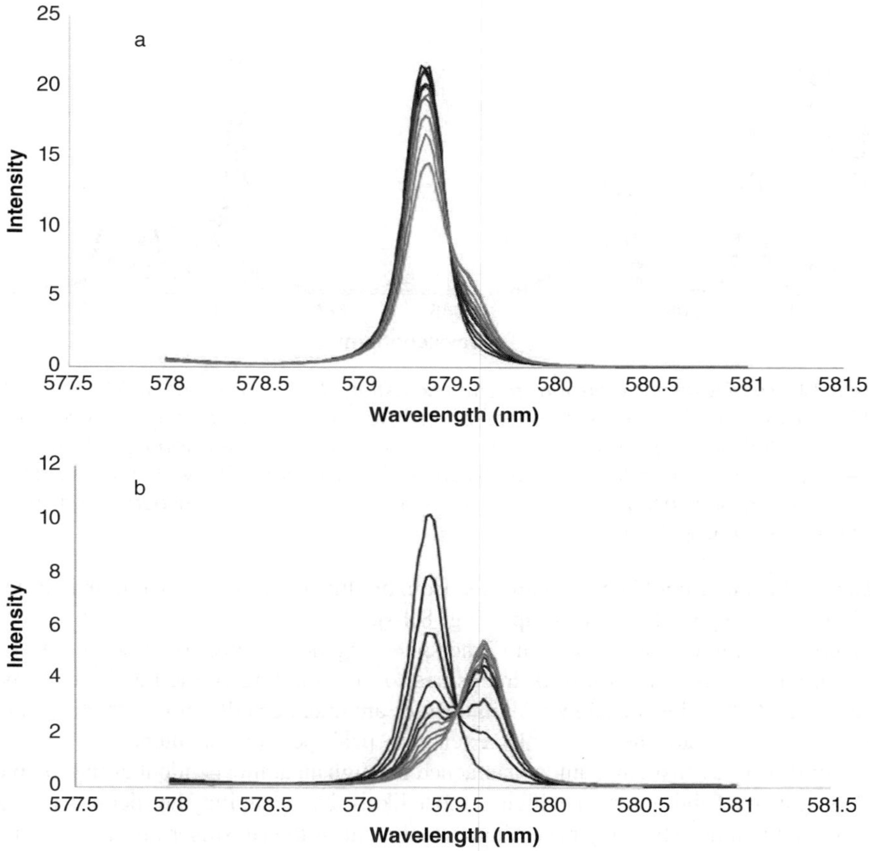


***Figure 8.13*** *The $^7F_0 \rightarrow {}^5D_0$, excitation spectra (emission at 628 ± 27 nm) of Eu(STHP) with 20 mM HEPES, 100 mM NaCl, pH 7.2 upon addition of: (a) lactate with new peak at 579.54 nm, or (b) citrate with new peak at 579.64 nm. Reproduced with permission from [8]. © 2011, American Chemical Society*

Both Eu(III) excitation spectroscopy and emission spectroscopy gave information about solution speciation as a function of anion concentration as well as the structure of the bound anions [8]. Carbonate and lactate bound the most weakly ($K_d = 35$ mM, 16 mM, respectively). The lactate and carbonate complexes participate in both innersphere and outersphere interactions as shown by the presence of both the original excitation peak of Eu(STHP)($OH_2$) and a new red-shifted peak for the innersphere complex (Fig. 8.13a). A non-integral $q$ number of 0.56 for lactate and 0.64 for carbonate is also consistent with a mixture of inner and outersphere complexes. Similarly, acetate, phosphate and methyl phosphate formed a mixture of outersphere and innersphere complexes depending on the conditions of the titrations. Only citrate appeared to form a predominantly innersphere complex with a formation constant of 17 μM (Fig. 8.13b).

Emission spectroscopy showed that carboxy anions bound to Eu(STHP) gave uniquely different emission peak ratios compared to phosphate anions bound to Eu(STHP). Thus,

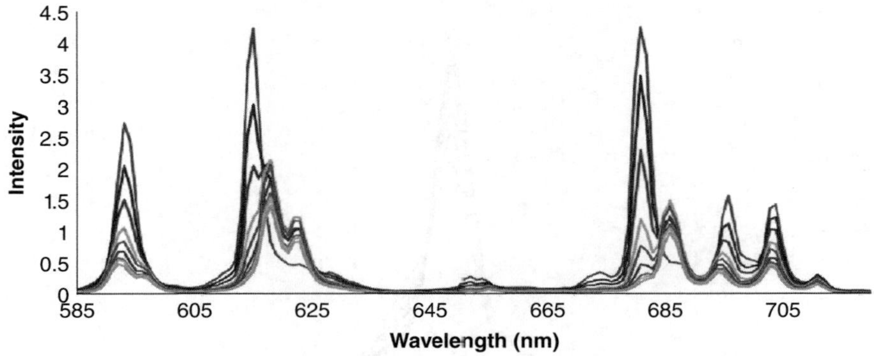


*Figure 8.14* *Luminescence emission spectra (excitation at 579.36 nm) of Eu(STHP) in 20 mM HEPES, 100 mM NaCl upon addition of phosphate (0–20 mM), at pH 7.2, with 100 μM Eu(STHP). Addition of phosphate leads to a decrease in the emission peak at 592 nm ($^5D_0 \rightarrow {}^7F_1$), a decrease in the major emission peak at 614 nm ($^5D_0 \rightarrow {}^7F_2$), and a decrease in the major emission band at 680 nm ($^5D_0 \rightarrow {}^7F_4$). Reproduced with permission from [8]. © 2011, American Chemical Society*

binding of phosphate could be distinguished from binding of carbonate or citrate in mixtures containing both types of donor groups (Fig. 8.14).

The combination of outersphere and innersphere ligand binding could be monitored by following changes in emission peak intensities to give binding isotherms for multimodal binding (Fig. 8.15). The emission peak intensities are unexpectedly more sensitive to these binding modes than are the $^7F_0 \rightarrow {}^5D_0$ excitation peak position or intensity.

The addition of a sensitising antenna attached through an amide pendent group to give Eu (THPC) destroyed the anion specificity, most likely by changing the alcohol hydrogen bonding interactions. This may necessitate attachment of the sensitiser through the macrocycle backbone. In all, this work showed that a combination of outersphere and innersphere ligand interactions may be important in determining both the binding selectivity and luminescence properties of the Eu(III) complexes [8].

## 8.5 Luminescence Studies of Ln(III) MRI Contrast Agents

### 8.5.1 Types of Ln(III) MRI Contrast Agents

Ln(III) contrast agents generally contain either macrocyclic ligands or multidentate chelates to form stable, and in some cases, kinetically inert complexes [65]. Direct excitation Eu(III) luminescence spectroscopy is especially important for the characterisation of MRI contrast agents, in part because there typically is no ligand antenna. Luminescence lifetime decays are used to determine $q$ numbers [59], an important parameter for $T_1$ agents in particular [66]. Eu(III) is readily used as a substitute for Gd(III), given its close proximity in the Ln(III) series. Also of critical importance is the number of available coordination sites for binding biologically relevant molecules as well as the solution chemistry and speciation of the MRI contrast agent. For example, the isomeric form of the contrast agent can influence the contrast efficiency and ionisation will likely influence the pH dependence of MRI contrast,

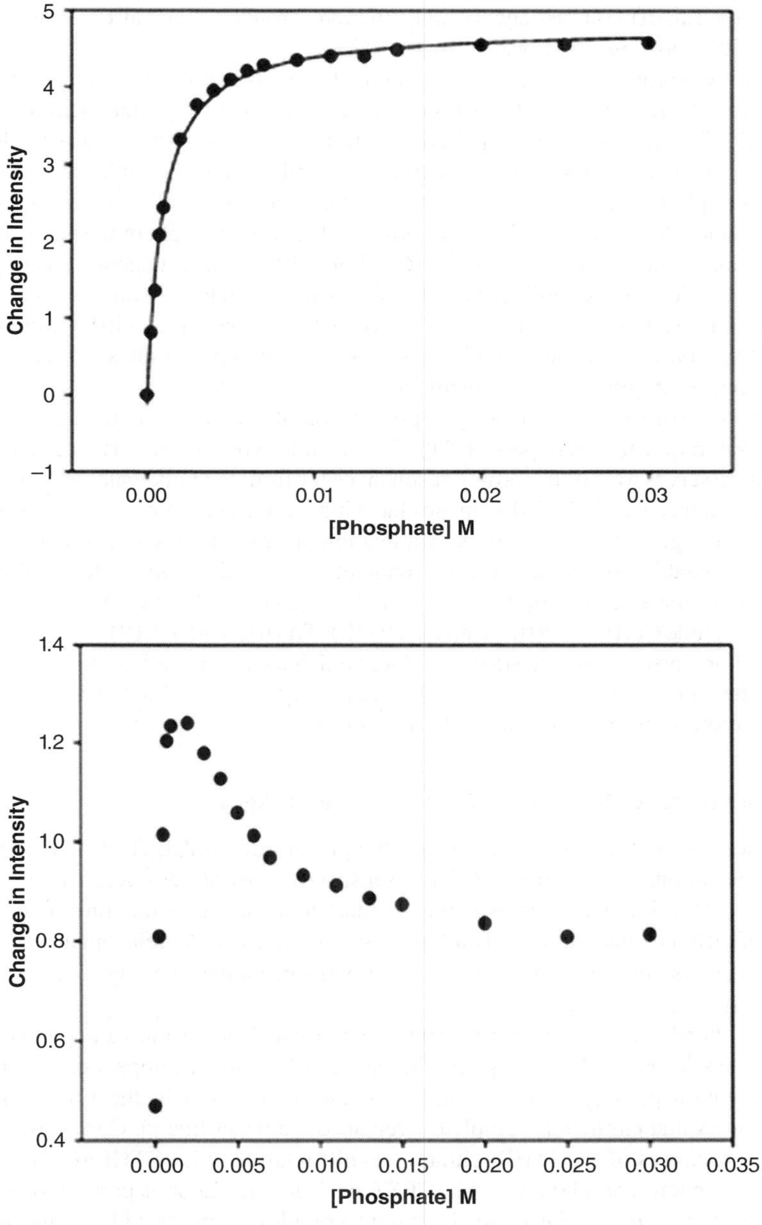


***Figure 8.15*** *Plot of the emission intensities of 0.10 mM Eu(STHP) in 20 mM HEPES, pH 7.2, 100 mM NaCl, ($\lambda_{ex}$ = 579.36 nm) with the addition of phosphate. Top: intensity at 593 nm, Bottom: intensity at 622 nm. The top curve is fitted to a 1:1 binding isotherm to give a $K_d$ of 0.3 mM. The bottom curve can be fitted to a 2:1 binding isotherm to give $K_d$ values of 0.3 mM and 3.0 mM. Reproduced with permission from [8]. © 2011, American Chemical Society*

especially for Ln(III) contrast agents that function through chemical exchange saturation transfer (CEST) as described below [67].

There are two prominent types of MRI contrast agent that use Ln(III) ions. The classical type of contrast agent includes Gd(III) complexes that function primarily as $T_1$ relaxation agents. Gd(III) contains seven unpaired electrons in a symmetric, relatively long-lived electronic state that exhibits high relaxivity [65,66]. Gd(III) complexes produce MRI contrast through shortening the longitudinal relaxation time constant ($T_1$) of protons in water molecules. Relaxation of the water protons through innersphere mechanisms is more substantial than outersphere mechanisms [68]. For Gd(III) contrast agents, it is important to balance available open coordination sites for water exchange while maintaining high thermodynamic stability and kinetic inertness of the complexes. Gd(III) contrast agents in clinical use typically have a single bound water although complexes with two bound water molecules are under development [69].

A second type of Ln(III) contrast agent produces contrast through paramagnetic chemical exchange saturation transfer (paraCEST) [70]. In this type of contrast, the paramagnetic Ln(III) centre serves to shift the proton resonances far from the bulk water protons. Contrast agents that produce paraCEST MR images have highly shifted proton resonances that are in chemical exchange with water protons. Application of a presaturation pulse at the frequency of the exchangeable proton saturates the magnetisation and, through chemical exchange, produces a decrease in the bulk water signal. Ln(III) which produce contrast through paraCEST include Ce(III), Nd(III), Eu(III), Dy(III), Tm(III), and Yb(III). These Ln(III) have relatively short-lived electronic states and large anisotropic magnetic moments that lead to highly shifted proton resonances [71]. The relaxivity of these Ln(III) is relatively low, leading to proton resonances that are fairly narrow.

### 8.5.2 Luminescence Studies of Ln(III) ParaCEST Agents

The chemical environment of the exchangeable protons of paraCEST agents is affected by solution speciation, anions and pH. This sensitivity towards environment makes them promising for development as responsive or smart contrast agents that turn on or off in the presence of different analytes [67]. Binding studies of biologically relevant anions to Ln(III) paraCEST agents show that both outersphere and innersphere anionic ligands affect the paraCEST spectrum [64,72,73].

Anionic ligands such as methylphosphate or citrate lead to the disappearance of the excitation peak for Eu(STHP)($OH_2$) and, in the case of citrate, the appearance of a new red-shifted excitation peak [8]. The *q* number of essentially zero in the Eu(STHP)(citrate) complex shows that citrate has completely replaced the water ligand. Correspondingly, the $^1$H NMR resonances of Eu(STHP)(citrate) all shift relative to Eu(STHP)($OH_2$), consistent with an innersphere complex. The paraCEST peak for the alcohol protons of Eu(STHP)(citrate) shifts by 3 ppm relative to the parent complex. This provides a mechanism to produce contrast in the presence of citrate. Methylphosphate (MP) binds to Eu(STHP) by changes in both luminescence and CEST spectra. At millimolar concentrations necessary for CEST experiments, an innersphere interaction predominates in solution. This is supported by a change in the chemical shift of the proton resonances of the Eu(III) complex upon addition of MP. In addition, the *q* number decreases from 1.0 to 0.36, as would be expected upon replacement of the water ligand. The excitation peak frequency does not

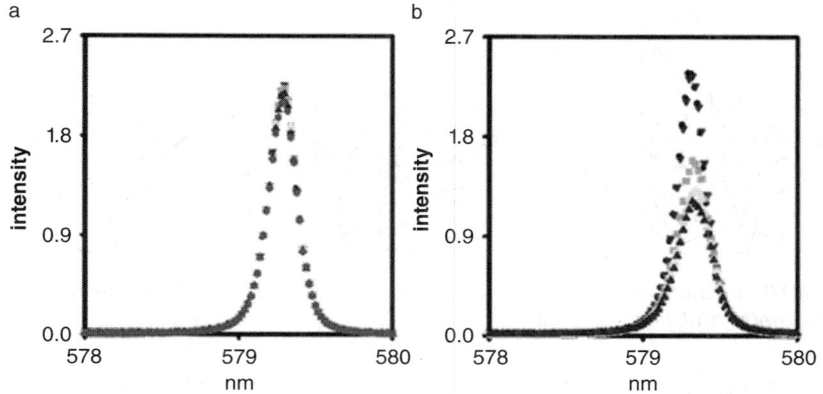


*Figure 8.16* Luminescence excitation spectra $^7F_0 \rightarrow {}^5D_0$ excitation at 579.32 nm (emission at 628 ± 27 nm) of 5.0 mM Eu(STHP) in 20.0 mM MES and 100 mM NaCl at pH 6.6, with addition of DEP showing little change of excitation peak (a) and a decrease in the excitation peak upon addition of MP (b). Reproduced with permission from [8]. © 2011, American Chemical Society

change markedly (Fig. 8.16). This is somewhat surprising given that bonding of MP anion would be anticipated to change the $^7F_0 \rightarrow {}^5D_0$ frequency.

The CEST spectra of both Eu(STHP) and Yb(STHP) are turned on at near neutral pH in the presence of diethylphosphate (DEP) [64], yet the $^7F_0 \rightarrow {}^5D_0$ excitation spectrum of Eu(STHP) at pH 6.5 is unchanged upon addition of excess DEP (Fig. 8.16). The luminescence lifetime of Eu(STHP) in the presence of DEP also does not change. Both of these experiments are consistent with the lack of formation of an innersphere complex by replacement of water with DEP.

However, addition of DEP to Eu(STHP) at pH 7.5 leads to disappearance of the excitation spectrum corresponding to the ionised species and appearance of the excitation spectrum of the non-ionised species (Fig. 8.17). The interaction of DEP with Eu(STHP)($OH_2$) essentially inhibits the ionisation of the Eu(III) complex (Fig. 8.18).

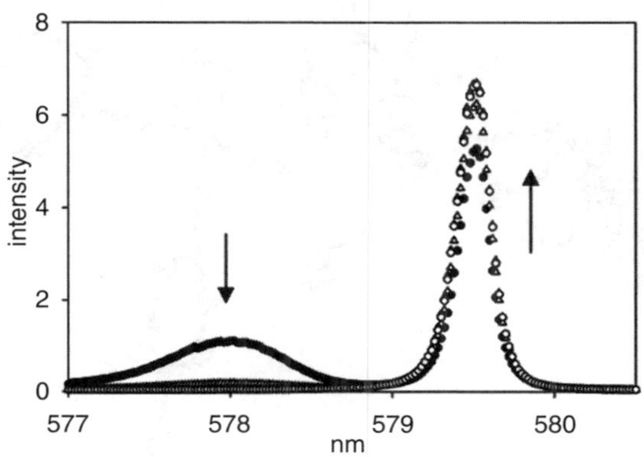


*Figure 8.17* $^7F_0 \rightarrow {}^5D_0$ excitation spectra ($^5D_0 \rightarrow {}^7F_2$ emission) of 10 mM Eu(STHP), 20 mM buffer, 100 mM NaCl with DEP (0–30 mM) at pH 8.0. Reproduced with permission from [64]. © 2010, American Chemical Society

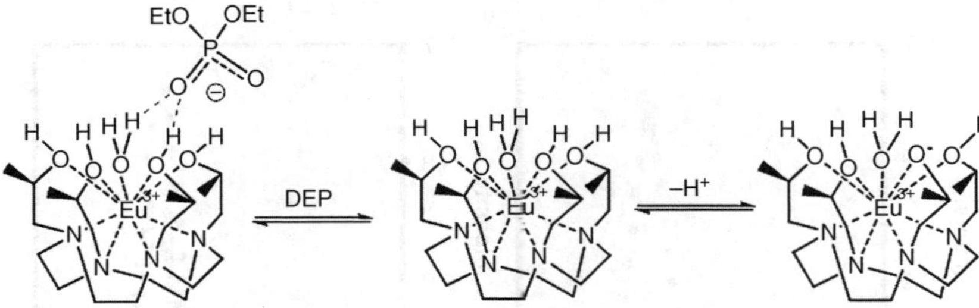


*Figure 8.18* Interaction of DEP with Eu(STHP). Reproduced with permission from [64]. © 2010, American Chemical Society

That this interaction occurs through outersphere contact of the phosphate ester anion and the hydroxyethyl groups of the macrocycle is shown by X-ray crystallographic studies and $^1$H NMR (Figs. 8.19 and 8.20). The crystal structure of $[Eu(STHP)(OH_2)](BNPP)_2(CF_3SO_3)$ (BNPP) = bis-4-nitrophenylphosphate) shows two phosphate diesters bound outersphere through hydrogen bonding interactions with the bound water protons and the hydroxyethyl protons of the macrocyclic complex (Fig. 8.19). Solution $^1$H NMR spectra show that the alcohol proton resonance grows in upon addition of DEP through slowing down exchange of the protons (Fig. 8.20). Correspondingly, CEST spectra and CEST images are turned on upon addition of DEP. Figure 8.21 shows a set of phantom images of tubes at different pH values.

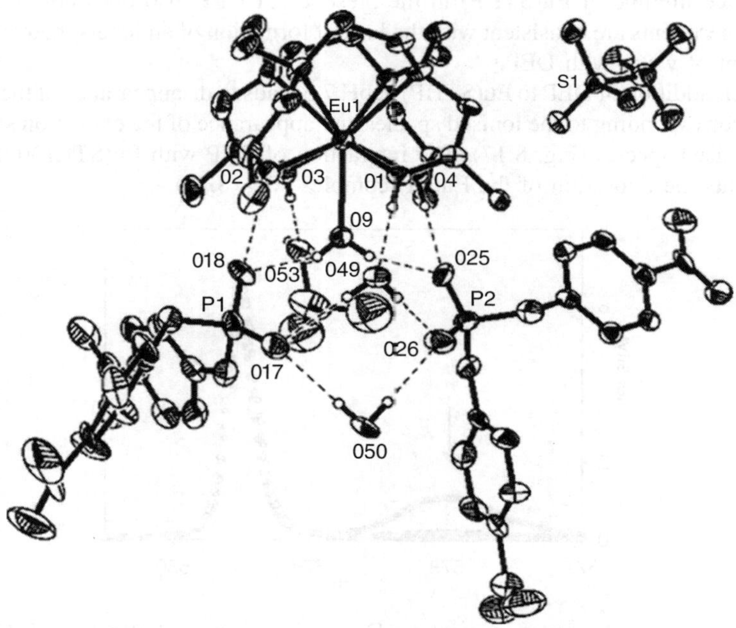


*Figure 8.19* Ortep diagram (50% ellipsoids) of $[Eu(STHP)(OH_2)][(O_2NPhO)_2PO_2^-]_2[CF_3SO_3]_3$ $(2\,H_2O)_3$ iPrOH. Reproduced with permission from [64]. © 2010, American Chemical Society

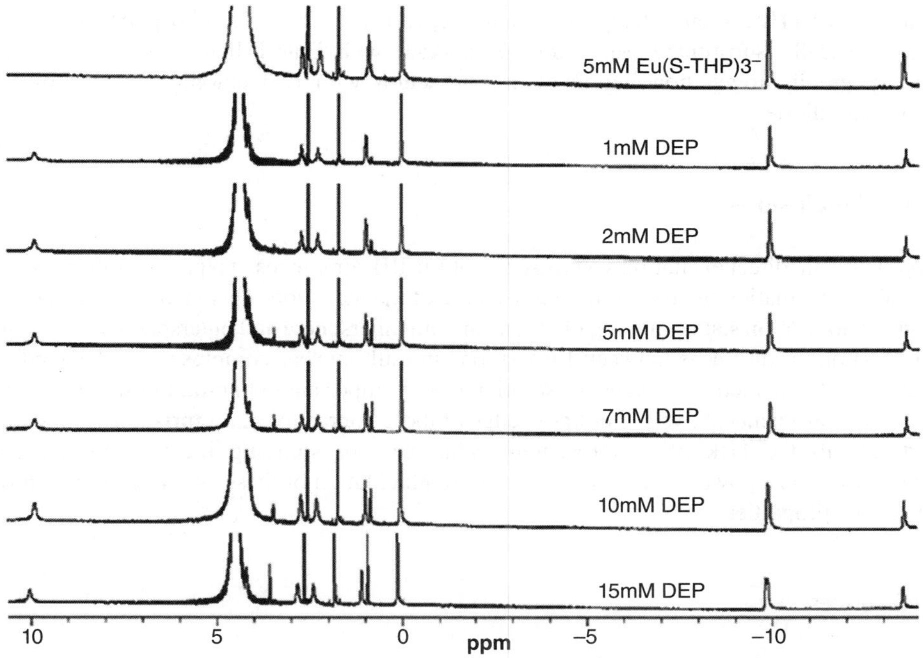


***Figure 8.20*** *$^{1}H$ NMR spectra of 5 mM Eu(STHP) titrated with DEP at pH 5.6. Reproduced with permission from [73]. © 2009, American Chemical Society*

Phantom images of Yb(STHP) at pH 4, 5, and 7 exhibit contrast only at the most acidic pH. However, addition of two equivalents of DEP (D, E, F) shows that contrast is now brightest for the tube at neutral pH [64].

Curiously, the outersphere interaction of phosphate diester with Eu(STHP) is selective for the type of phosphate diester. Only symmetrically substituted esters such as BNNP or DEP

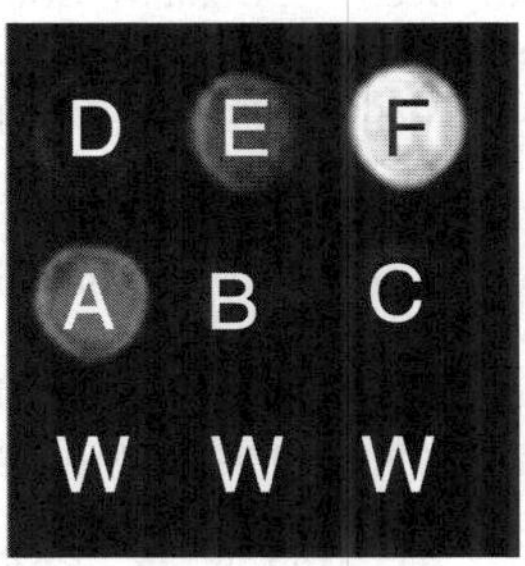


***Figure 8.21*** *CEST phantom images. Lower row: W is water for control. A–F contain 5 mM Yb(STHP), 20 mM buffer, 100 mM NaCl. Middle row: A at pH 4.6, B at pH 5.5, C at pH 7.0. Top row: after addition of 10 mM DEP, D at pH 4.6, E at pH 5.5, F at pH 7.0. Reproduced with permission from [64]. © 2010, American Chemical Society*

bind to Eu(STHP) in an outersphere interaction and turn CEST on. Phosphorothioates or asymmetrically substituted esters do not interact with Eu(STHP). This suggests that outersphere hydrogen bonding involves interaction with two anionic oxygens of the phosphate diester.

## 8.6 Conclusions

In conclusion, direct excitation spectroscopy of Eu(III) complexes in aqueous solution gives valuable information on the coordination sphere of the metal ion. This includes detection of isomers, ionisation states, binding of anions in both innersphere and outersphere modes, and information on distances between Ln(III) ions in multinuclear complexes. Understanding such delicate interactions in aqueous solution is very important in fine-tuning smart contrast agents and efficient catalysts. Our knowledge of these interactions is surprisingly limited, in part due to the lack of spectroscopic techniques for studying them. Luminescence spectroscopy is a powerful and versatile tool for elucidation of these essential coordination chemistry properties.

## References

1. Bünzli J-CG. Benefiting from the unique properties of lanthanide ions. Acc. Chem. Res. 2006; 39 (1):53–61.
2. Horrocks WD Jr, Sudnick DR. Lanthanide ion luminescence probes of the structure of biological macromolecules. Acc. Chem. Res. 1981; 14:384–392.
3. Horrocks WD Jr, Albin M. Lanthanide ion luminescence in coordination chemistry and biochemistry. Progress in Inorganic Chemistry. 1984; 31:1–104.
4. Amin S, Voss DA Jr, Horrocks WD Jr, Morrow JR. Restoration of catalytic activity by replacement of a coordinated amide group: synthesis and laser-induced luminescence studies of the phosphate diester transesterification catalyst $[Eu(NBAC)]^{3+}$. Inorg Chem. 1996; 35(26):7466–7467.
5. Nwe K, Andolina CM, Morrow JR. Tethered dinuclear Eu(III) catalysts for the cleavage of RNA. J Am Chem Soc. 2008; 130(44):14861–14871.
6. Dickins RS, Aime S, Batsanov AS, Beeby A, Botta M, Bruce JI, et al. Structural, Luminescence, and NMR Studies of the Reversible Binding of Acetate, Lactate, Citrate, and Selected Amino Acids to Chiral Diaqua Ytterbium, Gadolinium, and Europium Complexes. J Am Chem Soc. 2002; 124(43):12697–12705.
7. Montgomery CP, Murray BS, New EJ, Pal R, Parker D. Cell-penetrating metal complex optical probes: targeted and responsive systems based on lanthanide luminescence. Acc. Chem. Res. 2009; 42(7):925–937.
8. Hammell J, Buttarazzi L, Huang C-H, Morrow JR. Eu(III) complexes as anion-responsive luminescent sensors and paramagnetic chemical exchange saturation transfer agents. Inorg Chem. 2011; 50:4857–4867.
9. Nwe K, Andolina CM, Huang C-H, Morrow JR. PARACEST properties of a dinuclear neodymium(III) complex bound to DNA or carbonate. Bioconjugate Chem. 2009; 20(7):1375–1382.
10. Eliseeva SV, Bünzli J-CG. Lanthanide luminescence for functional materials and bio-sciences. Chem. Soc. Rev. 2010; 39(1):189–227.

11. Moore EG, Samuel APS, Raymond KN. From antenna to assay: lesson learned in lanthanide luminescence. Acc. Chem. Res. 2009; 42:542–552.
12. Horrocks WD Jr., Luminescence Spectroscopy. In: Riordan JF, Vallee BL, editors. Methods in Enzymology. San Diego, CA: Academic Press Inc.; 1993. pp. 495–5 8.
13. Choppin GR, Peterman DR. Applications of lanthanide luminescence spectroscopy to solution studies of coordination chemistry. Coord. Chem. Rev. 1998; 174:283–299.
14. Sastri VS, Bünzli J-CG, Rao R, Rayudu GVS, Perumareddi JR Modern Aspects of Rare Earths and Their Complexes. 1st edn. Amsterdam, Netherlands: Elsevier B. V.; 2003.
15. Supkowski RM, Horrocks WD Jr., On the determination of the number of water molecules, q, coordinated to europium(III) ions in solution from luminescence decay lifetimes. Inorg Chim Acta. 2002; 340:44–48.
16. Beeby A, Clarkson IM, Dickins RS, Faulkner S, Parker D, Royle L, et al. Non-radiative deactivation of the excited states of europium, terbium and ytterbium complexes by proximate energy-matched OH, NH and CH oscillators: an improved luminescence method for establishing solution hydration states. J Chem Soc, Perkin Trans. 2. 1999; 1999(3):493–504.
17. Andolina CM, Holthoff WG, Page PM, Mathews RA, Morrow JR, Bright FV. Spectroscopic system for direct lanthanide photoluminescence spectroscopy with nanomolar detection limits. Appl Spectrosc. 2009; 63(5):483–493.
18. Morrow JR, Andolina CM. Spectroscopic investigations of Lanthanide ion binding to Nucleic acids. Dordrecht, The Netherlands: Springer B.V; 2011.
19. Stein G, Wurzberg E. Energy Gap Law in the Solvent Isotope Effect on Radiationless Transitions of Rare Earth Ions. J Chem Phys. 1975; 62(1):208–213.
20. Werts VMH. Luminescent Lathanide Complexes. Amsterdam, Netherlands: University of Amsterdam; 2000.
21. Haas Y, Stein G, Würzberg J. Radiationless transitions in solutions: isotopes and proximity effects on $Dy^{3+}$ by C–H and C–N bonds. J Chem Phys. 1974; 60:258.
22. Frey ST, Horrocks WD Jr., On correlating the frequencey of the $^7F_0 \Rightarrow {}^5D_0$ transition in $Eu^{3+}$ complexes with the sum of 'nephelauxetic parameters' for all of the coordinating atoms. Inorg Chim Acta. 1995; 229(1/2): 383–390.
23. Choppin GR, Wang ZM. Correlation between ligand coordination number and the shift of the $^7F_0 \rightarrow {}^5D_0$ transition frequency in europium(III) complexes. Inorg Chem. 1997; 36(2):249–252.
24. Albin M, Horrocks WD Jr., Europium(III) luminescence excitation spectroscopy. quantitative correlation between the total charge on the ligands and the $^7F_0 \Rightarrow {}^5D_0$ transition frequency in Europium (III) complexes. Inorg Chem. 1985; 24(6):895–900.
25. Werts MHV, Jukes RTF, Verhoeven JW. The emission spectrum and the radiative lifetime of $Eu^{3+}$ in luminescent lanthanide complexes. Phys Chem Chem Phys. 2002; 4(9):1542–1548.
26. Bretonniere Y, Cann MJ, Parker D, Slater R. Design, synthesis and evaluation of ratiometric probes for hydrogencarbonate based on europium emission. Org. Biomol. Chem. 2004; 2(11):1624–1632.
27. Pal R, Parker D. A ratiometric optical imaging probe for intracellular pH based on modulation of europium emission. Org. Biomol. Chem. 2008; 6(6):1020–1033.
28. Kielar F, Montgomery CP, New EJ, Parker D, Poole RA, Richardson SL, et al. A mechanistic study of the dynamic quenching of the excited state of europium(III) and terbium(III) marcocyclic complexes by charge- or electron transfer. Org. Biomol. Chem. 2007; 5:2975–2982.
29. Poole RA, Kielar F, Richardson SL, Stenson PA, Parker D. A ratiometric and non-enzymatic luminescence assay for uric acid: differential quenching of lanthanide excited states by anti-oxidants. Chem. Commun. 2006; 2006(39):4084–4086.
30. Selvin PR, Rana TM, Hearst JE. Luminescence Resonance Energy Transfer. J Am Chem Soc. 1994; 116(13):6029–6030.

31. Selvin PR. The renaissance of fluorescence resonance energy transfer. Nat Struct Biol. 2000; 7:730–734.
32. Selvin PR. Principles and biophysical applications of lanthanide-based probes. Annu Rev Biophys Biomol Struct. 2002; 31:275–302.
33. Xiao M, Selvin PR. Quantum yields of luminescent lanthanide chelates and far-red dyes measured by resonance energy transfer. J Am Chem Soc. 2001; 123(29):7067–7073.
34. Horrocks WD Jr, Collier WE. Lanthanide ion luminescence probes. Measurement of distance between intrinsic protein fluorophores and bound metal ions: quantitation of energy transfer between tryptophan and terbium(III) or europium(III) in the calcium-binding protein parvalbumin. J Am Chem Soc. 1981; 103(10):2856–2862.
35. Horrocks WD Jr, Rhee M-J, Snyder AP, Sudnick DR. Laser-induced metal ion luminescence: interlanthanide ion energy transfer distance measurements in the calcium-binding proteins, parvalbumin, and thermolysin. Metalloprotein models address a photophysical problem. J Am Chem Soc. 1980; 102(10):3650–3652.
36. Zhang X, Chang CA, Brittain HG, Garrison JM, Telser J, Tweedle MF. pH Dependence of relaxivities and hydration numbers of gadolinium(III) complexes of macrocyclic amino carboxylates. Inorg Chem. 1992; 31(26):5597–5600.
37. Brittain HG. Intermolecular energy transfer between lanthanide complexes in aqueous solution. 4. Stereoselectivity in the transfer from Terbium(III) to Europium(III) complexes of aspartic acid. Inorg Chem. 1978; 18(7):1740–1745.
38. Spaulding L, Brittain HG. Intermoleculare Energy Tansfer between Lanthanide Complexes. 9. Tb(III) Donor and Eu(III) acceptor Complexes of Amino Polycarboxylic Acids. Inorg. Chem. 1983; 22:3486–3488.
39. Andolina CM, Morrow JR. Luminescence resonance energy transfer in heterodinuclear Ln(III) complexes for sensing biologically relevant anions. Eur J Inorg Chem. 2011; 2011:154–164.
40. Selvin PR. Lanthanide-based resonance energy transfer. IEEE J Sel Top Quantum Electron. 1996; 2(4):1077–1087.
41. Reifernberger JG, Ge P, Selvin PR. Progress in lanthanides as luminescent probes. Reviews in Fluorescence. 2005; 2:399–431.
42. Yuan F, Greenbaum NL. A FRET method to identify locations of site-bound metal ions in the splicesomal U2-U6 snRNA complex. Spectrum. 2007; 20(2):14–17.
43. Yuan F, Griffin L, Phelps L, Buschmann V, Weston K, Greenbaum NL. Use of a novel Forster resonance energy transfer method to identify locations of site-bound metal ions in the U2-U6 snRNA complex. Nucleic Acids Res. 2007; 35(9):2833–2845.
44. Parker D. Excitement in f block: structure, dynamics and function of nine-coordinate chiral lanthanide complexes in aqueous media. Chemical Society Reviews. 2004; 33(3):156–165.
45. Parker D, Dickins RS, Puschmann H, Crossland C, Howard JAK. Being Excited by Lanthanide Coordination Complexes: Aqua Species, Chirality, Excited-State Chemistry, and Exchange Dynamics. Chem Rev. 2002; 102(6):1977–20 8
46. Amin S, Voss DA Jr, Horrocks WD Jr, Lake CH, Churchill MR, Morrow JR. Laser-induced luminescence studies and crystal structure of the europium(III) complex of 1,4,7,10-tetrakis (carbamoylmethyl)-1,4,7,10-tetraazacyclododecane. The link between phosphate diester binding and catalysis by lanthanide(III) macrocyclic complexes. Inorg Chem. 1995; 34(12):3294–3300.
47. Nwe K, Richard JP, Morrow JR. Direct excitation luminescence spectroscopy of Eu(III) complexes of 1,4,7-tris(carbamoylmethyl)-1,4,7,10- tetraazacyclododecane derivatives and kinetic studies of their catalytic cleavage of an RNA analog. Dalton Transactions. 2007; 2007 (44):5171–5178.
48. Brittain HG, Choppin GR, Barthelemy P.P. pH-dependence of the metal ion hydration state in lanthanide complexes of polyaminopolycarboxylate ligands. J Coord Chem. 1992; 26:143–153.

49. Dunand FA, Merbach AE. UV-Vis observation of different $^7F_0 \rightarrow {}^5D_0$ transitions for both isomeric species in Eu(III) DOTA-like complexes. Inorg Chem Commun. 2001; 4(12):719–722.
50. Billard I. Lanthanide and actinide solution chemistry as studied by time-resolved emission spectroscopy. Handbook on the Physics and Chemistry of Rare Earths. Amsterdam, Netherlands: Elsevier; 2003. pp. 465–514.
51. Nehlig A, Elhabiri M, Billard I, Albrecht-Gary A-M, Lutzenkirchen K. Photoexcitation of europium(III) in various electrolytes: dependence of the luminescence liftetimes on the type of salts and the ionic strength. Radiochimica Acta. 2003; 91(1):37–43.
52. Lis S, Choppin GR. Luminescence lifetimes of aqueous Europium perchlorate chloride and nitrate solutions. Materials Chemistry and Physics. 1992; 31(1/2):159–161.
53. Andolina CM, Mathews RA, Morrow JR. Solution chemistry of europium(III) aqua Ion at micromolar concentrations as probed by direct excitation luminescence spectroscopy. Helvetica Chimica Acta. 2009; 92(11):2330–2348.
54. Cernochova K, Mathur JN, Choppin GR. Chemical speciation of Am, Cm and Eu with EDTA at high ionic strength: thermodynamics and laser fluorescence spectroscopy studies. Radiochimica Acta. 2005; 93(12):733–739.
55. Wu SL, Horrocks WD Jr., General method for the determination of stability constants of lanthanide ion chelates by ligand ligand competition: Laser-excited Eu3+ luminescence excitation spectroscopy. Anal Chem. 1996; 68(2):394–401.
56. Frey ST, Chang CA, Carvalho JF, Varadarajan A, Schultze LM, Pounds KL, et al. Characterization of Lanthanide Complexes with a Series of Amide-Based Macrocycles, Potential Mri Contrast Agents, Using Eu3+ Luminescence Spectroscopy and Molecular Mechanics. Inorg Chem. 1994; 33(13):2882–2889.
57. Horrocks WD Jr, Arkle VK, Liotta FJ, Sudnick DR. Kinetic parameters for a system at equilibrium from the time course of luminescence emission: a new probe of equilibrium dynamics. Excited-state europium(III) as a species label. J Am Chem Soc. 1983; 105(11):3455–3459.
58. Horrocks WD Jr, Sudnick DR. Time-resolved europium(III) excitation spectroscopy – luminescence probe of metal-Ion binding-sites. Science. 1979; 206(4423):1194–1196.
59. Supkowski RM, Horrocks WD Jr., Displacement of inner-sphere water molecules from $Eu^{3+}$ analogues of $Gd^{3+}$ MRI contrast agents by carbonate and phosphate anions: dissociation constants from luminescence data in the rapid-exchange limit. Inorg Chem. 1999; 38(24):5616–5619.
60. Chappell LL, Voss DA Jr, Horrocks WD Jr, Morrow JR. Effect of mixed pendent groups on the solution and catalytic properties of europium(III) macrocyclic complexes: bifunctional and monofunctional amide and alcohol pendents in septadentate or octadentate ligands. Inorg Chem. 1998; 37:3989–3998.
61. Epstein DM, Chappell LL, Khalili H, Supkowski RM, Horrocks WD Jr, Morrow JR. Eu(III) macrocyclic complexes promote cleavage of and bind to models for the 5′-cap of mRNA. effect of pendent group and a second metal ion. Inorg Chem. 2000; 39(10):2130–2134.
62. Sanchez-Lombardo I, Andolina CM, Morrow JR, Yatsimirsky AK. Speciation of Eu(III) hydroxo complexes in aqueous DMSO studied by direct excitation luminescence spectroscopy and their catalytic activity in phosphodiester cleavage. Dalton Trans. 2010; 39(3):864–873.
63. Morrow JR, Amyes TL, Richard JP. Phosphate binding energy and catalysis by small and large molecules. Acc. Chem. Res. 2008; 41(4):539–548.
64. Huang C-H, Hammell J, Ratnakar SJ, Sherry AD, Morrow JR. Activation of a PARACEST agent for MRI through selective outersphere interactions with phosphate diesters. Inorg Chem. 2010; 49(13):5963–5970.
65. Caravan P, Ellison JJ, McMurry TJ, Lauffer RB. Gadolinium(III) chelates as MRI contrast agents: structure, dynamics, and applications. Chem Rev. 1999; 99(9):2293–2352.

66. Terreno E, Delli CD, Viale A, Aime S. Challenges for molecular magnetic resonance imaging. Chem Rev. 2010; 110:3019–3042.
67. De L-RLM, Lubag AJM, Malloy CR, Martinez GV, Gillies RJ, Sherry AD. Responsive MRI agents for sensing metabolism in vivo. Acc. Chem. Res. 2009; 42:948–957.
68. Allen MJ, Meade TJ. Magnetic resonance contrast agents for medical and molecular imaging. Metal Ions in Biological Systems. 2004; 42:1–38.
69. Datta A, Raymond KN. Gd-Hydroxypyridinone (HOPO)-Based High-Relaxivity Magnetic Resonance Imaging (MRI) Contrast Agents. Acc. Chem. Res. 2009; 42:938–947.
70. Viswanathan S, Kovacs Z, Green KN, Ratnakar SJ, Sherry AD. Alternatives to Gadolinium-Based Metal Chelates for Magnetic Resonance Imaging. Chem. Rev. 2010; 110:2960–3018.
71. Geraldes CFGC, Luchinat C. Lanthanides as shift and relaxation agents in elucidating the structure of proteins and nucleic acids. Metal Ions in Biological Systems (Lanthanides and Their Interrelations with Biosystems). 2003; 40:513–588.
72. Aime S, Delli Castelli D, Fedeli F, Terreno E. A Paramagnetic MRI-CEST Agent Responsive to Lactate Concentration. J Am Chem Soc. 2002; 124(32):9364–9365.
73. Huang C-H, Morrow JR. A PARACEST agent responsive to inner- and outer-sphere phosphate ester interactions for MRI applications. J Am Chem Soc. 2009; 131(12):4206–4207.

# 9

# Heterometallic Complexes Containing Lanthanides

*Stephen Faulkner and Manuel Tropiano*

*Chemistry Research Laboratory, University of Oxford, UK*

## 9.1 Introduction

Lanthanide complexes containing more than one metal ion can have clearly defined properties that derive from their structure, and that can be controlled and exploited in a wide variety of ways. However, before addressing these, we must consider how the nature of the chemistry of the *f*-block elements defines approaches to the preparation of heterometallic systems. The study of the coordination chemistry of *f*-block ions in the solid phase and in solution has advanced enormously over the course of the last half-century, and the key principles that define its use have developed over the course of time – illustrating in microcosm the underpinning principles of coordination chemistry and supramolecular chemistry.

The shielded nature of *f*-orbitals directly affects the bonding properties of lanthanide ions to the extent that interactions with other elements, the overlap with ligand orbitals being practically negligible, can be regarded as mainly ionic in nature. Trivalent lanthanide ions have high coordination numbers with 8–10 donor atoms usually arranged in a prismatic geometry around the central metal. Indeed, unlike transition metal ions, where ligand field effects largely influence the coordination geometry, lanthanide ions are usually unaffected by the ligand environment. This is also reflected in magnetic and luminescence properties unaltered by the type of ligand. This, together with the lanthanide contraction and the most common +3 oxidation state throughout the series, makes the chemistry of lanthanides very similar with only occasional subtle changes.

*Luminescence of Lanthanide Ions in Coordination Compounds and Nanomaterials*, First Edition.
Edited by Ana de Bettencourt-Dias.

## 9.2 Properties of a Heteromultimetallic Complex

For a heteromultimetallic complex to have clearly defined properties, it is essential that each metal ion resides in its own binding pocket at all times with minimal dissociation between metal and ligand, hence forming a kinetically stable assembly. The family of lanthanide β-diketonate complexes is known to be kinetically labile to the point that different species are present in solution as a consequence of fast ligand exchange. Attempts to form heterometallic complexes in solution would hence result in fast transmetallation and statistical mixtures of different species [1]. Lanthanide chelates of linear and macrocyclic polyamino carboxylate ligands are instead kinetically and thermodynamically stable in solution. They can effectively be applied as robust building blocks to form multimetallic architectures through covalent or supramolecular interactions, in which the metal environment remains closely associated with the ligand during the course of the reaction and beyond. The thermodynamic stabilities of some representative lanthanide complexes are tabulated in Table 9.1. EDTA and DTPA show increasing stability upon increase of the charge density on the lanthanide ion. The linear increment along the series for EDTA is due to its structure flexibility. On the other hand, a plateau in stability is reached with $Tb^{3+}$ and $Ho^{3+}$ for DTPA, where the charge density is counterbalanced by the lack of rigidity of the ligand and hence difficulty to wrap around the metal ion. The overall higher stability of the latter is simply due to its higher denticity. Complexes formed with DOTA are at least two orders of magnitude more stable than the correspondent DTPA complexes. The preorganised macrocyclic structure and the higher basicity of donor groups have both been shown to play an important role in the overall thermodynamic stability of DOTA complexes along the lanthanide series [2,3].

Coordination compounds formed by hard and charged donors with lanthanides, like those of d- and s-block metal ions, are usually thermodynamically stable – especially if coordinated by chelating ligands. Table 9.1 shows a comparison of thermodynamic constants for some key polydentate ligand systems in common with lanthanides.

The kinetic stability of the lanthanide complexes with polydentate aminocarboxylate ligands has been extensively studied mainly for gadoliunium complexes given their use as MRI contrast agents. In this case the literature clearly shows a relationship between the rigidity of the ligand and the kinetics of formation and dissociation of the metal complexes. Due to their flexible structure, the complexation of lanthanide ions by EDTA and DTPA is a fast process. In the case of DTPA and its derivatives, the kinetics of complexation is

***Table 9.1*** *Thermodynamic stability constants for representative linear and macrocyclic polyamino carboxylate complexes of lanthanide ions*

| | Log $K_a$ ($H_2O$, $I=0.1$ M, 25 °C) | | |
|---|---|---|---|
| | EDTA [4] | DTPA [4] | DOTA [2] |
| $La^{3+}$ | 15.5 | 19.5 | 22.9 |
| $Eu^{3+}$ | 17.3 | 22.5 | 23.5 |
| $Gd^{3+}$ | 17.4 | 22.5 | 24.7 |
| $Tb^{3+}$ | 17.9 | 22.7 | 24.2 |
| $Lu^{3+}$ | 19.9 | 22.4 | 25.4 |

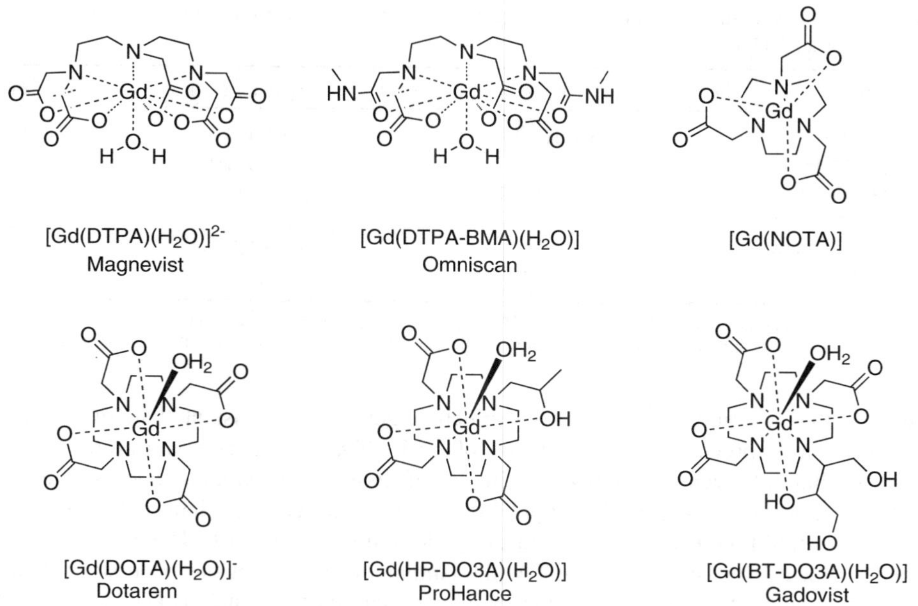


***Figure 9.1*** *Structures of the gadolinium chelates discussed in this chapter*

characterised by a saturation behaviour, indicating rapid formation of an intermediate and its slow rearrangement to give the final product [5].

This is not the case for lanthanide complexes with DOTA. The formation of a lanthanide-DOTA complex has been postulated to occur via a three-step mechanism; Moreau and co-workers have sought to identify the key intermediates in this process, describing coordination of the acetate arms and ligand (de)solvation processes that lead to formation of the complex [6]. The metal ion is firstly coordinated by the acetate arms outside the cavity in a fast process to form $LnH_2DOTA(H_2O)_5$. Coordination to two nitrogen atoms and concerted proton transfer to two carboxylate arms follow in the first slow rate determining step forming $LnH_2DOTA(H_2O)_3$. After fast deprotonation of the carboxylate arms the remaining nitrogen atoms coordinate to the lanthanide ion to give the final complex LnDOTA in the second rate-determining step (Fig. 9.1). Such processes are clearly influenced by the local pH, and the rate of lanthanide complexation is optimal in the pH range just below 7.0, which maximises the chance of deprotonating intermediate species while avoiding the prospect of forming lanthanide hydroxide precipitates and colloids.

When considering the kinetic stability of a lanthanide chelate, dissociation and trans-metallation reactions are the main processes that must be taken into account. For complexes formed with EDTA, the proton-assisted dissociation dominates over direct attack by another metal ion. Indeed, the structure does not leave space for direct attack, because none of the acetate arms remains exposed [7]. This can be contrasted with the DTPA case where the two residual negative charges on $Ln(DTPA)^{2-}$ make the direct attack reaction more likely. A study by Brücher et al. [8] demonstrated that at physiological pH the transmetallation mechanism dominates and that the kinetics of the process largely depends on the attacking

***Table 9.2*** *Dissociation kinetics of some gadolinium complexes discussed in this chapter ($k_{obs} = k_0 + k_1 [H^+]$)*

| Compound | $k_0$ ($s^{-1}$) | $k_1$ ($M^{-1}s^{-1}$) | Reference |
|---|---|---|---|
| $[Gd(DOTA)]^-$ | $<5 \times 10^{-8}$ | $5 \times 10^{-6}$ | [9] |
| $[Gd(DTPA)]^{2-}$ | —[a] | $\sim 1.2 \times 10^{-3}$ | [11] |
| [Gd(HP-DO3A)] | $-7(\pm 20) \times 10^{-10}$ | $2.6 \times 10^{-4}$ | [12] |
| [Gd(BT-DO3A)] | $2 \times 10^{-10}$ | $2.8 \times 10^{-5}$ | [12] |
| [Gd(DTPA-BMA)] | —[a] | $>2 \times 10^{-3}$ | [11] |
| [Gd(NOTA)] | $8.3 \times 10^{-6}$ | $2.3 \times 10^{-2}$ | [13] |

[a] No study of pH dependent and independent rates of lanthanice exchange has been reported for DTPA or DTPA-BMA. The values for $k_1$ are estimates, based on the assumption that the pH dependent pathway will dominate at pH = 1.0.

metal, with $Cu^{2+}$ and $Zn^{2+}$ possessing much larger kinetic constant than $Eu^{3+}$ for the transmetallation process. The dissociation of $Ln(DOTA)^-$ complexes is a proton assisted process, although even at low pH the kinetics of dissociation remains much slower than DTPA analogues (see Table 9.2). At pH 7.0 the acid catalysed dissociation of $Ln(DOTA)^-$ makes a negligible contribution to the rate of gadolinium exchange [9]. The transmetallation kinetics of $Zn^{2+}$ for DOTA and DOTA-like lanthanide complexes has been studied and a qualitative comparison with open chain polyamino carboxylate, such as DTPA, shows the higher kinetic stability for complexes derived from macrocyclic structures with regard to transmetallation with $Gd(DOTA)^-$ showing no reaction after four days [10].

Given the kinetic inertness of low spin $d^6$ transition metal complexes, especially with low spin metal ions, our focus will be mainly on the kinetic stability of lanthanide chelates and their derivatives described in this chapter.

Knowledge of the kinetic and thermodynamic stability of the starting building blocks and the final assembly is of crucial importance when designing heteromultimetallic architectures, and must be carefully considered.

At the simplest, an anionic lanthanide complex can interact with a cation to form a labile heterometallic complex in solution. Such complexes have been known for many years, though they have not always been recognised as such. For instance, Bryden et al. [14] showed that $^{23}Na$ shifts were observed for sodium chloride solutions containing lanthanide complexes of DOTA. Fig. 9.2 shows the variation in observed $^{23}Na$ chemical shift between

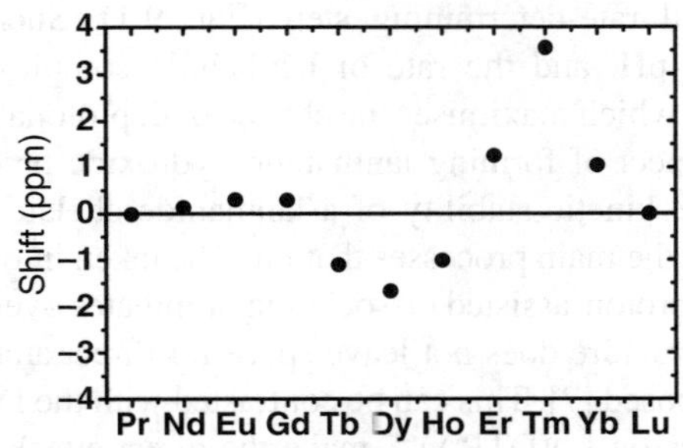


***Figure 9.2*** *$^{23}Na$ NMR shifts in 0.1 M $[Ln(DOTA)]^-$ solutions. Reproduced with permission from [14]. © 1981, American Chemical Society*

different complexes. Since the paramagnetic lanthanide ions give rise to pseudocontact shifts that vary from one lanthanide to another, these shifts vary across the series. The data reported of the interaction of sodium ions with $[Ln.DOTA]^-$ can be explained by binding of sodium either to the axially coordinated solvent molecule on DOTA, or directly to the carboxylates [14]. Though not identified as a heterometallic complex at the time, this report can be viewed with hindsight as the first in the field – meaning that heterometallic lanthanide complexes have now been known for more than 30 years.

Increasing the negative charge on the complex increases the electrostatic interaction with $Na^+$. This has been exploited with $[Tm(TTHA)]^{3-}$ and $[Dy(TTHA)]^{3-}$ to give large $^{23}Na$ shifts (though still in fast exchange regime): such complexes have been used to discriminate between intra- and extra-cellular sodium in cardiac and brain MRS imaging [15,16].

The preparation of kinetically inert structures is much more challenging. Strategies include differential binding of *d*- and *f*-block metals (when preparing a *d-f* hybrid) or the covalent linkage of stable building blocks. Over the last 15 years, we and others have been developing strategies to exploit bridging ligands in the controlled preparation of heterometallic systems. The rest of this chapter deals with the current state of this art.

## 9.3 Lanthanide Assemblies in the Solid State

The solid phase lends itself to the preparation of lanthanide containing heterometallic materials. In this case the choice of lanthanide complexes is ample since it is not only limited to kinetically inert complexes. The solid state acts as kinetic trap so that building blocks formed by β-diketonates are widely applied for designing multimetallic architectures such as the one represented in Fig. 9.3 [17–19].

Bridging organic units have been used as ligands to form heterometallic arrays between transition metals and lanthanide ions. The transition metal complex will act as antenna for lanthanide sensitization via its $^3$MLCT state so that the final solid will display lanthanide-centred luminescence in the NIR region. For instance, the bifunctional ligand 1H-benzimidazole-5-carboxylic acid has been employed as a bridging unit to construct heterometallic coordination polymers (Fig. 9.4c). Here the carboxylate group selectively coordinates the lanthanide ion while the nitrogen donors display a higher affinity for

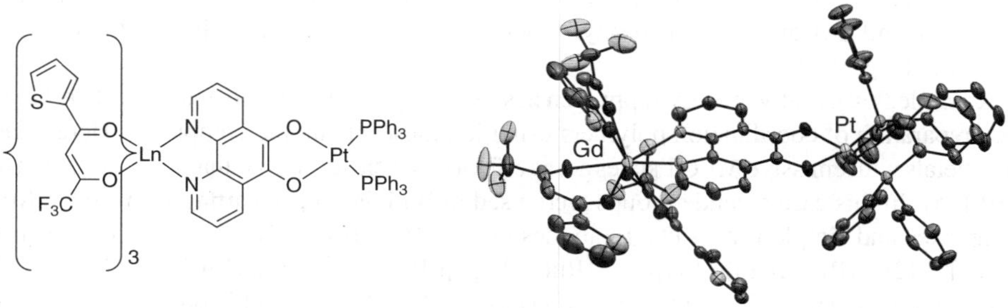


***Figure 9.3*** *Example of a solid state structure of a lanthanide β-diketonate* d-f *hybrid. Created using the Mercury 2.4 software from the crystallographic data in Reference [17]. (See colour plate section)*

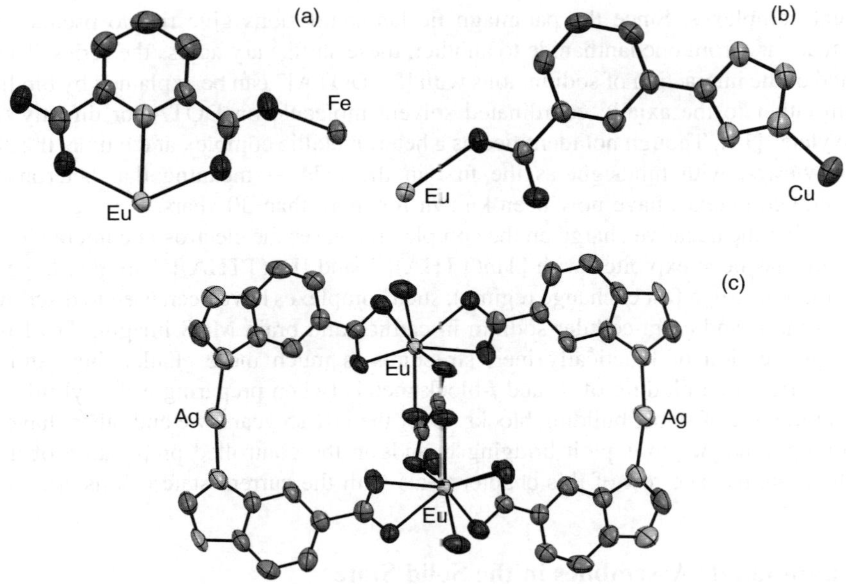


***Figure 9.4*** *Examples of solid state heterometallic architectures obtained using bridging organic ligands. Created using the Mercury 2.4 software from the crystallographic data in in the respective article (see text for references). (See colour plate section)*

transition metal ions such as $Ag^+$ and $Cu^+$ [20]. This approach shows advantages in that the coordination network forms by simple self-assembly in the solid state and is entirely predictable on the basis of the ligand structure and the metal ions employed. The same principle has been applied by Liang et al. [21] in the preparation of 3*d*-4*f* coordination networks based on tetrazolyl and carboxylate donors. Upon mixing of 3-cyanobenzoic acid, $NaN_3$ and the appropriate lanthanide and copper salts under hydrothermal conditions, *in situ* tetrazole formation occurs where the tetrazole selectively coordinates the transition metal while the lanthanide is coordinated exclusively by carboxylate groups (Fig. 9.4b). Other examples include the use of 1H-benzimidazole-5,6-dicarboxylic acid to form 3*d*-4*f* heterometallic coordination polymers with $Ag^+$, $Cu^+$ and $Ni^{2+}$ as transition metal ions [22] and pyridine-2,6-dicarboxylic acid to connect $Ln^{3+}$ and $Fe^{2+}$ [23] or $Mn^{2+}$ ions [24] (Fig. 9.4a).

A simpler but equally effective approach has been employed by Ward and coworkers [25] in the preparation of coordination polymers using luminescent anionic complexes. Here transition metals with emissive MLCT states act as effective sensitisers for lanthanide emission in the NIR [25]. In this case cyanide groups were used as bridging units starting from stable $Ru^{3+}$ complexes and simple $Ln^{3+}$ salts. Examples include $[Ru(Bipy)(CN)_4]^{2-}$ [26,27], $[Ru(Phen)(CN)_4]^{2-}$ [28], $[Ru(Bpym)(CN)_4]^{2-}$, $\{[Ru(CN)_4]_2(\mu\text{-Bpym})\}^{4-}$ [29], $[Ru(Hat)(CN)_4]^{2-}$ [25], $\{[Ru(CN)_4]_3(\mu^3\text{-Hat})\}^{6-}$, $[\{Ru(CN)_4\}_2(\mu^2\text{-Hat})]^{4-}$, $[Cr(CN)_6]^{3-}$, and $[Co(CN)_6]^{3-}$ [30]. The advantage of this method is that the building blocks are already kinetically stable in solution and the solid structure is dictated by the coordination number adopted by the lanthanide ion (Fig. 9.5).

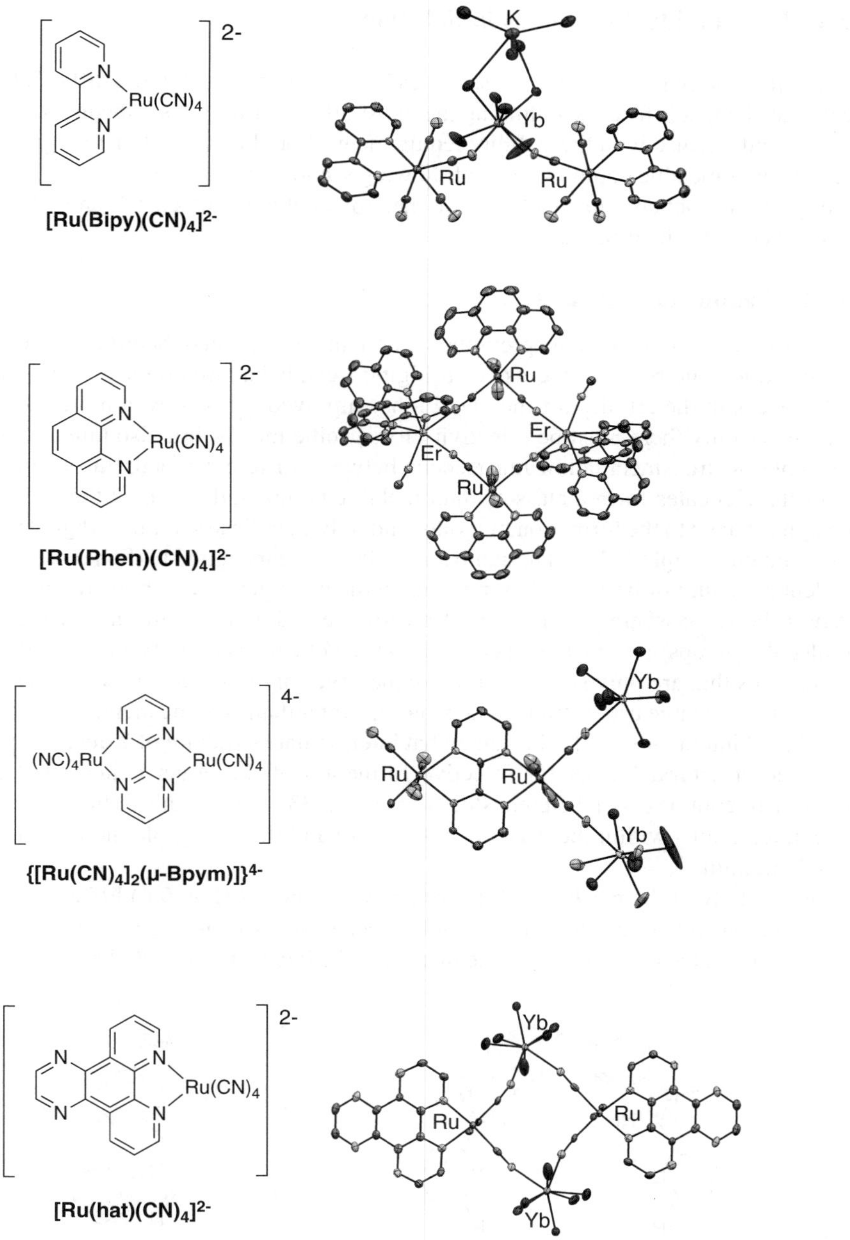


***Figure 9.5*** *Building blocks and solid structures of luminescent coordination polymers derived from complex anions. Created using the Mercury 2.4 software from the crystallographic data in the respective articles (see text for references). (See colour plate section)*

## 9.4 Lanthanide Assemblies in Solution

The formation of heterometallic assemblies in solution is a much more difficult task. In this case the chemist is faced with finding the, most often non-trivial, synthetic route to the final compound but also with the stability requirements, both kinetic and thermodynamic, of the target heterometallic complex. In the following section we will discuss strategies applied in the years to obtain kinetically stable heterometallic architectures and the properties associated with them in solution.

### 9.4.1 Lanthanide Helicates

As we have pointed out in the previous section, the formation of heterometallic complexes in the solid state requires the use of bridging ligands having donor groups with marked differences in the affinity for the metal ions employed. This is even more so in solution, where not only there must be selectivity for specific metals, but also kinetic stability, for example no transmetallation must occur between different metals in different binding pockets. Helicates form a class of multimetallic compound in which the accurate ligand design can lead to the formation of mono- and polymetallic assemblies. Piguet reported the first dinuclear triple-helical lanthanide complex 20 years ago [31] (Scheme 9.1). Here the tridentate nature of both binding domains in the pro-ligand **1,** combined with a $Eu^{3+}$ ion, drives the self-assembly formation of a triple helical architecture in which three ligand molecules wraps around two metal ions in a helical fashion. While ligand **1** leads to complexes that are only stable in non-competitive solvents, such as acetonitrile, ligand **3** forms highly stable triple stranded helicates in water thanks to the increased stability of the Ln-ligand interaction [32]. The somewhat intermediate structure of **2** does not possess the required structural factors for selective formation of the target helicate but leads to a mixture of complexes of different stoichiometries [33]. Clearly this is due to the absence of π–π interactions within the organic architecture and the negligible increase in Ln-ligand bond strength.

In the study of the mechanism of formation of helicates (Fig. 9.6) Ehlabiri et al. [34,35] have demonstrated that the self-assembly mechanism is mainly governed by cooperative electrostatic interactions between the metal and the ligand **4**. The initial coordination of two

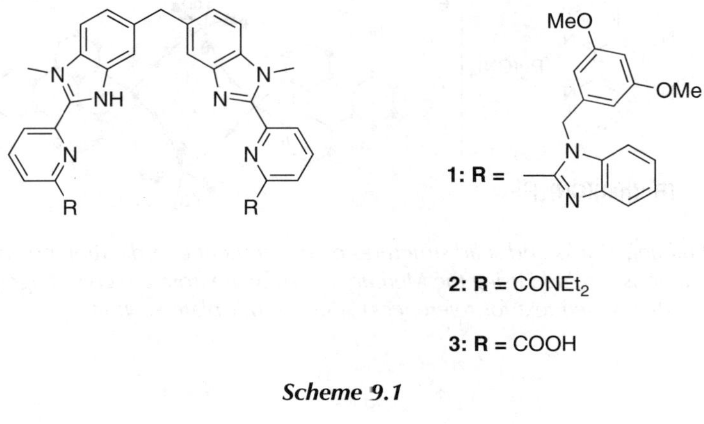


*Scheme 9.1*

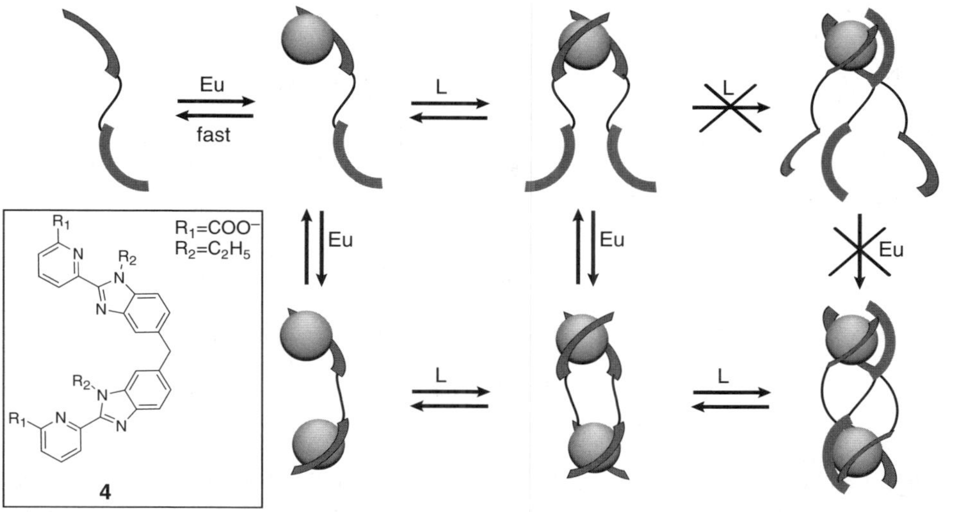


***Figure 9.6*** *Self-assembly mechanism for the formation of triple-stranded helicates as discussed by Elhabiri et al. [35]. Adapted with permission from [35]. Copyright © 2004, John Wiley. (See colour plate section)*

ligand strands around one metal ion prevents the approach of a third ligand by electrostatic repulsion of the negative carboxylate groups. The mechanism is hence forced toward the complexation of a second lanthanide ion followed by one last ligand molecule to form the final $Ln_2L_3$ helical structure.

Piguet and coworkers [36] have demonstrated that the rationalization of the possible electrostatic interactions occurring during the self-assembly process can help predict the final dominant structure in solution. With an accurate choice of experimental conditions, the formation of tetra- and penta- nuclear helicates is then possible as reported by Zeckert et al. and El Aroussi et al. [37,38]. The knowledge of the assembly mechanism and the underlying inter- and intramolecular interactions, allows a fine tuning of targeted properties through the controlled modification of the ligand building block. For instance, grafting of the benzimidazole moieties with polyoxoethylene groups renders the structure water-soluble, without preventing the formation of the triple-stranded helicate. Such compounds lend themselves to further modification and have been successfully used as luminescent bioprobes for cellular imaging [39,40].

The formation of heteromultimetallic complexes by self-assembly requires a more precise selection of ligands. While they must preserve their general structure, that underlies the overall helical architecture through weak interactions, they must also possess two, ideally orthogonally different, binding sites. Heterobimetallic complexes comprising a lanthanide ion and a transition metal are a class of compounds that has received increasing attention in the last fifteen years for ability of the transition metal to function as an antenna group, especially for NIR emitting lanthanide. The coordination geometry of a specific transition metal helps rationalising the ligand design and therefore facilitates the prediction of the final self-assembled architecture. The transition metal hence acts as a template around which the structure builds up.

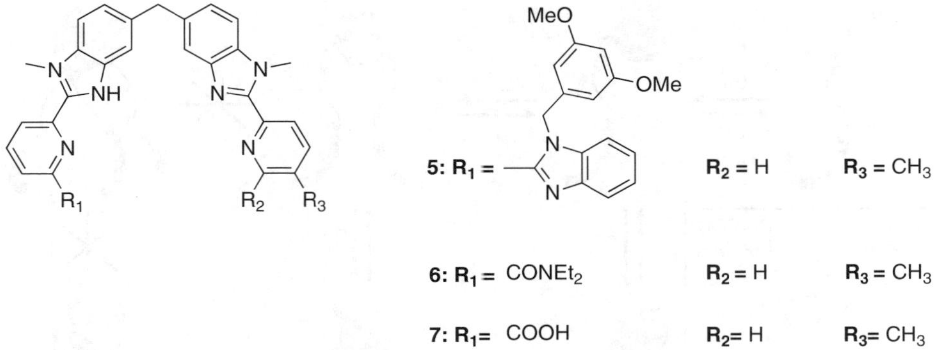


*Scheme 9.2*

Piguet and coworkers [41] reported the formation of the first heterodinuclear *d-f* triple-stranded helicate more than 15 years ago using the receptor ligand **5** (Scheme 9.2). Precise definition of the donor sites on the ligand was necessary in order to meet the coordination requirements of both metals (tricapped trigonal prismatic and octahedral for $Ln^{3+}$ and $Zn^{2+}$ respectively) and ensure that one helicate (i.e. that which combined three tridentate binding domains around one metal centre while simultaneously providing three bidentate domains at the other) is favoured over other regioisomers. Increasing the affinity of the ligand building block for the lanthanide ion through the use of oxygen donors increased the selectivity of formation of the triple stranded helicate under less strict stoichiometric conditions [42,43].

Indeed, minor modifications on the receptor can greatly modify the structural and electronic properties of the *d-f* assembly. While ligand **5** has a strong tendency to form homometallic double stranded helicates with $Zn^{2+}$, modification of the tridentate binding site to include an oxygen donor, markedly favoured the formation of the head to head to head (HHH) pure *d-f* helicate. It is noteworthy that the starting building blocks are designed almost entirely on the basis of thermodynamic considerations – and it should probably be stressed that the nature of potential kinetic traps during the self-assembly process (whether involving high activation energy barriers or arising from low solubility) must be considered in detail during the development and application of these systems. The use of labile metal ions is hence an indispensable requisite to let the system explore the entire potential energy surface, although it does not guarantee that the final conformation be the most kinetically and thermodynamically stable. However, the use of inert metals would greatly increase the kinetic stability of the helicate, thus opening it to possible applications in biomedical imaging. To overcome this problem post-modifications of the assembly have been used, e.g. by oxidation of a labile $Co^{2+}$ into its inert $Co^{3+}$ form [44,45] or using the $Cr^{2+}/Cr^{3+}$ redox couple [46].

The preparation of *f-f* heterometallic helicates is obviously more challenging. While for *d-f* assemblies the prediction of the structure relies on the metals and on the ligand properties, the small change in ionic radii among the lanthanides series remains the only discriminating factor. Perhaps the most representative example of the preparation of such assembly is given by Bünzli and coworkers [47] in which two binding groups, with orthogonal coordination affinities depending on the lanthanide ionic radius, are combined

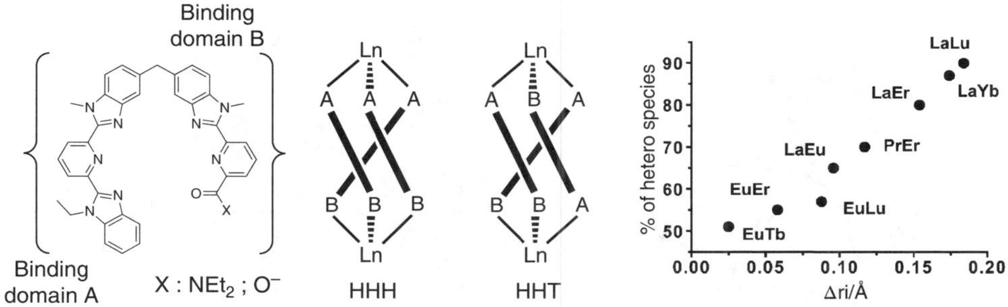


*Figure 9.7* *Ligand employed and conformational isomers obtained by Bünzli et al. [47] in the selective preparation of heterobimetallic helicates. The graph shows the selectivity of the formation of the heterometallic assembly upon increasing the* $\Delta r$ *of the lanthanide couple. Adapted with permission from [47]. Royal Society of Chemistry*

together on the typical bis-benzimidazole scaffold. For the Ln = La and Ln′ = Lu couple the HHH isomer dominates in solution (ca. 90%), while the head to head to tail (HHT) species gradually increases with decreasing $\Delta r$ to reach a statistical mixture for the Eu/Tb couple (Fig. 9.7) [47].

### 9.4.2 Non-helicate Structures

Several other polymetallic structures formed by self-assembly have been reported in the literature. For instance the class of metallacrowns [48] has received increasing attention over the last few years. These inorganic analogues of crown ethers are self-assembled from simple building blocks and transition metal ions, mainly $Cu^{2+}$ and $Fe^{2+}$ (Fig. 9.8). The lanthanide ion, situated in the central cavity, acts as a template in the formation of the heterometallic structure. The first lanthanide-containing structures reported were made of a hexanuclear cluster in which the lanthanide ion is surrounded by five transition metals

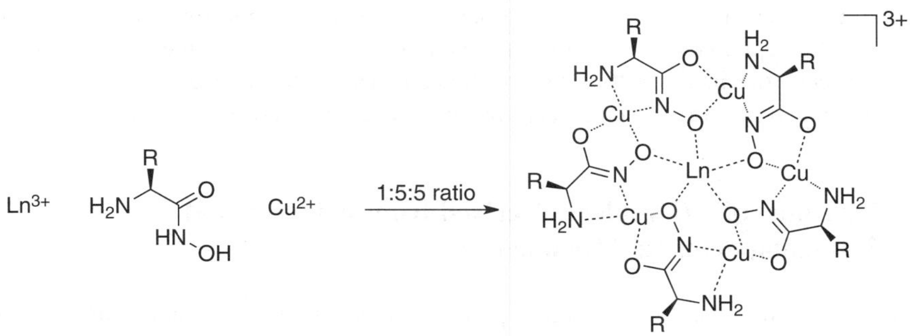


*Figure 9.8* *Selective formation of 15-MC-5 by self-assembly of lanthanide and transition metal ions with aminohydroxamate ligands*

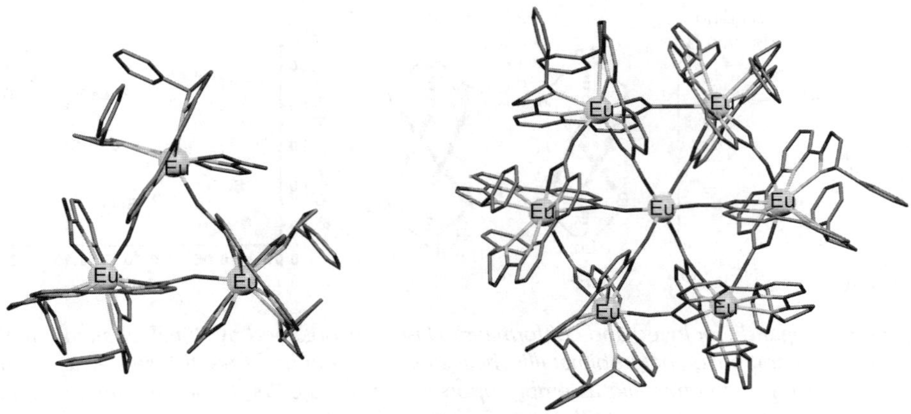


*Figure 9.9* *Representation of the solid-state structure for the europium molecular triangle and wheel obtained by Mazzanti and co-workers [52,53]. Created using the Mercury 2.4 software from the crystallographic data in the respective articles (see text for references). (See colour plate section)*

forming the family of 15-MC-5 metallacrowns. Since then more complex architectures have been designed and isolated, wherein the structure has expanded to incorporate an increasing number of metal centres up to 60-MC-20 architectures [48]. Their structures are entirely predictable based on the coordination angle of the aminohydroxamate ligands and the transition metal ions [49].

They have been shown to bind carboxylate anions through coordination to the lanthanide and they lend themselves to further functionalization through chemical modification of peripheral ligands [50,51]. This property could be exploited in the formation of supramolecular architectures through interactions with other kinetically stable building blocks to form heteromultimetallic arrays. Another example of how intermolecular interactions play a crucial role in the self- assembly of supramolecular architectures is reported by Mazzanti and coworkers [52,53] in the stereoselective preparation of a chiral heptanuclear europium wheel and a trinuclear europium triangle (Fig. 9.9). Steric constraints on the ligands surrounding each lanthanide ion cause alternating Δ and Λ isomers around the external $Eu^{3+}$ centres. In the heptanuclear assembly the seventh $Eu^{3+}$ ions sits in the central cavity and plays a crucial role in the self-assembly. Indeed, while an identical architecture can be obtained using $Nd^{3+}$, the heptanuclear species is not observed with the smaller $Yb^{3+}$.

## 9.5 Heterometallic Complexes Derived from Bridging and Multi-compartmental Ligands

Organic transformations can be used to prepare architectures from stable starting components. We initially used this approach in the first controlled preparation of *f*-*f*′ heterometallic systems, using stable 4-aminobenzylDO3A as building block complexes and reacting them with DTPA anhydride (Fig. 9.10), before adding a second lanthanide ion which binds in the DTPA diamide pocket [54].

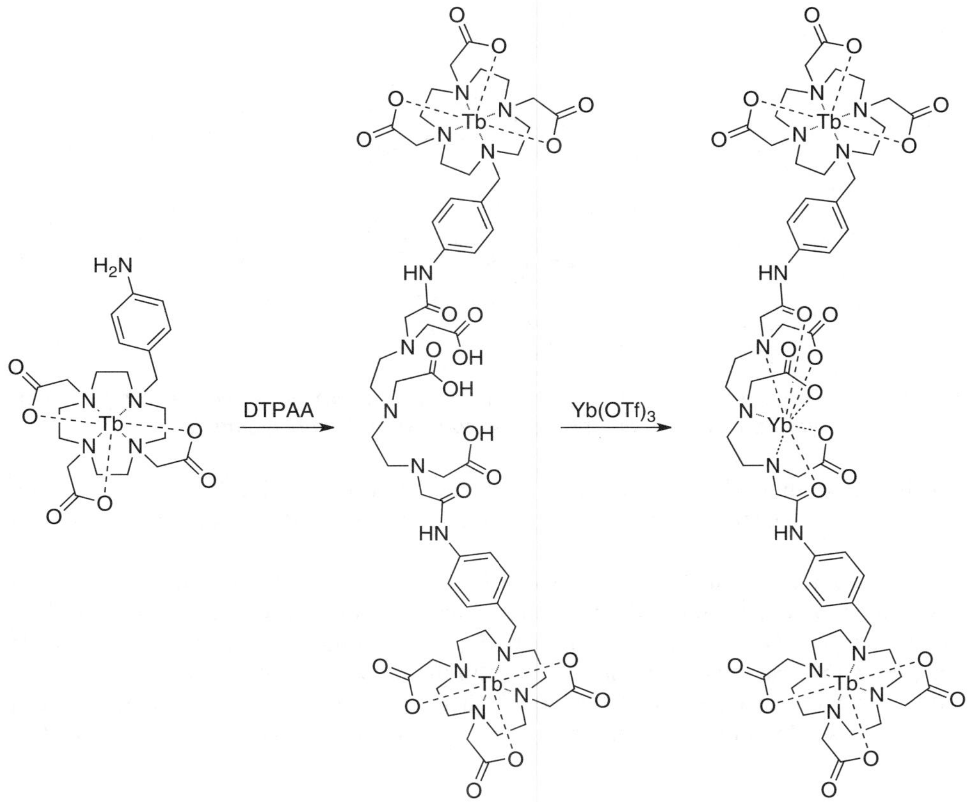


***Figure 9.10*** *Strategy for the preparation of pure f-f′ heterometallic systems obtained by organic transformation on kinetically stable building blocks as reported by Faulkner et al. [54]*

Tremblay and Sames [55] extended this approach to systems in which two binding sites are both coordinated to a lanthanide, before exploiting the lability of DTPA under acidic conditions to sequester one lanthanide and replace it with a second one (Fig. 9.11), while leaving the DOTA pocket untouched.

Related approaches can be used to prepare *d-f* hybrid complexes and *s-f* hybrids by linking stable complexes to additional binding sites. In the case of the former, stable *d*-block complexes can be reacted with azamacrocycles to yield stable *d-f* hybrids (Fig. 9.12) [56]. In the case of *s-f* hybrid arrays, a stable lanthanide complex must be linked to a binding site for alkali metals (Fig. 9.13) [57].

In the case of *f-f′* heterometallic systems, an orthogonal protection approach has been used to demonstrate a route to forming otherwise symmetrical heterometallic systems (Fig. 9.14) [58]. In this case, the synthetic strategy is dictated by the nature of the bound lanthanide ion: the order of the deprotection steps is of critical importance, since bound lanthanide ions catalyse decomposition of the ligand backbone through hydrolysis of the bound carboxamide. This application of protecting group chemistry probably has limited use beyond the proof of concept – the overall yield of the extended synthesis is sufficiently

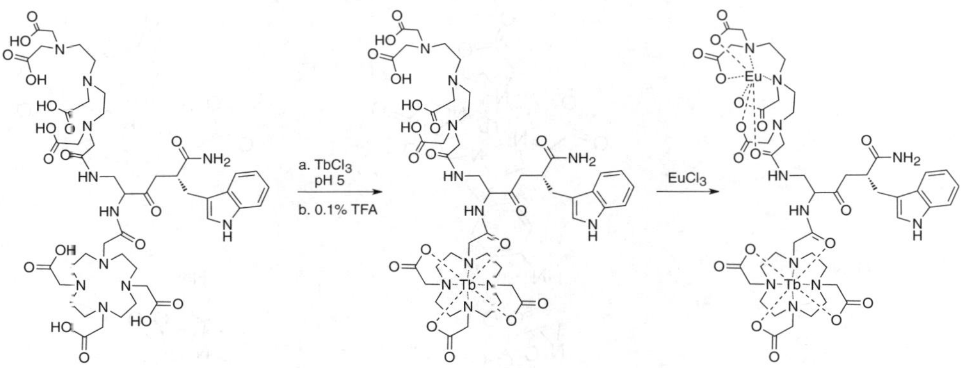


*Figure 9.11* *Formation of pure* f-f′ *heterometallic system through sequential complexation/decomplexation of a DTPA binding pocket as reported by Tremblay and Sames [55]*

low that preparation of similar structures would be just as effective if achieved by separation of LnL from a statistical mixture of L, LnL and $Ln_2L$, followed by addition of a different lanthanide ion.

Alternatively, two stable complexes can be linked together through the formation of covalent or coordinate bonds. For instance, we have employed click chemistry methods to link propargyl DO3A complexes together (Fig. 9.15) to form homobinuclear lanthanide complexes [59], and extended this approach to the preparation of *d-f* hybrids containing redox active ferrocene chromophores [60]. *f-f′* heterometallic systems can also be produced by linking stable components together, for instance in the preparation of tetranuclear

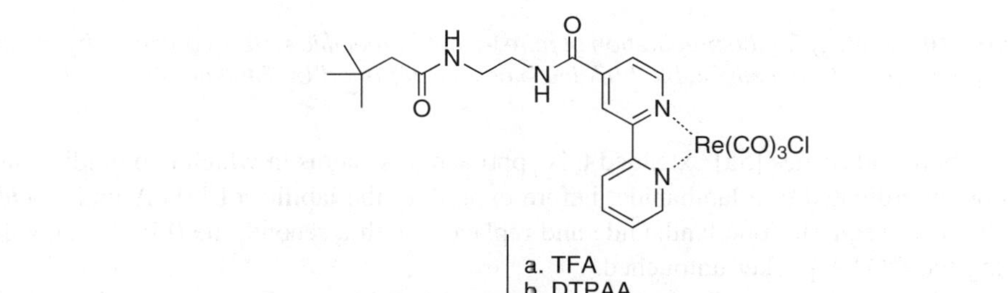


*Figure 9.12* *Preparation of* d-f *hybrid complexes starting from kinetically stable lanthanide and transition metal building blocks*

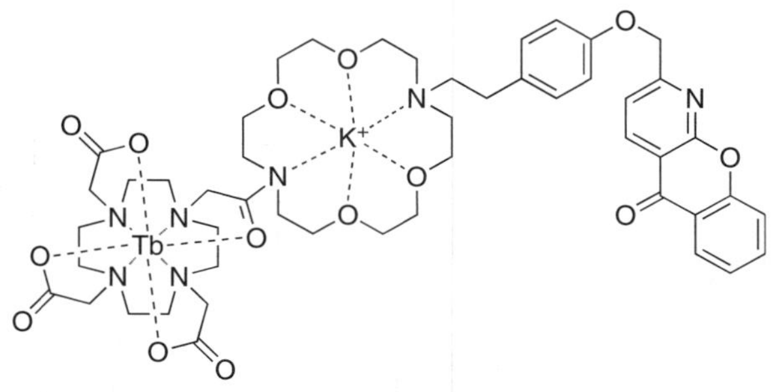

***Figure 9.13*** *Example of s-f hybrid complex in which an alkali metal binding site is linked with a lanthanide containing pocket*

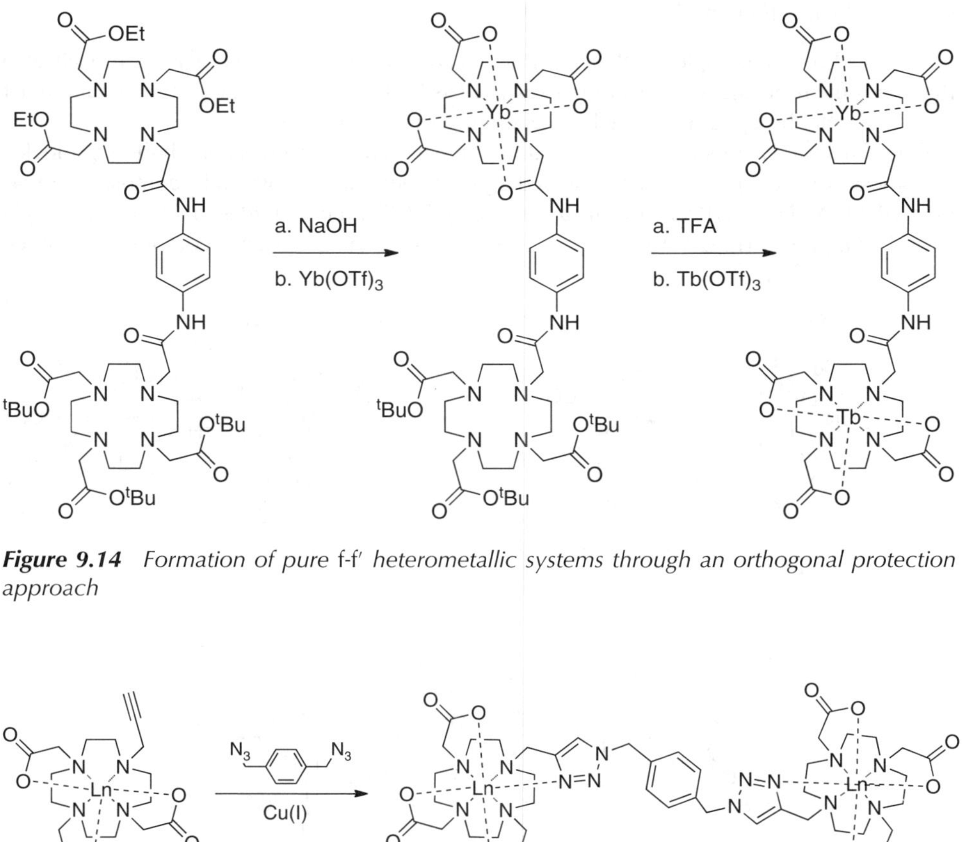


***Figure 9.14*** *Formation of pure f-f′ heterometallic systems through an orthogonal protection approach*

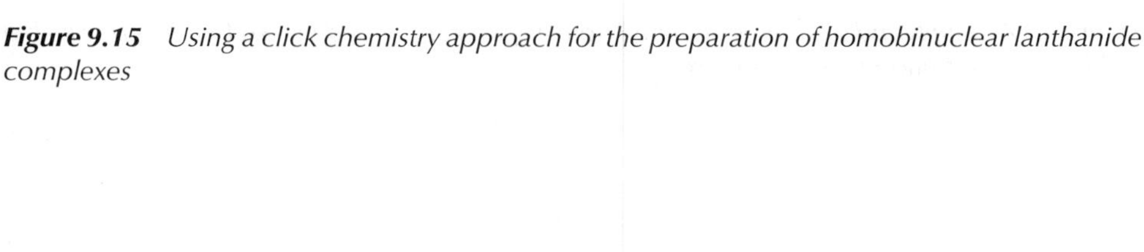


***Figure 9.15*** *Using a click chemistry approach for the preparation of homobinuclear lanthanide complexes*

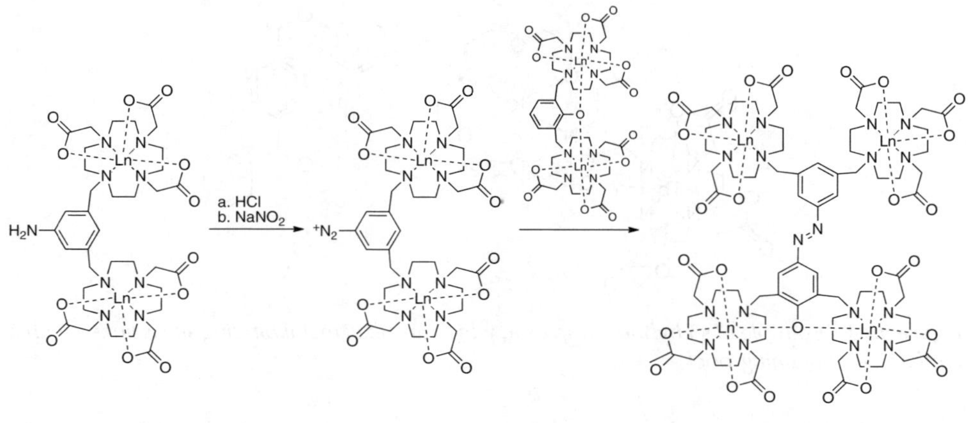


*Figure 9.16* Formation of f-f′ heterometallic systems obtained from building block that are stable to the reaction condition employed

bimetallic $Ln_2Ln'_2$ complexes through diazotization (Fig. 9.16) [61]. This approach also illustrates the robustness of the complexes used as building blocks, since they remain intact despite the acidic conditions used in generating diazonium salts.

Coordination chemistry can be used to prepare arrays from stable building blocks. Metathesis of Re(Bpy)$(CO)_3$Cl followed by treatment with lanthanide complexes of 4-picolylDO3A, TriazolylDO3A or triazole bearing DOTA monoamides yields a series of *d-f* hybrid complexes (Fig. 9.17, a−c). The Gd-Re bimetallic systems combine long-lived

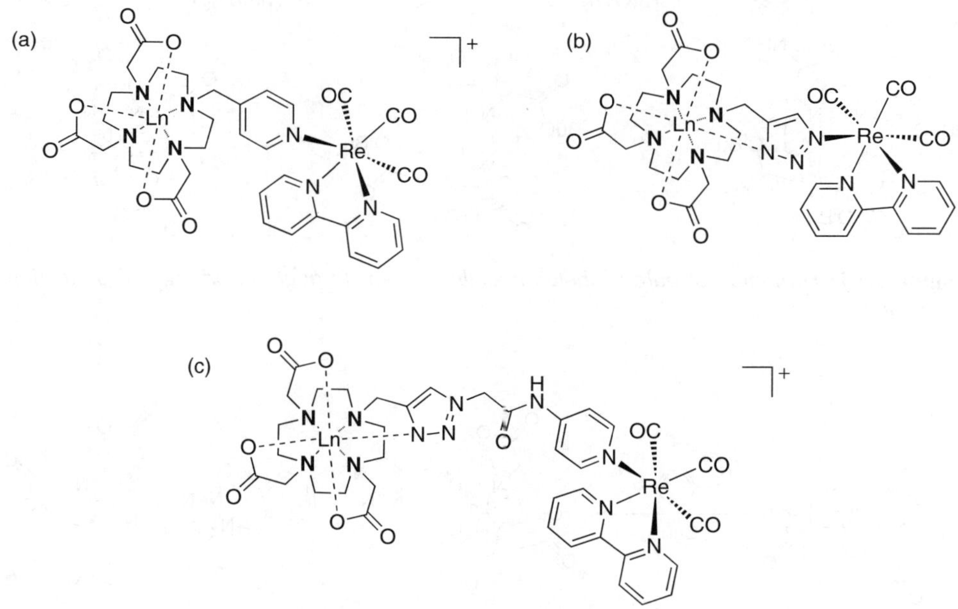


*Figure 9.17* d-f Hybrid structures obtained from a ligand appended DO3A building block able to bind to a transition metal complex

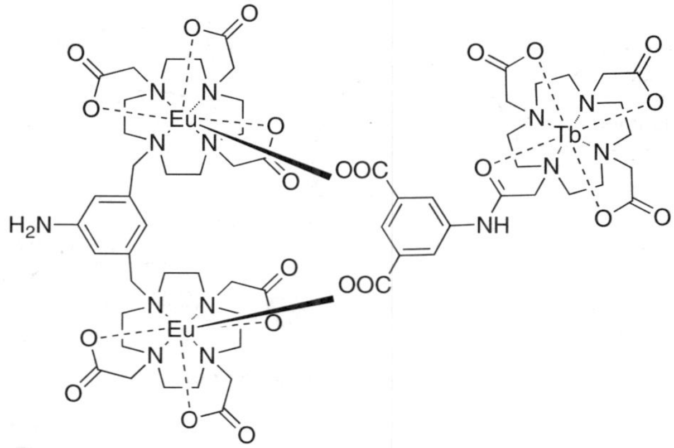


*Figure 9.18* Supramolecular structure obtained by the interaction an isophthalate appended DOTA monoamide complex and a seven coordinate dimetallic europium complex

luminescence (from the $^3$MLCT states) with the MRI properties of the gadolinium centres to give bimodal imaging agents that can be addressed both by optical methods and MRI [62,63].

The differences in structure between the complexes result in different properties: the relaxivity of complex a is greatly reduced in phosphate buffer, as might be expected given the heptadentate sites of the ligand, while that of complex c (with an octadentate lanthanide binding site) is unaffected by phosphate. Despite the fact that the gadolinium complex of triazoloDO3A contains an octadentate ligand and does not bind phosphate, the relaxivity of complex b is influenced by phosphate, indicating that rhenium coordination by the triazole reduces its affinity for the lanthanide.

Coordination to lanthanide centres can also be exploited in the preparation of arrays in solution. For instance, *d-f* hybrids can be prepared by adding bipyridylcarboxylate bearing complexes of rhenium(I), ruthenium(II) and osmium(II) to lanthanide complexes of DO3A; such systems have been used to screen sensitising chromophores for use with lanthanides [64]. Despite the fact that benzoic acid functionalised DO3A derivatives have been shown to aggregate in solution [65], more complicated systems are required if heterometallic complexes containing only lanthanides are to be prepared. Indeed, self-assembly of such heterometallic lanthanide complexes has only recently been observed, through exploiting the affinity of m-xylyl bridged DO3A derivatives for an isophthalate appended DOTA monoamide (Fig. 9.18) [66]. This system is remarkable in the high affinity of the host for the guest even in competitive media: the binding constant is unaltered in competitive media such as phosphate buffered saline, suggesting that self-assemblies could have potential *in vivo*.

## 9.6 Energy Transfer in Assembled Systems

The population of the excited states of a lanthanide ion can be achieved via several mechanisms. The diagram in Fig. 9.19 shows only some of the pathways leading to lanthanide-centred luminescence (see also Chapter 1). In the early years of the study of

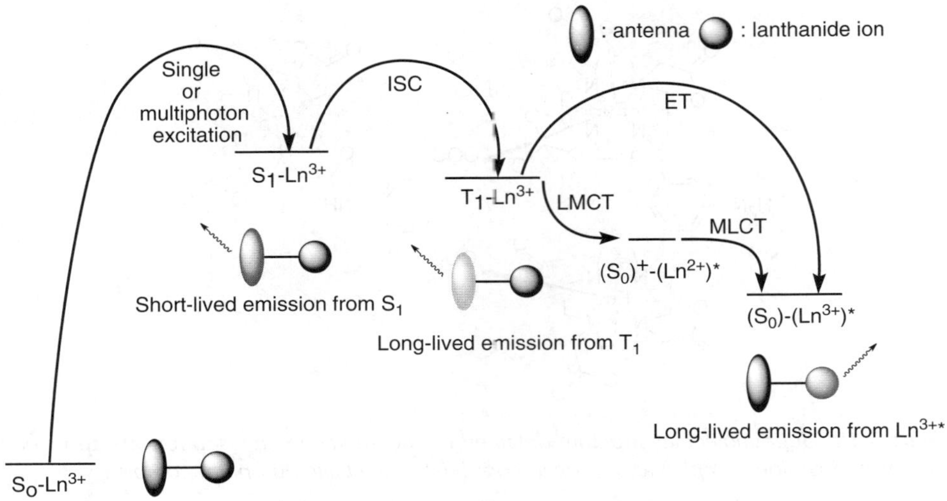


*Figure 9.19 Possible energy transfer processes that can occur in a lanthanide antenna assembly leading to luminescence from the lanthanide centre. (See colour plate section)*

luminescent complexes containing lanthanide ions, organic chromophores were widely exploited since high energy emitting lanthanides, such as terbium and europium, were mainly studied (for their high intrinsic luminescence and ease of detection). In the last 10 years, transition metal complexes have formed a new class of sensitising chromophores and are now becoming the most used antennas for NIR emitting lanthanides [67]. They offer higher photostability when compared to organic counterparts and usually long luminescence lifetime that allows a more accurate study of the energy transfer. Indeed, the study of *d-f* hybrids has incredibly increased our knowledge on the possible lanthanide sensitization mechanisms.

The population of the antenna excited states can be achieved through a single or multiphoton excitation process. Although direct two-photon excitation of the metal is in principle a Laporte allowed process, it has recently been shown that, whenever an organic chromophore is present, it will dominate the two-photon behaviour of the system [68]. Two-photon excitation of europium, via excited states centred on the antenna, has been demonstrated by Pålsson et al. [69] and offers great promise for biomedical applications (see also Chapter 5). The authors showed that the energy transfer cascade remained unaffected and the imaging probe retained its responsive behaviour towards carbonate [69]. Once the antenna chromophore has been populated, inter-system crossing to the chromophore-centred triplet state takes place and from there formation of the lanthanide-centred excited state can occur either by Förster or Dexter mechanism, depending on donor-acceptor energy overlap and chemical structure. This is only an overall description of a simplified mechanism and other possible processes must not be excluded *a priori*, for example sensitization from the antenna singlet state, as illustrated for the Ru-Ln dyads shown in Fig. 9.20 where energy transfer from a singlet and an $^3$MLCT state can occur simultaneously [70].

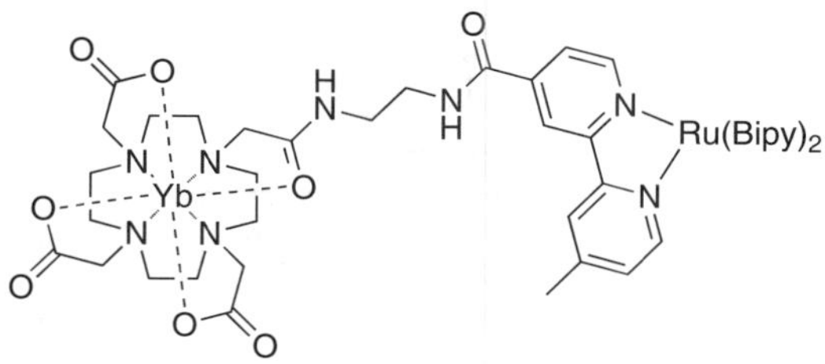


***Figure 9.20*** *A* d-f *hybrid complex in which the energy transfer is mediated by both the singlet and triplet MLCT states*

A sequential double electron transfer mechanism can also be invoked where it is thermodynamically feasible. The free energy dependence of the photoinduced electron transfer can be expressed (in eV) by the following equation [71]:

$$\Delta G_{\mathrm{ET}} = (E_{\mathrm{ox}} - E_{\mathrm{red}}) - E_s - \frac{e_0^2}{\varepsilon a}$$

where $E_{\mathrm{ox}}$ and $E_{\mathrm{red}}$ are the oxidation and reduction potentials of the donor and acceptor respectively. $E_s$ is the singlet state energy of the donor and $e_0^2/\varepsilon_a$ is the energy of interaction between the radical ion pair in the excited state. Among the lanthanides, only ytterbium and europium have a reduction potential low enough to make the forward electron transfer process energetically possible. However, while such electron transfer process yields the emissive state for ytterbium containing systems, the LMCT state for europium is usually lower in energy than the emissive $^5D_0$ state. The occurrence of such a mechanism was clearly demonstrated for an ytterbium complex containing a phenanthridine chromophore; in this system, sensitization of the $^2F_{5/2}$ excited state through a double electron transfer pathway is observed at high pH, but it becomes thermodynamically unfavourable upon protonation of the antenna (leaving slow triplet mediated energy transfer as the only pathway to sensitise the excited state) [72,73].

In an extensive study on the energy transfer between transition metal chromophores and lanthanide centres, Ward and coworkers [74] have demonstrated that the energy transfer in this class of dyads is mainly governed by a superexchange mechanism mediated by the bridging ligand whenever such a mechanism is possible. A Coulombic mechanism may contribute only partially whenever saturated linkers are present. Indeed the authors showed that even at a separation of about 20 Å between the metal centres, the efficiency of energy transfer in the Ru-Nd donor-acceptor dyad was still as high as 44% [74]. The change of the accepting lanthanide ion decreased the energy transfer efficiency due to the decrease in the donor−acceptor energy overlap following the trend Nd > Er > Yb. Changing the donor metal to $Os^{2+}$ did not alter the energy transfer mechanism although it decreased its efficiency substantially. In fact, while it can be said that the mechanism of energy transfer is most often mediated by a superexchange process, the rationalization of its efficiency cannot be based solely on the donor atom. We will discuss some illustrative examples in the following paragraphs.

Subtle aspects during the design of the *d-f* hybrid architecture can lead to a fine-tuning of the lanthanide centred luminescence. By exploiting this principle Piguet, Bünzli and

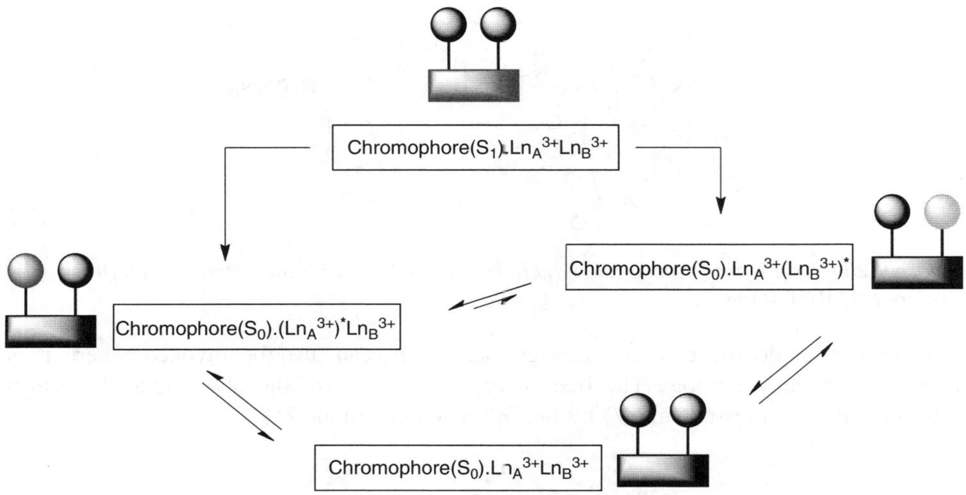


*Figure 9.21 Schematic representation of the possible energy transfer pathways in a bimetallic array. (See colour plate section)*

coworkers [75] reported a series of multimetallic helicates containing $Cr^{3+}$-$Ln^{3+}$ couples (Ln = Yb, Nd). Here, the distance between the metal centres limits the efficiency of energy transfer, to the point that $k_{ET}(CrLn) \ll k_{lum}$. This, combined with a rate constant of $Cr^{3+} \rightarrow Ln^{3+}$ energy transfer comparable with the decay rate constant of the $Cr^{3+}$ excited state, $k_{ET}(CrLn) \approx k_d$, leads to $Nd^{3+}$ and $Yb^{3+}$ NIR luminescence in the milliseconds range. Strategies for the optimization of such processes leading to long-lived NIR-emitting $Ln^{3+}$ states have recently been published. The use of two $Cr^{3+}$centres has shown to increase the luminescence quantum yield without perturbation of the overall mechanism [76].

Although the preparation of heteromultimetallic *f-f* complexes is challenging (see previous discussion), the properties arising from such a metal assembly could be exploited for various applications such as the development of dual emissive probes, bimodal imaging agents and for upconversion. The accurate understanding of the energy transfer mechanism within the complex ensemble is thus fundamental. If we consider the simplest scenario in which two different lanthanide ions are present in the same ligand backbone, either in identical or different binding pockets, the diagram in Fig. 9.21 represents the energetic pathways that the system can follow once its antenna-centred excited states have been populated.

The donor–acceptor spectral overlap and the overall chemical structure of the complex will determine the preferential energy transfer pathway from the antenna to the excited state of one of the lanthanide ions, thus influencing the Förster and Dexter contributions. Assuming that the two metal centres are independent, that is, no interaction between the two occurs, the emission of the assembly will be an average of the emission of the single ion, weighted for the efficiency of energy transfer from the common antenna, and for the non-radiative deactivation of their excited states. In the event of effective spectral overlap, energy transfer between the lanthanide centres can take place, provided that the kinetics of energy transfer are of the same order of non-radiative processes. For instance Faulkner and Pope reported a trinuclear complex in which two $Tb^{3+}$ centres act as sensitising units for

$Yb^{3+}$. This is the first example of communication between two lanthanide ions present in the same molecule. Indeed the energy transfer is assisted by a favourable energy overlap and by the chemical structure that allows the two metal centres to be in close proximity. Here, selective excitation of the $^5D_4$ state of $Tb^{3+}$ resulted in sensitised luminescence from $Yb^{3+}$ at 980 nm [77].

Finally it is worth mentioning that detrimental processes such as quenching or thermal repopulation of excited states can further influence this intricate energy transfer cascade. Indeed, the difference in energy between the chromophore and the lanthanide manifolds can be accurately engineered towards a fine-tuning of the ensemble properties. The overall intensity of luminescence from the lanthanide centre will then depend on the efficiency of the energy transfer of each step highlighted in Fig. 9.19, so that the emission of the array can be finely modulated by several means and strategies; some of these will be elucidated in the following section.

## 9.7 Responsive Multimetallic Systems

Through the accurate design of the molecular architecture, the overall luminescence behaviour of a lanthanide-containing probe can be finely tuned to respond to external changes in analyte concentration. The main strategies used are either the direct perturbation of the energy transfer cascade, through quenching or population of an excited state, or by indirect methods such as chemical structure modifications that lead to a rearrangement of the overall geometry of the complex and hence a change in the luminescence properties. Heteromultimetallic complexes offer the possibility of having two signalling units, hence allowing for ratiometric detection where the luminescence of one unit is uninfluenced by the presence of the analyte. Unfortunately, while several examples of monometallic complexes that respond to pH, $pO_2$, cations and anions concentrations are reported in the literature, the synthetic challenges arising from the synthesis of heteromultimetallic complexes have somewhat thwarted their development and use as ratiometric probes.

We recently reported a *d-f* hybrid (Fig. 9.22) in which the luminescence from europium could be modulated by changing the oxidation state of the ferrocene unit. Indeed, the

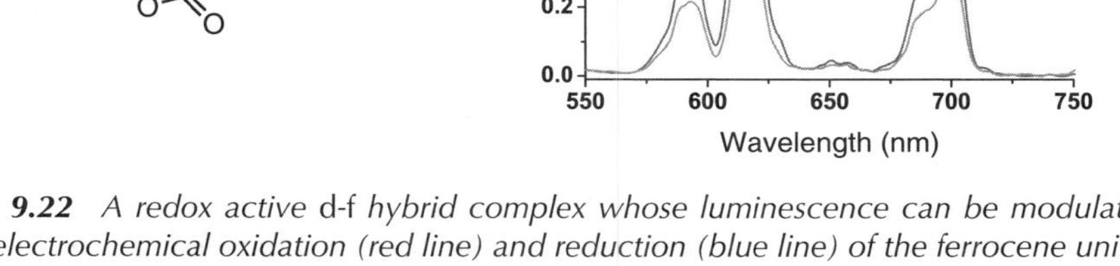


***Figure 9.22*** *A redox active* d-f *hybrid complex whose luminescence can be modulated by upon electrochemical oxidation (red line) and reduction (blue line) of the ferrocene unit. (See colour plate section)*

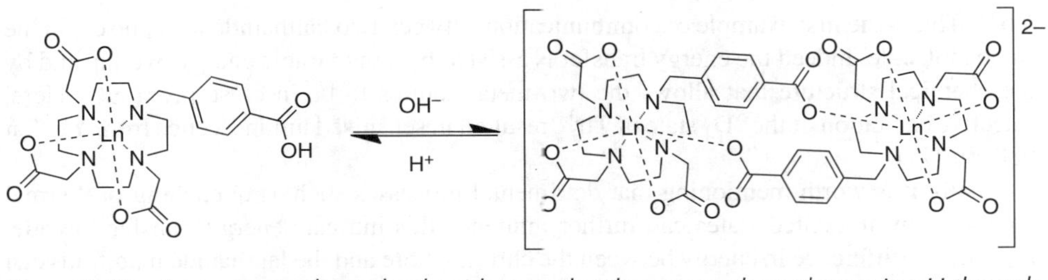


***Figure 9.23*** *A seven coordinate lanthanide complex that responds to change in pH through aggregation when in the deprotonated state, triggering an increase in the luminescence signal*

spectral overlap between the europium donor and ferrocenium acceptor is such that effective Förster energy transfer takes place only when the ferrocene unit is oxidised and does not affect the luminesce of the assembly when in its reduced state.

This is the first example of electrochemical tuning of lanthanide luminescence and opens up new opportunities for signal manipulation and responsive imaging [60]. The seven-coordinate lanthanide complex with an appended carboxylate-containing arm (Fig. 9.23) has been used as pH responsive probe through change in the molecular structure in solution. An increase in pH deprotonates the carboxylic moiety that in turn binds to the unsaturated lanthanide. This affords dimeric entities in solution where the displacement of solvent molecules triggers an increase in emission intensity proportional to pH [65].

Anion binding can also be used to template the synthesis of supramolecular assemblies such as the pseudorotaxane shown in Fig. 9.24. A macrocycle containing a transition

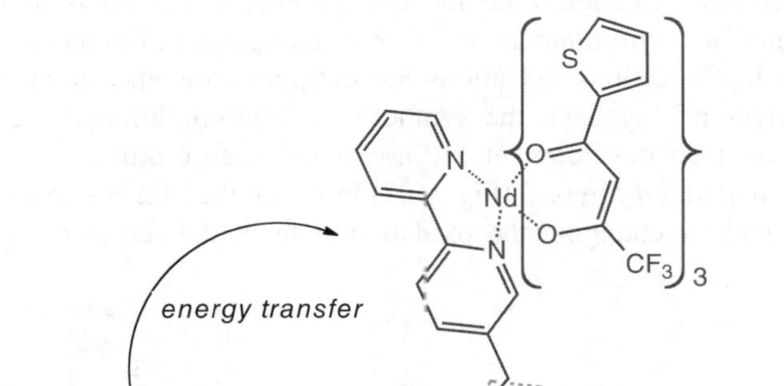


***Figure 9.24*** *Pseudorotaxane formation from a transition metal appended macrocycle and a lanthanide appended thread. (See colour plate section)*

metal and a lanthanide containing thread organise around a chloride anion through favourable hydrogen bonding and $\pi-\pi$ stacking interactions. The formation is evidenced by efficient through-space energy transfer between the transition metal and the lanthanide ion [78].

The coupling of two or more functionalities in the same assembly allows the use of the properties of the single components in a single probe or can result in drastically altered features originated from intramolecular interactions that can perturb the energy transfer cascade in response to an external stimulus such as change in pH, $pO_2$ or analyte concentration.

## 9.8 Summary and Prospects

We have now reached the stage where there are routes available to access a very wide range of metallic systems containing lanthanides by exploiting both kinetic control and thermodynamic selectivity. The former approach lends itself to the preparation of a wide variety of heterometallic systems, including *d-f*, *f-f′* and *s-f* hybrid complexes, while the latter is much less effective for the preparation of pure *f-f′* hybrid heterometallic systems. Through control of intermetallic separation, ligand structure and the nature of the chromophore it is possible to achieve a wide range of effects. Things have now reached the stage where such complexes can be exploited in a range of applications, ranging from electrochemical switching of luminescence to medical diagnosis. However, much remains to be done. Development of heterometallic patterned surface architectures is almost entirely unexplored, and though we have hinted at the possibilities for self-assembly of heterometallic architectures *in vivo*, there is much that remains to be done before the biological properties of heterometallic architectures can be exploited in full. The coming years will see new developments in structure and function of lanthanide containing arrays, but it is clear that these must be predicated upon a detailed understanding of kinetic and thermodynamic properties of their constituent parts, and upon the ability to measure and control speciation of, for example, isomeric forms of a given building block. This raises spectroscopic as well as synthetic challenges, since most measurements on heterometallic systems will give a weighted average (for instance an MRI measurement will always favour the isomer with the most hydration at the metal centre, while luminescence will always favour the least hydrated form). There is therefore an urgent need to improve our current understanding of the analytical methods used to assess solution structure, as well as an equal need to develop new strategies and components that can be used in building complex architectures. These requirements ensure that there is much still to be discovered.

## References

1. Binnemans K., in *Handbook on the Physics and Chemistry of Rare Earths*, J. C. B. K.A. Gschneidner, V. K. Pecharsky, Eds. (Elsevier, 2005), vol. 35, pp. 107–272.
2. Cacheris W. P., S. K. Nickle, A. D. Sherry, Thermodynamic study of lanthanide complexes of 1,4,7-triazacyclononane-N,N′,N″-triacetic acid and 1,4,7,10-tetraazacyclododecane-N,N′,N″,N‴-tetraacetic acid. *Inorg. Chem.*, **26**, 958–960 (1987).

3. Kumar K., C. A. Chang, M. F. Tweedle, Equilibrium and kinetic studies of lanthanide complexes of macrocyclic polyamino carboxylates. *Inorg. Chem.*, **32**, 587–593 (1993).
4. Smith R. M., A. E. Martell, Critical stability constants, enthalpies and entropies for the formation of metal complexes of aminopolycarboxylic acids and carboxylic acids. *Sci. Total Environ.*, **64**, 125–147 (1987).
5. Balogh E., R. Tripier, P. Fouskova, F. Reviriego, H. Handel, E. Toth, Monopropionate analogues of $DOTA^{4-}$ and $DTPA^{5-}$: kinetics of formation and dissociation of their lanthanide(iii) complexes. *Dalton Trans.* **2007**, 3572–3581 (2007).
6. Moreau J., E. Guillon, J.-C. Pierrard, J. Rimbault, M. Port, M. Aplincourt, Complexing Mechanism of the Lanthanide Cations $Eu^{3+}$, $Gd^{3+}$, and $Tb^{3+}$ with 1,4,7,10-Tetrakis(carboxymethyl)-1,4,7,10-tetraazacyclododecane (dota)—Characterization of Three Successive Complexing Phases: Study of the Thermodynamic and Structural Properties of the Complexes by Potentiometry, Luminescence Spectroscopy, and EXAFS. *Chem. Eur. J.* **10**, 5218–5232 (2004).
7. Brücher E. P. Szarvas, Kinetic study of the isotope-exchange reactions of the central ions of the lanthanide ethylenediaminetetraacetate and trans-1,2-diaminocyclohexanetetraacetate complexes. *Inorg. Chim. Acta* **4**, 632–636 (1970).
8. Sarka L., L. Burai, E. Brücher, The Rates of the Exchange Reactions between $[Gd(DTPA)]^{2-}$ and the Endogenous Ions $Cu^{2+}$ and $Zn^{2+}$: A Kinetic Model for the Prediction of the In Vivo Stability of $[Gd(DTPA)]^{2-}$, Used as a Contrast Agent in Magnetic Resonance Imaging. *Chem. Eur. J.* **6**, 719–724 (2000).
9. Tóth É., E. Brücher, I. Lazar, I. Tóth, Kinetics of Formation and Dissociation of Lanthanide(III)-DOTA Complexes. *Inorg. Chem.* **33**, 4070–4076 (1994).
10. Laurent S., L. V. Elst, R. N. Muller, Comparative study of the physicochemical properties of six clinical low molecular weight gadolinium contrast agents. *Contrast Media Mol. Imaging* **1**, 128–137 (2006).
11. Wedeking P., K. Kumar, M. F. Tweedle, Dissociation of gadolinium chelates in mice: Relationship to chemical characteristics. *Magnetic Resonance Imaging* **10**, 641–648 (1992).
12. Tóth É., R. Király, J. Platzek, B. Radüchell, E. Brücher, Equilibrium and kinetic studies on complexes of 10-[2,3-dihydroxy-(1-hydroxymethyl)-propyl]-1,4,7,10-tetraazacyclododecane-1,4, 7-triacetate. *Inorg. Chim. Acta* **249**, 191–199 (1996).
13. Tircsó G., Z. Kovács, A. D. Sherry, Equilibrium and Formation/Dissociation Kinetics of Some $Ln^{III}$PCTA Complexes. *Inorg. Chem.* **45**, 9269–9280 (2006).
14. Bryden C. C., C. N. Reilley, J. F. Desreux, Multinuclear nuclear magnetic resonance study of three aqueous lanthanide shift reagents: complexes with EDTA and axially symmetric macrocyclic polyamino polyacetate ligands. *Anal. Chem.* **53**, 1418–1425 (1981).
15. Pike M. M., J. C. Frazer, D. F. Dedrick, J. S. Ingwall, P. D. Allen, C. S. Springer Jr, T. W. Smith, $^{23}Na$ and $^{39}K$ nuclear magnetic resonance studies of perfused rat hearts. Discrimination of intra- and extracellular ions using a shift reagent. *Biophys. J.* **48**, 159–173 (1985).
16. Naritomi H., M. Kanashiro, M. Sasaki, Y. Kuribayashi, T. Sawada, In vivo measurements of intra- and extracellular $Na^{+}$ and water in the brain and muscle by nuclear magnetic resonance spectroscopy with shift reagent. *Biophys. J.* **52**, 611–616 (1987).
17. Shavaleev N. M., L. P. Moorcraft, S. J. A. Pope, Z. R. Bell, S. Faulkner, M. D. Ward, Sensitised near-infrared emission from lanthanides using a covalently-attached Pt(II) fragment as an antenna group. *Chem. Commun.* **2003**, 1134–1135 (2003).
18. Shavaleev N. M., G. Accorsi, D. Virgili, Z. R. Bell, T. Lazarides, G. Calogero, N. Armaroli, M. D. Ward, Syntheses and crystal structures of dinuclear complexes containing d-block and *f*-block luminophores. Sensitization of NIR luminescence from Yb(III), Nd(III), and Er(III) centers by energy transfer from Re(I)- and Pt(II)-bipyrimidine metal centers. *Inorg. Chem.*, **44**, 61–72 (2005).

19. Shavaleev N. M., L. P. Moorcraft, S. J. A. Pope, Z. R. Bell, S. Faulkner, M. D. Ward, Sensitised near-infrared emission from complexes of Yb-III, Nd-III and Er-III by energy-transfer from covalently attached Pt-II-based antenna units. *Chemistry-a European Journal* **9**, 5283–5291 (2003).
20. Peng G., Y.-C. Qiu, Z.-H. Liu, B. Liu, H. Deng, Spontaneous Assembly of d-f Coordination Frameworks: Syntheses, Structures, and Photoluminescence. *Cryst. Growth Des.* **10**, 114–121 (2009).
21. Liang L., G. Peng, L. Ma, L. Sun, H. Deng, H. Li, W. Li, A New Family of 3d-4f Heterometallic Tetrazole-based Coordination Frameworks: In Situ Tetrazole Ligand Synthesis, Structure, Luminescence, and Magnetic Properties. *Cryst. Growth Des.* **12**, 1151–1158 (2012).
22. Sun Y.-G., G. Xiong, M.-Y. Guo, F. Ding, S.-J. Wang, P. F. Smet, D. Poelman, E.-J. Gao, F. Verpoort, Hydrothermal synthesis, crystal structure and properties of Ni(II)-4f complexes based on 1H-benzimidazole-5,6-dicarboxylic acid. *Dalton Trans.* **41**, 7670–7680 (2012).
23. Zhao B., X.-Y. Chen, Z. Chen, W. Shi, P. Cheng, S.-P. Yan, D.-Z. Liao, A porous 3D heterometal-organic framework containing both lanthanide and high-spin Fe(II) ions. *Chem. Commun.* **2009**, 3113–3115 (2009).
24. Zhao B., X.-Y. Chen, P. Cheng, D.-Z. Liao, S.-P. Yan, Z.-H. Jiang, Coordination Polymers Containing 1D Channels as Selective Luminescent Probes. *J. Am. Chem. Soc.* **126**, 15394–15395 (2004).
25. Herrera J., S. Pope, A. Meijer, T. Easun, H. Adams, W. Alsindi, X. Sun, M. George, S. Faulkner, M. Ward, Photophysical and Structural Properties of Cyanoruthenate Complexes of Hexaaza-triphenylene. *J. Am. Chem. Soc.* **129**, 11491–11504 (2007).
26. Miller T., J. Jeffery, M. Ward, H. Adams, Photoinduced Ru–Yb energy transfer and sensitised near-IR luminescence in a coordination. *Dalton Trans.* **2004**, 491–504 (2004).
27. Davies G. M., S. J. Pope, H. Adams, S. Faulkner, M. D. Ward, Structural and photophysical properties of coordination networks combining $[Ru(bipy)(CN)_4]^{2-}$ anions and lanthanide(III) cations: rates of photoinduced Ru-to-lanthanide energy transfer and sensitized near-infrared luminescence. *Inorg. Chem.*, **44**, 4656–4665 (2005).
28. Baca S. G., H. Adams, D. Sykes, S. Faulkner, M. D. Ward, Three-component coordination networks based on $[Ru\ (phen)(CN)_4]^{2-}$ anions, near-infrared luminescent lanthanide (III) cations, and ancillary oligopyridine ligands: structures and photophysical properties. *Dalton Trans.* **2007**, 2419–2430 (2007).
29. Herrera J., S. J. Pope, H. Adams, S. Faulkner, M. D. Ward, Structural and photophysical properties of coordination networks combining [Ru(Bpym)(CN)4]2- or [[Ru(CN)4]2(mu-bpym)]4- anions (bpym = 2,2′-bipyrimidine) with lanthanide(III) cations: sensitised near-infrared luminescence from Yb(III), Nd(III), and Er(III) following Ru-to-lanthanide energy transfer. *Inorg. Chem.* **45**, 3895–3904 (2006).
30. Lazarides T., G. M. Davies, H. Adams, C. Sabatini, F. Barigelletti, A. Barbieri, S. J. A. Pope, S. Faulkner, M. D. Ward, Ligand-field excited states of hexacyanochromate and hexacyanocobaltate as sensitisers for near-infrared luminescence from Nd(III) and Yb(III) in cyanide-bridged d-f assemblies. *Photochem. Photobiol. Sci.* **6**, 1152–1157 (2007).
31. Piguet C., A. F. Williams, G. Bernardinelli, The First Self-Assembled Dinuclear Triple-Helical Lanthanide Complex: Synthesis and Structure. *Angew. Chem. Int. Ed.* **31**, 1622–1624 (1992).
32. Elhabiri M., R. Scopelliti, J.-C. G. Bünzli, C. Piguet, Lanthanide Helicates Self-Assembled in Water: A New Class of Highly Stable and Luminescent Dimetallic Carboxylates. *J. Am. Chem. Soc.* **121**, 10747–10762 (1999).
33. Martin N., J.-C. G. Bünzli, V. McKee, C. Piguet, G. Hopfgartner, Self-Assembled Dinuclear Lanthanide Helicates: Substantial Luminescence Enhancement upon Replacing Terminal Benz-imidazole Groups by Carboxamide Binding Units. *Inorg. Chem.* **37**, 577–589 (1998).

34. Elhabiri M., R. Scopelliti, J.-C. G. Bünzli, C. Piguet, The first lanthanide-containing helicates self-assembled in water. *Chem. Commun.* **1998**, 2347–2348 (1998).
35. Elhabiri M., J. Hamacek, J.-Claude G. Bünzli, A.-M. Albrecht-Gary, Lanthanide Homobimetallic Triple-Stranded Helicates: Insight into the Self-Assembly Mechanism. *Eur. J. Inorg. Chem.* **2004**, 51–62 (2004).
36. Piguet C., M. Borkovec, J. Hamacek, K. Zeckert, Strict self-assembly of polymetallic helicates: the concepts behind the semantics. *Coord. Chem. Rev.* **249**, 705–726 (2005).
37. Zeckert K., J. Hamacek, J.-M. Senegas, N. Dalla-Favera, S. Floquet, G. Bernardinelli, C. Piguet, Predictions, Synthetic Strategy, and Isolation of a Linear Tetrametallic Triple-Stranded Lanthanide Helicate. *Angew. Chem. Int. Ed.* **44**, 7954–7958 (2005).
38. Aroussi B. E., S. Zebret, C. Besnard, P. Perrottet, J. Hamacek, Rational Design of a Ternary Supramolecular System: Self-Assembly of Pentanuclear Lanthanide Helicates. *J. Am. Chem. Soc.* **133**, 10764–10767 (2011).
39. Chauvin A.-S., S. Comby, B. Song, C. D. B. Vandevyver, F. Thomas, J.-C. G. Bünzli, A Polyoxyethylene-Substituted Bimetallic Europium Helicate for Luminescent Staining of Living Cells. *Chemistry A European Journal* **13**, 9515–9526 (2007).
40. Deiters E., B. Song, A.-S. Chauvin, C. D. B. Vandevyver, F. Gumy, J.-C. G. Bünzli, Luminescent Bimetallic Lanthanide Bioprobes for Cellular Imaging with Excitation in the Visible-Light Range. *Chem. Eur. J.* **15**, 885–900 (2009).
41. Piguet C., G. Hopfgartner, A. F. Williams, J.-C. G. Bünzli, Self-assembly of the first heterodinuclear d-f triple helix in solution. *J. Chem. Soc., Chem. Commun.* **1995**, 491–493 (1995).
42. Piguet C., J.-C. G. Bünzli, G. Bernardinelli, G. Hopfgartner, S. Petoud, O. Schaad, Lanthanide Podates with Predetermined Structural and Photophysical Properties: Strongly Luminescent Self-Assembled Heterodinuclear d-f Complexes with a Segmental Ligand Containing Heterocyclic Imines and Carboxamide Binding Units. *J. Am. Chem. Soc.* **118**, 6681–6697 (1996).
43. Edder C., C. Piguet, J.-C. G. Bünzli, G. Hopfgartner, A water-stable and strongly luminescent self-assembled non-covalent lanthanide podate. *J. Chem. Soc. Dalton Trans.* **1997**, 4657–4664 (1997).
44. Piguet C., C. Edder, S. Rigault, G. Bernardinelli, J.-C. G. Bünzli, G. Hopfgartner, Isolated d-f pairs in supramolecular complexes with tunable structural and electronic properties. *J. Chem. Soc. Dalton Trans.* **2000**, 3999–4006 (2000).
45. Rigault S., C. Piguet, G. Bernardinelli, G. Hopfgartner, Lanthanide-Assisted Self-Assembly of an Inert, Metal-Containing Nonadentate Tripodal Receptor. *Angew. Chem. Int. Ed.* **37**, 169–172 (1998).
46. Cantuel M., G. Bernardinelli, D. Imbert, J.-C. G. Bünzli, G. Hopfgartner, C. Piguet, A kinetically inert and optically active $Cr^{III}$ partner in thermodynamically self-assembled heterodimetallic non-covalent d-f podates. *J. Chem. Soc. Dalton Trans.* **2002**, 1929–1940 (2002)
47. André N., R. Scopelliti, G. Hopfgartner, C. Piguet, J.-C. G. Bünzli, Discriminating between lanthanide ions: self-assembly of heterodimetallic triple-stranded helicates. *Chem. Commun.* **2002**, 214–215 (2002).
48. Mezei G., C. M. Zaleski, V. L. Pecoraro, Structural and functional evolution of metallacrowns. *Chem. Rev.* **107**, 4933–5003 (2007).
49. Tegoni M., M. Remelli, Metallacrowns of copper(II) and aminohydroxamates: Thermodynamics of self assembly and host-guest equilibria. *Coord. Chem. Rev.* **256**, 289–315 (2012).
50. Tegoni M., M. Tropiano, L. Marchio, Thermodynamics of binding of carboxylates to amphiphilic $Eu^{3+}/Cu^{2+}$ metallacrown. *Dalton Trans.* **2009**, 6705–6708 (2009).
51. Tegoni M., M. Furlotti, M. Tropiano, C. S. Lim, V. L. Pecoraro, Thermodynamics of Core Metal Replacement and Self-Assembly of Ca(2+) 15-Metallacrown-5. *Inorg. Chem.* **49**, 5190–5201 (2010).

52. Bozoklu G., C. Gateau, D. Imbert, J. Pécaut, K. Robeyns, Y. Filinchuk, F. Memon, G. Muller, M. Mazzanti, Metal-Controlled Diastereoselective Self-Assembly and Circularly Polarised Luminescence of a Chiral Heptanuclear Europium Wheel. *J. Am. Chem. Soc.* **134**, 8372–8375 (2012).
53. Bozoklu G., C. Marchal, C. Gateau, J. Pécaut, D. Imbert, M. Mazzanti, Diastereoselective Self-Assembly of a Homochiral Europium Triangle from a Bipyoxazoline-Carboxylate Ligand. *Chem. Eur. J.* **16**, 6159–6163 (2010).
54. Faulkner S., S. J. A. Pope, Lanthanide-sensitised lanthanide luminescence: Terbium-sensitised ytterbium luminescence in a trinuclear complex. *J. Am. Chem. Soc.* **125**, 10526–10527 (2003).
55. Tremblay M. S., D. Sames, Synthesis of luminescent heterometallic bis-lanthanide complexes via selective, sequential metallation. *Chem. Commun.* **2006**, 4116–4118 (2006).
56. Pope S. J., B. J. Coe, S. Faulkner, Re(I) sensitised near-infrared lanthanide luminescence from a hetero-trinuclear $Re_2Ln$ array. *Chem. Commun.* **2004**, 1550–1551 (2004).
57. Thibon A, V. C. Pierre, A Highly Selective Luminescent Sensor for the Time-Gated Detection of Potassium. *J. Am. Chem. Soc.* **131**, 434–435 (2009).
58. Natrajan L. S., A. J. L. Villaraza, A. M. Kenwright, S. Faulkner, Controlled preparation of a heterometallic lanthanide complex containing different lanthanides in symmetrical binding pockets. *Chem. Commun.* **2009**, 6020–6022 (2009).
59. Jauregui M., W. S. Perry, C. Allain, L. R. Vidler, M. C. Willis, A. M. Kenwright, J. S. Snaith, G. J. Stasiuk, M. P. Lowe, S. Faulkner, Changing the local coordination environment in mono- and bi-nuclear lanthanide complexes through "click" chemistry. *Dalton Trans.* **2009**, 6283–6285 (2009).
60. Tropiano M., N. L. Kilah, M. Morten, H. Rahman, J. J. Davis, P. D. Beer, S. Faulkner, Reversible Luminescence Switching of a Redox-Active Ferrocene-Europium Dyad. *J. Am. Chem. Soc.* **133**, 11847–11849 (2011).
61. Placidi M. P., A. J. L. Villaraza, L. S. Natrajan, D. Sykes, A. M. Kenwright, S. Faulkner, Synthesis and spectroscopic studies on azo-dye derivatives of polymetallic lanthanide complexes: using diazotization to link metal complexes together. *J. Am. Chem. Soc* **131**, 9916–9917 (2009).
62. Koullourou T., L. S. Natrajan, H. Bhavsar, S. J. A. Pope, J. Feng, J. Narvainen, R. Shaw, E. Scales, R. Kauppinen, A. M. Kenwright, S. Faulkner, Synthesis and spectroscopic properties of a prototype single molecule dual imaging agent comprising a heterobimetallic rhenium-gadolinium complex. *J. Am. Chem. Soc.* **130**, 2178–2179 (2008).
63. Tropiano M., C. J. Record, E. Morris, H. S. Rai, C. Allain, S. Faulkner, Synthesis and Spectroscopic Study of d-f Hybrid Lanthanide Complexes Derived from triazolylDO3A. *Organometallics* **31**, 5673–5676 (2012).
64. Pope S. J. A., B. J. Coe, S. Faulkner, E. V. Bichenkova, X. Yu, K. T. Douglas, Self-assembly of heterobimetallic d-f hybrid complexes: Sensitization of lanthanide luminescence by d-block metal-to-ligand charge-transfer excited states. *J. Am. Chem. Soc.* **126**, 9490–9491 (2004).
65. Faulkner S., B. P. Burton-Pye, pH Dependent self-assembly of dimetallic lanthanide complexes. *Chem. Commun.* **2005**, 259–261 (2005).
66. Tilney J. A., T. J. Sorensen, B. P. Burton-Pye, S. Faulkner, Self-assembly between dicarboxylate ions and a binuclear europium complex: formation of stable adducts and heterometallic lanthanide complexes. *Dalton Trans.* **40**, 12063–12066 (2011).
67. Klink S. I., H. Keizer, F. C. J. M. van Veggel, Transition metal complexes as photosensitizers for near-infrared lanthanide luminescence. *Angew. Chem. Int. Ed.* **39**, 4319–4321 (2000).
68. Sørensen T. J., O. A. Blackburn, M. Tropiano, S. Faulkner, Direct two-photon excitation of $Sm^{3+}$, $Eu^{3+}$, $Tb^{3+}$, $Tb.DOTA^-$, and Tb.propargylDO3A in solution. *Chem. Phys. Lett.* **541**, 16–20 (2012).
69. Palsson L.-O., R. Pal, B. S. Murray, D. Parker, A. Beeby, Two-photon absorption and photoluminescence of europium based emissive probes for bioactive systems. *Dalton Trans.* **37**, 5726–5734 (2007).

70. Perry W. S., S. J. Pope, C. Allain, B. J. Coe. A. M. Kenwright, S. Faulkner, Synthesis and photophysical properties of kinetically stable complexes containing a lanthanide ion and a transition metal antenna group. *Dalton Trans.* **39**, 10974–10983 (2010).
71. Weller A., Electron-transfer and complex formation in the excited state. *Pure Appl. Chem* **16**, 115–123 (1968).
72. Beeby A., S. Faulkner, D. Parker, J. Williams, Sensitised luminescence from phenanthridine appended lanthanide complexes: analysis of triplet mediated energy transfer processes in terbium, europium and neodymium complexes. *J. Chem. Soc., Perkin Trans. 2* **2001**, 1268–1273 (2001).
73. Beeby A., S. Faulkner, J. A. G. Williams, pH Dependence of the energy transfer mechanism in a phenanthridine-appended ytterbium complex. *J. Chem. Soc., Dalton Trans.* **2002**, 1918–1922 (2002).
74. Lazarides T., D. Sykes, S. Faulkner, A. Barbieri, M. D. Ward, On the Mechanism of d-f Energy Transfer in RuII/LnIIIand OsII/LnIIIDyads: Dexter-Type Energy Transfer Over a Distance of 20 Å. *Chem. Eur. J.* **14**, 9389–9399 (2008).
75. Imbert D., M. Cantuel, J.-C. G. Bünzli, G. Bernardinelli, C. Piguet, Extending lifetimes of lanthanide-based near-infrared emitters (Nd, Yb) in the millisecond range through Cr(III) sensitization in discrete bimetallic edifices. *J. Am. Chem. Soc.* **125**, 15698–15699 (2003).
76. Aboshyan-Sorgho L., H. Nozary, A. Aebischer, J.-C. G. Bünzli, P.-Y. Morgantini, K. R. Kittilstved, A. Hauser, S. V. Eliseeva, S. Petoud, C. Piguet, Optimizing Millisecond Time Scale Near-Infrared Emission in Polynuclear Chrome(III)-Lanthanide(III) Complexes. *J. Am. Chem. Soc.* **134**, 12675–12684 (2012).
77. Faulkner S., S. Pope, Lanthanide-sensitized lanthanide luminescence: terbium-sensitized ytterbium luminescence in a trinuclear complex. *J. Am. Chem. Soc* **125**, 10526–10527 (2003).
78. Sambrook M. R., D. Curiel, E. J. Hayes, P. D. Beer, S. J. A. Pope, S. Faulkner, Sensitised near infrared emission from lanthanides via anion-templated assembly of d–f heteronuclear [2] pseudorotaxanes. *New J. Chem.* **30**, 1133–1136 (2006).

# Index

**A**

acousto-optic modulator (AOM) 137
aliphatic polyaminocarboxylate and carboxylate probes 154–63
  lab-on-a-chip technology 162–3
  molecular probes and conjugates 155–60
  nanoparticle probes 160–163
alkali/alkaline earth metal ion sensors 237– 40
ancillary components 67
angular overlap model (AOM) 25–7
angular wave functions 7
anion binding affinities 235–6
antenna effect 40–44
  chemosensors 232, 236
  energy transfer 40–44
  heterometallic complexes 347–8
  intersystem crossing 40–42
  singlet and triplet states 40–42
  spectroscopic techniques 53, 55
  two-photon absorption 207–14, 220–221, 224–5
anti-epidermal growth factor 223
AOM *see* acousto-optic modulator; angular overlap model
aromatic carboxylate anions 255–6
arteriovenous malformation (AVM) 147
autofluorescence 131, 200
AVM *see* arteriovenous malformation
aza crown ethers 237–9, 255
azamacrocyclic probes 163–5

**B**

*Bacillus* spp. 134, 171
back energy transfer 40–41, 45, 310
barium 239–40
Bethe's notation 29–31
β-cell autoantibodies 159
β-diketonate probes 144–54
  inorganic–organic hybrid probes 154
  metal-enhanced luminescent probes 152–4
  molecular probes and conjugates 146–51
  multi-photon excitation imaging 154
  nanoscopic probes 151–2
bioimaging 125–96
  aliphaticpolyaminocarboxylate and carboxylate probes 154–63
  animal studies 287–90
  β-diketonate probes 144–54
  cell studies 283–7
  chemosensors 245, 247–9, 260–264
  circularly polarised luminescence 109–14
  classical optical microscopy 127–8
  commercial instruments 143–4
  concepts and definitions 125–7, 283
  deep-tissue optical imaging 291–3
  dendrimers 178–81
  direct excitation Ln(III) luminescence spectroscopy 318–20
  early instrumental developments 134–40
  heterometallic complexes 348
  hydrogen peroxide assays 178–9
  inorganic–organic hybrid probes 154
  lab-on-a-chip technology 162–3
  lanthanide luminescent bioprobes and bioconjugates 127, 133, 134–40, 144–82
  luminescence microscopy 127–44
  macrocyclic probes 163–71
  metal-enhanced luminescent probes 152–4
  molecular β-diketonate probes and conjugates 146–51
  molecular polyaminocarboxylate and carboxylate probes and conjugates 146–51
  multi-photon excitation imaging 154

*Luminescence of Lanthanide Ions in Coordination Compounds and Nanomaterials*, First Edition.
Edited by Ana de Bettencourt-Dias.

bioimaging (*Continued*)
nanoparticle polyaminocarboxylate and carboxylate probes 160–163
nanoprobes for single-object imaging 222–3
nanoscopic β-diketonate probes 151–2
optical sectioning 271, 283, 291–2
optimization of instrumentation 140–144
optimised bioprobe design considerations 217–21
principle of luminescence microscopy 128–31
principle of time-resolved luminescence microscopy 131–4
self-assembled triple helix bioprobes 171–7
spectroscopic techniques 67–8
time-resolved luminescence microscopy 127, 131–44
two-photon absorption 200–201, 203–5, 217–23
up-converting nanoparticles 127, 180–181, 271, 283–93
biotinylated concanavalin A 138
biphotonic microscopy imaging 215–25
bioimaging 200–201, 203–5, 217–23
nanoprobes for single-object imaging 222–3
NIR-to-NIR imaging 223–5
optimised bioprobe design considerations 217–21
proof of concept 215–17
bipyridine-based ligands 156
bis(benzimidazole)pyridines 171
bis-bipyridine-phosphineoxide ligands 250–251
bis-carboxylate anions 258
bis-pyrazolyl-triazine ligands 210, 213
Bohr Magneton 20
branching ratios 37–8
bridging ligands 335–6, 342–7, 349
bright field microscopy 128, 171
bromide 253, 259

**C**

cadmium 246
calcium 237–40
cationic chelates 177
CCDs *see* charge-coupled devices
CD *see* circular dichroism
central field approximation 4–9
cerium 289, 294, 322
CEST *see* chemical exchange saturation transfer
charge-coupled devices (CCDs) 57, 63, 66, 127–8, 136–8, 141–3, 149
charge transfer state mediated sensitization 209–11
chemical exchange saturation transfer (CEST) 322–6
chemosensors 231–67
alkali metal ions 237–40
bioimaging 245, 247–9, 260–264
covalent versus self-assembled receptor design 235–7
d-block metal ions 240–247
design principles 233–60
emission based sensors 231–2
halides and pH 250–260
hypersensitivity 233
interactions with DNA and biological systems 260–264
luminescence lifetime 232–3
luminescent labels and molecular beacons 247–9
photophysical properties of Ln(III) sensors 231–3
ratiometric peaks 233
receptor site and antenna component design 234–5
sensors for anions 249–60
sensors for cations 237–49
spectral form 233
chirality
biological relevant systems 109–14
circularly polarised luminescence 77–8, 93–115
concepts and definitions 77–8
lanthanide(III) complexes with achiral ligands 94–8
lanthanide(III) complexes with chiral ligands 99–109
chitosan 150–151
chloride 251, 253, 259
chopper wheels 141
chromium 336, 340, 349–50
circular dichroism (CD) 78–9, 84
circularly polarised luminescence (CPL) 77–123
applications 93–109
artifacts in CPL measurements 90–91

biological relevant systems 109–14
calibration and standards 88–91
chirality 77–8, 93–115
chiroptical tools 78–9
concepts and definitions 77–9
DPA-based achiral ligands 95–8
general theory 79–82
1,2-HOPO and IAM-based chiral ligands 99–100
instrumentation 84–8
lanthanide(III) complexes with achiral ligands 94–8
lanthanide(III) complexes with chiral ligands 99–109
lanthanide(III)-containing systems 93–4
macrocycle-type chiral ligands 106–9
measurements 84–93
phosphineoxide or phosphate-based chiral ligands 102–4
pinene-bipyridine based chiral ligands 104
proposed instrumental improvements 91–3
Pybox-derived chiral ligands 100–102
2,6-pyridinedicarboxamide-based chiral ligands 104–6
racemic mixtures: following circularly polarised excitation 94–5
racemic mixtures: perturbation of the equilibrium 95–8
sensing through coordination to antenna/receptor groups 112–14
sensing through coordination to metal centre 109–12
spectral characteristics and interpretation 82–4
theoretical principles 79–84
clathrin/caveolae-independent cytosis 168
click chemistry 344–6
cobalt 242, 336, 340
computed tomography (CT) 293–5
confocal microscopy 130–131, 168
conformational isomerism 340–341
continuous-wave excitation 54, 65, 85–7
copper 240–242, 244–6, 334, 336, 341
core nanoparticles 277–8
core–shell nanoparticles 277–9, 282–3
Coulomb interactions 9–10
coumarins 209–10
covalently linked chemosensors 235–7
CPL *see* circularly polarised luminescence
cross-polarised light illumination 128
cryptates
chemosensors 239–40
macrocyclic probes 169–71
*Cryptosporidium* spp. 133, 143, 148–9
crystal field parameters
angular overlap model 25–6
estimation of 21–7
point charge electrostatic model 21–5
simple overlap model 26–7
symmetry 14–18
*see also* Stark effect
CT *see* computed tomography
cyclams 237–9
cyclens
chemosensors 237–9, 255–60, 262
macrocyclic probes 165–9
cytotoxicity 165, 170–171, 174, 284, 290
Czerny–Turner optical design 64

**D**

dark-field illumination 128
*Daturainnoxia* 135
d-block metals
heterometallic complexes 334–7, 340–341, 344–9
ion sensors 240–247
deep-tissue optical imaging 291–3
dendrimers 178–80, 181
deoxyribose nucleic acid (DNA) 205, 260–264
Dexter energy transfer 43, 348, 350
diabetes mellitus 159
diastereomeric isomers 311–12
*Dictyostelium discoidenum* 138, 155
differential photon counter (DPC) 86
differential transition probability 80–81
diffraction gratings 64
diffuse reflectance 71–2
dipicamides 210
dipicolinates (DPA)
bioimaging 171
two-photon absorption 207–8, 210–214, 218–20, 223–5
dipole–dipole energy transfer 42–4
direct excitation Ln(III) luminescence spectroscopy 303–30
anion binding in catalysis and sensing 317–20
bioimaging 318–20
defining the Ln(III) coordination sphere 303–4, 311–17

direct excitation Ln(III) luminescence spectroscopy (*Continued*)
Eu(III) complex speciation in solution 311
excitation spectra of geometric isomers 311–12
inner-sphere coordination of anions 312–14, 319–20, 322–3, 326
ligand ionization 314–17
Ln(III) catalysts 303–4, 317–18
Ln(III) emission spectroscopy 307–8
Ln(III) excitation spectroscopy 306–7
luminescence properties of lanthanide ions 304–6
luminescence resonance energy transfer 310, 316, 318
MRI contrast agents 303–4, 320–326
number of excitation peaks 311
optical sensors 303–4, 318–20
phosphate ester binding and cleavage 317–18
time-resolved Ln(III) luminescence spectroscopy 308–10, 317
displacement assays 257
DNA *see* deoxyribose nucleic acid
DPA *see* dipicolinates
DPC *see* differential photon counter
dysprosium
circularly polarised luminescence 83, 93, 99
direct excitation Ln(III) luminescence spectroscopy 306–10, 322
heterometallic complexes 335
up-converting nanoparticles 295

**E**
ECD *see* electronic circular dichroism
ED *see* electric dipole
EDCD *see* emission detected circular dichroism
EDS *see* energy-dispersive X-ray spectroscopy
EELS *see* electron energy-loss spectroscopy
effective brightness 218
effective magnetic moment 20
Einstein's coefficient of spontaneous emission 38
electric dipole (ED) transitions 27–32, 34–9
electron energy-loss spectroscopy (EELS) 279–80, 283
electronic circular dichroism (ECD) 78
electron-multiplying charge-coupled devices (EMCCDs) 141–3, 149
electron transfer 237–8, 310, 348–9
electrostatic repulsion energy 8
EMCCDs *see* electron-multiplying charge-coupled devices
emission-based sensors 231–2
emission detected circular dichroism (EDCD) 93
emission efficiency 39–42, 70–72
endocytosis 167–8, 174
energy-dispersive X-ray spectroscopy (EDS) 279–81
energy transfer
antenna effect 40–41, 42–4
back energy transfer 40–41, 45, 310
bioimaging 169
Dexter energy transfer 43, 348, 350
fluorescence resonance energy transfer 248, 289
Förster resonant energy transfer 42–4, 126–7, 131, 348, 350
heterometallic complexes 347–51
luminescence resonance energy transfer 289, 310, 316, 318
non-radiative quenching 45
two-photon absorption 209
energy transfer up-conversion (ETU) 207
environmental samples 133
eosinophils 155–7
epifluorescence microscope 128–9, 137–8
erbium
bioimaging 181
direct excitation Ln(III) luminescence spectroscopy 306
heterometallic complexes 349
spin–orbit coupling 11–12
two-photon absorption 209
up-converting nanoparticles 270–277, 283–91
ESA *see* excited state absorption
*Escherichia coli* 134–5, 158
ETU *see* energy transfer up-conversion
europium
allowed and forbidden *f-f* transitions 33
antenna effect 41
bioimaging 134–6, 138, 146–80
chemosensors 233, 237, 241–5, 247, 249–60
circularly polarised luminescence 83, 88–93, 95, 99–115
crystal field parameters 27

direct excitation Ln(III) luminescence spectroscopy 304–26
emission efficiency and rate constants 40
heterometallic complexes 332, 334, 336, 341–2, 344, 347
Judd–Ofelt theory 36
non-radiative quenching 44–5
sensitised emission 2
spectroscopic techniques 53–4, 56, 73
Stark effect 16–17
transition probabilities and branching ratios 38
two-photon absorption 198, 206–14, 216, 218–23
EXAFS *see* extended X-ray absorption fine structure
excitation light source 65–6, 140–141
excited state absorption (ESA) 207
excited state lifetimes 73
expectation values 14
extended X-ray absorption fine structure (EXAFS) 282

**F**
ferrocenes 344, 351–2
ferro-electric liquid crystal (FLC) 138–9
field programmable gate array (FPGA) design 87
figure of merit (FOM) for brightness 291, 295
FLC *see* ferro-electric liquid crystal
FLIM *see* fluorescence lifetime imaging
fluorescence 40–42
fluorescence lifetime imaging (FLIM) 131
fluorescence resonance energy transfer (FRET) 248, 289
fluoride 251, 253
FOM *see* figure of merit
Förster resonant energy transfer (FRET) 42–4, 126–7, 131, 348, 350
frequency-upconverted pulsed Nd:YAG lasers 66
FRET *see* fluorescence resonance energy transfer; Förster resonant energy transfer

**G**
gadolinium
bioimaging 165–7
chemosensors 244–5
circularly polarised luminescence 83
direct excitation Ln(III) luminescence spectroscopy 306, 320–322
heterometallic complexes 332–5, 346–7
two-photon absorption 209–10
up-converting nanoparticles 290, 293–4
gated autosynchronous luminescence detector (GALD) 142–3
gene silencing 284
geometric isomers 311–12
*Giardia lamblia* 133, 140, 143, 148–9, 153, 161
GM *see* Göppert-Mayer
gold nanoparticles 161–2, 257–8
Göppert-Mayer (GM) parameter 198–9

**H**
HAADF *see* high-angle annular dark field
halide ion sensors 250–260
Hamiltonian operator
central field approximation 4–9
circularly polarised luminescence 80
Judd–Ofelt theory 34
Stark effect 13, 16, 18–19
Zeeman effect 20–21
head to head to head (HHH) isomer 340–341
HEK cells 170–171
HeLa cells
lanthanide luminescent bioprobes and bioconjugates 159–60, 168–9, 174–5
two-photon absorption 216–17, 220, 224
up-converting nanoparticles 284–7, 289–90
helicates 338–41
Her2/*neu see* human epidermal growth factor receptor 2
heterometallic complexes 331–58
antenna effect 347–8
bioimaging 348
bridging ligands 335–6, 342–7, 349
concepts and definitions 331
energy transfer in assembled systems 347–51
helicates 338–41
kinetic stability 332–4
multi-compartmental ligands 342–7
non-helicate structures 341–2
properties of heteromultimetallic complexes 332–5
responsive multimetallic systems 351–3

heterometallic complexes (*Continued*)
solid state lanthanide assemblies 335–7
solution state lanthanide assemblies 338–42
thermodynamic stability 332
transmetallation reactions 333–4
two-photon absorption 348–9
HHH *see* head to head to head
high-angle annular dark field (HAADF) imaging 279–281
high-resolution transmission electron microscopy (HR-TEM) 278–81
HNL *see* human neutrophil lipocalin
holmium
direct excitation Ln(III) luminescence spectroscopy 306
heterometallic complexes 332
up-converting nanoparticles 284, 289, 295
holographic gratings 64
HPV *see* human papillomavirus
HR-TEM *see* high-resolution transmission electron microscopy
human epidermal growth factor receptor 2 (Her2/*neu*) 162–3, 176, 249
human neutrophil lipocalin (HNL) 157
human papilloma virus (HPV) 138
hydrogen bonding 314, 318–20, 324–6
hydrogen peroxide assays 178–9
hyperpolarizability 201–2
hypersensitivity 38–9, 233, 237

**I**
ICCDs *see* image-intensifier charge-coupled devices
ILCT *see* intra-ligand charge transfer state
image-intensifier charge-coupled devices (ICCDs) 141
immunocytochemical applications 134–5, 155–7
indirect excitation 85
induced electric dipole transitions 1–2
InGaAs diodes 66–7
Inner-sphere coordination of anions 312–14, 319–20, 322–3, 326
inorganic–organic hybrid probes 154
integrating sphere 72–3
intersystem crossing (ISC) 40–41, 209
intra-ligand charge transfer state (ILCT) 209–11
intrinsic emission efficiency 40
iodide 259
iron 242, 247, 336, 341
ISC *see* intersystem crossing
islet cell autoantibodies 159

**J**
Jablonski diagram 40
Judd–Ofelt theory 34–7, 39

**K**
Kramers' ions
allowed and forbidden *f-f* transitions 32
Coulomb interactions 9
Stark effect 16
Zeeman effect 19–20
Kronecker delta symbol 12

**L**
lab-on-a-chip technology 162–3, 248–9
Laguerre polynomials 6–7
Landé's interval rule 12–13, 20
Langmuir–Blodgett films 105–6
lanthanide luminescent bioprobes and bioconjugates (LLBs)
aliphatic polyaminocarboxylate and carboxylate probes 154–63
β-diketonate probes 144–54
bioimaging 127, 133, 134–40, 144–80
dendrimers 178–80
hydrogen peroxide assays 178–9
inorganic–organic hybrid probes 154
lab-on-a-chip technology 162–3
luminescence microscopy 133, 134–40
macrocyclic probes 163–71
metal-enhanced luminescent probes 152–4
molecular β-diketonate probes and conjugates 146–51
molecular polyaminocarboxylate and carboxylate probes and conjugates 146–51
multi-photon excitation imaging 154
nanoparticle polyaminocarboxylate and carboxylate probes 160–163
nanoscopic β-diketonate probes 151–2
self-assembled triple helix bioprobes 171–7
lanthanum
allowed and forbidden *f-f* transitions 28–9
bioimaging 160, 175
circularly polarised luminescence 102
crystal field parameters 26–7

heterometallic complexes 332, 341
Laplace operator 4
Laporte rule 29–30, 34, 206
laser diodes 65–6
LED *see* light-emitting diode
Legendre functions 7, 8
leukocytes 155–7
ligand ionization 314–17
light-emitting diode (LED) 65–6, 140
liquid-nitrogen temperature emission spectra 57
lithium 237–40
LLBs *see* lanthanide luminescent bioprobes and bioconjugates
LRET *see* luminescence resonance energy transfer
luminescence emission spectra 49–51, 58, 307–8
luminescence excitation spectra 49–51, 58–9
  *see also* direct excitation Ln(III) luminescence spectroscopy
luminescence lifetime measurements 59–60, 232–3
luminescence microscopy 127–44
  classical optical microscopy 127–8
  commercial instruments 143–4
  early instrumental developments 134–40
  lanthanide luminescent bioprobes and bioconjugates 133, 134–40
  optimization of instrumentation 140–144
  principles 128–34
  time-resolved luminescence microscopy 127, 131–44
luminescence principles
  4*f* orbitals 2
  allowed and forbidden *f-f* transitions 27–33
  angular overlap model 25–6, 27
  angular wave functions 7
  antenna effect 40–44
  common emissive *f-f* transitions 6
  Coulomb interactions 9–10
  crystal field parameters and symmetry 14–18
  electronic configuration of +III oxidation state 2–4
  emission efficiency and rate constants 39–42, 45
  empirical parameters describing optical transitions 28–9
  energies of crystal field split terms 18–20
  energy levels for Ln(III) ions 5
  energy level term symbols 2–4
  estimation of crystal field parameters 21–7
  ground state term symbols for Ln(III) ions 4
  Hamiltonian in central field approximation 4–9
  historical development 1–2
  hypersensitive transitions 38–9
  intensity of induced electric dipole transitions 34–7
  intensity of magnetic dipole transitions 32–3
  Judd–Ofelt theory 34–7, 39
  nature of *f-f* transitions 4–40
  non-radiative quenching 40–41, 44–5
  one-electron wave functions 5–7
  parity rule 29–30, 34
  point charge electrostatic model 21–5, 27
  sensitization mechanism 40–45
  simple overlap model 26–7
  spin rule 29
  spin–orbit coupling 9–13
  Stark effect 9–10, 13–20
  symmetries of the terms 30–32
  transition probabilities and branching ratios 37–8
  Zeeman effect 9–10, 19–21
luminescence resonance energy transfer (LRET) 289, 310, 316, 318
luminescent lanthanide labels 247–9
luthetium 102, 332

**M**

macrocyclic probes 163–71, 220–221
macropinocytosis 168
magnesium 237–40
magnetic dipole (MD) transitions 27–33, 38, 40
magnetic field effect *see* Zeeman effect
magnetic resonance imaging (MRI)
  bioimaging 181
  contrast agents 303–4, 320–326
  direct excitation Ln(III) luminescence spectroscopy 303–4, 320–326
  up-converting nanoparticles 289–90, 293–5
magnetic spin quantum number 7
manganese 336
master oscillator power oscillator (MOPO) 304, 306
MCF-7 cells 220, 285, 289
MD *see* magnetic dipole
mercury 240–241, 244–6

metal-enhanced luminescent probes 152–4
metallacrowns 342
metal-organic frameworks 181
metal oxide/p-doped silicon (MOS) capacitors 130
metal-to-ligand charge transfer transitions (MLCT) 209, 335–6, 348–9
Michler ketone 210, 213
microemulsions 222
microfluidic devices 162–3
MLCT *see* metal-to-ligand charge transfer transitions
molecular beacons 247–9
molecular engineering 201–5
molecular imaging 126
molecular probes and conjugates
  aliphaticpolyaminocarboxylate and carboxylate probes 155–60
  β-diketonateprobes 146–51
monocapped square antiprism 25
monochromators 63–5, 85–6
monoclonal antibodies 148, 157, 175–6, 223
MOPO *see* master oscillator power oscillator
MOS *see* metal oxide/p-doped silicon
MRI *see* magnetic resonance imaging
multi-compartmental ligands 342–7
multiphotonic imaging 201–5, 216–17
multi-photon excitation imaging 154

**N**

nanoparticles
  aliphaticpolyaminocarboxylate and carboxylate probes 160–163
  β-diketonate probes 151–2
  chemosensors 248–9, 257–8
  lab-on-a-chip technology 162–3, 248–9
  spectroscopic techniques 68
  surface modification 278
  synthesis of core and core–shell nanoparticles 277–8
  two-photon absorption 222–3
  *see also* up-converting nanoparticles
nanoprobes for single-object imaging 222–3
nano-rattles 287
near-field optical microscopy 68
near infrared excitation/emission (NIR-NIR) imaging 180–181, 223–5
neodymium
  allowed and forbidden *f-f* transitions 33
  bioimaging 181
  chemosensors 239–40, 245–6, 256, 261–2
  circularly polarised luminescence 93
  crystal field parameters 27
  direct excitation Ln(III) luminescence spectroscopy 310, 312, 316, 322
  heterometallic complexes 342, 349–50
  Judd–Ofelt theory 37
  spectroscopic techniques 73
  spin–orbit coupling 11
  two-photon absorption 209
neutrophils 155–7
nickel 336
NIR-NIR *see* near infrared excitation/emission
NMR *see* nuclear magnetic resonance
nonlinear optics (NLO) 197–205
nonlinear transmission method 199
non-radiative quenching
  direct excitation Ln(III) luminescence spectroscopy 308–14, 318–20, 322–6
  sensitization mechanism 40–41, 44–5
nuclear magnetic resonance (NMR) spectroscopy 78, 334–5

**O**

oblique illumination 128
oligochitosan 150–151
one-electron wave functions 5–7
optical parametric oscillator (OPO) 66
optical sectioning 271, 283, 291–2
optical sensors 303–4
opto-electronic shutters 141
organic luminophores 126
osmium 347, 349
outer-sphere quenching 309–10, 314, 318–20, 322–6

**P**

palladium 155–7
PALM *see* photoactivation localization microscopy
PAMAM *see* polyamidoamine
paramagnetic chemical exchange saturation transfer (paraCEST) 322–6
parity rule 29–30, 34, 206
PCEM *see* point charge electrostatic moment
PD *see* power density
PDMS *see* polydimethylsiloxane
PDZ domain 158–9

PEM *see* photo-elastic (or elasto-optic) modulator
PeT *see* photoinduced electron transfer
PET *see* positron emission tomography
Pfeiffer effect 96–8
phantom images 324–5
phase contrast microscopy 128
pH low insertion peptide (pHLIP) 161–2
phosphate ester binding and cleavage 317–18
phosphorescence
  antenna effect 40–42
  bioimaging 131–2, 138
photoactivation localization microscopy (PALM) 68
photo-avalanche mechanism 275
photobleaching 129, 201
photo-elastic (or elasto-optic) modulator (PEM) 86–8, 90–91
photoinduced electron transfer (PeT) 237–8, 348–9
photoluminescence 274
photomultiplier tubes (PMTs) 66–7
  circularly polarised luminescence 86–7
  luminescence microscopy 128, 130
  up-converting nanoparticles 276
pH sensors 250–260
plasmonics 152–4
platelets 161–2
platinum 155–7
PMTs *see* photomultiplier tubes
point charge electrostatic model (PCEM) 21–5
  comparison with AOM and SOM methods 27
  monocapped square antiprism 25
  square antiprism 23–4
  tricappedtrigonal prism 24
polyamidoamine (PAMAM) 178–80
polyaminocarboxylates *see* aliphatic polyaminocarboxylate and carboxylate probes
polydimethylsiloxane (PDMS) microfluidics 162–3
porphyrins 135–6, 156, 214–15
positron emission tomography (PET) 290, 293
potassium 237–40
power density (PD) 270–272, 275–6, 283–4, 287–91
PPGA *see* field programmable gate array
praseodymium
  allowed and forbidden *f-f* transitions 33
  crystal field parameters 26
  Judd–Ofelt theory 36
  radial charge density distribution 2–3
  spin–orbit coupling 13
  Stark effect 19
pseudorotaxanes 352–3
pulsed excitation/gated detection 54, 65
pulsed nitrogen lasers 66

**Q**

quantum dots (QDs) 126
quantum yield 39
  absolute measurement with integrating sphere 72–3
  bioimaging 153, 168–9
  chemosensors 232
  measurement against standard in solid-state 71–2
  measurement against standard in solution 70–71
  spectroscopic techniques 69–73
  two-photon absorption 203, 210
  up-converting nanoparticles 275–6, 291

**R**

radial charge density distribution 2–3
rate constants 39–40, 45
ratiometric peaks 233
repulsive Coulomb energy 8
responsive multimetallic systems 351–3
rhenium 344–7
rotation formula 30
Russell–Saunders quantum number 2–3
  *see also* spin–orbit coupling
ruthenium 261, 336–7, 347–9

**S**

samarium
  bioimaging 163, 174–5
  circularly polarised luminescence 83, 93, 99
  direct excitation Ln(III) luminescence spectroscopy 306–10
  spectroscopic techniques 73
SBMC *see* streptavidin-based macromolecular complex
scanning microscopy 216, 222
scanning monochromators 63–4
Schrödinger equation 4, 7
  *see also* Hamiltonian operator
self-assembled chemosensors 235–7

self-assembled triple helix bioprobes 171–7
  cell penetration and imaging 174–5
  cytotoxicity 174
  heterometallic complexes 338–41
  photophysical properties 173–4
  specific targeting with bioconjugated luminescent probes 175–6
  structure, thermodynamic stability and kinetic inertness 171–3
sensitization mechanism 40–45
  antenna effect 40–44, 207–10
  charge transfer state mediated 209–11
  chemosensors 231–2
  energy transfer 40–44
  intersystem crossing 40–42
  non-radiative quenching 40–41, 44–5
  singlet and triplet states 40–42
  two-photon absorption 207–11
sensitised emission 2
signal saturation 129
signal to noise ratio (SNR) 131–2, 248
silver 336
SIM *see* structured illumination microscopy
simple overlap model (SOM) 26–7
single-object imaging 222–3
Slater two-electron radial integrals 8
slit width 64–5
SNR *see* signal to noise ratio
sodium 237–40
sol-gel 214, 222
solid-state inorganic ion exchange matrices 56
SOM *see* simple overlap model
SOS *see* sum-over states
spark spectroscopy 1
spectrographs 63–4
spectroscopic techniques 49–75
  absolute measurement with integrating sphere 72–3
  adapting experiments for diverse samples 52–3
  ancillary components 67
  antenna effect 53, 55
  challenges in experiment design and interpretation 52–7
  common luminescence experiments 57–60
  components and characteristics 63–7
  concepts and definitions 49–52
  design elements and configurations 61–3
  detectors 66–7
  excitation sources 65–6
  excited state lifetimes 73
  in-solution measurements 70–71
  instrumentation in luminescence spectroscopy 52–69
  instrument modules 61–2
  luminescence emission spectra 49–51, 58
  luminescence excitation spectra 49–51, 58–9
  luminescence lifetime measurements 59–60
  monochromators 63–4
  number of coordinated solvent molecules 73
  quantum yield measurement 69–73
  recent advances in instrumentation 67–9
  resolving and assigning spectral bands 55–7
  resolving weak emission from intense background 53–5
  solid-state measurements 71–2
  spectral range for Ln(III) ions 49–51
  standards 70–72
  time-resolved emission and excitation spectra 60
  typical absorption and emission spectra 49–50
spin coordinates 7
spin–orbit coupling 9–13
spin rule 29
splitting energy 20–21
square antiprism 23–4
Stark effect 9–10, 13–20
  crystal field parameters and symmetry 14–18
  energies of crystal field split terms 18–20
  expectation values 14
STED *see* stimulated emission depletion
Stern–Volmer constants 167
stimulated emission depletion (STED) 68
stochastic optical reconstruction microscopy (STORM) 68
streptavidin-based macromolecular complex (SBMC) 138
structured illumination microscopy (SIM) 68
sum-over states (SOS) method 201–2
super-resolution techniques 68
symmetry
  allowed and forbidden *f-f* transitions 30–32
  crystal field parameters 14–18
synchrotron radiation 280–282

T
TCSPC *see* time-correlated single photon counting

TEM *see* transmission electron microscopy
terbium
bioimaging 155–60, 163–7, 169–77
chemosensors 237–9, 241–3, 245, 249, 251–6, 258–60, 262
circularly polarised luminescence 83–4, 95, 99–100, 104, 108–9, 111–12
direct excitation Ln(III) luminescence spectroscopy 306–10
heterometallic complexes 332, 343–5, 347, 350–351
non-radiative quenching 45
spectroscopic techniques 55, 73
two-photon absorption 206–8, 212, 214–18
up-converting nanoparticles 294
terpyridine-based ligands 156, 178–9
thalidomide 77
theranostics 125
three-level model 202–3
thulium
bioimaging 181
direct excitation Ln(III) luminescence spectroscopy 306, 322
heterometallic complexes 335
up-converting nanoparticles 274–6, 284–90, 292
time-correlated single photon counting (TCSPC) 69, 131
time-gated spectroscopy 214–15
time-resolved amplified cryptate emission (TRACE) 170
time-resolved circularly polarised luminescence (CPL) 87
time-resolved detection (TRD)
bioimaging 126–7, 131–4, 136–8, 148, 160–163, 171
direct excitation Ln(III) luminescence spectroscopy 308–10, 317
emission and excitation spectra 60, 69
time-resolved luminescence microscopy (TRLM) 127, 131–44
commercial instruments 143–4
early instrumental developments 134–40
lanthanide luminescent bioprobes and bioconjugates 133, 134–40
optimization of instrumentation 140–144
principles 131–4
tip enhanced optical microscopy 68
total internal reflection fluorescence microscopy (TIRFM) 139–40
TPEF *see* two-photon excited fluorescence
TPUWFM *see* two-photon upconversion wide field microscopy
TRACE *see* time-resolved amplified cryptate emission
transition probabilities 37–8
transmetallation reactions 333–4
transmission electron microscopy (TEM) 278–81, 283–7
transparency window 200, 223, 271
TRD *see* time-resolved detection
tricapped trigonal prism 24
TRLM *see* time-resolved luminescence microscopy
tumour detection
lanthanide luminescent bioprobes and bioconjugates 162–3, 165, 176
time-resolved luminescence microscopy 135–6
two-photon absorption 222–3
up-converting nanoparticles 285–93
two-photon absorption 197–230
antenna effect 207–14, 220–221, 224–5
bioimaging 200–201, 203–5
biphotonic microscopy imaging 200–201, 203–5, 215–25
breakthrough experiments 205–6
brightness trade-off 211–14
charge transfer state mediated sensitization 209–11
concepts and definitions 197–8
excitation of *f-f* transitions 206–7
excited luminescence in solid matrix 214
experimental determination of 2PA efficiency 199–200
heterometallic complexes 348–9
molecular engineering for multiphotonic imaging 201–5, 216–17
nanoprobes for single-object imaging 222–3
NIR-NIR imaging 223–5
nonlinear optics 197–205
nonlinear transmission method 199
optimization of molecular two-photon cross section 211–14
sensitization mechanism 207–11
spectroscopic evidence 205–15

two-photon absorption (*Continued*)
theoretical and historical background 198–9, 205–6
time-gated spectroscopy 214–15
two-photon excited fluorescence procedure 199–200
two-photon fluorescence microscopy 200–201, 203–5, 215–25
Z-scan method 199
two-photon excited fluorescence (TPEF) procedure 199–200
two-photon fluorescence microscopy *see* biphotonic microscopy imaging
two-photon scanning microscopy 216, 222
two-photon upconversion wide field microscopy (TPUWFM) 292

**U**

up-converting nanoparticles (UCNPs) 269–302
animal studies 287–90
bioimaging 127, 180–181, 271, 283–93
cell studies 283–7
characterization 278–83
concepts and definitions 269–71, 283
deep-tissue optical imaging 291–3
figure of merit for brightness 291, 295
magnetic resonance imaging 289–90, 293–5
optical sectioning 271, 283, 291–2
physical properties of Ln(III) ions 272
principles of upconversion 272–6
surface modification 278
synthesis of core and core–shell nanoparticles 277–8

**V**

vibrational circular dichroism (VCD) 78

**W**

water counting studies 314, 316, 318

**X**

xenon flash lamps 65
X-ray absorption near-edge fine structure (XANES) 282
X-ray photo-electron spectroscopy (XPS) 281–2
X-ray (powder) diffraction (XRD) 278

**Y**

ytterbium
allowed and forbidden *f-f* transitions 31–2
bioimaging 135–6, 174, 181
chemosensors 245–6, 255–6, 261–2
circularly polarised luminescence 83, 93
direct excitation Ln(III) luminescence spectroscopy 306, 322
heterometallic complexes 342–3, 345, 349–51
spectroscopic techniques 57, 73
two-photon absorption 209, 212–15, 223–25
up-converting nanoparticles 270–277, 283–91

**Z**

Zeeman effect 9–10, 19–21
zinc 159, 177, 240–244, 246, 334, 340
*Zonula occludens-1* 158–9
Z-scan method 199